The Evolution of Multicellularity

T0321361

Evolutionary Cell Biology

Series Editors

Brian K. Hall – Dalhousie University, Halifax, Nova Scotia, Canada
Sally A. Moody – George Washington University, Washington DC, USA

Editorial Board

Michael Hadfield – University of Hawaii, Honolulu, USA
Kim Cooper – University of California, San Diego, USA
Mark Martindale – University of Florida, Gainesville, USA
David M. Gardiner – University of California, Irvine, USA
Shigeru Kuratani – Kobe University, Japan
Nori Satoh – Okinawa Institute of Science and Technology, Japan
Sally Leys – University of Alberta, Canada

Science publisher

Charles R. Crumly – CRC Press/Taylor & Francis Group

For more information about this series, please visit: www.crcpress.com/Evolutionary-Cell-Biology/book-series/CRCEVOCELBIO

The Evolution of Multicellularity

Edited by

Matthew D. Herron, Peter L. Conlin,
and William C. Ratcliff

CRC Press
Taylor & Francis Group
Boca Raton London New York

CRC Press is an imprint of the
Taylor & Francis Group, an **informa** business

COVER ART by Pedro Márquez-Zacarías
"This is a piece of generative art that combines deterministic and stochastic processes of growth. The network component represents the growth of a Wolfram Physics model, in which simple substitution rules make a system develop into an intricate network of causal dependencies. The cell cluster was generated with a self-avoiding random walk, where each cell can stochastically become one of two phenotypes. This piece represents the interplay between constraints, contingency, and symmetry breaking that produce the unity and diversity of multicellular organisms."

First edition published 2022
by CRC Press
6000 Broken Sound Parkway NW, Suite 300, Boca Raton, FL 33487-2742

and by CRC Press
2 Park Square, Milton Park, Abingdon, Oxon, OX14 4RN

© 2022 selection and editorial matter, Matthew D. Herron, Peter L. Conlin, and William C. Ratcliff; individual chapters, the contributor

CRC Press is an imprint of Taylor & Francis Group, LLC

This book contains information obtained from authentic and highly regarded sources. While all reasonable efforts have been made to publish reliable data and information, neither the author[s] nor the publisher can accept any legal responsibility or liability for any errors or omissions that may be made. The publishers wish to make clear that any views or opinions expressed in this book by individual editors, authors or contributors are personal to them and do not necessarily reflect the views/opinions of the publishers. The information or guidance contained in this book is intended for use by medical, scientific or health-care professionals and is provided strictly as a supplement to the medical or other professional's own judgement, their knowledge of the patient's medical history, relevant manufacturer's instructions and the appropriate best practice guidelines. Because of the rapid advances in medical science, any information or advice on dosages, procedures or diagnoses should be independently verified. The reader is strongly urged to consult the relevant national drug formulary and the drug companies' and device or material manufacturers' printed instructions, and their websites, before administering or utilizing any of the drugs, devices or materials mentioned in this book. This book does not indicate whether a particular treatment is appropriate or suitable for a particular individual. Ultimately it is the sole responsibility of the medical professional to make his or her own professional judgements, so as to advise and treat patients appropriately. The authors and publishers have also attempted to trace the copyright holders of all material reproduced in this publication and apologize to copyright holders if permission to publish in this form has not been obtained. If any copyright material has not been acknowledged please write and let us know so we may rectify in any future reprint.

With the exception of Chapter 13, no part of this book may be reprinted or reproduced or utilised in any form or by any electronic, mechanical, or other means, now known or hereafter invented, including photocopying and recording, or in any information storage or retrieval system, without permission in writing from the publishers.

Chapter 13 of this book is available for free in PDF format as Open Access at www.taylorfrancis.com. It has been made available under a Creative Commons Attribution-NonCommercial-NoDerivatives (CC-BY-NC-ND) 4.0 International license.

Trademark notice: Product or corporate names may be trademarks or registered trademarks and are used only for identification and explanation without intent to infringe.

ISBN: 9780367356965 (hbk)
ISBN: 9781032207797 (pbk)
ISBN: 9780429351907 (ebk)

DOI: 10.1201/9780429351907

The Open Access version of chapter 13 was funded by Howard Hughes Medical Institute.

Contents

SECTION 1   Theory and Philosophy

SECTION 2   Aggregative Multicellularity

SECTION 3   Clonal Multicellularity

SECTION 4   Life Cycles and Complex Multicellularity

SECTION 5   *Synthesis and Conclusions*

Foreword by Andrew H. Knoll

Multicellularity. On the face of it, the concept seems simple and unambiguous: some organisms package everything needed for metabolism, behavior, locomotion and reproduction into a single cell, whereas others contain multiple cells, commonly with varying functions. This is true enough, but it fails to account for the remarkable variety of multicellular organisms, or the nuances of their origins, development, and function. Accordingly, biologists commonly preface "multicellularity" with adjectives that acknowledge various axes of diversity. Multicellular organisms can be aggregative or clonal, emphasizing a fundamental distinction in life cycle. They may be obligate or facultative, changing in response to environmental signals. Or they may be viewed as simple or complex, with complexity variously defined by three-dimensionality, number of cell types, size, or function.

Thinking about this a decade ago, I threw my lot in with function, albeit an aspect of function that correlates with size, three-dimensionality and the differentiation of distinct cell types. As multicellular organisms begin to develop in three dimensions, cells in the interior of the organism become progressively distanced from the ambient environment – their source of food, oxygen and external molecular signals. With increasing size come expanding opportunities for feeding, defense, and environmental occupation, but also a greater need for mechanisms beyond diffusion to support interior cells. For this reason, my attempt to define complex multicellularity focused on the presence of tissues or organs that circumvent the limitations of diffusion. That works pretty well for bilaterian animals, plants, kelps and macroscopic fungi, but, admittedly, presents problems for florideophyte red algae, which pass tests for three-dimensionality, macroscopic size and cell differentiation with flying colors, but fare less well in terms of diffusion and its circumvention.

In fact, all attempts to categorize multicellularity have their strengths and limitations, for the simple reason that multicellular organisms are so various. Although commonly left out of the conversation, bacteria exhibit simple multicellularity in a number of clades; obligately multicellular cyanobacteria have existed for at least 2.5 billion years, and cyanobacteria capable of cell differentiation have been found as fossils in 1.5-billion-year-old rocks. Moreover, myxobacteria aggregate to form macroscopic structures. Among eukaryotes, simple multicellular clades have evolved repeatedly through time, with fossils of simple (presumed) macrophytes found in rocks as old as 1800–1600 million years, and more complex forms entering the record around 600 million years ago, as the age of animals began. In terms of complexity, there does seem to be something different about animals, plants, mushrooms and kelps, but all contain examples that flout the rules. Is *Trichoplax*, for example, complexly multicellular? Were the earliest known animals, preserved in Ediacaran rocks, complex, even though they probably consisted of little more than upper and lower epithelia surrounding a fluid- or gel-filled interior space? The answers may depend on your definition of complexity.

Ways of thinking about multicellularity have expanded along with our inventory of multicellular clades, and many of these exciting new approaches can be found in this volume. Better phylogenies and improved sampling, especially among protists and fungi, now show just how many times multicellular organisms have evolved from unicellular ancestors. And the application of the various 21st-century "omics" is illuminating the molecular basis of multicellular development and function, while providing clues to the origins of multicellularity in diverse lineages. Integrating genomics with phylogeny, a number of labs are showing the virtues of molecular comparisons between multicellular clades and a relatively wide range of their unicellular relatives. For example, genomes of the closest sister groups to plants and animals appear to exhibit gene losses as well as gains, evident when a broader range of streptophytes and holozoans is examined.

When does a colony of unicells become a multicellular organism? Such a question emphasizes the importance of individuality in thinking about this subject. At a fundamental level, multicellularity reflects cooperative behavior among cells, but what mechanisms actually underlie the emergence of multicellular structures as evolutionary individuals? And how did nascent multicellular organisms deal with the confounding problem of cheaters, constant threats to the integrity of the system? In these pages, Rick Michod argues that only group selection – long debated by evolutionary biologists – "can make a cell suboptimal were it to leave the group" (Chapter 3).

While increasingly detailed studies of multicellularity emphasize the distinct evolutionary paths traveled by different multicellular clades, some commonalities persist. The evolution of clonal multicellularity – and, therefore, by far the greatest diversity of multicellular species – requires mechanisms of adhesion to ensure that cells remain in place following division. (This, in turn, also puts emphasis on intracellular mechanisms for regulating the geometry of cell division.) Communication among cells is equally key. A number of authors have underscored the evolution of plasmodesmata in photosynthetic clades and gap junctions in animals as key innovations in cell-cell communication. Such structures not only facilitate the passage of signals and other molecular traffic between cells, but provide a mechanism for signaling that is spatially specific. Related to this, the regular morphologies of multicellular organisms require developmental programs in which the fates of individual cells are determined by signals from their neighbors.

Although inherently obvious, it bears repetition that multicellular clades evolved from unicellular ancestors. Therefore, a better understanding of single-celled sister groups can do much to illuminate the bases of multicellularity. Unicells have sophisticated mechanisms for the transduction of signals from the environment and from conspecific neighbors, eliciting a variety of cellular responses. Homologous signaling mechanisms occur in their multicellular descendants, and so knowing signal's function in both ancestors and descendants provides clues to the causes, consequences and mechanisms of multicellular evolution. Life cycles also illuminate aspects of the multicellular program. It has long been known that genes active in the regulation of cell differentiation during the development of *Volvox* also play a role in cell differentiation during the life cycle of its close unicellular sisters. Evidence summarized in this volume also supports the hypothesis that cAMP regulation of multicellular aggregation and differentiation in *Dictyostelium* reflects an earlier evolution of starvation-induced encystment in its unicellular ancestors (Chapters 5 and 8). Comparable insights are flowing from detailed studies of choanoflagellates and other holozoan relatives of animals (Chapter 13), as well as streptophyte sister groups of land plants (Chapter 16). Ancestors also impose constraints on multicelluarity, for example, the rigid cell walls of streptophytes and terrestrial fungi.

Finally, one of the greatest hurdles to the emergence of multicellularity lies in the fitness of early stages in the evolutionary path toward complexity. It is easy enough to appreciate the functional advantages of complex organisms with billions of cells, but what about the early stages of multicellular evolution, populated by organisms with perhaps a few dozen cells? Does predator avoidance furnish both a fitness advantage and scaffolding for the emergence of complexity? To what extent does the coordinated behavior of multiple cells enhance feeding, and does this, as well, provide a sufficient seed for complex multicellularity? Here is where ecology can help. We know little about how many simple multicellular taxa work in nature, and while experiments are providing welcome illumination, more is needed.

Plants, animals, mushrooms and seaweeds shape the ecosystems of our common experience, but they have characterized only the most recent fifteen percent of life's history. How did life expand so remarkably so late in the evolutionary day? And how did innovations in different multicellular clades feed back onto one another? In the pages that follow, readers will find a wealth of new observations, experiments and models that collectively enrich our understanding of multicellularity in all its manifestations. They provide welcome opportunity to learn – and an invitation to think further.

Andrew H. Knoll
Harvard University, Cambridge, MA
aknoll@oeb.harvard.edu

Acknowledgments

We wish to express our sincere appreciation to the contributing authors for sharing their knowledge and deep insight into the evolution of multicellularity. It has been a privilege working with them to assemble this collection, and we appreciate their patience. We think their chapters will serve as a great resource for both newcomers to the field and more experienced researchers for a long time to come.

Each chapter was reviewed by two other chapter contributors and at least one of the editors. We are extremely grateful to all referees for taking their time to read the submitted contributions and to provide thoughtful, critical, and helpful suggestions for improving the chapters. We also wish to thank Andrew Knoll for contributing the foreword to this book.

We think it's worth mentioning that most of the work on this book, by us and by the authors, was done during the COVID-19 pandemic. It has become a cliché to talk about "these challenging/difficult/unprecedented times," but that shouldn't distract from the havoc of the last year. Everyone has been affected to some degree, ranging from inconvenience, isolation, and anxiety for the most privileged among us to loss and death for the less privileged and the unlucky. In spite of the resulting unexpected and unwelcome disruptions, our authors persevered, without exception, to deliver outstanding chapters. We know about some of the extraordinary challenges they faced, but we have no doubt that there was more we don't know about.

P.L.C. is grateful for support through a NASA Postdoctoral Program Fellowship and funding from the David and Lucile Packard Foundation. W.C.R. gratefully acknowledges generous support from the National Science Foundation (DEB-1845363) and through a Packard Fellowship for Science and Engineering. This material is based upon work while M.D.H. was serving at the National Science Foundation.

Contributors

Rebecka Andersson
Department of Mathematics and
   Mathematical Statistics
and
Integrated Science Lab
Umeå University
Umeå, Sweden

Pierrick Bourrat
Philosophy Department
Macquarie University

and

Department of Philosophy & Charles
   Perkins Centre
The University of Sydney

John L. Bowman
School of Biological Sciences
Monash University
Melbourne, Australia

Liam N. Briginshaw
School of Biological Sciences
Monash University
Melbourne, Australia

Cathleen Broersma
School of Natural Sciences
Massey University
Auckland, New Zealand

Thibaut Brunet
Howard Hughes Medical Institute and
   the Department of Molecular and
   Cell Biology
University of California, Berkeley
California, USA

and

Department of Cell Biology and
   Infection
Institut Pasteur
Paris, France

J. Mark Cock
Laboratory of Integrative Biology of
   Marine Models
Station Biologique de Roscoff
Sorbonne Université
Roscoff, France

Susana M. Coelho
Department of Algal Development and
   Evolution
Max Planck Institute for Developmental
   Biology
Tübingen, Germany

Peter L. Conlin
School of Biological Sciences
Georgia Institute of Technology
Atlanta, Georgia, USA

Guilhem Doulcier
Institut de Biologie de l'Ecole Normale
   Supérieure (IBENS)
Université Paris Sciences et Lettres,
   CNRS, INSERM
Paris, France

and

Department of Evolutionary Theory
Max Planck Institute for Evolutionary
   Biology
Plön, Germany

Roberta M. Fisher
Department of Biology
University of Copenhagen
Copenhagen, Denmark

Marco La Fortezza
Institute for Integrative Biology
ETH Zürich
Zürich, Switzerland

Jordi van Gestel
Department of Microbiology and
    Immunology
University of California San Francisco
San Francisco, California, USA

Katrin Hammerschmidt
Institute of Microbiology
Kiel University
Kiel, Germany

Matthew D. Herron
School of Biological Sciences
Georgia Institute of Technology
Atlanta, Georgia, USA

Hanna Isaksson
Department of Mathematics and
    Mathematical Statistics
and
Integrated Science Lab
Umeå University
Umeå, Sweden

Israt Jahan
Department of Biology
Washington University in St. Louis
St. Louis, Missouri, USA

Stefania E. Kapsetaki
Biodesign Institute
Arizona State University
Tempe, Arizona, USA

Nicole King
Howard Hughes Medical Institute and
    the Department of Molecular and
    Cell Biology

Andrew H. Knoll
Department of Organismic and
    Evolutionary Biology
Harvard University
Cambridge, Massachusetts, USA

Tyler Larsen
Department of Biology
Washington University in St. Louis
St. Louis, Missouri, USA

Michelle M. Leger
Institute of Evolutionary Biology
    (CSIC-Universitat Pompeu Fabra)
Barcelona, Spain

Eric Libby
Department of Mathematics and
    Mathematical Statistics
Umeå University
Umeå, Sweden
and
Integrated Science Lab

and

Santa Fe Institute
Santa Fe, New Mexico, USA

Alexander N. May
Research Casting International
Quinte West, Canada

Richard E. Michod
Department of Ecology and
    Evolutionary Biology
University of Arizona
Tucson, Arizona, USA

László G. Nagy
Synthetic and Systems Biology Unit
Biological Research Centre
Szeged, Hungary

and

Department of Plant Anatomy
Institute of Biology
Eötvös Loránd University
Budapest, Hungary

Aurora M. Nedelcu
Biology Department
University of New Brunswick
Fredericton, Canada

Maureen A. O'Malley
School of History and Philosophy of
    Science
University of Sydney
Sydney, NSW Australia

Elizabeth A. Ostrowski
School of Natural Sciences
Massey University
Auckland, New Zealand

David C. Queller
Department of Biology
Washington University in St. Louis
St. Louis, Missouri, USA

William C. Ratcliff
School of Biological Sciences
Georgia Institute of Technology
Atlanta, Georgia, USA

Iñaki Ruiz-Trillo
Institute of Evolutionary Biology
    (CSIC-Universitat Pompeu Fabra)
Barcelona, Spain

Kaitlin A. Schaal
Institute for Integrative Biology
ETH Zürich
Zürich, Switzerland

Pauline Schaap
School of Life Sciences
University of Dundee
Dundee, UK

Merlijn Staps
Department of Ecology and
    Evolutionary Biology
Princeton University
Princeton, New Jersey, USA

Joan E. Strassmann
Department of Biology
Washington University in St. Louis
St. Louis, Missouri, USA

Corina E. Tarnita
Department of Ecology and
    Evolutionary Biology
Princeton University
Princeton, New Jersey, USA

Gregory J. Velicer
Institute for Integrative Biology
ETH Zürich
Zürich, Switzerland

1 Introduction
The Evolution of Multicellularity in Context

Matthew D. Herron
School of Biological Sciences, Georgia Institute
of Technology, Atlanta, Georgia, USA

Peter L. Conlin
School of Biological Sciences, Georgia Institute
of Technology, Atlanta, Georgia, USA

William C. Ratcliff
School of Biological Sciences, Georgia Institute
of Technology, Atlanta, Georgia, USA

CONTENTS

> If indeed it is true that all living bodies are productions of nature, we are driven to the belief that she can only have produced them one after another and not all in a moment. Now if she shaped them one after another, there are grounds for thinking that she began exclusively with the simplest, and only produced at the very end the most complex organisations both of the animal and vegetable kingdoms.
>
> *(Lamarck, 1809, p. 129)*

1.1 INTRODUCTION

Barbara Kingsolver called it "audacious" to send a piece of writing out into a world that "already contains *Middlemarch*" (Kingsolver, 2001). To ask readers to spend time on your creation, when they could instead choose from a raft of powerful, wise, and profound novels that already exist, it had, she concluded, "better be important." In the context of this collection, our *Middlemarch* is the outstanding work that has

DOI: 10.1201/9780429351907-1

already been done on the evolution of multicellularity. Excellent books by David Kirk (1998), John Tyler Bonner (2000), and Richard Kessin (2001), as well as collections edited by David Whitworth (2008), by Iñaki Ruiz-Trillo and Aurora Nedelcu (2015), and by Karl Niklas and Stuart Newman (2016), have been dedicated to the topic, not to mention thousands of scholarly papers.

Our audacity comes from the conviction that an open niche still exists, a sort of book on multicellularity that hasn't previously been written. Our goal has been to organize a set of chapters that would collectively serve as an in-depth review of the subfield of evolutionary biology that deals with the origins of multicellularity. We intend the book to serve as a jumping-off point, stimulating further research by summarizing the topics that students and researchers of the evolution of multicellularity should be familiar with.

We hope that it will provide a sufficient overview so that a reader unfamiliar with the relevant literature (a beginning graduate student, for example) will come away with an understanding of the major issues. What types of multicellular organisms exist? What are their evolutionary relationships? What processes led to their origins and subsequent evolution? In what conceptual frameworks can their evolution be understood? Crucially, what questions remain to be answered (see Chapter 18 for a detailed discussion)? In addition to providing an overview for newcomers to the field, we hope the book will serve as a reference for more established researchers.

1.2   BACKGROUND

The idea that multicellular animals and plants evolved from single-celled organisms has been around for as long as there has been a coherent theory of evolution. Jean-Baptiste Lamarck, for example, believed that (mostly) unicellular "infusoria" were constantly arising through spontaneous generation and evolving into more complex forms due to the motion of fluids within their bodies (Lamarck, 1809, 1815) (more details on Lamarck's views can be found in Chapter 13). Although Charles Darwin considered Lamarck's ideas about spontaneous generation "superfluous (and groundless)" (Darwin, 1887, p. 210), he agreed that animals and plants likely descend from "some one primordial form" (Darwin, 1859, p. 425).

In the post-Darwin world, descent of multicellular organisms from unicellular ancestors has by and large been taken as a given. Furthermore, plants and animals have long been considered to have independently evolved multicellularity. Ernst Haeckel, for example, proposed that animals descended from protozoa and plants from protophyta (Haeckel, 1894) (more details on Haeckel's views can be found in Chapter 13). Henry Cadwalader Chapman thought it "probable that Monera in past time divided into animal and vegetal Monera," which gave rise to the animals and plants (including red, green, and brown algae), respectively (Chapman, 1873, p. 83). August Weismann agreed that animals and plants descended from distinct unicellular ancestors (Weismann, 1889).

As the big picture of phylogenetic relationships among kingdoms and phyla began to emerge, it became clear that two origins of multicellularity would not suffice. For example, the fundamental distinction between cells with and without nuclei, recognized by Haeckel (1869) and formalized in the taxonomy of Copeland (1938), necessitates an independent origin in the filamentous cyanobacteria. This did not, however, resolve the extreme heterogeneity of Copeland's Kingdom Protoctista ("Nucleate organisms not of the characters of plants [including green algae] and animals" [Copeland, 1956, p. 4]). The recognition of fundamental differences among the phyla within the kingdom, including, for example, red algae, brown algae, and ciliates, further implied that the multicellular members of each of these groups represent at least one additional independent origin of multicellularity.

Further advances in phylogenetic systematics have shown that even within some of these taxa, multicellularity has evolved more than once. This is almost certainly the case in the green algae (Chapter 9), the fungi (Chapter 14), and the Amoebozoa (Chapter 5), for example. In 2007, Grosberg and Strathmann estimated "at least 25" independent origins of multicellularity (Grosberg and Strathmann, 2007, p. 622), but this is very likely a serious underestimate. Recent phylogenetic reconstructions based on whole transcriptome data suggest that there may have been this many independent origins of multicellularity in the green algal lineage alone (One Thousand Plant Transcriptomes Initiative, 2019). Furthermore, we should not forget that essentially all estimates of the number of origins are based exclusively on extant taxa; there is no telling how many species may have evolved multicellularity and subsequently gone extinct without leaving much of a fossil record.

In the second half of the twentieth century, the evolution of multicellularity began to be seen as one example of a broader category of transitions leading to new, more inclusive biological units. John Tyler Bonner, for example, wrote of "cases where in one jump a new level of complexity is reached" (Bonner, 1974, p. 58), including the origins of life, of eukaryotic cells, of multicellularity, and of social groupings. Leo Buss interpreted the hierarchy of life, from genes to species, as resulting from a series of transitions from less to more inclusive units of selection (Buss, 1987). In their foundational book, John Maynard Smith and Eörs Szathmáry treated the evolution of multicellularity as an example of a "Major Transition in Evolution," events in which new levels of biological organization evolved (Maynard Smith and Szathmáry, 1995).

Maynard Smith and Szathmáry's book established the Major Transitions as a subfield of evolutionary biology, which has expanded greatly in the last 25 years. In both biology and the philosophy of biology, the evolution of multicellularity has been viewed through this lens. Subsequent authors have revised the list of transitions and continue to do so (Herron, 2021, and references therein), but every version we are aware of has included the evolution of multicellularity.

1.3 RATIONALE FOR THE STRUCTURE OF THIS BOOK

Aside from the introductory and concluding chapters, we have organized the book into four sections. The first, Theory and Philosophy, addresses the ways in which the topic of the evolution of multicellularity has informed and been informed by the philosophy of biology (Chapter 2), the theory of multilevel selection (Chapter 3), and the evolution of life cycles (Chapter 4). The evolution of multicellularity has long played a central role in discussions of the nature of biological individuality, which biological units are the bearers of fitness, predictability *versus* contingency in evolution, how complexity is defined and how it evolves, biological hierarchies, the evolution of cooperation, and the diversity of life cycles, among other topics.

Multicellular life cycles are remarkably diverse. In eukaryotes, though, nearly all involve an alternation of haploid and diploid generations, with fertilization establishing the diploid phase and meiosis restoring the haploid condition. Either phase, or both, may undergo mitosis, and the products of mitosis may form multicellular structures in neither, either, or both phases. In some cases, those multicellular structures result from the products of mitosis failing to separate; in others, free-living unicells aggregate to form a multicellular structure. We call these situations clonal and aggregative multicellularity, respectively, and this is a fundamentally important distinction.

In clonal multicellularity, as the name suggests, all of the cells in the multicellular structure are clones of the progenitor cell (zygote, spore, or other propagule) and related to each other by $r \approx 1$. In aggregative multicellularity, relatedness among the cells in an aggregate can take any value, depending on the spatial structure of the population and the efficacy of kin recognition during group formation. Social evolution theory predicts that very different outcomes will result from this difference in relatedness, namely that altruistic cooperation will evolve more easily and to higher levels in the clonal case, where relatedness among cells is consistently high. These predictions are consistent with observations of multicellular complexity, which is only known to reach high levels in clonal multicellularity.

Because aggregative and clonal multicellularity have such fundamental differences, we have devoted a section of the book to each (Sections 2 and 3, respectively). Each of these sections begins with an overview of the diversity and phylogeny of independent origins of multicellularity of the appropriate type (Chapters 5 and 9), followed by chapters on group formation (Chapters 6 and 10), group maintenance (Chapters 7 and 11), and group transformation (Chapters 8 and 12). This organization was inspired by Andrew Bourke's book *Principles of Social Evolution*, which proposes the formation, maintenance, and transformation of social groups as a general framework for understanding the Major Transitions (Bourke, 2011). While we don't view social group formation, maintenance, and transformation as necessarily sequential stages of a Major Transition in Evolution – because the challenges of social group maintenance may persist/arise even after group transformation has occurred, for example – we feel that Bourke's stages nicely illustrate the primary barriers to a successful transition in individuality. We hope the parallel structure of Sections 2 and 3 will facilitate comparison of aggregative and clonal multicellular organisms, helping readers to identify both similarities and differences between these two modes of multicellular development.

Section 4 addresses the evolution of multicellularity through the lenses of life cycle evolution and the evolution of multicellular development. Where the previous two sections focus largely on the early steps in the evolution of multicellularity across a broad range of taxa, the chapters in this section are largely focused on subsequent steps in particular taxa: animals (Chapter 13), fungi (Chapter 14), algae (Chapter 15), and plants (Chapter 16), which collectively comprise the so-called complex multicellular taxa, those with more than a handful of cell types and that exhibit intercellular communication, three-dimensional tissue structure, and genetically regulated tissue differentiation

(Knoll, 2011). The final chapter in this section (Chapter 17) considers the case of "organisms" made up of multiple species, such as lichens and other intimate symbioses.

1.4 CONCLUSION

Our goal for this book was a single resource that summarizes what we know about the evolution of multicellularity. We aimed high and sent invitations to some of the most influential authors in this field, fully expecting that most would decline. To our delight and amazement, nearly all accepted.

The results have far exceeded our expectations. We feel that the chapters cover their respective topics exceptionally well, combining broad overviews with compelling examples and historical context with recent discoveries. Taken as a whole, we think they will serve the roles we had in mind for the book – primer, reference, and catalyst for further research – admirably.

We live in exciting times for the evolution of multicellularity. New theoretical advances, new model systems coming online, and new capabilities for fast, cheap genome sequencing are combining to produce new discoveries at an unprecedented pace and to provide new perspectives for understanding them. Microbial evolution experiments allow direct observations of transitions to multicellular life and tests of adaptive hypotheses. The expansion of CRISPR/Cas9 genome editing into non-model organisms has immense potential to identify genes and gene functions related to the evolution of multicellularity. A new appreciation for the importance of microbial symbionts expands our perspective, not only on how multicellularity can evolve but on what it means to be a multicellular organism. All of these developments mean that our understanding of the evolution of multicellularity is likely to grow at an accelerating pace, and we are excited to see what lies ahead.

ACKNOWLEDGMENTS

This material is based upon work while M.D.H. was serving at the National Science Foundation. P.L.C. was supported by a NASA Postdoctoral Program Fellowship and funding from the David and Lucile Packard Foundation. W.C.R. was supported by NSF DEB-1845363 and a Packard Fellowship for Science and Engineering.

REFERENCES

Bonner, J. T. (1974). *On Development: The Biology of Form*. Boston, Harvard University Press.

Bonner, J. T. (2000). *First Signals: The Evolution of Multicellular Development*. Princeton, Princeton University Press.

Bourke, A. F. G. (2011). *Principles of Social Evolution*. Oxford, Oxford University Press.

Buss, L. W. (1987). *The Evolution of Individuality*. Princeton, Princeton University Press.

Chapman, H. C. (1873). *Evolution of Life*. Philadelphia, JB Lippincott.

Copeland, H. F. (1938). The kingdoms of organisms. *The Quarterly Review of Biology*, 13:383–420.

Copeland, H. F. (1956). *The Classification of Lower Organisms*. Palo Alto, California, Pacific Books.

Darwin, C. (1859). *The Origin of Species* (6th edition, 1872). London, Reprinted by D. Appleton and Company.

Darwin, C. (1887). Letter to C. Lyell, October 11th, 1859. In Darwin, F., editor, *The Life and Letters of Charles Darwin*, Volume 2, pages 208–215. John Murray, London.

Grosberg, R. K. and Strathmann, R. R. (2007). The evolution of multicellularity: A minor major transition? *Annual Review of Ecology Evolution and Systematics*, 38:621–654.

Haeckel, E. (1869). Monograph of Monera by Ernst Haeckel. *Quarterly Journal of Microscopical Science*, 9:27–42.

Haeckel, E. (1894). *Systematische Phylogenie: Systematische Phylogenie der Protisten und Pflanzen*, volume 1. Berlin, Verlag von Georg Reimer.

Herron, M. D. (2021). What are the major transitions? *Biology & Philosophy*, 36:1–19.

Kessin, R. H. (2001). *Dictyostelium: Evolution, Cell Biology, and the Development of Multicellularity*, Volume 38. Cambridge, UK, Cambridge University Press.

Kingsolver, B. (2001). Foreword. In Kingsolver, B. and Kenison, K., editors, *The Best American Short Stories, 2001: Selected from US and Canadian Magazines*. New York, Houghton Mifflin Harcourt.

Kirk, D. L. (1998). *Volvox: A Search for the Molecular and Genetic Origins of Multicellularity and Cellular Differentiation*. Cambridge, UK, Cambridge University Press.

Knoll, A. H. (2011). The multiple origins of complex multicellularity. *Annual Review of Earth and Planetary Sciences* 39, 217–239.

Lamarck, J.-B. (1809). *Philosophie Zoologique ou Eexposition des Considérations Relatives à l'Histoire Naturelle des Animaux* (1963 translation by H. Elliot), Volume 1. Paris, F. Savy.

Lamarck, J.-B. (1815). *Histoire Naturelle des Animaux sans Vertèbres, Présentant les Caractères Généraux et Particuliers de ces Animaux, Leur Distribution, Leurs Classes, Leurs Familles, Leurs Genres et la Citation des Principales Espèces qui s'y Rapportent, précédée d'une Introduction Offrant la Détermination des Caractères Essentiels de l'Animal, sa Distinction du végétal et des Autres Corps Naturels, Enfin, l'Exposition des Principes Fondamentaux de la Zoologie*. Brussels, Meline, Cans et Compagnie.

Maynard Smith, J. and Szathmáry, E. (1995). *The Major Transitions in Evolution*. Oxford, UK, Oxford University Press.

Niklas, K. J. and Newman, S. A. (2016). *Multicellularity: Origins and Evolution*. Cambridge, MA, MIT Press.

One Thousand Plant Transcriptomes Initiative (2019). One thousand plant transcriptomes and the phylogenomics of green plants. *Nature*, 574:679.

Ruiz-Trillo, I. and Nedelcu, A. M. (2015). *Evolutionary Transitions to Multicellular Life: Principles and Mechanisms*. Dordrecht, Springer.

Weismann, A. (1889). *Essays Upon Heredity and Kindred Biological Problems (Authorised Translation)*. Oxford, Clarendon Press.

Whitworth, D. E. (2008). *Myxobacteria: Multicellularity and Differentiation*. Washington, DC, ASM Press.

Section 1

Theory and Philosophy

2 Getting at the Basics of Multicellularity

Maureen A. O'Malley
School of History and Philosophy of Science
University of Sydney, Sydney, NSW Australia

CONTENTS

2.1 INTRODUCTION

All scientific investigations are underpinned by basic assumptions, and the study of multicellularity is no exception. Asking questions about these foundations can help illuminate the field's subject matter and investigative tools. This chapter addresses three general questions about multicellularity:

1. What is multicellularity?
2. How is multicellularity explained?
3. How is multicellularity valued?

Multicellularity is a broad label that covers an important attribute of living systems. However, because multiple phenomena are covered by the term multicellularity, any simple categorization scheme is likely to fail to capture how multicellularity manifests across the tree of life.

DOI: 10.1201/9780429351907-3

9

Taking the various dimensions of multicellularity into account has implications for how multicellularity is theorized and explained. In particular, a single overarching evolutionary explanation of multicellularity as an adaptive phenomenon might in certain respects be inadequate and misleading. A range of different, highly tailored explanations might be more appropriate for the multiple versions of multicellularity on the evolutionary record.

Tangled up with both these issues of categorizing and explaining multicellularity is another very basic one to do with how multicellularity is valued. This chapter asks whether multicellularity is any 'better' than unicellularity in any evolutionary or ecological sense.

2.2 CATEGORIZING MULTICELLULARITY

Like the attempt to define 'life', defining multicellularity is a tenuous endeavor, often clouded by anthropocentrism

(Lyons and Kolter 2015, p. 21)

The diversity of origins and forms of multicellularity across the domains of life is well-recognized (e.g., Parfrey and Lahr 2013). Multicellularity is often characterized as a 'repeated invention' and thus an instance of convergent evolution (Rokas 2008, p. 472; Szathmáry 2015). However, there are numerous cross-cutting distinctions made about multicellularity that show it to be a multi-faceted phenomenon: clonal versus aggregative, simple versus complex, obligate versus facultative, environmentally versus internally driven.

2.2.1 CLONAL VERSUS AGGREGATIVE MULTICELLULARITY

A fundamental way of categorizing extant multicellularity is to divide it into 'aggregative' and 'clonal' forms (e.g., Grosberg and Strathmann 2007). The former reproduce separately but cluster together by various means; the latter forms arise from a series of dividing cells that remain closely connected and may differentiate into further cell types after additional division.

Clonal multicellularity is well-known in animals, plants, algae, and fungi, but it also occurs in microbes that are normally defined as unicellular. Cyanobacteria form filaments, some with differentiated cell types, by not separating completely after cell division (Schirrmeister et al. 2011; Claessen et al. 2014). *Streptomyces* and many other actinobacteria also form differentiated multicellular bodies, but via coenocytes that form hyphal networks and eventually culminate in spore production (Claessen et al. 2014; Barka et al. 2016). *Zoothamnium*, an aquatic ciliate, forms large branching colonies with a common stalk formed from a syncytial fusion of cells (Clamp and Williams 2006). Some amoeboid organisms also use coenocytic and syncytial multicellular structures from which they eventually release amoebal progeny (e.g., *Ichthyosporea*, *Syssomonas*). And although magnetotactic bacteria are usually referred to as 'aggregates', some of them undergo synchronized cloning to replicate as hollow spherical assemblies (Abreu et al. 2014), as do their non-magnetotactic multicellular relatives (LeFèvre et al. 2010).

Aggregative forms of multicellularity with cell differentiation are seen in the fruiting body phenotypes of organisms such *Dictyostelium*, the ciliate *Sorogena*, and a number of other eukaryotic groups (Kaiser 2003; Brown et al. 2012; Schilde and Schaap 2013). Many aggregative multicellular eukaryotes do not carry out everyday activities of feeding and movement in their differentiated multicellular states, which are primarily formations for collective spore dispersal (Cavalier-Smith 2017). In Bacteria, myxobacteria and *Bacillus subtilis* are the classic exemplars of aggregative multicellularity (Kroos 2007). Other bacterial taxa also form multicellular structures by aggregative means (e.g., *Salmonella, Escherichia coli, Proteus mirabilis*) and some of them feature differentiation in cell types (e.g., *Pseudomonas aeruginosa*). In Archaea, several methanogens form tight aggregations, in which unicellular dispersals lead to new aggregative formations (Robinson et al. 1985; Kern et al. 2015). The lifecycles of these apparently obligatory aggregations dictate individual cellular morphologies.

2.2.2 COMPLEX VERSUS SIMPLE MULTICELLULARITY

Another way of distinguishing forms of multicellularity is to place each instance of it on a gradient of complexity. This is often estimated by numbers of cell types (e.g., Rokas 2008; Bonner 2004; Knoll 2011). Particular versions of more differentiated multicellularity sometimes accomplish 'three-dimensional' organization (e.g., Nagy, Kovács et al. 2018), which is a property mostly attributed to animals, plants, and fungi. On this spectrum of simple-to-complex multicellularity, bacterial and archaeal multicellularity is not that different from 'simple' eukaryotic multicellularity. These less differentiated multicellularities are considered to be much more limited in evolutionary and phenotypic innovation than the 'complex multicellularity' exhibited by plants, algae, animals, and fungi (e.g., Knoll 2011).

Is thinking of complexity in this way helpful? Attempts to quantify complexity levels mostly map on to organisms already believed to be complex (Bell and Mooers 1997; Cock and Collén 2015). Such quantifications sideline the immensely complicated lifecycles that many unicellular organisms with facultative multicellular stages undergo (Herron et al. 2013). But because complexity is often discussed with examples like animals in mind, the intricacies of lifecycles rarely count. Nor does metabolic complexity, which is at least as (if not more) likely to be found in less morphologically complex organisms (Poole et al. 2003). Other evaluations argue that due to the many steps and mechanisms involved (i.e., more complexity), 'aggregative multicellularity might be … more difficult to evolve', and yet, at the same time, these 'simple' aggregations might have 'more limited evolutionary potential' (Brunet and King 2017, pp. 127–128).

How is such limitation known? Again, largely by not measuring up to the morphological standards of animals and plants. And even then, some huge but unicellular algae (coenocytes) display great morphological complexity, developmental phases, and bodily regions (Mandoli 1998; Ranjan et al. 2015). It might seem even more remarkable to achieve these capacities with one – albeit large – cell than with numerous differentiated cells. If complexity were measured by lifecycle complexity or sheer numbers of reproductive strategies, many plants and fungi would score reasonably well (Herron et al. 2013; Tripp and Lendemer 2017; Naranjo-Ortiz and Gabaldón 2020), along with a huge range of unicellular organisms that only occasionally opt for multicellular states. Many animals would score very poorly on this measure of complexity, which would mean that equating all instances of clonal multicellularity with greater innovation might not be justifiable.

2.2.3 OBLIGATE VERSUS FACULTATIVE MULTICELLULARITY

A further distinction is that of obligate and facultative multicellularity: whether an organism (or offspring in a lifecycle) persists in a unicellular state or not. This dimension can cut across the clonal-aggregative divide. Although it is often thought that obligate multicellularity is always clonal (e.g., Fisher et al. 2013, p. 1120), and that aggregative multicellularity is mostly facultative – because it occurs in multigenerational life cycles of organisms that have unicellular stages (Brown et al. 2012) – this is not always the case. For instance, some aggregative multicellularity can occur entirely without a unicellular stage. Although multicellularity is almost always discussed in relation to single-species arrangements, lichen propagules – containing their photobiont – are an example of multispecies multicellular reproduction of obligate symbiotic aggregations (Tripp and Lendemer 2017). These lifelong commitments to multicellularity might be considered, therefore, exceptions to general observations that 'no multicellular organisms that develop by aggregation also produce multicellular propagules' (Grosberg and Strathmann 2007, p. 626).

And although a great deal of clonal multicellularity in large eukaryotes is obligatory, at least some forms of clonal multicellularity are facultative (e.g., choanoflagellates, plasmodial slime molds). Moreover, some clonal organisms reproduce from multicellular propagules, meaning a multicellular organism gives rise to another multicellular organism (e.g., vegetatively propagating plants, modular clonal animals, mycelial fragmentation in fungi, choanoflagellate colonies and rosettes, experimental snowflake yeast). In other words, these multicellular entities need not have an obligatorily unicellular stage, which is often taken to be a hallmark of the clonal multicellularity they represent (e.g., Grosberg and Strathmann 1998; Niklas and Newman 2013).

Making the situation more complicated are multicellular magnetotactic bacteria that have unviable unicells and always reproduce as multicellular units. These organisms are thus the most obligatorily multicellular of all (Keim et al. 2004; Abreu et al. 2014). This strategy falls outside standard definitions of clonal multicellularity (serial cell division from a single cell progenitor). Moreover, recent analysis shows genetic variation between cells in the multicellular magnetotactic assemblage (Schaible 2020). The general lesson is that it is difficult to be too definitive about how obligatory or not any form of multicellularity will be, or even whether clonal multicellularity absolutely requires a unicellular bottleneck.

2.2.4 ENVIRONMENTALLY VERSUS INTERNALLY TRIGGERED MULTICELLULARITY

A final distinction is sometimes made between multicellularity that is triggered environmentally or by signals internal to the organism. Many facultatively multicellular organisms require some sort of environmental trigger to initiate the developmental programs of aggregative and even clonal multicellularity. Dictyostelia are an example of the former, for which the classic trigger is starvation (Schilde and Schaap 2013), as are some algae, for which aggregative multicellularity may be invoked by the presence of predators (Kapsetaki and West 2019). Choanoflagellates can be an example of the latter when bacterial signals trigger the serial cell divisions that result in multicellular development (Alegado et al. 2012). But obligate clonal multicellularity, especially the versions leading to many cell types and organs, appears to be driven internally even if environmental conditions affect aspects of the multicellularity that is attained (Fisher and Regenberg 2019).

Although internally triggered multicellularity is usually discussed in relation to plants and animals, some cyanobacteria that develop into clonal filaments do so without environmental triggers (Herrero et al. 2016). Obligate multicellularity without a unicellular stage, as in magnetotactic bacteria, also appears to be internally driven rather than environmentally cued (Keim et al. 2004). But in important respects, this distinction is not very helpful. 'Internal' triggers originate from the cellular constituents of the developing organism, whereas external signals can come from other organisms in close proximity. In the latter case, such signals are what prompt aggregation and eventually the internalization of those very same signals. For example, when *Pseudomonas* aggregates begin the differentiation of cell structures (e.g., 'mushroom' formations), at least some of the signals to differentiate are at that point 'internal' to the developing multicellular organism (Klausen et al. 2003).

2.2.5 SUMMARIZING MULTICELLULARITY DISTINCTIONS

Table 2.1 shows how three distinctions (clonal/aggregative, obligate/facultative, internal/environmental) produce eight loosely defined states of multicellularity. All of them have representatives in the living world today. Each cell in Table 2.1 is represented by an example discussed in the chapter. The table could be further complicated by adding a complex/simple distinction, but since this would involve judgments favoring either complex lifecycles or cell differentiation (see 1.2), or some other factor (McShea 1996), that dimension would not be informative without considerable qualification.

In addition to the cross-cutting nature of these distinctions, phylogenetic reconstructions make clear how many very different phenomena are lumped beneath the label of 'multicellularity', even in close evolutionary quarters (e.g.,

TABLE 2.1
Eight States of Multicellularity

	Facultative	Obligate	Internal Trigger	Environmental Cue
Clonal	Plasmodial slime molds, streptomyces	Cyanobacteria	Metazoans	Choanoflagellates
Aggregative	*Dictyostelium*	Methanobacteria	*Pseudomonas* (mushroom structures)	Myxobacteria

Schirrmeister et al. 2011; Brown et al. 2012; Sebé-Pedrós et al. 2017). And beyond the possible permutations of clonal, aggregative, facultative, obligatory, genetically driven, and environmentally cued development lie a range of very different mechanisms and cellular behaviors. Incomplete separation, sticky adherence, coenocyte growth, coenocyte cellularization, syncytial fusion, and baroque lifecycles are just some of the phenomena that distinguish the varieties of multicellularity distributed across the tree of life. Talking about multicellularity in general and especially '*the* evolution of multicellularity' might be misleading, because it would obscure the different mechanisms and evolutionary histories of such multicellular states. The overall plurality of multicellular states has implications for how relevant phenomena are investigated.

2.3 EXPLAINING THE EVOLUTION OF MULTICELLULARITY

Researchers interested in multicellularity as a broad phenomenon study it not just to understand its various forms but also to gain insight into its evolution. Although these investigations explore a range of phenomena and use different methods, they address similarly basic questions.

2.3.1 EXPLAINING CONVERGENCE AND DIVERGENCE

The diversity of manifestations of multicellularity raises questions about the evolutionary relationships between different versions. Do certain forms give rise to others? Or are they separate evolutionary innovations sharing only some general selection pressures? Although aggregative multicellularity is often thought of as a bit of a deadend, in that it does not explore a great deal of phenotypic (i.e., morphological) space (e.g., Brunet and King 2017), clonal multicellularity is usually described as leading to extensive innovation and impressive morphological diversity (e.g., Szathmáry 2015). Does this mean that the mechanisms and selection pressures of clonal multicellularity are different from those of aggregative versions?

An example that sheds light on this question involves unicellular sister groups to metazoans and choanoflagellates (together forming the clade of Holozoa; see Figure 2.1). Members of these groups are thought to illuminate how metazoans originated and subsequently evolved. The closest group, filastereans, displays a sophisticated form of aggregative multicellularity and development (Sebé-Pedrós et al. 2013). Yet the next closest group, ichthyosporeans (the sister to animals, choanoflagellates, and filastereans), deploys yet another means of reaching multicellularity, in that the multicellular form emerges from the cellularization of a coenocyte, with released cells heading off to establish their own multicellular body (Suga and Ruiz-Trillo 2013; Sebé-Pedrós et al. 2017). These organisms are therefore clonally multicellular.

Choanoflagellates themselves, the closest group to metazoans, exhibit a facultative form of clonal multicellularity that sometimes depends on bacterial signaling (Fairclough et al. 2010; Alegado et al. 2012). Does this suggest that the ancestor of choanoflagellates and metazoans discovered a clonal form of multicellularity, with no ties to these other two forms of multicellularity (aggregation and coenocyte cellularization), and it was such a good discovery (or at least expensive to opt out of)

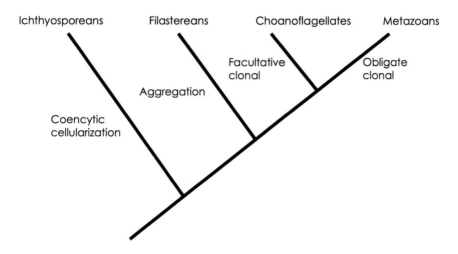

FIGURE 2.1 Holozoan experiments in multicellularity. (Credit: Peter L. Conlin.)

that it became obligate? But many choanoflagellate lineages do not engage in multi-cellularity of any kind, and their ancestral ventures into this lifestyle may not be the basis for metazoan multicellularity (Suga and Ruiz-Trillo 2013). It might only be jus-tified to say that all these lineages – ichthyosporeans, filastereans, choanoflagellates, metazoans, plus others that appear to be closely related and using yet other strategies (see Tikhonenkov et al. 2020) – were intensively exploring multicellularity and its potential, and all of them preserve their very different achievements to this day.

2.3.2 Unifying Versus Particular Explanations

Increasing comparative work shows that looking for common explanations of similar multicellular strategies is not necessarily a helpful strategy. For instance, although *Dictyostelium* multicellularity shares some similar genetic and architectural features with animal multicellularity, the phylogenetic distance between these forms points to convergence, rather than a shared ancestral state of multicellularity (Parfrey and Lahr 2013). Nevertheless, dictyostelid and myxobacterial capacities for multicellu-larity mean they are often used as models for morphogenesis (Kaiser 2003).

Experimental investigations of the origins of multicellularity necessarily focus on particular tractable organisms (e.g., Ratcliff et al. 2012; 2013; Hammerschmidt et al. 2014). Such studies are designed to be extrapolated: what is found in yeast, for instance, is suggested to have broader implications for how 'readily' and 'rapidly' multicellularity can evolve (Ratcliff et al. 2012, p. 109; Koschwanez et al. 2013). But just as intriguing is the fact that these newly emerged multicellular yeasts reproduce as multicellular units, without a unicellular stage (Ratcliff and Travisano 2014). In other experiments, the co-option of unicellular stages for minimizing genetic varia-tion challenges the standard view that unicellular bottlenecks are adaptively central to the origin of clonal multicellularity (Ratcliff et al. 2013).

Although these experiments cannot recapitulate history, they seek generality by abstracting away from the experimental details of lab-evolved multicellularity.

These abstractions might involve the relative importance of ecological versus genetic constraints (Ratcliff et al. 2013), mechanisms for cooperation (Bastiaans et al. 2016), and resource efficiencies (Koschwanez et al. 2013). But it is clear that such experiments produce novel occurrences of multicellularity that arise as the result of particular selection pressures, historical backgrounds, genetic capacities, and phenotypic proclivities. Such novelties might further undermine hopes of a unifying explanation for any multicellular phenomenon

2.3.3 Adaptive Versus Non-Adaptive Explanations

Although it is well-known that multicellularity has many independent origins and historical trajectories, there is still a tendency to seek general explanations of how multicellularity has emerged and persisted. A common feature of such accounts is an explanatory focus on adaptiveness. A standard list of common advantages for any form of multicellular organization includes predator resistance, new food sources (i.e., other organisms), reproductive efficiency, resource storage, recruitment of symbionts, sequestration of incompatible features (e.g., metabolic pathways, motility, and reproduction), and phenotypic flexibility (see Grosberg and Strathmann 2007; Umen 2014; Cavalier-Smith 2017; Nagy, Tóth et al. 2018). And if at least some multicellularity with differentiated cells does not require a high genetic budget to originate and maintain (Nagy, Kovács et al., 2018), then the existence of many independent origins might be explained quite simply (i.e., low-cost exploration of phenotypic space, with potentially high payoff).

Although many experimental and natural-historical studies of multicellularity focus on explanations that emphasize advantages, at least as much if not more can be learned about multicellularity by considering the implications of disadvantages for both the origin and persistence of any particular form (Cavalier-Smith 2017). Non-adaptive accounts of multicellularity highlight biophysical processes that lead to lineages 'drifting' into multicellular explorations and leading in some cases to the co-option of features such as adhesion (Niklas and Newman 2020). Committed multicellularity, in which differentiation is spatial rather than temporal, means that the flexibility to switch between unicellular and multicellular states is lost, which may be risky in changing environments.

Relatedly, if indeed some lineages had a pre-existing toolkit for other functions that was co-opted for novel multicellular functions (Parfrey and Lahr 2013; Ratcliff et al. 2013; Niklas and Newman 2020), it might mean that some multicellularity takes an evolutionary road that is hard to back out of. Once certain tools get locked into the clonal construction of new forms of interdependent cellular organization, it might be difficult to put those tools back to work in old independent functions. Small population sizes of facultatively multicellular organisms might incur less stringent selection and thus persist despite modestly detrimental outcomes (Lynch 2007). Subsequently, if certain environments favor multicellularity even briefly, there could be disadvantages to maintaining unicellular options throughout the lifecycle of generations, and so these stages are lost and prove difficult to re-evolve (Libby et al. 2016). Despite some lineages managing to give up on clonal multicellularity, they are vastly outnumbered by the plants, animals, fungi, and algae that were obliged to continue with a lifecycle of reduced complexity but morphology of greater complexity.

2.3.4 Explanatory Implications of Losing Multicellularity

Although metazoans are sometimes referred to as possessors of 'unconditional multicellularity' (Schilde and Schaap 2013, p. 7), meaning all metazoans are multicellular, this universality has a few interesting aberrations. Microscopic myxozoans are the classic examples of animals giving up on complex body plans (but not complex lifecycles), where not only organs have been discarded along the evolutionary way, but also aerobic mitochondria (Chang et al. 2015; Yahalomi et al. 2020). Transmissible cancers are an extreme example of reversed multicellularity in a range of metazoans (e.g., dogs, Tasmanian devils, shellfish), in which clonal unicellular lineages have made their escape from obligatory multicellular origins (Murgia et al. 2006; Rebbeck et al. 2009). Some commentators go so far as to suggest human cancer cultures, selected and nurtured by biomedical researchers for generations, are also reduced metazoans, now living as 'protists' (Strathmann 1991).

Elsewhere, yeasts have convergently given up on multicellularity (e.g., *Saccharomyces*, *Schizosaccharomyces*), although they can recover it in certain conditions, meaning that perhaps they have managed to reverse a previously obligate state of multicellularity to a merely facultative one (Dickinson 2005; Nagy et al. 2018). Some unicellular green algae show evidence of derivation from multicellular lineages (Delwiche and Cooper 2015). Cyanobacterial lineages have frequently lost and occasionally rediscovered multicellularity (Schirrmeister et al. 2011). These instances of reversion of multicellularity, although wide-ranging, are nevertheless fewer than all the independent origins, which makes multicellularity look like a trait that is harder to lose than to gain.

But rather than thinking only about the loss of multicellularity as a phenomenon that needs explaining, we might also want to think about the retention of unicellular stages in some facultatively multicellular lineages. What holds them back from a full-blown commitment to multicellularity? Why not give in to obligate clonal multicellularity? Even though lifecycle complexity in organisms with facultative multicellularity seems like hard work to maintain, it has persisted in some lineages for longer than the switch to simple lifecycles with a unicellular zygotic or spore stage. But seeing extensive cell differentiation in clonally multicellular organisms as some sort of advance is potentially problematic especially if the long-term future of life on Earth is considered. In all mass extinction events, large obligatorily multicellular forms are the most prone to being wiped out (Mata and Bottjer 2012). Nevertheless, many discussions of the evolution of multicellularity proceed as if obligate clonal multicellularity is a state worth attaining, despite its vulnerability to major environmental changes.

2.4 WHY DO WE VALUE MULTICELLULARITY?

It is a curious phenomenon that despite most life on Earth being unicellular, a great deal of value is placed on the emergence of multicellular life forms, and of those, very particular versions of multicellularity. For instance,

> One of the most remarkable events in evolutionary history was the emergence and radiation of eukaryotic multicellular organisms. *(Niklas and Newman 2013, p. 41)*
>
> Although multicellularity arose more than 20 times, the 'spectacular' forms arose only in plants, animals and fungi. *(Szathmáry 2015, p. 10108)*

A number of justifications are given for why certain forms of multicellularity are so important and interesting:

> the evolution of complex multicellularity … is clearly where key questions of diversity and ecological success lie [rather than the origin of simple multicellularity. *(Knoll 2011, p. 221)*
>
> Of all the transitions between levels of organization, the advent of multicellularity is perhaps the most interesting … Multicellularity completely redefines the concept of what is an individual organism, and has occurred independently dozens of times across all domains of life. Multicellularity represents a transition from the microscopic to the macroscopic world [and thus different responses to] physical laws. Multicellularity also enabled vast phenotypic expansion and diversification, primarily via cell differentiation and temporal development of morphological structures within an organism. And finally, multicellularity is most likely a necessary step along the evolutionary path to intelligence and consciousness. *(Lyons and Kolter 2015, p. 21)*

Quotes such as these suggest several reasons for valuing multicellularity and transitions to it: increasing diversity, adaptive complexity, and macroscopic size.

2.4.1 DIVERSITY

The most basic reason for valuing multicellularity is that there are many instances of it (or even of just 'complex' multicellularity) and that it leads to further diversification in the form of more speciose lineages and greater exploration of phenotypic space. However, the same could be said of unicellularity, especially in terms of thermodynamic 'space' (Nealson and Conrad 1999; Poole et al. 2003), and we might think unicellular diversity has a higher chance of persisting when drastic changes occur to the Earth (Mata and Bottjer 2012). But in response, it might then be argued that multicellularity is more adaptive in important respects.

2.4.2 ADAPTIVE COMPLEXITY

> Multicellularity is a major evolutionary innovation … [that] is a requisite for the development of adaptive complexity. *(Herrero et al. 2016, p. 832)*

But here we might ask whether there is less adaptive complexity in, say, unicellular organisms that have a range of metabolic capacities, all of which are rarely achieved by standard multicellular organisms. For example, many unicellular organisms can switch from autotrophy to heterotrophy, or from aerobic to anaerobic respiration (Kelly and Wood 2006). They can oxidize methane aerobically for breakfast and hydrogen anaerobically for dinner (Carere et al. 2017). Other populations of microorganisms might be defined by their proclivity to drink hydrogen, but have hidden tendencies to snack on glucose and sip alcohols (Schichmann and Müller 2016). More generally, numerous microbes can make extensive use of inorganic energy substrates, which is beyond the scope of large eukaryotes (Nealson and Conrad 1999). This metabolic diversity and flexibility is why microbes are dominant players in most biogeochemical cycles. Versatility in one sphere (i.e., metabolism) but not others (i.e., size and shape) might be discounted if one sphere is intuitively valued more highly.

2.4.3 MACROSCOPIC SIZE

Although most of the individual organisms living on Earth today are still unicellular, if all multicellular eukaryotes suddenly vanished from Earth, our planet would appear as barren as Mars. *(Kirk 2005, p. 299)*

[If we were to] remove the multicellular land plants … [and] get rid of … animals [and] multicellular fungi … [there would be an] empty landscape of bare soil dotted by microscopic photosynthetic bacteria and algae. *(Ratcliff and Travisano 2014, p. 383)*

The 'emptiness' postulated by such views seems to depend heavily on projections of familiar landscapes. It is highly likely that prior to the visual dominance of plants and animals, the Earth was a teeming mass of phototrophic microorganisms of every shade of green and red, forming mats in the shallows and turbid blooms in open waters (Butterfield 2015). The supposedly 'bare soil' was crusted, filmed, and threaded by a variety of unicellular lifeforms, in all sorts of shades and shapes, with considerable biogeochemical impact (Horodyski and Knauth 1994; Labandeira 2005; Wellman and Strother 2015). Some were probably coenocytes, and their large clusters would have added to the visual diversity. Rocks were dramatically stained by other microbial life and ebbing and flowing glaciers dappled by conglomerations of photosynthesizers and other cold-tolerant microbes (Vincent et al. 2000; Beraldi-Campesi 2013). Admittedly, these many types of colorful and intricate communities were not visually similar to today's forests or grasslands, but the landscape would not have been devoid of displays of life or even beauty. And these lifestyles are as biome-defining and niche-creating as the existence of large plants, herbivores, and predators. In other words, it is not clear that claims about 'barrenness' can be justified beyond the fact that only certain visible lifeforms meet standard human expectations.

2.5   CONCLUSION

In short, using the term 'multicellularity' as if it refers to a particular phenomenon looks problematic. We know about many different origins and instances of multicellularity. We may be attuned to features they have in common rather than their differences and even uniqueness. This proclivity in turn might make overly broad any explanation of particular achievements of multicellularity, whether these achievements are historical or experimental. We are inclined to over-value certain forms of multicellularity, partly because of visibility and obvious impact on us as humans. However, there is no denying that the world we inhabit and perceive is one of large forms with extensive environmental impact and that advancing theory and interpreting evidence for how such forms evolved is an utterly central project in evolutionary studies. The more that is learned about all the different instances of multicellularity and their origins, the more complete our understanding of evolution will be.

REFERENCES

Abreu, F., Morillo, V., Nascimento, F. F. et al. 2014. Deciphering unusual uncultured magnetotactic multicellular prokaryotes through genomics. *ISME Journal* 8: 1055–1068.

Alegado, R. A., Brown, L. W., Cao, S. et al. 2012. A bacterial sulfonolipid triggers multicellular development in the closest living relatives of animals. *eLife* 1: e00013.

Barka, E. A., Vatsa, P., Sanchez, L. et al. 2016. Taxonomy, physiology, and natural products of Actinobacteria. *Microbiology and Molecular Biology Reviews* 80: 1–43.

Bastiaans, E., Debets, A. J. M., and Aanen, D.K. 2016. Experimental evolution reveals that high relatedness protects multicellular cooperation from cheaters. *Nature Communications* 7: 11435.

Bell, G. and Mooers, A. O. 1997. Size and complexity among multicellular organisms. *Biological Journal of the Linnaean Society* 60: 345–363.

Beraldi-Campesi, H. 2013. Early life on land and the first terrestrial ecosystems. *Ecological Processes* 2(1).

Bonner, J. T. 2004. Perspective: the size-complexity rule. *Evolution* 58: 1883–1890.

Brown, M. W., Kolisko, M., Silberman, J. D. and Roger, A. J. 2012. Aggregative multicellularity evolved independently in the eukaryotic supergroup Rhizaria. *Current Biology* 22: 1123–1127.

Brunet, T. and King, N. 2017. The origin of animal multicellularity and cell differentiation. *Developmental Cell* 43: 124–140.

Butterfield, N. J. 2015. Proterozoic photosynthesis – a critical review. *Palaeontology* 58: 953–972.

Carere, C. R., Hards, K., Houghton, K. M., et al. 2017. Mixotrophy drives niche expansion of verrucomicrobial methanotrophs. *ISME Journal* 11: 2599–2610.

Cavalier-Smith, T. 2017. Origin of animal multicellularity: precursors, causes, consequences – the choanoflagellate/sponge transition, neurogenesis and the Cambrian explosion. *Philosophical Transactions of the Royal Society London B* 372: 20150476.

Chang, E. S., Neuhof, M., Rubinstein, N. D. et al. 2015. Genomic insights into the evolutionary origin of Myxozoa within the Cnidaria. *Proceedings of the National Academy of Sciences USA* 112: 14912–14917.

Claessen, D., Rozen, D. E., Kuipers, O. P., Søgaard-Andersen, L., and Van Wezel, G. P. 2014. Bacterial solutions to multicellularity: a tale of biofilms, filaments and fruiting bodies. *Nature Reviews Microbiology* 12: 115–124.

Clamp, J. C. and Williams, D. 2006. A molecular phylogenetic investigation of *Zoothamnium* (Ciliophora, Peritrichia, Sessilida). *Journal of Eukaryotic Microbiology* 53: 494–498.

Cock, J. M. and Collén, J. 2015. Independent emergence of complex multicellularity in the brown and red algae. In *Evolutionary Transitions to Multicellular Life* ed. I. Ruiz-Trillo and A. M. Nedelcu, 335–361. Dordrecht: Springer.

Delwiche, C. F. and Cooper, E. D. 2015. The evolutionary origin of a terrestrial flora. *Current Biology* 25: R899–R910.

Dickinson, J. R. 2005. Are yeasts free-living unicellular eukaryotes? *Letters in Applied Microbiology* 41: 445–447.

Fairclough, S. R., Dayel, M. J., and King, N. 2010. Multicellular development in a choanoflagellate. *Current Biology* 20: R875–R876.

Fisher, R. M., Cornwallis, C. K., and West, S. A. 2013. Group formation, relatedness, and the evolution of multicellularity. *Current Biology* 23: 1120–1125.

Fisher, R. M. and Regenberg, B. 2019. Multicellular group formation in *Saccharomyces cerevisiae*. *Proceedings of the Royal Society London B* 286: 20191098.

Grosberg, R. K. and Strathmann, R. R. 1998. One cell, two cell, red cell, blue cell: the persistence of a unicellular stage in multicellular life histories. *Trends in Ecology & Evolution* 13: 112–116.

Grosberg, R. K. and Strathmann, R. R. 2007. The evolution of multicellularity: a minor major transition? *Annual Review of Ecology, Evolution, and Systematics* 38: 621–654.

Hammerschmidt, K., Rose, C. J., Kerr, B., and Rainey, P. B. 2014. Life cycles, fitness decoupling and the evolution of multicellularity. *Nature* 515: 75–79.

Herrero, A., Stavans, J., and Flores, E. 2016. The multicellular nature of filamentous heterocyst-forming cyanobacteria. *FEMS Microbiology Reviews* 40: 831–854.

Herron, M. D., Rashidi, A., Shelton, D. E., and Driscoll, W. W. 2013. Cellular differentiation and individuality in the 'minor' multicellular taxa. *Biological Reviews* 88: 844–861.

Horodyski, R. J. and Knauth, L. P. 1994. Life on land in the Precambrian. *Science* 263: 494–498.

Kaiser, D. 2003. Coupling cell movement to multicellular development in myxobacteria. *Nature Reviews Microbiology* 1: 45–54.

Kapsetaki, S. E. and West, S. A. 2019. The costs and benefits of multicellular group formation in algae. *Evolution* 73(6): 1296–1308.

Kelly, D. P. and Wood, A. P. 2006. The chemolithotrophic prokaryotes. *Prokaryotes* 2: 441–456.

Keim, C. N., Martins, J. L., Abreu, F. et al. 2004. Multicellular life cycle of magnetotactic prokaryotes. *FEMS Microbiology Letters* 240: 203–208.

Kern, T., Linge, M., and Rother, M. 2015. *Methanobacterium aggregans* sp. nov., a hydrogenotrophic methanogenic archaeon isolated from an anaerobic digester. *International Journal of Systematic and Evolutionary Microbiology* 65: 1975–1980.

Kirk, D. L. 2005. A twelve-step program for evolving multicellularity and a division of labor. *BioEssays* 27: 299–310.

Klausen, M., Aaes-Jørgensen, A., Molin, S., and Tolker-Nielsen, T. 2003. Involvement of bacterial migration in the development of complex multicellular structures in *Pseudomonas aeruginosa* biofilms. *Molecular Microbiology* 50: 61–68.

Koschwanez, J. H., Foster, K. R., and Murray, A. W. 2013. Improved use of a public good selects for the evolution of undifferentiated multicellularity. *eLife* 2: e00367.

Knoll, A. H. 2011. The multiple origins of complex multicellularity. *Annual Review of Earth and Planetary Science* 39: 217–239.

Kroos, L. 2007. The *Bacillus* and *Myxococcus* developmental networks and their transcriptional regulators. *Annual Review of Genetics* 41: 13–39.

Labandeira, C. C. 2005. Invasion of the continents: cyanobacterial crusts to tree-inhabiting arthropods. *Trends in Ecology and Evolution* 20: 253–262.

Lefèvre, C. T., Abreu, F., Lins, U., and Bazylinski, D. A. 2010. Nonmagnetotactic multicellular prokaryotes from low-saline, nonmarine aquatic environments and their unusual negative phototactic behavior. *Applied and Environmental Microbiology* 76: 3220–3227.

Libby, E., Conlin, P. L., Kerr, B., and Ratcliff, W. C. 2016. Stabilizing multicellularity through ratcheting. *Philosophical Transactions of the Royal Society London B* 371: 20150444.

Lynch, M. 2007. The frailty of adaptive hypotheses for the origins of organismal complexity. *Proceedings of the National Academy of Sciences USA* 104 (Suppl. 1): 8597–8604.

Lyons, N. A. and Kolter, R. 2015. On the evolution of bacterial multicellularity. *Current Opinion in Microbiology* 24: 21–28.

Mata, S. A. and Bottjer, D. J. 2012. Microbes and mass extinctions: paleoenvironmental distribution of microbialites during time of biotic crisis. *Geobiology* 10: 3–24.

Mandoli, D. F. 1998. Elaboration of body plan and phase change during development of *Acetabularia*: how is the complex architecture of a huge unicell built? *Annual Review of Plant Physiology and Plant Molecular Biology* 49: 173–198.

McShea, D. W. 1996. Metazoan complexity and evolution: is there a trend? *Evolution* 50: 477–492.

Murgia, C., Pritchard, J. K., Kim, S. Y., Fassati, A., and Weiss, R. A. 2006. Clonal origin and evolution of a transmissible cancer. *Cell* 126: 477–487.

Nagy, L. G., Kovács, G. M., and Krizsán, K. 2018. Complex multicellularity in fungi: evolutionary convergence, single origin, or both? *Biological Reviews* 93: 1778–1794.

Nagy, L. G., Tóth, R., Kiss, E., Slot, J., Gácser, A., and Kovacs, G. M. 2018. Six traits of fungi: their evolutionary origins and genetic bases. *The Fungal Kingdom* 1: 35–56.

Naranjo-Ortiz, M. A. and Gabaldón, T. 2020. Fungal evolution: cellular, genomic and metabolic complexity. *Biological Reviews* 95: 1198–232.

Nealson, K. H. and Conrad, P. G. 1999. Life: past, present and future. *Philosophical Transactions of the Royal Society London B* 354: 1923–1939.

Niklas, K. J. and Newman, S. A. 2013. The origins of multicellular organisms. *Evolution & Development* 15: 41–52.

Niklas, K. J. and Newman, S. A. 2020. The many roads to and from multicellularity. *Journal of Experimental Botany* 71: 3247–3253.

Parfrey, L. W. and Lahr, D. J. G. 2013. Multicellularity arose several times in the evolution of eukaryotes. *Bioessays* 35: 339–347.

Poole, A. M., Phillips, M. J., and Penny, D. 2003. Prokaryote and eukaryote evolvability. *BioSystems* 69: 163–185.

Ranjan, A., Townsley, B. T., Ichihashi, Y., Sinha, N. R., and Chitwood, D. H. 2015. An intracellular transcriptomic atlas of the giant coenocyte *Caulerpa taxifola*. *PLOS Genetics* 11: e1004900.

Ratcliff, W. C., Denison, R. F., Borrello, M., and Travisano, M. 2012. Experimental evolution of multicellularity. *Proceedings of the National Academy of Sciences USA* 109: 1595–1600.

Ratcliff, W. C., Herron, M. D., Howell, K., Pentz, J. T., Rosenzweig, F., and Travisano, M. 2013. Experimental evolution of an alternating uni- and multicellular life cycle in *Chlamydomonas reinhardtii*. *Nature Communications* 4: 2742.

Ratcliff, W. C. and Travisano, M. 2014. Experimental evolution of multicellular complexity in *Saccharomyces cerevisiae*. *BioScience* 64: 383–393.

Rebbeck, C. A., Thomas, R., Breen, M., Leroi, A. M., and Burt, A. 2009. Origins and evolution of a transmissible cancer. *Evolution* 63(9): 2340–2349.

Robinson, R. W., Aldrich, H. C., Hurst, S. F., and Bleiweis, A.S. 1985. Role of the cell surface of *Methanosarcina mazei* in cell aggregation. *Applied and Environmental Microbiology* 49: 321–327.

Rokas, A. 2008. The molecular origins of multicellular transitions. *Current Opinion in Genetics & Development* 18: 472–478.

Schaible, G. 2020. Multicellular magnetotactic bacteria: organized complexity in the Domain Bacteria. *AAAS Annual Meeting*, Seattle WA, Feb. 13–16.

Schichmann, K. and Müller, V. 2016. Energetics and application of heterotrophy in acetogenic bacteria. *Applied and Environmental Microbiology* 82: 4056–4069.

Schilde, C. and Schaap, P. 2013. The Amoebozoa. In *Dictyostelium discoideum Protocols*, ed. L. Eichnger L and F. Rivero F. Methods in Molecular Biology 983 (pp. 1–15). Springer, New York.

Schirrmeister, B. E., Antonelli, A., and Bagheri, H. C. 2011. The origin of multicellularity in cyanobacteria. *BMC Evolutionary Biology* 11: 45.

Sebé-Pedrós, A., Irimia, M., Del Campo, J., Parra-Acero, H., Russ, C., Nusbaum, C., Blencowe, B. J., and Ruiz-Trillo, I. 2013. Regulated aggregative multicellularity in a close unicellular relative of metazoa. *eLife* 2: e01287.

Sebé-Pedrós, A., Degnan, B. M., and Ruiz-Trillo, I. 2017. The origin of Metazoa: a unicellular perspective. *Nature Reviews Genetics* 18: 498–512.

Strathmann, R. R. 1991. From metazoan to protist via competition among cell lineages. *Evolutionary Theory* 10: 67–70.

Suga, H. and Ruiz-Trillo, I. 2013. Development of ichthyosporeans sheds light on the origin of metazoan multicellularity. *Developmental Biology* 377: 284–292.

Szathmáry, E. 2015. Toward major evolutionary transitions theory 2.0. *Proceedings of the National Academy of Sciences USA* 112: 10104-10111.

Tikhonenkov, D. V., Hehenberger, E., Esaulov, A. S., Belyakova, O. I., Mazei, Y. A., Mylnikov, A. P., and Keeling, P. J. 2020. Insights into the origin of metazoan multicellularity from predatory unicellular relatives of animals. *BMC Biology* 18: 39.

Tripp, E. A. and Lendemer, J. C. 2017. Twenty-seven modes of reproduction in the obligate lichen symbiosis. *Brittonia* 70: 1–14.

Umen, J. G. 2014. Green algae and the origins of multicellularity in the plant kingdom. *Cold Spring Harbor Perspectives in Biology* 6: a016170.

Vincent, W. F., Gibson, J. A. E., Pienitz, R., and Villeneuve, V. 2000. Ice shelf microbial ecosystems in the high Arctic and implications for life on snowball Earth. *Naturwissenschaften* 87: 137–141.

Wellman, C. H. and Strother, P. K. 2015. The terrestrial biota prior to the origin of land plants (embryophytes): a review of the evidence. *Palaeontology* 58: 601–627.

Yahalomi, D., Atkinson, S. D., Neuhof, M., Chang, E. S., Philippe, H., Cartwright, P., Bartholomew, J. L., and Huchon, D. 2020. A cnidarian parasite of salmon (Myxozoa: *Henneguya*) lacks a mitochondrial genome. *Proceedings of the National Academy of Sciences USA* 117: 5358–5363.

3 Multi-Level Selection of the Individual Organism

Richard E. Michod
Department of Ecology and Evolutionary Biology,
University of Arizona, Tucson, AZ, USA

CONTENTS

3.1 INTRODUCTION

"Is there anything in evolution that can't be answered by individual selection, that needs to be explained by selection acting on groups?" asks Jerry Coyne, an evolutionary geneticist at the University of Chicago. "I can't think of any." *(Morrell, 1996)*

Although rhetorical, this remark reflects a common view in evolutionary biology that most questions can be addressed by viewing organisms as the sole unit of selection. For billions of years there were only unicellular organisms on earth; where did multicellular organisms come from? The answer, of course, is that multicellular organisms evolved from unicellular organisms when unicellular organisms started forming cell groups. In this chapter, I review work concerning the role of multi-level selection (MLS) in evolutionary transitions in individuality (ETIs), with focus on theoretical work on the transition from single cells to multicellular organisms. As we will see, MLS is needed to explain the origin of the multicellular organism, that very entity that is supposed to deny the need for MLS in evolutionary biology. The central question I wish to address with these models is, how do groups of individuals become a new kind of individual, or, with respect to the evolution of multicellularity, how do groups of cells become an individual multicellular organism?

3.2 BACKGROUND

3.2.1 EVOLUTIONARY TRANSITIONS IN INDIVIDUALITY (ETIs)

In the present chapter, I discuss the use of multilevel selection theory (MLS) to study evolutionary transitions in individuality, or ETIs, with focus on the transition from unicellular to multicellular organisms. ETIs are changes in the unit of selection and adaptation, that is, changes in the evolutionary individual. Examples of ETIs include the evolution of the cellular genome from replicating molecules, the evolution of complex eukaryotic cells from groups of bacterial and archaeal cells, the evolution of multicellular organisms from unicellular organisms, and the evolution of eusocial societies from solitary organisms. ETIs are rare events having happened dozens of times during the history of life. While rare, they have contributed to one of life's most fundamental characteristics, its hierarchical structure. While not all levels in the hierarchy of life are evolutionary individuals (for example, tissues or organs), evolutionary individuals constitute levels in the hierarchy of life (for example, bacterial cells, eukaryotic cells, multicellular organisms, eusocial insect societies). ETIs involve the evolution of a group of existing individuals into a new kind of individual, such as the evolution of a group of cells into a multicellular organism. ETIs involve multilevel selection and the evolution of traits, such as conflict mediators discussed below, that modify the development of the groups to enhance the individuality of the group.

3.2.2 WRIGHT'S SHIFTING BALANCE THEORY

MLS has been part of population genetics since the foundations of population genetics in the early part of the last century in the work of Sewell Wright and his shifting-balance theory of evolution (1932, 1977, Chapter 13). In Wright's view, a large global population partially subdivided into local groups is the most favorable for continued evolution. The groups in this theory are local subpopulations partially isolated from other such subpopulations. In the terminology of MLS theory introduced below; Wright's shifting balance theory is in the realm of MLS1. Wright's shifting balance theory does not focus on the fitness interactions that can occur within groups, and the groups in Wright's theory are not evolving into new kinds of evolutionary individuals. Nevertheless, Wright appreciated the significance of group structure and MLS to evolution. Stochastic variation in local subpopulations allows groups to explore different fitness peaks. This within-group selection leads to an ensemble of groups, each attracted to local and likely different fitness optima. If one of the local optima is also a global optimum, then between-group selection mediated by, for example, differential migration among subpopulations (Wright termed this group selection phase "asymmetric diffusion") could result in transformation of the larger population or entire species. This group selection phase of Wright's theory builds upon individual selection; in the language of MLS theory introduced below, individual effects are filtering up to the group level; there are no true group effects. Still, in a partially subdivided population, local adaptation is possible, and so is mass transformation of the species. Large homogeneous populations without subdivision become trapped on local optima unable to explore the fitness surface, and small isolated populations suffer inbreeding and deleterious effects of genetic drift. Partially subdivided populations are more favorable for continued evolutionary change or movement towards fitness maxima according to Wright.

The group selection Phase 3 of Wright's shifting balance process depends on differences among groups in their average fitness. The average fitness of the group describes its growth and output into the global population through a set of equations that Wright developed (Wright, 1931, 1932, 1969, 1977). The discussion here is primarily based on the second volume of his 4-volume treatise (Wright, 1969). When fitness is constant, the average fitness of a group also controls within group change resulting from selection among individuals within the group. Consequently, when selection is constant, there is a harmony between selection at two levels in a selection hierarchy, in the sense that traits that increase the fitness of individuals also increase the average fitness of groups. When selection is constant, there is no conflict between the two levels in the selection hierarchy, and Wright's shifting balance process can operate with the third phase of group selection building upon the gene frequency change occurring during the second phase of within group selection.

Frequency-dependent selection based on fitness interactions within the group changes all this because the phase of within-group selection and the phase of between-group selection are determined by different mathematical functions leading to the possibility of conflict between levels of selection in the direction of gene frequency change. Wright's (1969, p. 121) "fitness function" is maximized by the dynamics of within-group change in both cases of frequency-dependent and constant selection. However, under frequency-dependent selection, Wright's fitness function no longer equals average individual fitness, as it does with constant selection. Population growth is still determined by average individual fitness with frequency-dependent selection, as is the case with constant selection. In Wright's theory, we are in the pre-ETI realm of MLS1; after an ETI the fitness of the group is decoupled from the fitnesses of its members, but that is not what Wright is concerned with in the shifting balance process.

Frequency-dependent selection and group selection can work in different directions. In frequency-dependent models, within-population change among organisms can lead to demise of the group and local extinction (for some simple examples, such as the evolution of spite, see Wright (1969, p. 127)). ETIs by their nature depend upon the frequency-dependent evolution of cooperative interactions within groups. The tension between the well-being of the group and selection dynamics among its members leading to conflict between levels is the basic problem that must be solved during an evolutionary transition to a new unit of selection and adaptation, a transition to a new kind of evolutionary individual (Michod, 1999).

Although the maximization of individual fitness no longer occurs generally under frequency-dependent selection, maximization principles may be developed in specific cases. For example, during multilevel selection in populations that are structured into family groups, Wright's fitness function (1969, p. 121) equals the average inclusive fitness effect which is maximized by the population dynamics (Michod and Abugov, 1980). Future work is needed to determine what property might be maximized during an ETI. In the modifier models discussed below, the ratio between selection at the group level and selection within groups appears to increase during the ETI (Figure 3.2C), but more work is needed to show if this is indeed a maximization process. Having a maximization principle for the evolution of individuality, even in simple limiting cases, would be extremely useful for understanding the concept of biological individuality. There is a large literature devoted to understanding biological individuality, with several collections of papers providing an overview of this exciting field (Bouchard and Huneman, 2013; Calcott and Sterelny, 2011; Gissis, Lamm, and Ayelet, 2017; Van Baalen and Huneman, 2014).

3.2.3 MULTILEVEL SELECTION (MLS)

MLS occurs in a population that is structured into subpopulations, or groups. I use the term group and subpopulation interchangeably in this chapter. Group structure has several consequences. Most relevant to our concerns in this chapter is group structure facilitates the evolution of cooperative interactions. Migration is reduced between groups, so interactions occur preferentially within the group. The environment of each group may differ leading to different forces of selection within each group that in turn leads to different traits and variation between groups in genetic composition. The population size is smaller in groups than in the global population with the possibility of increased genetic drift producing different gene frequencies in each group and increased variation between groups. As a result of variation between groups, there is a possibility for selection at the group level, such as when some groups persist longer, survive better, or produce more offspring and migrants than other groups. In a group-structured population, there may be selection among individuals within a group and selection between groups.

Although present in the foundational work of Wright (1932, 1977, Chapter 13), MLS did not emerge as a subfield within evolutionary biology until the 1970s, along with interest in the evolution of social and anti-social behavior. Interest in MLS has continued more recently because of its role in ETIs, beginning especially with the work of Buss (1987). Contemporary MLS theory began with the foundational work of Price, who developed a covariance approach to selection (1970) in the MLS context (1972) that is discussed in more detail below. Wilson (1975) developed an MLS model, termed the "trait-group" model, that has been extremely influential in the study of the evolution of cooperative and social behavior. Like Wright, Wilson distinguished two phases of selection, within- and between-group selection, but Wilson was interested in the effects of this multilevel selection on the evolution of fitness-affecting interactions within the group such as the evolution of cooperation. During the within-group phase, a cooperative trait will usually decline in frequency, because of the costs paid by cooperative individuals relative to non-cooperative or defecting individuals in the same group. During the between-group selection phase, groups with more cooperation survive at higher rates or output more offspring to the next generation than groups with less cooperation. Wilson showed how between-group selection in favor of cooperation can overcome within group selection against cooperation. In other foundational works, Heisler and Damuth (1987) developed a contextual analysis approach to study selection in structured populations and Damuth and Heisler (1988) distinguished between two kinds of MLS. MLS-type 1 (termed MLS1 here) occurs when the focal entities are the individuals within the group and the group provides context for selection on the individuals. MLS-type 2 (termed MLS2 here) occurs when the groups themselves are the focus and the groups differentially survive and reproduce as groups. As discussed more below, ETIs have been characterized as a transition between MLS1 and MLS2 (see, for example, Okasha, 2005). There are several general models of MLS (Frank, 2012; Gardner, 2015; Gardner and Grafen, 2009). The best introduction to MLS in evolutionary biology is Okasha's book (2006).

We have used MLS theory to study the development of cell groups during the evolution of multicellular organism (Michod, 1996, 1999; Michod, Nedelcu, and Roze, 2003; Michod and Roze, 1997a, 1997b, 1999, 2001; Roze and Michod, 2001). There are three analytical tools or modeling approaches we use in our work. The first tool is Price's covariance approach to selection (1970) which he developed in MLS context (1972). The second tool involves kin selection and the study of evolution in genetically structured populations (Hamilton, 1964a, 1964b; Michod, 1982; Michod and Abugov, 1980). The third tool involves game theory and its use in the study of the evolution of cooperation and conflict in structured populations. Kin selection is implicit in the MLS models discussed here because the models involve groups that develop clonally from a single cell propagule (Figure 3.1). I begin with Price's equation since its analysis in an MLS context leads to the notion of counterfactual fitness that is useful in quantifying evolution through an ETI.

3.2.4   Price Equation and Counterfactual Fitness

Following Darwin, a population evolves by natural selection when there is heritable variation in fitness. The Price equation or Price's theorem (Price, 1970, 1972) can be thought of as a mathematical version of this "conditions approach" to natural selection (Okasha, 2006, pp. 36–37), in which the Darwinian conditions are represented in equation form (Shelton and Michod, 2020). For an overview of the Price equation in evolutionary biology, see a recent collection of papers on this topic (Lehtonen, Okasha, and Helanterä, 2020). The logic of natural selection is general (Lewontin, 1970), and Darwin's conditions may apply at several hierarchically nested levels at the same time. The occurrence of selection at multiple levels is MLS. Price (1972) and Hamilton (1975) showed how the Price equation can be applied recursively to

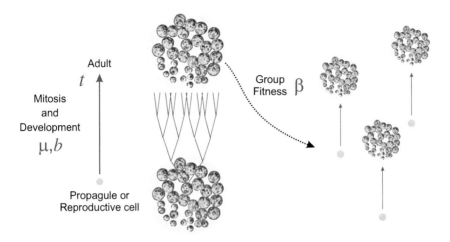

FIGURE 3.1   Development and multi-level selection in a model of a multicellular organism. Mitotic cell divisions produce an adult cell group from a propagule. There can be selection at both the group level and the within group or cell level during development of the adult group. Solid blue lines indicate development, dashed black lines indicated group productivity which is determined by the frequency of cooperative cells in the group. Parameters μ, b, t in red refer to the parameters of development: μ is the mutation rate (from cooperate to defect), b cell division rate, and t is the time available for cell division and development, respectively. The parameter β refers to the beneficial group fitness effect of cooperation on the number of propagules produced by the group. Other parameters may be studied such as the survival of cells during development and cell size, but these characters are not considered here. There are, however, additional parameters related to the nature of modifier locus and how it modifies the developmental parameters μ, b, t. Several different kinds of modifiers have been considered: cell policing, programmed cell death, determinant group size, and germ line modifier alleles (Michod, 2003). For example, the germ line modifier considered in Figure 3.2 reduces the development time of the germ line by amount δ; in other words, the soma develops for time t and the germ line develops for time $t - \delta$, so there is less time available for deleterious mutation in the germ line relative to the somatic line. In the germ line model, the timing of sequestration of the germ line is another parameter that has been considered (Michod et al., 2003).

represent selection at different levels giving Equation 3.1, in which Δq and Δq_s are the frequencies of an allele of interest in the global population and subpopulation s, respectively, and $\bar{W}$ and $\bar{W}_s$ are the average fitnesses of the global population and subpopulations (indexed by s), respectively. The derivation of Equation 3.1, or similar forms of the Price equation, can be found in many places; here I follow Michod (1999, Chapter 4).

$$\Delta q \bar{W} = Cov\left[\bar{W}_s, q_s\right] + E\left[\bar{W}_s \Delta q_s\right].$$ (3.1)

Hamilton (1975) interpreted Equation 3.1 as accomplishing a "formal separation of levels of selection," in which the first term on the right-hand-side represented the effects on gene frequency change of between-group selection (or just "group selection") and the second term represented the effects of within-group or individual selection. Unfortunately, further analysis of the Price partition has shown the causal situation is not so simple, and different partitions are needed to formally separate group and individual selection in explicit MLS models. See Okasha (Okasha, 2006) and Shelton and Michod (2020) for a fuller discussion of these issues.

The basic issue that must be addressed with the conditions approach represented in Equation 3.1 may be termed the "pseudo-group" problem (Shelton and Michod, 2014). Pseudo-groups meet Darwin's conditions of heritable variation in fitness, without there being "true" group effects on fitness. Following Williams (1966b), consider a herd of fleet deer, that is, a population of deer in which there is variation in the running speed of individual deer. If this population is subdivided into groups, by chance, groups will contain different compositions of deer and the groups will vary in average group running speed. This between-group variation in average running speed, likely an important component of group fitness, is not due to any interaction among the deer within the group and so it is not a "true" group effect. Rather, the between group variation in average running speed is determined solely by sampling individual properties of the deer, properties the deer have in isolation from one another. The situation in which the evolution of a group-level trait is influenced only by natural selection at the individual-level and not by natural selection at the group level has been termed a "fortuitous benefit" (Williams, 1966b) or "cross-level byproduct" (Okasha, 2006). In this chapter, I use the term "pseudo-group" for these kinds of groups; in pseudo-groups, the fitness variation at the individual level "filters up" to the group level (Shelton and Michod, 2014).

Beginning with Darwin, as represented in, for example, Lewontin (1970), it was assumed that the conditions for natural selection were sufficient for understanding natural selection and the partitioning of the contribution of each level in a selection hierarchy to overall genetic change. It is now clear that there is more to the problem of partitioning "group selection" in MLS scenarios than the conditions approach alone can resolve, and explicit mathematical models are needed to clarify the causation of selection, for example, whether selection is caused by true group effects in which interactions within groups play a causal role.

An approach based on counterfactual fitness allows for a clear partition between levels of selection and a possible resolution of the pseudo-group problem (Shelton and Michod 2020). Another approach to resolve the pseudo-group problem is to redefine the term "group" and restrict it to cases in which there are true group effects (Clarke, 2016). Counterfactual fitness is the fitness an individual would have were it to leave the group. Shelton and Michod assumed that in MLS models without genetic drift, pleiotropy, and epistasis, that only "group selection" can lower counterfactual fitness, that is, only group selection can lower the fitness of a cell were it to leave the group. Based on this assumption, Michod and Shelton (2014, 2020) developed a partition of group and individual selection that attributes the degree of group and individual selection in problem cases like when there are pseudo groups. Unlike the Price approach, the counterfactual approach has the desirable feature that the degree of inferred group-specific selection increases continuously with the degree of group effect. Furthermore, as the ETI proceeds, group fitness becomes decoupled from counterfactual fitness, in the sense that fitness in a group may be quite high,

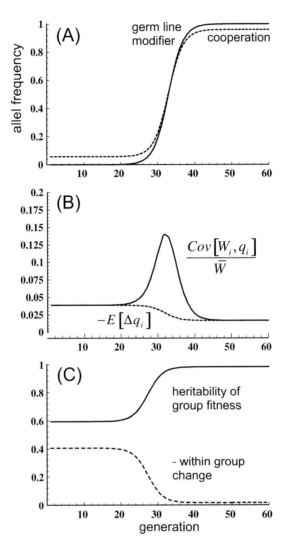

FIGURE 3.2 Components of an evolutionary transition to multicellular individuality as modeled by a two-locus MLS modifier model given in Figure 3.1 and Table 3.1. Figure modified from Michod and Roze (1997). The figure considers the case of a transition between equilibrium 3 and 4 (see Table 3.1 and associated text) for a costly germ line modifier of within organism change (mathematical model presented in Michod [1996]). The modifier is assumed to decrease the development time for the germ line (when compared to the soma) by amount δ. The parameter values studied are $\mu = 0.003$, $\beta = 30$, $t = 40$, $\delta = 35$ and $b = 1.1$ (b is the replication rate of defecting cells relative to unity for cooperating cells). The parameter δ is the reduction of time for cell division in the germ line compared to the somatic line. In other words, the somatic line divides for time t, while the germ line divides for time $t - \delta$, so there is less time available for deleterious mutation in the germ line. The x-axis for all panels is time in organism generations. The y-axis in panel (A) is gene frequency (frequency of either the C allele or the M allele); the y-axis in panels (B) and (C) correspond to the different curves as labeled. Modifier M alleles may increase and sweep through the population

even as counterfactual fitness decreases to zero. In this way, counterfactual fitness can be used to quantify progression through an ETI. Although more work needs to be done on this problem, our analyses to date based on counterfactual fitness suggest that there are at least three kinds of selection that can be occurring at the same time in MLS models: group-specific selection along with two kinds of individual selection, within-group selection and global individual selection. "Global individual selection" refers to the aspects of individual selection that are independent of interactions within a group (more on global individual selection below).

3.3 MLS MODELS OF THE EVOLUTION OF ORGANISMAL DEVELOPMENT

3.3.1 OVERVIEW

We have used MLS to study the development of cell groups and the conditions under which modifiers of development evolve that increase the heritability of group fitness, reduce within group selection, and lead to a decoupling of fitness between levels (Michod, 1996, 1999; Michod and Roze, 1997, 1999, 2001; Roze and Michod, 2001). These models are hypotheses for the transition between MLS1 and MLS2 and the origin of multicellular individuality. These models are intended as heuristic devices for understanding the evolutionary transition to multicellularity and how a genotype-phenotype map could be reconstructed at the group level when initially it is present only at the cell level. The basic setup is given in Figure 3.1, for concreteness a volvocine green alga colony is shown, but the models are abstract and general population genetics models; they are not specific to the volvocine algae. Using these models, we have

FIGURE 3.2 *(continued)*

as shown by the solid curve of panel (A) leading to an increase in cooperativity among cells (dashed curve in panel [A]). As shown in panel (B), the underlying cause of the modifier's success during the transition is the fact that the heritable covariance in fitness for the organism, $\frac{Cov(W_i, q_i)}{\bar{W}}$ (solid curve), is greater than the average within organism change, $E(\Delta q_i)$ (dashed curve). The derived gene frequency dynamics under multilevel selection can be represented in terms of Price's Equation 3.1 which becomes in this case $\Delta q = \frac{Cov(W_i, q_i)}{\bar{W}} + E(\Delta q_i)$, with $q_i, \Delta q_i$ being the frequency, and change in frequency, respectively, of the C (cooperate) allele within zygotes of type i, and $q, \Delta q$ the frequency, and change in frequency, respectively, of the C gene in the total population ($i = 1, 2, 3, 4$ for CM, Cm, DM and Dm zygotes respectively). As occurs in panel (B), the two components of the Price equation must be equal at both equilibria, before (at equilibrium 3) and after the transition when the population returns to equilibrium (now equilibrium 4). As a consequence of the modifier's success and fixation, the level of cooperation in the population increases dramatically from nearly zero initially to greater than 0.90 after the transition (panel [A] dashed curve), and the heritability of fitness at the emerging organism level increases from approximately 0.6 to close to unity (panel [C], solid curve), while the within organism change in groups with cooperating cells drops from about 0.4 to near zero (panel [C], dashed curve). In this figure, heritability of fitness at the group level is measured by the regression of offspring fitness on adult fitness. See text for further explanation.

studied the evolution of developmental mechanisms by which cooperative cell groups are constructed during the clonal cell divisions that create the adult group from a zygote or propagule.

The model assumes that the development of the cell group starts with a propagule. The propagule may be a single cell, as is the case in many multicellular organisms that develop from a fertilized egg, like the volvocine green alga shown for example in the figure. The propagule may be produced asexually or sexually. Alternatively, the model has been used to study the evolution of the single cell bottleneck that is so common in the development of multicellular groups by considering propagules comprised of multiple cells sampled in various ways from adult groups of the previous generation (Michod and Roze, 2000; Roze and Michod, 2001). In the model, the concept of a genotype-phenotype map at the group level involves the mapping between genetic traits present in the propagule and those present in the adult group stage. In particular, the models study how various ways of constructing adult groups, such as using a germ line or cell policing, affect this genotype-phenotype map.

The models embed a cooperate/defect game within a two-locus, multilevel selection framework to study how modifiers of development evolve at a second locus in response to mutation and selection at the primary cooperate/defect locus. Before the second modifier locus is considered, the primary cooperate/defect locus embodies a standard MLS1 group selection model (see, for example, Michod, 1997a, 1997b). For simplicity, haploidy is assumed, except for a transient diploid stage during sex. Development involves the conversion of a cell propagule into an adult cell group through cell division described by a variety of parameters given in Figure 3.1. As already mentioned, propagules contain one or more cells sampled from an adult group in the previous generation (or from several adult groups in the case of aggregation). Sex may occur in the case of single celled propagules that fuse with propagules from other groups to start a new group.

The genotype-phenotype map is a mapping between the propagule's genotype and the phenotype of the adult cell group derived from the propagule. The main group phenotype of interest is the degree to which cells in the adult group cooperate with each other to benefit the group.

The model is an abstract population genetics model; for concreteness, a colonial volvocine green algal species like *Pleodorina starrii* is shown in the figure, a species in which the adult group develops from a single cell propagule. Shown in Figure 3.1 is the adult cell group created by a single reproductive cell in a parental *P. starrii* colony. The adult group has non-specialized reproductive cells (they participate in both reproductive and survival functions) and smaller specialized somatic cells appearing at the bottom left of the colony image (the somatic cells specialize in somatic functions like flagellar action).

3.3.2 Cooperation and Conflict

The cell behavior locus is assumed to have two alleles, C and D, which causes cells to express cooperation and defection, respectively, when interacting in the group. Cooperation benefits other cells in the adult group at a cost to the cooperating cell paid during the mitotic divisions that produce the adult group. Cooperating cells pay

a cost of cooperation by either replicating more slowly or surviving less often compared to defecting cells. Defecting cells do not cooperate and do not pay a cost, but
they may receive benefits from cooperating cells present in their adult group because
groups with cooperating cells are more productive. During development of the adult
group, there is recurrent mutation from C to D at each cell division; back mutation
is ignored on the assumption that there are many more ways to lose a functional trait
like cooperation than to gain it. These defector mutations disrupt the functioning of
the adult cell group by reducing the level of cooperation. Mutation increases the variance and opportunity for selection at the within-group or cell level during the mitotic
divisions that create the adult stage. After the adult group is formed, a propagule is
made either asexually or sexually. The output of propagules depends on the degree
of cooperation at the adult group stage according to a parameter β. Depending on
the parameters of development (mutation rate, cell replication rate, time for cell division), the costs and benefits of cooperation, and the mode of propagule formation, a
polymorphism may be maintained at the C/D locus by mutation-selection balance.
This polymorphism sets the stage for the evolution of modifiers of development
assumed to be encoded by a second locus.

3.3.3 MUTATION AND MULTILEVEL SELECTION

Mutation from cooperate to defect is assumed at rate μ, while back mutation is
ignored. Back mutation from defection to cooperation is ignored, relative to mutation
leading to the loss of cooperation, on the assumption that it is much more likely to
lose a complex trait like cooperation than gain it through random mutation. Because
of the hierarchical nature of selection within and between organisms, there are two
levels of selection at which to consider mutational effects: the cell and the cell group.
This leads to a classification scheme, +/+, +/–, –/+, –/–, with the effect of the mutation
on the cell given on the left and the effect of the mutation on the group or emerging
organism given on the right. Uniformly advantageous mutations (+/+) which benefit
both the fitness of cells and the fitness of the whole organism will sweep through the
population: there is little reason to model them explicitly, given the deterministic
assumptions of the model (the effects of finite population sampling are ignored).
Likewise for uniformly deleterious mutations which detract from the fitness of both
levels (–/–), except they will be lost from the population. There is some evidence for
the –/– kind or effect (Demerec, 1936). In this case (–/–), the occurrence of selection among cells within the organism may have the benefit of lowering the overall
mutation load in the population of organisms and this effect has been considered by
several authors (Crow, 1970; Otto and Orive, 1995; Whitham and Slobodchikoff,
1981). Mutations that benefit the cell's replication rate but detract from organism fitness (+/–) are the case of interest here since they arise when C cells mutate to D cells
during mitotic cell division. Considerable evidence exists for this kind of mutation in
animals – most notably cancer mutants (Aktipis, 2020). The other class of mutations
which harm the cell but benefit the organism (–/+) can be addressed by an adjustment
of the parameters in the models given below but have not been studied in detail.

In short, we embed a standard mutation selection balance model in a two-locus
modifier model to study the evolution of individuality through the modification of

development of cell groups. This approach is different from the standard mutation selection model of population genetics. Two levels of selection are assumed in the developmental modifier model and so there are two levels of mutational effects that must be considered as discussed above. Mutation may be uniformly deleterious, or uniformly beneficial, in that it decreases, or increases, the fitness at both the cell and the group levels simultaneously. On the other hand, the effects of mutation at the two levels could go in different directions, such as is the case with an altruistic mutation that increases the fitness at the group level, while decreasing the fitness at the cell level. We now consider the second modifier locus assumed in the model.

3.3.4 Modifiers of Development

A second modifier locus is considered that modifies the parameters and mode of development of the adult cell group and how cells are sampled to create propagules for the next generation (gametes in the case of sexual reproduction). For example, in the case of a germ line modifier, we assume the propagules for the next generation are sampled from a group of cells that is separate from the somatic cell line, and that has a lower mutation rate and/or less time available for cell division. As a consequence, the mutation rate parameter μ and/or the development time parameter t in the germ line are reduced relative to what these parameters are in the somatic line. Another means of reducing conflict among cells is by cells actively policing and regulating the benefits of defection. To model cell policing, we assume that cooperating cells expressing policing spend time and energy monitoring other cells and reducing the advantages of defecting at a cost to the cell group. As a result of policing, the benefits of cooperation to the adult group are reduced, while the advantages to cells of defecting are reduced. In similar ways, we have considered modifiers creating a unicellular propagule (Michod and Roze, 2000; Roze and Michod, 2001), programmed cell death (the modifier reduces the survival and replication rate of defecting cells), and determinate adult group size, which fixes the size of the adult group and has the effect of reducing the opportunity for within group change (Li and Michod, unpublished). The different kinds of modifiers studied have been reviewed by Michod (2003).

By modifying the parameters of development and mode by which cells are sampled to create the next generation, the modifier locus molds the genotype-phenotype map and the degree to which the propagule produced by an adult resembles the propagule that founded the group. This resemblance is a measure of heritability at the group level and is used to study the heritability of group traits and interpret the results of the models. By molding development, the modifier (so-called M allele below) increases heritability at the group level and the capacity of the groups to reproduce themselves. Reproduction of the cell group is the essential function of development (Griesemer, 2001). Modifier alleles create "higher level" functions, in the sense that their traits are selected by virtue of their tilting the balance in favor of selection at the group level and away from selection at the individual level.

3.3.5 Evolutionary Transitions in Individuality

The results of these models show that cooperation among cells in a proto-organism may be vulnerable because traits that benefit the cell and harm the cell group can

TABLE 3.1
Equilibria and Their Interpretation in the Two-Locus Modifier Model
Introduced in Figure 3.1

Equil.	Genotype	Description of Loci	Interpretation
1	D, m	no cooperation; no modifier	*Single cells, no organism*
2	D, M	no cooperation; modifier fixed	Not of biological interest, never stable
3	$C/D, m$	polymorphic for cooperation and defection; no modifier	*Group of cooperating cells or proto-organism*: no higher-level functions
4	$C/D, M$	polymorphic for cooperation and defection; modifier fixed	*Individual organism*: integrated group of cooperating cells with higher-level functions mediating within organism change

Note: The cooperate/defect locus has two alleles, C, for cooperate, and D, for defect. The modifier locus has two alleles, m and M, that affect development. The non-modifier m allele has the basic parameters of development μ, b, t shown in Figure 3.1, while the modifier allele M causes changes in these and other parameters depending on the kind of modifier considered. For example, the germ line modifier considered in Figure 3.2 reduces the development time of the germ line by amount δ; the soma develops for time t and the germ line develops for time $t - \delta$. As a result, there is less opportunity for mutation in the germ line.

increase within each cell group during the mitotic proliferation that forms the adult group. This is similar to issues of cooperation and conflict that have been much discussed in a synchronic context in the sociobiology literature. The diachronic context in these models comes from explicit consideration of the evolution of modifier traits that affect development. Modifier traits affect development of the multicellular group and can tip the balance in favor of cooperation by changing aspects of development that affect the interplay of levels of selection. By subverting within-group natural selection, modifiers can set the stage for enhanced cooperation and elaborate integration of cell-groups into adaptive wholes and multicellular individuals.

Why do these modifier M alleles evolve and how do they lead to the capacity of a group to reproduce itself? In the Michod and Roze (1997) model, four possible equilibria were studied as described in Table 3.1. Equilibrium 1 has only Dm cells; there is no cooperation and no group fitness. Equilibrium 2 has only DM cells and is not stable or biologically interesting. Equilibrium 3 is polymorphic at the primary cooperate/defect locus and has Cm and Dm cells. Groups at equilibrium 3 may be viewed as a proto-organism, a cell group with cooperation and fitness variance, but no higher-level functions. Finally, equilibrium 4 is polymorphic at the primary locus, fixed for the modifier M allele, and so has CM and DM cells (Michod, 1999, p. 114). At equilibrium 4, the population has transitioned to existing as groups of cooperating cells with higher-level group functions that mediate conflict at the lower level. Consequently, we refer to the groups at equilibrium 4 as individual organisms. Equilibrium 1 occurs when the advantage of defection is high. In this model, within-group selection favors defector (D) cells and between-group selection favors cooperator (C) cells. Thus, for there to be a population that is polymorphic for C and D (equilibriums 3 and 4), selection at the two levels must be in balance. This balance means that the first and second terms of the right-hand side of the Price equation (Equation 3.1) are equal in magnitude. At equilibrium 3, the population is fixed for the non-modifier allele (allele m). The C/D polymorphism consists of cooperating cells being maintained at relatively low frequencies (Michod, 1999, p. 123, see Figure 6–3 in that reference). The exact level of cooperation depends on several parameter values, but the general observation of lower cooperation in a population fixed for the m allele holds. For a population in equilibrium 4, again cell- and group-level selection are in balance. However, the frequency of the cooperator (C) allele in this equilibrium can be much higher as the cell groups have higher-level group-specific functions that suppress lower-level selection for defection.

The mutation-selection balance equilibrium at the C/D locus implies that C alleles are fitter than D alleles, to compensate for mutation from C to D. The models show that, under certain conditions, alleles at the modifier locus evolve due to hitchhiking with the fitter C allele. This has the effect of increasing the between-group variance and decreasing the within-group variance, thereby increasing the level of cooperation and the fitness of the group. Examples of conflict modifiers studied by this approach include germ-soma specialization, reduced mutation rate, policing, programmed cell death, passing the life cycle through a single-cell zygote stage, and fixed group size (reviewed in Michod [2003]). By increasing the variance at the group level and decreasing the variance at the cell level, the modifiers have the effect of decoupling the fitness at the group level from cell fitness (i.e., the counterfactual fitness of cells) as well as enhancing the capacity of the group to reproduce itself.

Although the Price equation may not partition selection correctly between levels because of the pseudo-group problem discussed above, it may be used to help us understand the transition to equilibrium 4 in the modifier model. In Figure 3.2(B), the two components of the Price covariance Equation 3.1 are plotted for the case of the modifier model. These components partition the total change in gene frequency into heritable fitness effects at the organism level (solid line) and within-organism change (dashed line). In the model studied here, within-organism change is always negative, because defecting cells replicate faster than cooperating cells and there is no back mutation from defection to cooperation. At equilibrium, before and after the transition, the two components of the Price equation must equal one another in magnitude, or else the population could not be in equilibrium (this is shown in Figure 3.2[B]). However, during the transition we see that the covariance of fitness with genotype at the emerging organism level (solid curve, Figure 3.2[B]) is greater than the average change at the cell level (dashed curve, Figure 3.2[B]). This greater heritable covariance in fitness at the higher level forces the modifier into the population. Note that after the transition, the within-organism change is smaller than before. Recall that the $E(\Delta q_i)$ term includes the effect of lower-level selection, which is seen as "property change", using the Price (1972) parlance, at the group level. Since the levels of property change are lower in equilibrium 4 compared to equilibrium 3, the population has evolved to have higher heritability of traits (and therefore higher heritable fitness) at the group level.

3.3.6 GROUP REPRODUCTION AND GROUP FITNESS HERITABILITY

Several approaches to quantifying "group reproduction" and group fitness heritability have been used in this work. The capacity of a group to reproduce itself may be measured by the degree to which a group created by a propagule resembles the group the propagule came from. Alternatively, since the group is made from a propagule, and the two-locus recurrence equations are in terms of the gene and genotype frequencies at the propagule stage, we may measure the capacity for reproduction and heritability of group traits as the degree to which the propagules produced by a group are similar to the propagule(s) that created the group. Heritability of fitness at the organism level may be measured in the standard way by the regression of offspring fitness on adult fitness (Michod, 1999, Chapter 6 and Appendix). Using this definition of heritability of fitness and the Price Equation 3.1, Figure 3.2 shows how the heritability of fitness increases during a model ETI (transition from Equilibrium 3 to 4 given in Table 3.1) involving the evolution of a developmental modifier that creates germ-soma specialization (Michod and Roze, 1997). Note that for the notation used in deriving Figure 3.2, the Price Equation 3.1 becomes $\Delta q = \frac{Cov(W_i, q_i)}{\overline{W}} + E(\Delta q_i)$, where i indexes cell groups, the proto-organisms.

Like any trait, heritability of fitness may be defined as the regression of offspring fitness on the fitness of parents. During the transition in the model in Figure 3.2, heritability of fitness at the group level increases. It can be shown that the evolutionary transition always leads to an increase in heritability of fitness (Michod and Roze, 1997, 1999). The heritability of fitness is further studied in Michod (1999, especially Appendix pp. 203–218).

In their review of the evolution of multicellularity, Rainey and de Monte (2014) observe that collective level or group heritability is a derived state, something that must be explained. They state, "Although MLS theory appropriately describes the state of a population of cells before and after the transition to multicellularity, it provides no explanation for how selection shifts from lower-level entities to collectives." I agree that collective level heritability is a derived state, a derived state that is explained by the modifier models considered here. I do not think their criticism of MLS applies to the explicitly diachronic setting of the MLS modifier models reviewed here. Indeed, these models provide a hypothesis for how MLS will act on modifiers of development, and, by so doing, create the capacity of the group to reproduce itself thereby increasing the fitness heritability at the group level. These diachronic MLS models explain the shift in selection from lower-level entities to higher-level collectives with the developmental capacity for group reproduction (Figures 3.1 and 3.2).

This section reviewed MLS models for the evolution of modifiers of the development of cell groups and propagules that are sampled from the adult cell groups to produce the next generation. These developmental modifiers create the first true group-level functions, such as specialized germ and somatic cell lines, that mold development to increase group heritability and the mapping of offspring genotype to adult phenotype. In this way, these models provide a hypothesis for how a group of individuals may become a new kind of individual, the central question posed by ETI theory.

These models take a decidedly diachronic view on levels-of-selection questions. As already mentioned, ETIs raise the questions of how the group level emerges and takes on properties of an evolutionary individual. The two-locus MLS modifier models present a hypothesis in which developmental features of the group-level reproductive system can themselves evolve by MLS, and the Price Equation 3.1 analysis helps to highlight the within-group versus between-group selective dynamics as shown in Figure 3.2(B).

The sampling of populations of Darwinian individuals easily creates groups with Darwinian properties giving rise to the problem of pseudo-groups and the need to distinguish true group effects, like those created by the genetic modifiers in the MLS models of development. There are at the least three kinds of selection that occur in MLS models, within-group individual selection, between-group selection, and global individual selection (Shelton and Michod, 2020). Michod and Roze (1999, p. 10) discuss the issues of pseudo-groups and the global individual selection that comes up in the MLS modifier models. Colonies with more defecting D cells (cells that replicate faster, all else equal) would be fitter than those with fewer D cells, even if there were no interactions between the cells within the groups. The high-replication-rate cells do better in competition within each cell group, and these are the same cells that (for the same reason) would do better without any group context at all. This issue has also been discussed by Okasha (2006, Chapter 8.4).

These models also contribute to our understanding of fitness during ETIs and how it is reorganized during an ETI. I now turn to the issue of fitness reorganization during the evolution of multicellularity and how it creates individuality at the group level.

3.4   FITNESS REORGANIZATION AND GERM-SOMA SPECIALIZATION

As the modifier allele increases in frequency in the population, heritable variation in fitness increases at the group level and decreases at the cell level (Figure 3.1[C]). In the case of the germ line modifier, as it spreads in the population, fitness becomes reorganized, in the sense that cells become specialized in the fitness components of the group. Fitness always has two basic components, survival and reproduction. These components must be rebuilt at the group level when initially they are present only at the cell level (Table 3.2). The evolution of germ-soma specialization accomplishes this because germ cells specialize in the reproduction component of fitness of the group and somatic cells specialize in the viability component. As already discussed, the germ-soma modifier allele builds a new genotype-phenotype map for fitness at the group level when initially this mapping is present at the cell or individual level.

By "reorganization of fitness," I mean the increase of fitness heritability at the group level and decrease at the cell level, with the specialization of lower-level units (cells) in the fitness components of the cell group (the new individual) (Michod, 2005, 2006, 2007). Because of this reorganization of fitness, fitness at the group level becomes decoupled from fitness at the cell level. "Decoupled" in the sense that the fitness of the group may be high, while the individual fitness of the specialized

TABLE 3.2
Reorganization of Fitness during ETIs.

Fitness components	Viability (vegetative/somatic functions).
	Fecundity (reproductive functions).
Definition of fitness reorganization	Transfer of fitness from lower to higher level. Lower levels specialize in fitness components. Heritability of fitness emerges at higher level.
Means of fitness reorganization	Stress responses. Fitness trade-offs.
	Somatic specialization. Group inseparability. Gene co-option.
Consequences of fitness reorganization	Individuality at higher level. Specialization at lower level. Complexity. Evolvability.

cells would be low were they to leave the group. A germ cell cannot survive well on its own and a somatic cell cannot reproduce, yet, together in a group, the group can survive and reproduce. That is counterfactual fitness for these specialized cells declines during an ETI.

The MLS modifier model is intended as a heuristic device for understanding the general issues involved and overlooks a variety of practical issues that are involved in the reorganization of fitness (Table 3.2), such as how the genes for fitness reorganization arise. Genes previously used for life-history stress responses in a unicellular ancestor may be co-opted for the evolution of soma and division of labor in the group (Nedelcu and Michod, 2006, 2020; Olson and Nedelcu, 2016). Cell division, previously the reproduction component of cell fitness is co-opted for growth of the group and organism body size (Nedelcu and Michod, 2003). A group life cycle must evolve from a cell cycle (Hanschen et al., 2016; Maliet, Shelton, and Michod, 2015; Shelton, Leslie, and Michod, 2017; Shelton and Michod, 2014), possibly through the coevolution of a cell life history trait (such as cell growth) and a group trait (such as time spent in the group) (Maliet et al., 2015; Shelton and Michod, 2014). Within the group, germ cells specialize in reproduction and somatic cells in viability of the group, as in the germ-soma modifier model reviewed above.

Specialization of somatic cells is a key stage in the remapping of fitness to the group level, for as individuals specialize in the fitness components of the group, they lose their individual (counterfactual) fitness outside of the group. Germ-soma specialization also increases the individuality of the group by making it indivisible. Germ and soma specialized cells have low fitness when removed from the context of the group, even as the fitness of the group may be quite high. In effect, the germ and soma specialized cells constitute a good team that together brings high fitness to the group even as their fitness when alone declines. Group fitness may also be decoupled from the individual fitness through the evolution of life-history traits; as individuals spend more time in the group, their individual properties will change from values optimal for living alone to values optimal for living in the group (Maliet et al., 2015; Shelton and Michod, 2014).

Trade-offs between fitness components have a special role to play in the reorganization of fitness. A simple trade-off at the individual level, say, between survival and reproduction, can lead to altruism when cells are in a group (Michod, 1999; Michod and Roze, 1999). Cells that put more effort into survival functions, if these same functions benefit the group, are behaving altruistically relative to cells that put less effort into survival. For example, in the volvocine green algae, flagellar motility is a significant survival component of both cells and cell groups. However, flagellar motility at the cell level interferes with the capacity of the cell to reproduce. As a consequence of this trade-off, when groups are first formed, cells that keep their flagella longer are behaving altruistically relative to cells that lose their flagella earlier.

The concept of fitness decoupling that arises from individuality modifiers like germ-soma specialization is similar to the ideas of MLS1 and MLS 2 developed by Damuth and Heisler (1988). In MLS1, group fitness is an aggregate property of individual fitness, while, in MLS2, fitness is a non-aggregate, or emergent, property of the group. In the case of cell groups, MLS1 would be the kind of group selection that occurs when group fitness is an average of cell fitness. With the evolution of germ-soma division of labor, the cell group enters the realm of MLS2, because group fitness is decoupled from the (counterfactual) fitness of cells (Okasha, 2005, 2006).

3.5   CRITICISMS AND COMMENTARIES ON THE MLS APPROACH TO ETIS

3.5.1   DARWINIAN PROPERTIES

A repeated concern with the MLS approach to ETIs has been the concern that MLS assumes the existence of groups with Darwinian properties, something that should be explained, not assumed (Clarke, 2014; Huneman, 2012; Rainey and Kerr, 2010; Rainey and De Monte, 2014). For example, de Monte and Rainey (2014) say, "… it is possible to fall into the trap, as we and others have emphasized, of invoking Darwinian properties as the cause of their own evolution."

The concern is important; a hypothesis should not be circular, that is, an explanation should not assume what is to be explained. It is a simple fact that groups formed by sampling Darwinian populations will often themselves have Darwinian properties at the group level to some extent. The groups formed by sampling a population of individuals will likely be comprised of different frequencies of types of individuals; there will often be differences between groups in group fitness, taken as the average of the fitnesses of members of the group.

The easy Darwinization of groups is the basis of the pseudo-group problem discussed above, which recognizes that groups of Darwinian individuals are themselves easily Darwinized. The Darwinian properties filter up so-to-speak from the individual level to the group level during the sampling process that creates the group.

The challenge is not explaining why groups may have Darwinian properties; this is rather easy to understand based on the sampling process that creates the groups. The challenge is with explaining how groups may gain properties of an evolutionary individual. The evolution of developmental modifiers discussed above is a hypothesis for how individual properties may evolve beginning with a group-structured population of cooperating individuals. These groups will often themselves have Darwinian properties bearing on the evolution of cooperation because of sampling a population of cooperating individuals.

I am belaboring this point because it has been misunderstood. In the evolutionary transition to multicellularity, it is not difficult to understand why cell groups have Darwinian properties to some degree. The problem is with explaining how increased Darwinian properties of groups may arise from true group effects, such as the effects created by the developmental modifier traits discussed above. These modifiers tweak the developmental processes that create cell groups, to enhance the evolutionary individuality of those groups. This is what panel C of Figure 3.2 and other analyses show.

Rainey and colleagues have argued for what they call a "take-nothing-for-granted account" (Black, Bourrat, and Rainey, 2019; De Monte and Rainey, 2014), in which they suggest that "Darwinian properties might emerge from non-Darwinian entities and, therefore, by non-Darwinian means" (Black et al., 2019). When one looks at the mathematical model for this "take-nothing-for-granted account" account, one finds the standard assumptions of MLS theory, most basically, a sampling process of Darwinian individuals that generates cell groups with Darwinian properties (termed "patches" in the model [Black et al., 2019]). These patches or groups possess Darwinian properties by virtue of sampling cells with Darwinian properties as occurs in all MLS models. The authors use the term "ecological scaffolding" for this process by which a sample of Darwinian individuals itself has Darwinian properties. I do not see how the model shows that "Darwinian properties might emerge from non-Darwinian entities and, therefore, by non-Darwinian means" (Black et al., 2019). The authors (Black et al., 2019) go on to argue that their model involves a "shift from levels to timescales [that] does much to clarify the kinds of conditions necessary to effect transitions in individuality." While a feature of their model is that it has two different time scales, it also has two different levels, the cell, and the patch or group. The general assumption in their model, common to all MLS models for ETIs, including the models discussed in this chapter, is that a group of Darwinian individuals may itself have Darwinian properties and these groups can, through further evolution, be molded into a new multicellular individual.

3.5.2 GROUP REPRODUCTION

The issue of explaining group reproduction has received special concern as well it should. Rainey (2007) states "The catch-22 is that selection is powerless to act at the group level because newly emerged groups are incapable of differential reproduction." As we have seen, groups likely have Darwinian properties, including differential reproduction, by virtue of sampling individuals with these Darwinian properties. The model of Black et al. (2019) discussed above is based on this assumption as are all MLS1 type models.

In describing the MLS modifier models presented here, Rainey and Kerr (2010) state "While such a scenario describes plausible changes, the model assumes that the capacity to leave group offspring is already in place. But how such a new level of reproduction emerges requires explanation." I agree that group reproduction requires explanation, but I do not agree that the models described here begin by assuming a new level of reproduction. In fact, as I have tried to explain, before variation is introduced at the modifier locus, there is a mutation-selection balance equilibrium at the first cooperate-defect locus at which some groups are more productive than others depending on the frequency of cooperation in the group. However, this is standard MLS1 type of variation and not a new level of reproduction. The new level of reproduction comes about because of variation introduced by the second modifier locus, the new modifier allele changes how groups develop to create a new level of reproduction. As I have already discussed, a feature of the MLS modifier models described here is that they explain the capacity of the group to leave offspring through the evolution of modifiers of development such as germ line modifiers and the effects of these changes in development on group heritability and the genotype-phenotype map (panel C of Figure 3.2).

The central issue in explaining group reproduction is understanding the evolution of the process by which adult cell groups are formed and produce propagules for the next generation. As we have seen, modifiers of development may mold both the development of adult groups and the sampling process by which cells are taken from the adult group to produce propagules for the next generation. These modifiers in turn increase heritability of fitness at the group level (panel C of Figure 3.2). The MLS modifier models discussed here assume that a generation starts with propagules sampled from adult groups in the previous generation, these propagules then give rise to the adult groups through development. "Development" in the model refers to the number and rate of mitotic cell divisions, cell specialization, and the timing and manner of sampling cells to produce the next generation. The propagules may be sampled randomly from the entire adult group, or the sample may come from a smaller group of cells set aside and sequestered at a certain stage in development as occurs with a germ line modifier (Michod et al., 2003; Roze and Michod, 2001). In the case of a germ line modifier, the propagule sample comes from a separate lineage of cells that may have fewer divisions or a lower mutation rate, because germ cells are separated from the metabolic activities present in the somatic line.

Reproduction of the group is quantified and explained in the modifier model by studying the mapping of group properties, especially fitness, from the propagule to adult. As the modifiers spread, they increase the heritability of group fitness (Figure 3.2[C]). The evolution of modifiers of development gives rise to increased heritability of fitness at the group level and by so doing gives rise to group reproduction.

The evolution of these modifiers of development accomplishes what is needed for the acquisition of group reproduction. Following Griesemer (2001) "Development from an evolutionary point of view can be thought of, in general, as the acquisition of the capacity to reproduce." The main feature of the modifier models reviewed here is their capacity to explain the evolution of group reproduction through the evolution of development.

3.5.3 FITNESS TRANSFER AND FITNESS DECOUPLING

A concern with the MLS models described above is that the concepts of "fitness decoupling" and "fitness transfer" are metaphorical and descriptive (Black et al.,

2019). Fitness decoupling and transfer have been used as descriptions of the results of the MLS modifier models reviewed here, such as the results given in Figure 3.2(C). There we see that, as the modifier increases, the fitness at the two levels diverges, with group fitness increasing and cell fitness decreasing. I have referred to this as fitness decoupling and/or fitness transfer. However, speaking of fitness "transfer" may suggest that fitness is a conserved quantity in the models which it is not. There is a mechanistic sense in which the evolution of altruistic forms of cooperation such as occurs in the model (Figure 3.1) can be seen as transferring fitness between levels (Shelton and Michod, 2020). An altruistic behavior is defined as having both a cost at the individual level and benefit at the group level. Consequently, as an altruistic allele spreads in a population, its costs decrease fitness at the cell level, while its benefits increase fitness at the group level. In this sense, the evolution of altruism transfers fitness between levels.

As discussed above, the idea that there is a decoupling of fitness between levels during an ETI relates to the idea that an ETI is a transition between MLS1 and MLS2. Libby and Rainey (2013) state that "The difficulty is that MLS theory fails to explain how the transition from MLS1 to MLS2 comes about." I hope it is clear from the presentation here that MLS theory can, when coupled with the evolution of developmental modifiers, explain the transition from MLS1 to MLS2. For example, after the germ line modifier evolves, the average fitness of the cell group is no longer an aggregate property of the cell level (counterfactual) fitnesses, and the heritability of the group fitness increases as a result. For this reason, the transition from Equil. 3 to Equil. 4 in Table 3.1 and Figure 3.2 may be viewed as a transition from MLS1 to MLS2.

3.5.4 Origin of Individuating Properties

Clarke (2014) warns of a possible "evolutionary chicken and egg" problem when discussing evolutionary transitions. "We must not presuppose the existence of higher-level organisms when offering evolutionary explanations [of higher-level organisms]," and she goes on to make clear that this warning applies not to just higher-level organisms themselves, but to the kinds of traits that define higher-level organisms and give rise to their individuality. Traits that define higher-level organisms Clarke calls "individuating mechanisms," traits like a germ line, single cell bottlenecks, policing; in short, the kind of modifiers of development considered in the two-locus modifier models discussed above and reviewed in (Michod, 2003). I agree that we must explain the existence of these traits and cannot assume them to be both causes and consequences of higher-level selection. I see the modifier models discussed here as explanatory hypotheses for the evolution of such individuating traits that define higher-level individual organisms.

In these models, before the evolution of the modifier allele, there is variation in fitness at the group level because different groups will contain different allele frequencies at the cooperate/defect locus, as discussed above. Initially, there is no developmental modifier allele M. The modifier allele is introduced at equilibrium 3 in Table 3.1. Equilibrium 3 is a mutation-balance selection equilibrium at the cooperate-defect locus; consequently, cooperation is more fit than defection, so as to compensate for mutation from C to D. Mechanistically, in a population genetic sense, the modifier allele, *M*, may hitchhike with the more fit *C* allele, along with the new kinds of groups it creates, and the population may transition from equilibrium 3 (*C/D*, *m*) to equilibrium 4 (*C/D*, *M*) (Table 3.1). This transition from equilibrium 3 to equilibrium 4 takes cooperation and heritable group fitness to higher levels as it creates groups with more individuality precisely because they now possess individuating mechanisms such as a germ live (Figure 3.1).

Is there a chicken and egg problem with this model as a hypothetical explanation for the ETI from unicellular to multicellular individuals? Does the model assume traits associated with higher-level individuals, like the germ line modifier, are both a cause and an effect of higher-level individuality? I do not think so. Before the evolution of the modifier, the fitness variances at the individual and group levels are a result of both cell division and the sampling creating groups (as diagrammed in Figure 3.1). In addition, there are assumptions related to the cooperate/defect game, the mutation/selection balance, and the standard assumptions of haploid two-locus population genetics.

Once introduced, the modifier allele coevolves with its effects on enhanced cooperation and group heritability (Figure 3.1), but this is what we would hope to see in an explanation of an ETI. For example, Sober and Wilson (1998, p. 97) say: "The coevolution of traits that influence population structure with traits that are favored by the new population structure can result in a feedback process that concentrates natural selection at one level of the biological hierarchy" (Sober and Wilson 1998, p. 97).

Concerning the MLS modifier models discussed above, Clarke (2014, p. 9) says that the modifiers, and the traits they cause, are present at the beginning of the model. I do not agree. The modifier *M* allele is not present at equilibrium 3 (Table 3.1) where it is introduced. A locus where the modifier allele *M* might arise is assumed, but, until the modifier is introduced, this locus has no effect on the model, the *m* allele assumed to reside there is neutral without any effect. On the contrary, it seems to me that the MLS modifier model is an excellent example of what Clarke advances in her paper as the remedy to the chicken and egg problem in explaining individuality. Clarke argues that individuality is built up over time with some aspects being present early in the process and other aspects arising later. In the MLS modifier model, cooperation, defection, and sampling into groups are present initially, with the evolution of developmental modifier traits caused by the modifier *M* allele coming later.

To assume a single gene locus encodes a complex trait, like a germ line, is, of course, an oversimplification of many steps. But, this limitation is found in all simple one and two-locus population genetic models and does not reduce the heuristic power of these models in explaining evolutionary processes (Michod, 1981).

3.6 CONCLUSIONS, OPEN QUESTIONS, AND FUTURE WORK

3.6.1 GENERAL

I have given a brief overview and history of MLS in population genetics and evolutionary biology. MLS was key to Wright's shifting balance theory (Wright, 1977, Figure 13.1) in which groups of individuals selected to a local fitness peak may output more individuals into the global population leading to transformation of the species (if the local fitness peak is also a global peak). Using the terminology developed in this chapter, we see that the MLS in Wright's theory is MLS1 and group fitness is a direct reflection of individual fitness. In Wright's process, the fitness of individuals filters up, so to speak, to create group fitness.

The determination of group fitness during ETIs requires something different. MLS always involves group fitness whether in Wright's theory or during ETIs. What is different in an ETI is how group fitness is comprised. The fitness of a group of individuals that have become a new kind of individual is no longer a simple average of the fitness of its members because the members will often specialize in different activities and components of fitness of the group. If alone, group members would have little fitness but together in a group the fitness can be quite high. An ETI begins with the group as a collection of individuals and ends with the group being a new kind of individual. This requires the specialization and integration of activities of members in service of group fitness, the survivorship and reproduction of the group. The specialization and integration of the group involve the evolution of new developmental processes.

I have reviewed how MLS may be used to develop hypotheses about the evolution of development during the transition from unicellular to multicellular life. We have seen how developmental modifiers may coevolve with group structure and create the first true group-level functions such as a sequestered germ line and cell policing. In effect, these modifiers take the population from MLS1 to MLS2, from groups of cooperating cells to groups of cooperating cells with higher group-level individuating functions, such as germ-soma separation, that mediate conflict within the group and enhance the heritability, reproduction, and individuality of the cell group. After the transition, fitness at the group level is no longer the average of cell fitness, since the group is comprised of specialized somatic and germ cells that would be deficient if they were to leave the group.

The case of germ soma specialization was presented here (Hanschen, Shelton, and Michod, 2015; Michod, 1999; Michod et al., 2003; Michod and Roze, 1997). The germ-soma model discussed above assumes that the modifier allele creates both germ and soma specialization. It would be useful to revisit these MLS modifier models to consider the evolution of somatic specialization before a specialized germ line. In the volvocine green algae lineage, somatically specialized cells evolved before germ-specialized cells (Herron and Michod, 2008; Herron et al., 2009). The reproductive cells in a species like *Pleodorina starrii* (shown in Figure 3.1) are not specialized at reproduction; they additionally participate in somatic activities like flagellar beating. In *P. starrii*, there are typically 64 cells with smaller specialized somatic cells (seen at the bottom left of each colony in the figure) and non-specialized reproductive cells that first have flagella before losing their flagella during cell division and reproduction (most of the cells in the colony image are these kind of non-specialized cells). The modifier models could be used to develop hypotheses about the soma first and germ later evolution observed in the volvocine clade.

3.6.2 MAXIMIZATION PRINCIPLES AND INDIVIDUALITY

Maximization principles are useful in science because they summarize complex dynamics in terms of a few variables and concepts. What might be maximized during an ETI? This measure could be used to quantify individuality, a concept that is difficult to understand as discussed above. In Figure 3.2(C), we see that as the ETI proceeds from equilibrium 3 to equilibrium 4 (Table 3.1), the fitness of the group increases, and the degree of within group change declines. Is some measure of the relative intensity of selection at the group level relative to the cell level maximized by the dynamics of the ETI? Within-group change can be seen as a kind of transmission error that lowers the heritability of the adult group phenotype (Frank, 2012; Michod, 1999). Additional analyses of the effect of the mutation rate on fitness after versus fitness before the ETI for the two levels of selection, group and cell level, are given in Michod and Roze (1999, Figures. 13–14). The increase in heritability at the group level relative to the cell level holds for both uniformly deleterious mutations ([−, −] in the notation given above), as well as for selfish mutations (+/−) assumed to result from mutation from cooperation to defection. So, after the ETI, fitness at the group level has increased relative to the cell level. Cell level fitness here refers to the replication rate of cells within the group. In the MLS modifier models, the rate of cell division depends only on cell genotype and does not depend on group context; the benefit of cooperation is assumed to affect the functionality of the adult group, not the replication rate of cells that make up the adult group (Figure 3.1). Further work should clarify what if anything is maximized by an ETI.

3.6.3 Fitness

Fitness is a unique and fundamental concept in biology. Lewontin remarked, "Natural selection of the character states themselves is the essence of Darwinism. All else is molecular biology (Lewontin, 1972)." It is a challenge to understand and be clear about the meaning of "fitness" when there is just one level of selection. We may expect increased difficulties when considering multiple, simultaneous levels of selection with the goal of understanding the transition from one unit or level of fitness to another.

"Fitness" is used in a variety of senses in this chapter. In the MLS modifier models, there are two aspects of cell fitness, the cell replication rate (which depends only on cell genotype) and the cell's cooperative behavior that contributes to group fitness of the adult cell group in a frequency-dependent manner. In these models, a cell's replication rate depends only on the genotype of the cell and does not depend on the composition of the group. This is because in the model cell replication creates the group in which the cooperative behavior is expressed, as shown in Figure 3.1. The composition of the group affects group fitness after the group has been made, but the group composition does not affect the replication rate of cells during the mitotic divisions that create the group. The overall fitness of a cell in the MLS models includes both its replication rate as well as its differential propagation through the differential output from groups, which is a frequency-dependent function of the frequency of cooperative cells in the adult group.

I have discussed "global individual selection" of cells which refers to the contributions to individual fitness that do not depend on group membership and give an example in the MLS modifier models of a defecting cell that has a higher replication rate regardless of context (see also, Okasha, 2006, Chapter 8.4 where this is further discussed). A defecting cell, in the MLS models, has two aspects to its fitness: a context-dependent component, by virtue of finding itself in groups with cooperators, and a context-independent component, such as a higher replication rate during cell division that gives rise to global individual selection. Although more work needs to be done, our analyses based on counterfactual fitness suggest that there are at least three kinds of selection that can be occurring at the same time in MLS models: group-specific between-group selection, along with two kinds of individual selection, within-group selection and global individual selection (Shelton and Michod, 2020).

"Counterfactual fitness" refers to the fitness a cell that is inside a group would have were it to leave the group. The degree to which counterfactual fitness differs from the fitness of unicells, that have not evolved in the context of the group, may be used to quantify progression through the ETI, however more work needs to be done.

There is also "fitness" in the sense of gene or allelic fitness, which considers the fitness effects of all the contexts the allele is in on the overall change in frequency of the gene. For example, I refer to gene fitness when I say that "cooperation is fitter than defection at the mutation-selection balance equilibrium." Likewise, one could refer to just "cell fitness," and ignore the different underlying fitness partitions that arise due to levels of selection. In this sense, cell fitness considers all sources of differential survival and reproduction of a cell. However, for a full understanding of evolution we would like to know if cell fitness is being caused by group effects. Likewise, group fitness may stem from different sources such as sampling of cell fitness or from "true group effects."

It is a challenge to get clear on "fitness" when levels of selection are changing, and modifier alleles are creating new kinds of groups and changing the degree to which selection is occurring at different levels. More work is needed to clarify fitness, how it is partitioned between levels, and its causal basis during ETIs. The power of individual selection and the primacy of organisms were often used to deny the need for MLS in evolutionary biology; however, the multicellular organism is a derived state and MLS theory is needed to explain its origin and evolution.

ACKNOWLEDGEMENTS

Comments from Dinah Davison, the editors and two reviewers greatly improved the manuscript. I thank Dinah Davison for her permission to use her image of *P. starrii* in Figure 3.1. This work was supported by NSF grant DEB-2029999.

REFERENCES

Aktipis, A. 2020. *The Cheating Cell*. Princeton: Princeton University Press.
Black, A. J., Bourrat, P., and Rainey, P. B. 2019. Ecological scaffolding and the evolution of individuality: the transition from cells to multicellular life. *Nature Ecology and Evolution* 4: 426–436.
Bouchard, F. and Huneman, P. Eds. 2013. *From Groups to Individuals: Evolution and Emerging Individuality*. Cambridge, MA: The MIT Press.
Buss, L. W. 1987. *The Evolution of Individuality*. Princeton: Princeton University Press.
Calcott, B. and Sterelny, K. Eds. 2011. *The Major Transitions in Evolution Revisited*. Cambridge, MA: The MIT Press.
Clarke, E. 2014. Origins of evolutionary transitions. *Journal of Biosciences*, 39: 303–317.
Clarke, E. 2016. A levels-of-selection approach to evolutionary individuality. *Biology and Philosophy* 31: 893–911.

Crow, J. F. 1970. Genetic loads and the cost of natural selection. In *Mathematical Topics in Population Genetics* Ed. K. Kojima, 128–177. New York: Springer-Verlag.

Damuth, J. and Heisler, I. L. 1988. Alternative formulations of multilevel selection. *Biology and Philosophy*, 34: 407–430.

De Monte, S. and Rainey, P. B. 2014. Nascent multicellular life and the emergence of individuality. *Journal of Biosciences* 392: 237–248.

Demerec, M. 1936. Frequency of "cell-lethals" among lethals obtained at random in the X-chromosome of *Drosophila melanogaster*. *Proceedings of the National Academy of Sciences, USA*, 22: 350–354.

Frank, S. A. 2012. Natural selection. III. Selection versus transmission and the levels of selection. *Journal of Evolutionary Biology*, 25: 227–243.

Gardner, A. 2015. The genetical theory of multilevel selection. *Journal of Evolutionary Biology*, 28: 1747–1751.

Gardner, A. and Grafen, A. 2009. Capturing the superorganism: a formal theory of group adaptation. *Journal of Evolutionary Biology* 22: 659–671.

Gissis, S. B., Lamm, E., and Ayelet, S. Eds. 2017. *Landscapes of Collectivity in the Life Sciences*. Cambridge Massachusetts: The MIT Press.

Griesemer, J. R. 2001. The units of evolutionary transition. *Selection* 1: 67–80.

Hamilton, W. D. 1964a. The genetical evolution of social behaviour. I. *Journal of Theoretical Biology* 7: 1–16.

Hamilton, W. D. 1964b. The genetical evolution of social behaviour. II. *Journal of Theoretical Biology* 7: 17–52.

Hamilton, W. D. 1975. Innate social aptitudes of man: an approach from evolutionary genetics. *Biosocial Anthropology* 53: 133–155.

Hanschen, E. R., Marriage, T. N., Ferris, P. J. et al. 2016. The *Gonium pectorale* genome demonstrates co-option of cell cycle regulation during the evolution of multicellularity. *Nature Communications* 7: 11370.

Hanschen, E. R., Shelton, D. E., and Michod, R. E. 2015. Evolutionary transitions in individuality and recent models of multicellularity. In *Evolutionary Transitions to Multicellular Life: Principles and Mechanisms* Ed. I. Ruiz-Trillo and A. M. Nedelcu, 165–188. Dordrecht: Springer.

Heisler, I. L. and Damuth, J. 1987. A method for analyzing selection in hierarchically structured populations. *The American Naturalist* 130: 582–602.

Herron, M. D., and Michod, R. E. 2008. Evolution of complexity in the volvocine algae: Transitions in individuality through Darwin's eye. *Evolution*, 62(2): 436–451. https://doi.org/10.1111/j.1558-5646.2007.00304.x

Herron, M. D., Hackett, J. D., Aylward, F. O., and Michod, R. E. 2009. Triassic origin and early radiation of multicellular volvocine algae. *Proceedings of the National Academy of Sciences, USA*, 106(9): 3254–3258. https://doi.org/10.1073/pnas.0811205106

Huneman, P. 2012. Adaptations in transitions: how to make sense of adaptation when beneficiaries emerge simultaneously with benefits? In *From groups to individuals: Evolution and emerging individuality* Ed. F. Bouchard and P. Huneman, 141–172. Cambridge, MA: The MIT Press.

Lehtonen, J., Okasha, S., and Helanterä, H. 2020. Theme issue "Fifty years of the Price equation." *Philosophical Transactions of the Royal Society B* 375: 20190350.

Lewontin, R. C. 1970. The units of selection. *Annual Review of Ecology and Systematics*, 1: 1–18.

Lewontin, R. C. 1972. Testing the theory of natural selection. *Nature* 236: 181–182.

Libby, E. and Rainey, P. B. 2013. A conceptual framework for the evolutionary origins of multicellularity. *Physical Biology* 10: 035001.

Maliet, O., Shelton, D. E., and Michod, R. E. 2015. A model for the origin of group reproduction during the evolutionary transition to multicellularity. *Biology Letters* 11: 20150157.

Michod, R. E. 1981. Positive heuristics in evolutionary biology. *British Journal for the Philosophy of Science* 32: 1–36.

Michod, R. E. 1982. The theory of kin selection. *Annual Review of Ecology and Systematics* 13: 23–55.

Michod, R. E. 1996. Cooperation and conflict in the evolution of individuality. II. Conflict mediation. *Proceedings of the Royal Society of London B, Biological Sciences* 263: 813–822.

Michod, R. E. 1997a. Coooperation and conflict in the evolution of individuality. I. Multilevel selection of the organism. *American Naturalist* 149: 607–645.

Michod, R. E. 1997b. Evolution of the individual. *The American Naturalist* 150: S5–S21.

Michod, R. E. 1999. *Darwinian Dynamics: Evolutionary Transitions in Fitness and Individuality*. Princeton, NJ: Princeton University Press.

Michod, R. E. 2003. Cooperation and conflict mediation during the origin of multicellularity. In *Genetic and Cultural Evolution of Cooperation* Ed P. Hammerstein, 261–307.

Michod, R. E. 2005. On the transfer of fitness from the cell to the multicellular organism. *Biology and Philosophy* 20: 967–987.

Michod, R. E. 2006. The group covariance effect and fitness trade-offs during evolutionary transitions in individuality. *Proceedings of the National Academy of Sciences, USA* 103: 9113–9117.

Michod, R. E. 2007. Evolution of individuality during the transition from unicellular to multicellular life. *Proceedings of the National Academy of Sciences, USA* 104: 8613–8618.

Michod, R. E. and Abugov, R. 1980. Adaptive topography in family-structured models of kin selection. *Science* 210: 667–669.

Michod, R. E., Nedelcu, A. M., and Roze, D. 2003. Cooperation and conflict in the evolution of individuality. IV. Conflict mediation and evolvability in *Volvox carteri*. *BioSystems* 69: 95–114.

Michod, R. E. and Roze, D. 1997. Transitions in individuality. *Proceedings of the Royal Society of London B, Biological Sciences* 264: 853–857.

Michod, R. E. and Roze, D. 1999. Cooperation and conflict in the evolution of individuality. III. Transitions in the unit of fitness. In *Mathematical and Computational Biology: Computational Morphogenesis, Hierarchical Complexity, and Digital Evolution* Ed. C. L. Nehaniv, 47–92. Providence: American Mathematical Society.

Michod, R. E. and Roze, D. 2000. Some aspects of reproductive mode and the origin of multicellularity. *Selection* 1: 97–109.

Michod, R. E. and Roze, D. 2001. Cooperation and conflict in the evolution of multicellularity. *Heredity* 81: 1–7.

Morrell, V. 1996. Genes versus teams. *Science* 273: 739–740.

Nedelcu, A. M. and Michod, R. E. 2003. Evolvability, modularity, and individuality during the transition to multicellularity in volvocalean green algae. In *Modularity in Development and Evolution* Ed. G. Schlosser and G. P. Chicago: Wagner, 66–489.

Nedelcu, A. M. and Michod, R. E. 2006. The evolutionary origin of an altruistic gene. *Molecular Biology and Evolution*, 23: 1460–1464.

Nedelcu, A. M. and Michod, R. E. 2020. Stress responses co-opted for specialized cell types during the early evolution of multicellularity. *BioEssays* 42: 2000029.

Okasha, S. 2005. Multilevel selection and the major transitions in evolution. *Philosophy of Science* 72: 1013–1025.

Okasha, S. 2006. *Evolution and the Levels of Selection*. Oxford: Oxford University Press.

Olson, B. J. S. C. and Nedelcu, A. M. 2016. Co-option during the evolution of multicellular and developmental complexity in the volvocine green algae. *Current Opinion in Genetics and Development* 39: 107–115.

Otto, S. P. and Orive, M. E. 1995. Evolutionary consequences of mutation and selection within an individual. *Genetics* 141: 1173–1187.

Price, G. R. 1970. Selection and covariance. *Nature* 227: 529–531.

Price, G. R. 1972. Extension of covariance selection mathematics. *Annals of Human Genetics* 35: 485–490.

Rainey, P. B. 2007. Unity from conflict. *Nature* 446: 616.

Rainey, P. B. and Kerr, B. 2010. Cheats as first propagules: a new hypothesis for the evolution of individuality during the transition from single cells to multicellularity. *BioEssays* 32: 872–880.

Rainey, P. B. and De Monte, S. 2014. Resolving conflicts during the evolutionary transition to multicellular life. *Annual Review of Ecology, Evolution, and Systematics*, 45: 599–620.

Roze, D. and Michod, R. E. 2001. Mutation, multilevel selection, and the evolution of propagule size during the origin of multicellularity. *The American Naturalist* 158: 638–654.

Shelton, D. E., Leslie, M. P., and Michod, R. E. 2017. Models of cell division initiation in *Chlamydomonas*: a challenge to the consensus view. *Journal of Theoretical Biology* 412: 186–197.

Shelton, D. E. and Michod, R. E. 2014. Group selection and group adaptation during a major evolutionary transition: insights from the evolution of multicellularity in the volvocine algae. *Biological Theory* 9: 452–469.

Shelton, D. E. and Michod, R. E. 2020. Group and individual selection during evolutionary transitions in individuality: meanings and partitions. *Proceedings of the Royal Society B* 375: 20190364.

Sober, E. and Wilson, D. S. 1998. *Unto Others: the Evolution and Psychology of Unselfish Behavior*. Cambridge, MA: Harvard University Press.

Van Baalen, M. and Huneman, P. 2014. Organisms as ecosystems/ecosystems as organisms. *Biological Theory* 9: 357–360.

Whitham, T. G. and Slobodchikoff, C. N. 1981. Evolution by individuals, plant-herbivore interactions, and mosaics of genetic variability: the adaptive significance of somatic mutations in plants. *Oecologia* 49: 287–292.

Williams, G. C. 1966a. *Adaptation and Natural Selection*. Princeton, NJ: Princeton Univ. Press.

Williams, G. C. 1966b. *Adaptation and Selection: A Critique of Some Current Evolutionary Thought*. Princeton, NJ: Princeton University Press.

Wilson, D. S. 1975. *A Theory of Group Selection*. Proceedings of the National Academy of Sciences, USA 72: 143–146.

Wright, S. 1931. Evolution in Mendelian populations. *Genetics* 16: 97–159.

Wright, S. 1932. The roles of mutation, inbreeding, crossbreeding and selection in evolution. *Proceedings of the Sixth International Congress of Genetics* 1: 356–366.

Wright, S. 1969. Evolution and the Genetics of Populations. Vol. 2. *The Theory of Gene Frequencies*. Chicago: Univ. Chicago Press.

Wright, S. 1977. Evolution and the Genetics of Populations. Vol. 3. *Experimental Results and Evolutionary Deductions*. Chicago: Univ. Chicago Press.

4 Life Cycles as a Central Organizing Theme for Studying Multicellularity

Merlijn Staps
Department of Ecology and Evolutionary Biology,
Princeton University, Princeton, NJ, USA

Jordi van Gestel
Department of Microbiology and Immunology, University
of California, San Francisco, San Francisco, CA, USA

Corina E. Tarnita
Department of Ecology and Evolutionary Biology,
Princeton University, Princeton, NJ, USA

CONTENTS

4.1 INTRODUCTION

Repeated evolutionary transitions from unicellular to multicellular life gave rise to an extraordinary diversity of multicellular life forms (Bonner 1998; Claessen et al. 2014; Grosberg and Strathmann 2007; Herron et al. 2013; Lyons and Kolter 2015; Ratcliff et al. 2017). Yet, our thinking of multicellularity is mostly shaped by a few paradigmatic, macroscopic examples. We begin this chapter by discussing several examples that showcase the broader diversity. Subsequently, we use the concept of the life cycle as a tool to systematically categorize multicellular diversity and study its evolutionary origin. We then review how recent advances in both empirical and theoretical research have improved our understanding of the evolution of multicellular life cycles, discuss the types of questions that still remain unanswered, and distinguish the conceptual approaches—bottom-up versus top-down—that can be used to investigate those questions. Finally, we show how a bottom-up approach can be employed theoretically to explore the origin of multicellular life.

DOI: 10.1201/9780429351907-5

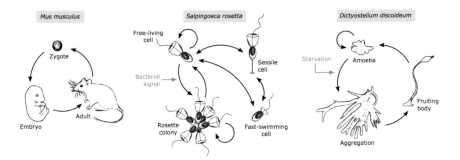

FIGURE 4.1   Extant multicellular organisms have diverse life cycles. Simplified depictions of the life cycles of the mouse *Mus musculus*, the choanoflagellate *Salpingoeca rosetta* (Dayel et al. 2011), and the cellular slime mold *Dictyostelium discoideum*. Arrows indicate life stage transitions. Environmentally induced group formation is indicated with a gray arrow for *S. rosetta* and *D. discoideum*; the dashed arrow indicates colony fission for *S. rosetta*. For simplicity, not all life stages are shown; for example, the *S. rosetta* life cycle does not show chain colonies or the sexual cycle.

When thinking about multicellularity, animals almost immediately jump to mind. Unsurprisingly, then, much of our intuition regarding multicellularity is derived from the animals. For example, among the textbook examples of multicellular development are the mammals (Figure 4.1). Mammals develop from a single cell (zygote) that repeatedly divides and ultimately gives rise to a mature multicellular individual. Reproduction takes place sexually: each individual produces gametes, which, after fusion, form a zygote capable of recapitulating the same developmental process. The mammalian form of multicellularity can be well understood as repeated cycles of development and reproduction: a zygote gives rise to a multicellular individual through repeated cell divisions (development), and this multicellular individual then generates single-celled gametes that again form a zygote (reproduction), thereby closing the cycle. There are, however, also animals for which multicellularity takes strikingly different forms than for the mammalian textbook example, such as the almost undifferentiated placozoa, the "moss animals" (Bryozoa), and the coral-forming "flower animals" (Anthozoa) (Buss 1983; 1987). Unlike mammals, these animals may reproduce asexually without a single-celled intermediate, form sessile colonies that blur the boundaries of individuality, or exhibit diverse morphological forms depending on environmental conditions (Harvell 1991; Hughes 1989; Simpson et al. 2020; Todd 2008). Thus, even though animals represent a single evolutionary transition to multicellularity (meaning that they share a common multicellular ancestor), they are multicellular in diverse ways (Cavalier-Smith 2017; Minelli and Fusco 2010; Moran 1994).

Right next door from the animal kingdom we find the choanoflagellate *Salpingoeca rosetta*, a marine eukaryote and one of the closest animal relatives (Figure 4.1) (Brunet and King 2017; Fairclough et al. 2010). *S. rosetta* exhibits a simple[1] form of multicellularity, consisting of small colonies of cells. Like animals, *S. rosetta* becomes multicellular via repeated cell divisions. Unlike animals, however, *S. rosetta* is facultatively multicellular, which means that it can also exist and reproduce as a unicellular organism. What's more, it can adopt three different unicellular life forms that are so different from one another that they were believed to belong to different genera until it was discovered that they were, in fact, different life stages of a single organism (Dayel et al. 2011). The different life stages are adapted to different conditions: one allows surface attachment whereas the other two are free-living. Of the two free-living forms, one is a fast swimmer specialized for dispersal and quick swimming towards patches of prey bacteria and the other is a slow swimmer that can become multicellular (Dayel et al. 2011; Koehl 2020; Miño et al. 2017). Two different types of multicellular structures can be formed: linear "chain colonies" and spherical "rosette colonies." Although much remains uncertain about the regulatory mechanisms through which *S. rosetta* switches between its different possible unicellular and multicellular forms, it has become clear that the environment plays an important role. For example, formation of rosette colonies from single cells can be caused by a specific environmental trigger: the detection of lipids secreted by bacteria in the environment (Alegado et al. 2012; Woznica et al. 2016). The reverse transition from colonies to single cells is less well understood, although it has been observed that colonies can split in two through fission (Dayel et al. 2011). This would suggest that while *S. rosetta* is strictly unicellular in the absence of the environmental trigger that induces colony formation, it may be strictly multicellular in its presence—with colonies giving rise to new colonies without the need for a single-celled intermediate. Thus, *S. rosetta* can express a range of different forms, both unicellular and multicellular, and some of these forms are induced by the environment.

Yet different types of multicellularity can be found in more distant branches of the tree of life, in organisms that evolved multicellularity independently from animals and choanoflagellates. The cellular slime molds provide one example (Figure 4.1). Cellular slime molds are soil-dwelling microbes that, like *S. rosetta*, are facultatively multicellular: they can live and reproduce as unicellular amoebae and only become multicellular in response to specific environmental conditions—in their case, the trigger for multicellularity is starvation (Bonner 2009; Schaap 2011). Whereas animals and choanoflagellates become multicellular by repeated cell divisions, cellular slime molds exhibit aggregative multicellularity, meaning that individual amoebae aggregate to form a multicellular organism. The multicellular stage is transient and short-lived, culminating in the formation of a fruiting body consisting of a stalk and a head filled with spores that is raised in the air (Bonner 1957; Schaap 2011). The fruiting body facilitates survival and dispersal. The starvation-resistant spores will germinate upon encountering favorable conditions and continue their life as unicellular amoebae. This type of multicellularity can be understood as an "emergency response" to harsh environmental conditions, allowing cells to quickly join forces to achieve efficient dispersal (Brunet and King 2017). Thus, the cellular slime molds are facultatively multicellular, with a short-lived aggregative multicellular stage that is induced by the environment.

Animals, choanoflagellates, and cellular slime molds are but a few examples of the immensely diverse ways to be multicellular. Here, we employ the concept of the life cycle to organize this diversity. Broadly speaking, an organism's life cycle describes the life stages between which an organism transitions in time. One of the prominent advocates for taking a life cycle-centered approach to the study of multicellularity was John Bonner, who argued that a dynamic view of all the life stages of an organism is more meaningful evolutionarily than a static view that emphasizes a specific stage of the life cycle (e.g., the adult at maturity) (Bonner 1965, 1993; van Gestel and Tarnita 2017). As the above examples highlight, these different life stages need not occur in a predictable succession and some life stages may only be expressed sporadically, under specific conditions. What all multicellular life cycles have in common, however, is that they involve *group formation* (the construction of multicellular groups, often starting from single cells) and *group propagation* (the process by which multicellular groups beget new multicellular groups). Therefore, these two processes—together with the intrinsic mechanisms (i.e., development) and extrinsic factors (e.g., environmental cues) that regulate them—offer a meaningful starting point for cataloging multicellular life cycles.

Group formation occurs in one of two principal ways (but hybrids of these are also possible, see [van Gestel and Tarnita 2017]): either groups are formed by cells staying together after division (also known as clonal multicellularity), as in animals or choanoflagellates, or groups are formed by cells coming together via aggregation (also known as aggregative multicellularity), as in the cellular slime molds (Bonner 1965, 1998; Grosberg and Strathmann 2007; Tarnita et al. 2013). The staying together—coming together dichotomy encapsulates a wide range of possible grouping mechanisms. For example, staying-together multicellularity can be achieved by means of a shared cell wall, an extracellular matrix, adhesive molecules on the cell surface, or coenocytic growth (Abedin and King 2010; Suga and Ruiz-Trillo 2013), while coming together can be achieved through aggregation via chemotaxis or binding to a common surface (van Gestel and Tarnita 2017). Both staying together and coming together have evolved multiple times and in evolutionarily distant clades (Bonner 1998; Fisher et al. 2013; Grosberg and Strathmann 2007). While staying-together multicellularity includes both facultative and obligate forms of multicellularity, coming together is always associated with facultative multicellularity, with the multicellular stage being transient and induced by specific (typically adverse) environmental conditions (Brown et al. 2012; Brunet and King 2017; Du et al. 2015; Fisher et al. 2013; Sebé-Pedrós et al. 2013, 2017).

Much like there are multiple mechanisms by which group formation can be achieved, there are also multiple ways for groups to propagate (Angert 2005; Pichugin et al. 2017). In many cases, group propagation requires reversal from the multicellular to the unicellular stage. For example, mammals have a very brief unicellular stage that comprises just the haploid gametes (sperm and egg) and the fertilized egg cell (zygote). Group propagation—the production and release of gametes—can occur repeatedly throughout the mature organism's reproductive lifespan. Cellular slime molds, in contrast, have a life cycle where the multicellular stage ends with a single group propagation event: the release of spores from the mature fruiting body. Finally, multicellular groups of *S. rosetta* propagate through fission, whereby the group splits in two (Dayel et al. 2011). Group propagation modes in which a group splits into two groups allow a life cycle to proceed without a single-cell intermediate.

Not only do life cycles differ in the mechanisms used to achieve group formation and group propagation, but they also differ in the *regulation* of these processes. For example, while mammals are obligately multicellular organisms with a life cycle that is under tight developmental control, the facultative multicellularity of choanoflagellates and cellular slime molds is more flexible. In these organisms, the multicellular stage is expressed conditionally on the presence of a specific environmental trigger, highlighting an important role for the environment in the regulation of the life cycle. Environmental triggers involved in life cycle regulation can be both abiotic (e.g., nutrient or drought stress [King 2004; Ritchie et al. 2008]), as in the case of cellular slime molds, or biotic (e.g., quorum-sensing, predator-prey interactions [Alegado and King 2014; Waters and Bassler 2005; Woznica et al. 2017]), as in the case of choanoflagellates, and they ensure that life stages (in particular, the multicellular stage) are only expressed under the appropriate environmental conditions (Woznica and King 2018). Sensitivity to environmental conditions is not unique to facultatively multicellular organisms; the life cycles of many obligate multicellular organisms also depend strongly on environmental conditions (Nagy et al. 2018; Schlichting 1986; Walbot 1996).

Two general points emerge from a broad outlook on multicellular life cycles (Figure 4.1). First, evolution has been extraordinarily creative in inventing different ways for multicellular organisms to develop and reproduce, leading to a plethora of multicellular life cycles. Second, a life cycle can only be properly understood in its ecological context, as the environment may provide crucial signals that govern life stage transitions. The extent to which the life cycle depends on the environment varies widely across life cycles: at one extreme are facultatively multicellular organisms, for which specific environmental conditions are required to trigger the multicellular stage; at the other extreme are life cycles that require limited environmental input, in which different life stages occur in a predictable succession (e.g., zygote → embryo → adult, for mammals).

4.2   LIFE CYCLE EVOLUTION

A life cycle-centered approach can also shed light on the evolutionary origins of multicellularity. A transition from unicellular to multicellular life requires the evolution of mechanisms by which cells can attach to each other (group formation) as well as the evolution of mechanisms by which multicellular groups can propagate themselves (group propagation) (Libby and Rainey 2013). Group formation alone is not sufficient for the evolution of multicellularity. For example, a mutation that blocks cell separation after division may quickly lead to the formation of multicellular groups, but will not lead to multicellular life if cells stay attached indefinitely—in this case, the multicellular group is an evolutionary dead end because it lacks the ability to propagate. Thus, while unicellular organisms may frequently undertake opportunistic experiments with multicellularity in which multicellular groups are formed, these experiments can only lead to the evolution of a multicellular lineage when a primitive multicellular life cycle is established. This renders the question of how multicellularity originated *identical* to the question of how a multicellular life cycle originated (Black et al. 2020; Bonner 1993; De Monte and Rainey 2014; van Gestel and Tarnita 2017).

The empirical understanding of how multicellular life cycles evolve has greatly expanded in the past two decades. First, work in comparative genomics has reshaped our thinking about the genomic underpinnings of transitions to multicellularity. Most strikingly, this research has revealed that the unicellular ancestors at the stems of multicellular lineages are, from a genomic perspective, more complex than previously appreciated and that much of the genetic toolkit for multicellularity predates the evolution of multicellularity (Brooke and Holland 2003; Glöckner et al. 2016; Hanschen et al. 2016; Prochnik et al. 2010; Rokas 2008; Sebé-Pedrós et al. 2017; Suga et al. 2013). The last unicellular ancestor of animals, for instance, was already equipped with genes employed in animal cell adhesion, with transcription factors used in animal development, and with some components of developmental signaling pathways—although the cell signaling repertoire also greatly expanded with the advent of multicellularity (King et al. 2008; Sebé-Pedrós et al. 2017, 2018; Srivastava et al. 2010). The fact that many multicellularity genes are of ancient, unicellular origin implies that understanding the unicellular ancestor is key to understanding the first emergence of a multicellular life cycle: multicellularity may more readily evolve from preadapted unicellular organisms equipped with genes or behaviors that can be co-opted for multicellular organization (King 2004). The evolution of facultative multicellularity, for example, is often contingent upon ancestral mechanisms that allow the ancestor to sense its external environment and to express different cellular behaviors in response (Kawabe et al. 2015; Ritchie et al. 2008).

Second, recent progress in experimental evolution has made it possible to study the emergence of multicellular life cycles in the laboratory (Boraas et al. 1998; Hammerschmidt et al. 2014;; Herron et al. 2019; Ratcliff et al. 2012, 2013). This approach makes use of experiments in which a unicellular organism is subjected to an artificial selective pressure that favors multicellularity, such as a selective pressure for increased size. For example, in an evolution experiment in which the unicellular green alga *Chlamydomonas reinhardtii*[2] was subjected to selection for rapid settling in liquid medium (which favors larger clusters of cells), a life cycle evolved that had alternating unicellular and multicellular life stages (Ratcliff et al. 2013). Experimental evolution studies reinforce the idea that transitions to multicellularity can readily be made, provided that a unicellular organism is equipped with the right preadaptations and that a selective pressure exists that favors multicellularity.

Despite such major empirical advances, there are also aspects of the evolution of multicellular life cycles that remain largely inaccessible empirically, at least for now. While we can use comparative genomics to make inferences about the genome of unicellular organisms that underwent transitions to multicellularity, characterizing the ancestral functions of these genes in the life cycle of the unicellular ancestor is challenging. And while the emergence of a multicellular life cycle can be studied experimentally under an artificial selective pressure, we have little idea of what selective environments drove the evolution of multicellularity in nature. These gaps in our empirical knowledge currently leave unanswered many questions about the evolution of multicellular life cycles. Why did multicellular life cycles emerge in some lineages, but not in others? Why did so many different types of life cycles evolve and how much of that diversity was present early on? And how does the life cycle facilitate (or impede) the emergence of multicellular innovations and the evolution of multicellular complexity?

Different conceptual approaches have been used to make progress on these open questions. For instance, a "top-down" approach starts by identifying the common features of many different multicellular life forms—such as high levels of cooperation and coordination between cells in multicellular groups, the individuality[3] of the multicellular organism, and the prevalence of a single-cell bottleneck in the life cycle (De Monte and Rainey 2014; Godfrey-Smith 2009; Queller and Strassmann 2009;

Ratcliff et al. 2017; West et al. 2015). Next, the top-down approach aims to explain how multicellular groups could have evolved these key features. For example, a clear pattern among extant multicellular organisms is that all examples of complex multicellularity, such as animals and plants, are associated with a staying-together life cycle (Fisher et al. 2020). In comparison, the multicellular complexity of cellular slime molds and other organisms with coming-together life cycles has remained limited; these organisms invariably have a transient and short-lived multicellular stage (Brunet and King 2017). A potential explanation for this pattern is that coming together can lead to genetically heterogeneous groups and hence evolutionary conflict (where the evolutionary interests of cells within a multicellular group are misaligned). In the absence of mechanisms to prevent genetically heterogeneous groups, such conflict could impede the evolution of multicellular complexity beyond a transient multicellular life stage (Michod and Herron 2006; Michod and Roze 2001; Queller 2000; Rainey and De Monte 2014). Conversely, in the absence of conflicts, staying together would allow for longer-lived groups and facilitate the evolution of multicellular innovations (e.g., the evolution of cell specialization and complex morphologies), which could lead to complex forms of multicellularity.

The top-down framework has been successful in providing potential evolutionary explanations that do not hinge on the particularities of any individual transition to multicellularity. However, there are also limitations to the top-down perspective. First, when identifying the features that multicellular organisms have in common (and could be of importance for their evolutionary origin), the top-down approach runs the risk of introducing biases towards the paradigmatic examples of complex multicellularity that we are most familiar with, even if the resulting collection of organisms is not representative of the actual diversity of multicellular life (van Gestel and Tarnita 2017). Second, within the top-down framework, it can be challenging to select among multiple competing explanations. For example, proposed top-down explanations for the prevalence of a single-cell bottleneck range from its ability to reduce conflicts of interests between cells (Grosberg and Strathmann 1998; Ratcliff et al. 2015; Roze and Michod 2001) to its ability to purge deleterious mutations (Grosberg and Strathmann 1998) or to ensure coherent development (Wolpert and Szathmáry 2002), and it is not immediately clear which of these explanations should carry the most weight. In fact, none of these explanations may be required at all, as empirical work shows that life cycles with a single-cell bottleneck can arise rapidly, without requiring selection, through co-option of the ancestral unicellular form (Ratcliff et al. 2013), and mathematical modeling suggests that a single-cell bottleneck may simply be favored because it maximizes population growth under a wide range of conditions (Pichugin et al. 2017). Such studies highlight the fact that the explanations provided by the top-down framework are best seen as working hypotheses that require independent testing, both empirically and theoretically.

The second conceptual framework is the bottom-up approach. In contrast to the top-down approach, which starts by considering extant multicellular organisms, the bottom-up approach asks what features of multicellularity might evolve given a certain unicellular ancestor and its environmental setting (van Gestel and Tarnita 2017). In doing so, it aims to obtain a mechanistic understanding of the way in which the unicellular ancestor and its ecology shape nascent multicellular life. The bottom-up approach can distinguish between features of multicellularity that emerge spontaneously, as a direct consequence of the way in which multicellularity arises from the unicellular ancestor, and features of multicellularity that require subsequent selection, beyond the origins of multicellularity, in favor of a particular multicellular organization. Moreover, by considering explicit evolutionary trajectories from a unicellular ancestor to a multicellular organism, the bottom-up approach is ideally suited to test specific hypotheses—such as the hypotheses that the top-down framework extracts. Relative to the top-down approach, a bottom-up perspective has the advantage that it explores the full potential of transitions to multicellularity and minimizes bias towards the actually realized or most paradigmatic multicellular lineages (van Gestel and Tarnita 2017). Just like the top-down approach, however, the bottom-up approach has its limitations. By explicitly considering the ancestral starting point and the mechanistic underpinnings of an evolutionary trajectory towards multicellularity, the scope of the bottom-up is more limited than that of the top-down approach, and the explanations it offers may generalize less broadly.

While the theoretical literature on the evolution of multicellularity has historically largely employed the top-down framework, these top-down insights have recently been complemented by bottom-up models that explore specific scenarios for the origin of multicellularity (Solé and Duran-Nebreda 2015; Solé and Valverde 2013). These scenarios range from public goods sharing (Biernaskie and West 2015) or collective motion (Garcia et al. 2014; 2015) in microbial populations to filament formation (Rossetti et al. 2011) or surface colonization (van Gestel and Nowak 2016) in bacteria. What makes studying the evolution of multicellularity with bottom-up models exciting is that these models enable us to do something that is very difficult to do empirically: directly study how the unicellular ancestor and the selective pressures it faces shape the evolution of multicellularity. And, by explicitly accounting for the cell-cell and cell-environment interactions that shape multicellularity, bottom-up models shed mechanistic light on the transition to multicellularity in the process.

4.3   A BOTTOM-UP APPROACH TO LIFE CYCLE EMERGENCE

To give a concrete example, we recently constructed a mechanistic model to study how a multicellular life cycle first emerges from a unicellular ancestor during a transition to multicellularity (Staps et al. 2019). Our goal was to explore (1) what kinds of multicellular life cycles could emerge and (2) what features of the ancestor and its environment would shape these life cycles at the origin of multicellularity. The evolutionary starting point is a unicellular organism that is able to sense its environment and express different genes in response. We assumed a simple (but non-constant) environment that fluctuates back and forth between two different states, mimicking, for example, the diurnal cycle faced by photosynthetic algae or the feast-and-famine cycles faced by soil-dwelling amoebae. Motivated by the empirical observation that co-option of ancestral functions underlies multiple transitions to multicellularity (Hanschen et al. 2016; Kawabe et al. 2015; Olson and Nedelcu 2016), we introduced a potential for multicellularity by allowing an ancestral gene to start causing some degree of cell adhesion, leading to daughter cells staying attached after division. We then investigated whether and what multicellular life cycles could arise through evolutionary changes in gene regulation. Thus, the bottom-up model makes the ancestral ecology and ancestral regulatory mechanisms explicit, and studies what life cycles may emerge from this starting point.

This bottom-up approach provided two important insights. First, despite the simple setup, this model produced a surprising diversity of life cycles. What life cycle evolved depended on two factors: the biophysical constraints on group formation (i.e., the strength of the attachments between cells, which determines the maximum group size that could theoretically be reached by cells that express the cell adhesion gene), and the benefits of multicellularity (specifically, the minimum size that groups need to reach for multicellularity to provide a benefit) (Figure 4.2a,b). Among the

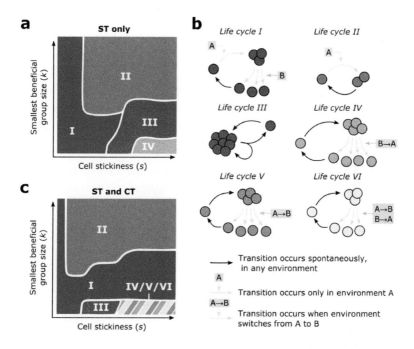

FIGURE 4.2 Modeling the emergence of multicellular life cycles. a, Evolved multicellular life cycles in simulations of the evolution of multicellularity. What life cycle evolves depends on the ability of cells to stay attached to each other (cell stickiness, s, ranging from virtually instantaneous detachment on the left to indefinite attachment on the right) and the minimum group size at which being in a group provides a benefit to constituent cells (smallest beneficial group size, k, range 2–10). Different colors indicate different life cycles. b, Cartoons depicting the evolved life cycles. Arrows indicate life-stage transitions; gray arrows indicate life stage transitions that are under environmental control. The environment switches back and forth between two states A and B. In life cycle I, groups are formed in environment A, and groups fall apart in environment B. In life cycle II, groups may occasionally form in environment A but groups fall apart immediately. In life cycle III, group formation is obligate, but occasionally cells detach from their group. In life cycles IV–VI group formation is also obligate, but groups propagate by dissolving into solitary cells in response to specific environmental triggers. c, The evolved multicellular life cycle in simulations in which coming together (CT) is possible in addition to staying together (ST). See (Staps et al. 2019) for a full description of the model and the original data figures.

several evolved life cycles, we recovered simple analogs of the mammalian one—an obligately multicellular life cycle in which groups are dependent for their propagation on the occasional release of cells from the group (Figure 4.2b, life cycle III)—and the cellular slime molds one—an environment-dependent life cycle in which groups regularly propagate by dissolving completely into solitary cells (Figure 4.2b, life cycle IV). The obligately multicellular life cycle arises when cell stickiness is sufficiently high for group formation to reap the benefits of multicellularity but low enough to have the occasional release of single-cell propagules that ensure group propagation. In contrast, the environment-dependent life cycle arises when high cell stickiness renders accidental detachment of cells unlikely and instead relies on environmentally induced active down-regulation of adhesion for the group to dissolve and ensure propagation. Thus, our simple bottom-up model reveals that diverse multicellular life cycles can evolve already at the origin of multicellularity through co-option of ancestral regulatory mechanisms.

Second, in an extended version of the model, we allowed cells to join groups via coming together, in addition to their capacity to form groups via staying together. Allowing for group formation via coming together favored the evolution of multicellular life cycles with a short-lived multicellular stage and frequent dissolution into solitary cells (Figure 4.2b, life cycles I, IV, V, VI). How can this be explained? Since coming together is not limited by the rate of cell division, it allows groups to assemble faster from single cells than staying together alone (Pentz et al. 2020). Faster group formation also enables groups to propagate more frequently, and thus results in a shorter life cycle. Thus, our simple bottom-up model provides an alternative explanation for the short-lived multicellular life stages observed in many organisms with aggregative multicellularity: it might not necessarily be that the potential for within-group conflict *prevents* longer-lived multicellularity, but rather that the potential for fast group formation and propagation *allows* for short-lived multicellular life stages.

Our life cycle model showcases the potential of the bottom-up approach. Explicitly accounting for the life cycle and ecological context of the unicellular ancestor can readily generate new insights: a surprising diversity of life cycles could have emerged at the origin of multicellularity and the transience of coming-together life cycles might be explained without needing to invoke conflicts among constituent cells. Interestingly, these insights emerge even though the model makes several simplifying assumptions about the ecology (the fluctuating environment) and development (the regulatory architecture underlying cell adhesion). This simplicity reflects purposeful abstraction as well as a (partially inevitable) ignorance about the life history and ecology of organisms that underwent transitions to multicellularity. Current empirical developments pave the way for a next generation of models that can more realistically account for the biology of early multicellular organisms and their unicellular predecessors. Indeed, studies of unicellular relatives of multicellular lineages enhance our understanding of the relevant regulatory mechanisms at the origin of multicellularity (Arenas-Mena 2017; Arenas-Mena and Coffman 2015; Levin et al. 2014; Sebé-Pedrós et al. 2018, 2016). At the same time, progress in biophysics and single-cell biology helps reveal the physical, metabolic, and behavioral constraints on early multicellular organisms (Dexter et al. 2019; Goodwin 1989; Kempes et al. 2012; Larson et al. 2020). And finally, largely overlooked knowledge from paleobiology can inform the environmental conditions that selected for transitions to multicellularity (Butterfield 2000, 2009; Erwin 1993; Valentine and Marshall 2015).

4.4   CONCLUSION

Life cycles are central to studying the evolution of multicellularity. They are a useful tool to catalog the variety of ways different branches of the tree of life have found to achieve multicellularity, and, because they encapsulate development and reproduction, they determine how evolution can act on nascent multicellular life. Understanding the evolution of multicellularity, and of multicellular organizing principles such as cell differentiation or morphogenesis, therefore requires life-cycle-centered approaches that explicitly consider the life cycle and its evolutionary origin. Future progress depends on the sustained dialog between two complementary approaches: a historically dominant top-down approach that aims to provide general evolutionary explanations that transcend the particularities of individual transitions to multicellularity, and a complementary bottom-up perspective that emphasizes specific evolutionary trajectories. The top-down approach can inform the bottom-up approach by identifying general organizing principles of multicellularity and by suggesting underlying evolutionary explanations. In turn, the bottom-up approach is able to inform the top-down approach by critically evaluating these explanations, adding mechanistic insight, and elucidating specific evolutionary trajectories as they unfold starting from a unicellular ancestor. It is by integrating the top-down and bottom-up perspectives that we can paint a fuller picture of the evolutionary transition to multicellularity: starting with how multicellular life arose when the first multicellular life cycle emerged from a unicellular ancestor and building towards an understanding of why multicellular life as we see it around us today looks the way it does.

NOTES

1. Multicellular complexity is difficult to define. Accordingly, there is no commonly accepted definition for what constitutes "simple" or "complex" multicellularity. Here, we use these terms to capture the difference between microscopic multicellular organisms with few differentiated cell types ("simple"), such as cyanobacteria, choanoflagellates, or volvocine algae and macroscopic multicellular organisms with many differentiated cell types ("complex"), such as most plants and animals. Many multicellular organisms do not fit neatly in either of these categories.
2. *C. reinhardtii* is a close relative of the multicellular genus *Volvox*, a model organism for germ-soma differentiation.
3. What constitutes a biological individual is a challenging conceptual question. Various definitions have been proposed, which may, for example, require indivisibility, genetic homogeneity, physiological autonomy, and/or satisfying the Lewontin conditions for evolution by natural selection (i.e. variation, reproduction and heritability) (Buss 1987; De Monte and Rainey 2014; Godfrey-Smith 2009; Michod 2007; Santelices 1999; van Gestel and Tarnita 2017).

REFERENCES

Abedin, Monika, and Nicole King. 2010. "Diverse Evolutionary Paths to Cell Adhesion." *Trends in Cell Biology, Special issue—CellBio-X*, 20 (12): 734–42. https://doi.org/10.1016/j.tcb.2010.08.002.

Alegado, Rosanna A, Laura W Brown, Shugeng Cao, Renee K Dermenjian, Richard Zuzow, Stephen R Fairclough, Jon Clardy, and Nicole King. 2012. A Bacterial Sulfonolipid Triggers Multicellular Development in the Closest Living Relatives of Animals. Edited by Peter Greenberg. *eLife* 1 (October): e00013. https://doi.org/10.7554/eLife.00013.

Alegado, Rosanna A., and Nicole King. 2014. "Bacterial Influences on Animal Origins." *Cold Spring Harbor Perspectives in Biology* 6 (11): a016162. https://doi.org/10.1101/cshperspect.a016162.

Angert, Esther R. 2005. "Alternatives to Binary Fission in Bacteria." *Nature Reviews Microbiology* 3 (3): 214–24. https://doi.org/10.1038/nrmicro1096.

Arenas-Mena, César. 2017. "The Origins of Developmental Gene Regulation." *Evolution & Development* 19 (2): 96–107. https://doi.org/10.1111/ede.12217.

Arenas-Mena, Cesar, and James A. Coffman. 2015. "Developmental Control of Transcriptional and Proliferative Potency during the Evolutionary Emergence of Animals." *Developmental Dynamics* 244 (10): 1193–1201. https://doi.org/10.1002/dvdy.24305.

Biernaskie, Jay M., and Stuart A. West. 2015. "Cooperation, Clumping and the Evolution of Multicellularity." *Proceedings of the Royal Society B: Biological Sciences* 282 (1813): 20151075. https://doi.org/10.1098/rspb.2015.1075.

Black, Andrew J., Pierrick Bourrat, and Paul B. Rainey. 2020. "Ecological Scaffolding and the Evolution of Individuality." *Nature Ecology & Evolution*, February, 1–11. https://doi.org/10.1038/s41559-019-1086-9.

Bonner, John Tyler. 1957. "A Theory of the Control of Differentiation in the Cellular Slime Molds." *The Quarterly Review of Biology* 32 (3): 232–46.

Bonner, John Tyler. 1965. *Size and Cycle: An Essay on the Structure of Biology*. Princeton University Press.

Bonner, John Tyler. 1993. *Life Cycles: Reflections of an Evolutionary Biologist*. Princeton University Press.

Bonner, John Tyler. 1998. "The Origins of Multicellularity." *Integrative Biology* 1 (1): 27–36.

Bonner, John Tyler. 2009. *The Social Amoebae: The Biology of Cellular Slime Molds*. Princeton University Press.

Boraas, Martin E., Dianne B. Seale, and Joseph E. Boxhorn. 1998. "Phagotrophy by a Flagellate Selects for Colonial Prey: A Possible Origin of Multicellularity." *Evolutionary Ecology* 12 (2): 153–64. https://doi.org/10.1023/A:1006527528063.

Brooke, Nina M, and Peter WH Holland. 2003. "The Evolution of Multicellularity and Early Animal Genomes." *Current Opinion in Genetics & Development* 13 (6): 599–603. https://doi.org/10.1016/j.gde.2003.09.002.

Brown, Matthew W., Martin Kolisko, Jeffrey D. Silberman, and Andrew J. Roger. 2012. "Aggregative Multicellularity Evolved Independently in the Eukaryotic Supergroup Rhizaria." *Current Biology* 22 (12): 1123–27. https://doi.org/10.1016/j.cub.2012.04.021.

Brunet, Thibaut, and Nicole King. 2017. "The Origin of Animal Multicellularity and Cell Differentiation." *Developmental Cell* 43 (2): 124–40. https://doi.org/10.1016/j.devcel.2017.09.016.

Buss, Leo W. 1983. "Evolution, Development, and the Units of Selection." *Proceedings of the National Academy of Sciences* 80 (5): 1387–91. https://doi.org/10.1073/pnas.80.5.1387.

Buss, Leo W. 1987. *The Evolution of Individuality*. Princeton University Press.

Butterfield, Nicholas J. 2000. *"Bangiomorpha pubescens* n. gen., n. sp.: Implications for the Evolution of Sex, Multicellularity, and the Mesoproterozoic/Neoproterozoic Radiation of Eukaryotes." *Paleobiology* 26 (3): 386–404. https://doi. org/10.1666/0094-8373(2000)026<0386:BPNGNS>2.0.CO;2.

Butterfield, Nicholas J. 2009. "Modes of Pre-Ediacaran Multicellularity." *Precambrian Research, World Summit on Ancient Microscopic Fossils,* 173 (1): 201–11. https://doi. org/10.1016/j.precamres.2009.01.008.

Cavalier-Smith, Thomas. 2017. "Origin of Animal Multicellularity: Precursors, Causes, Consequences—the Choanoflagellate/Sponge Transition, Neurogenesis and the Cambrian Explosion." *Philosophical Transactions of the Royal Society B: Biological Sciences* 372 (1713): 20150476. https://doi.org/10.1098/rstb.2015.0476.

Claessen, Dennis, Daniel E. Rozen, Oscar P. Kuipers, Lotte Søgaard-Andersen, and Gilles P. Van Wezel. 2014. "Bacterial Solutions to Multicellularity: A Tale of Biofilms, Filaments and Fruiting Bodies." *Nature Reviews Microbiology* 12 (2): 115–24. https:// doi.org/10.1038/nrmicro3178.

Dayel, Mark J., Rosanna A. Alegado, Stephen R. Fairclough, Tera C. Levin, Scott A. Nichols, Kent McDonald, and Nicole King. 2011. "Cell Differentiation and Morphogenesis in the Colony-Forming Choanoflagellate *Salpingoeca rosetta.*" *Developmental Biology* 357 (1): 73–82. https://doi.org/10.1016/j.ydbio.2011.06.003.

De Monte, Silvia, and Paul B Rainey. 2014. "Nascent Multicellular Life and the Emergence of Individuality." *Journal of Biosciences,* April. https://doi.org/10.1007/ s12038-014-9420-5.

Dexter, Joseph P., Sudhakaran Prabakaran, and Jeremy Gunawardena. 2019. "A Complex Hierarchy of Avoidance Behaviors in a Single-Cell Eukaryote." *Current Biology* 29 (24): 4323-4329.e2. https://doi.org/10.1016/j.cub.2019.10.059.

Du, Qingyou, Yoshinori Kawabe, Christina Schilde, Zhi-hui Chen, and Pauline Schaap. 2015. "The Evolution of Aggregative Multicellularity and Cell–Cell Communication in the Dictyostelia." *Journal of Molecular Biology* 427 (23): 3722–33. https://doi. org/10.1016/j.jmb.2015.08.008.

Erwin, Douglas H. 1993. "The Origin of Metazoan Development: A Palaeobiological Perspective." *Biological Journal of the Linnean Society* 50 (4): 255–74. https://doi. org/10.1111/j.1095-8312.1993.tb00931.x.

Fairclough, Stephen R., Mark J. Dayel, and Nicole King. 2010. "Multicellular Development in a Choanoflagellate." *Current Biology* 20 (20): R875–76. https://doi.org/10.1016/j. cub.2010.09.014.

Fisher, R. M., J. Z. Shik, and J. J. Boomsma. 2020. "The Evolution of Multicellular Complexity: The Role of Relatedness and Environmental Constraints." *Proceedings of the Royal Society B: Biological Sciences* 287 (1931): 20192963. https://doi.org/10.1098/ rspb.2019.2963.

Fisher, Roberta M., Charlie K. Cornwallis, and Stuart A. West. 2013. "Group Formation, Relatedness, and the Evolution of Multicellularity." *Current Biology* 23 (12): 1120–25. https://doi.org/10.1016/j.cub.2013.05.004.

Garcia, Thomas, Leonardo Gregory Brunnet, and Silvia De Monte. 2014. "Differential Adhesion between Moving Particles as a Mechanism for the Evolution of Social Groups." *PLoS Computational Biology* 10 (2): e1003482. https://doi.org/10.1371/jour-nal.pcbi.1003482.

Garcia, Thomas, Guilhem Doulcier, and Silvia De Monte. 2015. "The Evolution of Adhesiveness as a Social Adaptation." *eLife* 4 (November). https://doi.org/10.7554/ eLife.08595.

van Gestel, Jordi, and Martin A. Nowak. 2016. "Phenotypic Heterogeneity and the Evolution of Bacterial Life Cycles." *PLOS Computational Biology* 12 (2): e1004764. https://doi. org/10.1371/journal.pcbi.1004764.

van Gestel, Jordi, and Corina E. Tarnita. 2017. "On the Origin of Biological Construction, with a Focus on Multicellularity." *Proceedings of the National Academy of Sciences* 114 (42): 11018–26. https://doi.org/10.1073/pnas.1704631114.

Glöckner, Gernot, Hajara M. Lawal, Marius Felder, Reema Singh, Gail Singer, Cornelis J. Weijer, and Pauline Schaap. 2016. "The Multicellularity Genes of Dictyostelid Social Amoebas." *Nature Communications* 7 (June): 12085. https://doi.org/10.1038/ncomms12085.

Godfrey-Smith, Peter. 2009. *Darwinian Populations and Natural Selection.* OUP Oxford.

Goodwin, Brian C. 1989. "Unicellular Morphogenesis." In *Cell Shape: Determinants, Regulation, And Regulatory Role*, 365–91.

Grosberg, Richard K., and R. R. Strathmann. 1998. "One Cell, Two Cell, Red Cell, Blue Cell: The Persistence of a Unicellular Stage in Multicellular Life Histories." *Trends in Ecology & Evolution* 13 (3): 112–16. https://doi.org/10.1016/S0169-5347(97)01313-X.

Grosberg, Richard K., and Richard R. Strathmann. 2007. "The Evolution of Multicellularity: A Minor Major Transition?" *Annual Review of Ecology, Evolution, and Systematics* 38 (1): 621–54. https://doi.org/10.1146/annurev.ecolsys.36.102403.114735.

Hammerschmidt, Katrin, Caroline J. Rose, Benjamin Kerr, and Paul B. Rainey. 2014. "Life Cycles, Fitness Decoupling and the Evolution of Multicellularity." *Nature* 515 (7525): 75–79. https://doi.org/10.1038/nature13884.

Hanschen, Erik R., Tara N. Marriage, Patrick J. Ferris, Takashi Hamaji, Atsushi Toyoda, Asao Fujiyama, Rafik Neme, et al. 2016. "The *Gonium pectorale* Genome Demonstrates Co-Option of Cell Cycle Regulation during the Evolution of Multicellularity." *Nature Communications* 7 (April): 11370. https://doi.org/10.1038/ncomms11370.

Harvell, C. Drew. 1991. "Coloniality and Inducible Polymorphism." *The American Naturalist* 138 (1): 1–14.

Herron, Matthew D., Joshua M. Borin, Jacob C. Boswell, Jillian Walker, I. Chen Kimberly Chen, Charles A. Knox, Margrethe Boyd, Frank Rosenzweig, and William C. Ratcliff. 2019. "De Novo Origins of Multicellularity in Response to Predation." *Scientific Reports* 9 (1): 2328. https://doi.org/10.1038/s41598-019-39558-8.

Herron, Matthew D., Armin Rashidi, Deborah E. Shelton, and William W. Driscoll. 2013. "Cellular Differentiation and Individuality in the 'Minor' Multicellular Taxa." *Biological Reviews* 88 (4): 844–61. https://doi.org/10.1111/brv.12031.

Hughes, Roger Neville. 1989. *Functional Biology of Clonal Animals.* Springer Science & Business Media.

Kawabe, Yoshinori, Christina Schilde, Qingyou Du, and Pauline Schaap. 2015. "A Conserved Signalling Pathway for Amoebozoan Encystation That Was Co-Opted for Multicellular Development." *Scientific Reports* 5 (1). https://doi.org/10.1038/srep09644.

Kempes, Christopher P., Stephanie Dutkiewicz, and Michael J. Follows. 2012. "Growth, Metabolic Partitioning, and the Size of Microorganisms." *Proceedings of the National Academy of Sciences* 109 (2): 495–500. https://doi.org/10.1073/pnas.1115585109.

King, Nicole. 2004. "The Unicellular Ancestry of Animal Development." *Developmental Cell* 7 (3): 313–25. https://doi.org/10.1016/j.devcel.2004.08.010.

King, Nicole, M. Jody Westbrook, Susan L. Young, Alan Kuo, Monika Abedin, Jarrod Chapman, Stephen Fairclough, et al. 2008. "The Genome of the Choanoflagellate *Monosiga brevicollis* and the Origin of Metazoans." *Nature* 451 (7180): 783–88. https://doi.org/10.1038/nature06617.

Koehl, M. A. R. 2020. "Selective Factors in the Evolution of Multicellularity in Choanoflagellates." *Journal of Experimental Zoology Part B: Molecular and Developmental Evolution.* https://doi.org/10.1002/jez.b.22941.

Larson, Ben T., Teresa Ruiz-Herrero, Stacey Lee, Sanjay Kumar, L. Mahadevan, and Nicole King. 2020. "Biophysical Principles of Choanoflagellate Self-Organization." *Proceedings of the National Academy of Sciences* 117 (3): 1303–11. https://doi.org/10.1073/pnas.1909447117.

Levin, Tera C, Allison J Greaney, Laura Wetzel, and Nicole King. 2014. "The Rosetteless Gene Controls Development in the Choanoflagellate *S. rosetta.*" *eLife* 3 (October). https://doi.org/10.7554/eLife.04070.

Libby, Eric, and Paul B Rainey. 2013. "A Conceptual Framework for the Evolutionary Origins of Multicellularity." *Physical Biology* 10 (3): 035001. https://doi.org/10.1088/1478-3975/10/3/035001.

Lyons, Nicholas A, and Roberto Kolter. 2015. "On the Evolution of Bacterial Multicellularity." *Current Opinion in Microbiology* 24 (April): 21–28. https://doi.org/10.1016/j.mib.2014.12.007.

Michod, R. E. 2007. "Evolution of Individuality during the Transition from Unicellular to Multicellular Life." *Proceedings of the National Academy of Sciences* 104 (Supplement 1): 8613–18. https://doi.org/10.1073/pnas.0701489104.

Michod, Richard E., and Matthew D. Herron. 2006. "Cooperation and Conflict during Evolutionary Transitions in Individuality." *Journal of Evolutionary Biology* 19 (5): 1406–9. https://doi.org/10.1111/j.1420-9101.2006.01142.x.

Michod, Richard E., and Denis Roze. 2001. "Cooperation and Conflict in the Evolution of Multicellularity." *Heredity* 86 (1): 1–7. https://doi.org/10.1046/j.1365-2540.2001.00808.x.

Minelli, A., and G. Fusco. 2010. "Developmental Plasticity and the Evolution of Animal Complex Life Cycles." *Philosophical Transactions of the Royal Society B: Biological Sciences* 365 (1540): 631–40. https://doi.org/10.1098/rstb.2009.0268.

Miño, Gastón L., M. A. R. Koehl, Nicole King, and Roman Stocker. 2017. "Finding Patches in a Heterogeneous Aquatic Environment: pH-Taxis by the Dispersal Stage of Choanoflagellates." *Limnology and Oceanography Letters* 2 (2): 37–46. https://doi.org/10.1002/lol2.10035.

Moran, Nancy A. 1994. "Adaptation and Constraint in the Complex Life Cycles of Animals." *Annual Review of Ecology and Systematics* 25: 573–600.

Nagy, László G., Gábor M. Kovács, and Krisztina Krizsán. 2018. "Complex Multicellularity in Fungi: Evolutionary Convergence, Single Origin, or Both?" *Biological Reviews* 93 (4): 1778–94. https://doi.org/10.1111/brv.12418.

Olson, Bradley JSC, and Aurora M Nedelcu. 2016. "Co-Option during the Evolution of Multicellular and Developmental Complexity in the Volvocine Green Algae." *Current Opinion in Genetics & Development* 39 (August): 107–15. https://doi.org/10.1016/j.gde.2016.06.003.

Pentz, Jennifer T., Pedro Márquez-Zacarías, G. Ozan Bozdag, Anthony Burnetti, Peter J. Yunker, Eric Libby, and William C. Ratcliff. 2020. "Ecological Advantages and Evolutionary Limitations of Aggregative Multicellular Development." *Current Biology*, September. https://doi.org/10.1016/j.cub.2020.08.006.

Pichugin, Yuriy, Jorge Peña, Paul B. Rainey, and Arne Traulsen. 2017. "Fragmentation Modes and the Evolution of Life Cycles." *PLOS Computational Biology* 13 (11): e1005860. https://doi.org/10.1371/journal.pcbi.1005860.

Prochnik, S. E., J. Umen, A. M. Nedelcu, A. Hallmann, S. M. Miller, I. Nishii, P. Ferris, et al. 2010. "Genomic Analysis of Organismal Complexity in the Multicellular Green Alga *Volvox carteri.*" *Science* 329 (5988): 223–26. https://doi.org/10.1126/science.1188800.

Queller, D. C., and Joan E. Strassmann. 2009. "Beyond Society: The Evolution of Organismality." *Philosophical Transactions of the Royal Society B: Biological Sciences* 364 (1533): 3143–55. https://doi.org/10.1098/rstb.2009.0095.

Queller, David C. 2000. "Relatedness and the Fraternal Major Transitions." *Philosophical Transactions of the Royal Society of London. Series B: Biological Sciences* 355 (1403): 1647–55. https://doi.org/10.1098/rstb.2000.0727.

Rainey, Paul B., and Silvia De Monte. 2014. "Resolving Conflicts During the Evolutionary Transition to Multicellular Life." *Annual Review of Ecology, Evolution, and Systematics* 45 (1): 599–620. https://doi.org/10.1146/annurev-ecolsys-120213-091740.

Ratcliff, William C., R. F. Denison, M. Borrello, and M. Travisano. 2012. "Experimental Evolution of Multicellularity." *Proceedings of the National Academy of Sciences* 109 (5): 1595–1600. https://doi.org/10.1073/pnas.1115323109.

Ratcliff, William C., Johnathon D. Fankhauser, David W. Rogers, Duncan Greig, and Michael Travisano. 2015. "Origins of Multicellular Evolvability in Snowflake Yeast." *Nature Communications* 6 (1): 6102. https://doi.org/10.1038/ncomms7102.

Ratcliff, William C., Matthew Herron, Peter L. Conlin, and Eric Libby. 2017. "Nascent Life Cycles and the Emergence of Higher-Level Individuality." *Philosophical Transactions of the Royal Society B: Biological Sciences* 372 (1735): 20160420. https://doi. org/10.1098/rstb.2016.0420.

Ratcliff, William C., Matthew D. Herron, Kathryn Howell, Jennifer T. Pentz, Frank Rosenzweig, and Michael Travisano. 2013. "Experimental Evolution of an Alternating Uni- and Multicellular Life Cycle in *Chlamydomonas reinhardtii*." *Nature Communications* 4 (1): 2742. https://doi.org/10.1038/ncomms3742.

Ritchie, A. V., S. Van Es, C. Fouquet, and P. Schaap. 2008. "From Drought Sensing to Developmental Control: Evolution of Cyclic AMP Signaling in Social Amoebas." *Molecular Biology and Evolution* 25 (10): 2109–18. https://doi.org/10.1093/molbev/msn156.

Rokas, Antonis. 2008. "The Molecular Origins of Multicellular Transitions." *Current Opinion in Genetics & Development* 18 (6): 472–78. https://doi.org/10.1016/j.gde.2008.09.004.

Rossetti, V., M. Filippini, M. Svercel, A. D. Barbour, and H. C. Bagheri. 2011. "Emergent Multicellular Life Cycles in Filamentous Bacteria Owing to Density-Dependent Population Dynamics." *Journal of the Royal Society Interface* 8 (65): 1772–84. https:// doi.org/10.1098/rsif.2011.0102.

Roze, Denis, and Richard E. Michod. 2001. "Mutation, Multilevel Selection, and the Evolution of Propagule Size during the Origin of Multicellularity." *The American Naturalist* 158 (6): 638–54. https://doi.org/10.1086/323590.

Santelices, Bernabé. 1999. "How Many Kinds of Individual Are There?" *Trends in Ecology & Evolution* 14 (4): 152–55. https://doi.org/10.1016/S0169-5347(98)01519-5.

Schaap, Pauline. 2011. "Evolutionary Crossroads in Developmental Biology: *Dictyostelium discoideum*." *Development* 138 (3): 387–96. https://doi.org/10.1242/dev.048934.

Schlichting, C. D. 1986. "The Evolution of Phenotypic Plasticity in Plants." *Annual Review of Ecology and Systematics* 17 (1): 667–93. https://doi.org/10.1146/annurev. es.17.110186.003315.

Sebé-Pedrós, Arnau, Cecilia Ballaré, Helena Parra-Acero, Cristina Chiva, Juan J. Tena, Eduard Sabidó, José Luis Gómez-Skarmeta, Luciano Di Croce, and Iñaki Ruiz-Trillo. 2016. "The Dynamic Regulatory Genome of *Capsaspora* and the Origin of Animal Multicellularity." *Cell* 165 (5): 1224–37. https://doi.org/10.1016/j.cell.2016.03.034.

Sebé-Pedrós, Arnau, Elad Chomsky, Kevin Pang, David Lara-Astiaso, Federico Gaiti, Zohar Mukamel, Ido Amit, Andreas Hejnol, Bernard M. Degnan, and Amos Tanay. 2018. "Early Metazoan Cell Type Diversity and the Evolution of Multicellular Gene Regulation." *Nature Ecology & Evolution* 2 (7): 1176–88. https://doi.org/10.1038/ s41559-018-0575-6.

Sebé-Pedrós, Arnau, Bernard M. Degnan, and Iñaki Ruiz-Trillo. 2017. "The Origin of Metazoa: A Unicellular Perspective." *Nature Reviews Genetics* 18 (8): 498–512. https:// doi.org/10.1038/nrg.2017.21.

Sebé-Pedrós, Arnau, Manuel Irimia, Javier del Campo, Helena Parra-Acero, Carsten Russ, Chad Nusbaum, Benjamin J Blencowe, and Iñaki Ruiz-Trillo. 2013. "Regulated Aggregative Multicellularity in a Close Unicellular Relative of Metazoa." *eLife* 2 (December): e01287. https://doi.org/10.7554/eLife.01287.

Simpson, Carl, Amalia Herrera-Cubilla, and Jeremy B. C. Jackson. 2020. "How Colonial Animals Evolve." *Science Advances* 6 (2): eaaw9530. https://doi.org/10.1126/sciadv. aaw9530.

Solé, Ricard V., and Salva Duran-Nebreda. 2015. "In Silico Transitions to Multicellularity." In *Evolutionary Transitions to Multicellular Life*, edited by Iñaki Ruiz-Trillo and Aurora M. Nedelcu, 2: 245–66. Dordrecht: Springer Netherlands. https://doi. org/10.1007/978-94-017-9642-2_13.

Solé, Ricard V., and Sergi Valverde. 2013. "Before the Endless Forms: Embodied Model of Transition from Single Cells to Aggregates to Ecosystem Engineering." *PLOS ONE* 8 (4): e59664. https://doi.org/10.1371/journal.pone.0059664.

Srivastava, Mansi, Oleg Simakov, Jarrod Chapman, Bryony Fahey, Marie E. A. Gauthier, Therese Mitros, Gemma S. Richards, et al. 2010. "The *Amphimedon queenslandica* Genome and the Evolution of Animal Complexity." *Nature* 466 (7307): 720–26. https:// doi.org/10.1038/nature09201.

Staps, Merlijn, Jordi Van Gestel, and Corina E. Tarnita. 2019. "Emergence of Diverse Life Cycles and Life Histories at the Origin of Multicellularity." *Nature Ecology & Evolution* 3 (8): 1197–1205. https://doi.org/10.1038/s41559-019-0940-0.

Suga, Hiroshi, Zehua Chen, Alex De Mendoza, Arnau Sebé-Pedrós, Matthew W. Brown, Eric Kramer, Martin Carr, et al. 2013. "The *Capsaspora* Genome Reveals a Complex Unicellular Prehistory of Animals." *Nature Communications* 4 (1): 1–9. https://doi. org/10.1038/ncomms3325.

Suga, Hiroshi, and Iñaki Ruiz-Trillo. 2013. "Development of Ichthyosporeans Sheds Light on the Origin of Metazoan Multicellularity." *Developmental Biology* 377 (1): 284–92. https://doi.org/10.1016/j.ydbio.2013.01.009.

Tarnita, Corina E., Clifford H. Taubes, and Martin A. Nowak. 2013. "Evolutionary Construction by Staying Together and Coming Together." *Journal of Theoretical Biology* 320 (March): 10–22. https://doi.org/10.1016/j.jtbi.2012.11.022.

Todd, Peter A. 2008. "Morphological Plasticity in Scleractinian Corals." *Biological Reviews* 83 (3): 315–37. https://doi.org/10.1111/j.1469-185X.2008.00045.x.

Valentine, James W., and Charles R. Marshall. 2015. "Fossil and Transcriptomic Perspectives on the Origins and Success of Metazoan Multicellularity." In *Evolutionary Transitions to Multicellular Life: Principles and Mechanisms*, edited by Iñaki Ruiz-Trillo and Aurora M. Nedelcu, 31–46. Advances in Marine Genomics. Dordrecht: Springer Netherlands. https://doi.org/10.1007/978-94-017-9642-2_2.

Walbot, Virginia. 1996. "Sources and Consequences of Phenotypic and Genotypic Plasticity in Flowering Plants." *Trends in Plant Science* 1 (1): 27–32. https://doi.org/10.1016/ S1360-1385(96)80020-3.

Waters, Christopher M., and Bonnie L. Bassler. 2005. "QUORUM SENSING: Cell-to-Cell Communication in Bacteria." *Annual Review of Cell and Developmental Biology* 21 (1): 319–46. https://doi.org/10.1146/annurev.cellbio.21.012704.131001.

West, Stuart A., Roberta M. Fisher, Andy Gardner, and E. Toby Kiers. 2015. "Major Evolutionary Transitions in Individuality." *Proceedings of the National Academy of Sciences* 112 (33): 10112–19. https://doi.org/10.1073/pnas.1421402112.

Wolpert, Lewis, and Eörs Szathmáry. 2002. "Multicellularity: Evolution and the Egg." *Nature* 420 (December): 745. https://doi.org/10.1038/420745a.

Woznica, Arielle, Alexandra M. Cantley, Christine Beemelmanns, Elizaveta Freinkman, Jon Clardy, and Nicole King. 2016. "Bacterial Lipids Activate, Synergize, and Inhibit a Developmental Switch in Choanoflagellates." *Proceedings of the National Academy of Sciences* 113 (28): 7894–99. https://doi.org/10.1073/pnas.1605015113.

Woznica, Arielle, Joseph P. Gerdt, Ryan E. Hulett, Jon Clardy, and Nicole King. 2017. "Mating in the Closest Living Relatives of Animals Is Induced by a Bacterial Chondroitinase." *Cell* 170 (6): 1175–1183.e11. https://doi.org/10.1016/j.cell.2017.08.005.

Woznica, Arielle, and Nicole King. 2018. "Lessons from Simple Marine Models on the Bacterial Regulation of Eukaryotic Development." *Current Opinion in Microbiology* 43 (June): 108–16. https://doi.org/10.1016/j.mib.2017.12.013.

Section 2

Aggregative Multicellularity

5 Eukaryote Aggregative Multicellularity

Phylogenetic Distribution and a Case Study of Its Proximate and Ultimate Cause in Dictyostelia

Pauline Schaap
School of Life Sciences, University of Dundee, Dundee, UK

CONTENTS

5.1   INTRODUCTION

The evolution of multicellularity provided organisms with a novel means to increase in size and to specialize cells to perform different roles. These abilities were apparently so useful that multicellularity evolved many times independently in both pro- and eukaryotes. All extant multicellular organisms spend at least a part of their life in unicellular form as gamete, zygote, or spore, but otherwise, a distinction can be made between organisms that feed in multicellular form and those that feed as single cells and come together in response to starvation. The first category comprises the animals, many photosynthetic pro- and eukaryotes, and fungi. A primary characteristic of these organisms is that after division of the initial progenitor cell, the daughter cells remain together and the organism feeds and spends the rest of its life in multicellular form. The second category comprises many prokaryotes and eukaryotes. They share as a primary characteristic that after cell division the daughter cells swarm out to feed individually. They only come together when deprived of

DOI: 10.1201/9780429351907-7

food to form a fruiting structure, inside of which all or most cells enter dormancy. This type of aggregative or facultative multicellularity has reached its highest level of complexity in the Dictyostelia, where the cells inside the multicellular agglomerate have, in addition to dormant spores, specialized into up to five somatic cell types, which lift and support the spore mass and protect it from bacterial infection. This chapter describes the known types of organisms with aggregative multicellularity and their phylogenetic distribution in eukaryotes. The evolutionary trajectory taken by the Dictyostelia to reach their current state of complexity is next discussed in more detail.

5.2 AGGREGATIVE MULTICELLULARITY IS EVOLUTIONARY DERIVED FROM UNICELLULAR ENCYSTATION

The phenotype of all extant organisms is largely the end result of their existential struggle with environmental stress. A common response to stress is dormancy, which in unicellular eukaryotes is most commonly achieved by encystment. Here the stressed cell encapsulates itself in a resilient wall and shuts down metabolism. After the stress condition, usually starvation or drought has passed, the cell leaves its capsule and starts feeding again.

Aggregative multicellularity is also triggered by starvation stress, but the aggregated cells form a fruiting structure to aerially lift the encapsulating cells, which are now usually called spores. The fruiting body is called a sorocarp and this type of multicellularity is therefore also called sorocarpic multicellularity. In prokaryotes, the myxobacteria display sorocarpic multicellularity with a wide variety of morphologically distinctive fruiting structures (Kaiser et al., 2010), while in eukaryotes sorocarpic multicellularity evolved independently in almost all major divisions (Figure 5.1).

In Excavates, the sorocarpic *Acrasis* and *Pocheina* amoebas form a monophyletic clade within the otherwise unicellular allovahlkampfid amoebas. Upon starvation, the amoebas either form globose cysts individually or aggregate to form a globular mound. In the mound, cells start to encapsulate as stalk cells, with others climbing on top and also encapsulating. The stalk cells assume an oblong or cuboidal shape and assemble into tiers of one or several cells wide. The ovoid spores assemble next; in *Acrasis* spp. as single or branched tiers and in *Pocheina* as a globular mass. Despite their different shapes, the cysts, spores, and stalk cells all germinate to yield amoebas, when food is available again (Brown et al., 2012b).

In Rhizaria, *Guttulinopsis vulgaris* is part of a small clade of unicellular *Rosculus* amoebae (Bass et al., 2016). When starved, individual amoebas collect into loose aggregates that transform into round mounds. During sorocarp formation amoebae secrete a slimy matrix, which congeals into sheets, inside which they become progressively trapped and then disintegrate. Cells that manage to reach the top of the structure differentiate into irregularly shaped spores that combine to form a globose sorus. Amoebas of the related species *G. nivea* can also encyst individually (Brown et al., 2012a; Raper et al., 1977).

In Stramenopiles, the spindle-shaped *Sorodiplophrys stercorea* amoebae feed on dung by osmotrophy (uptake of dissolved organic compounds by osmosis) and move using long branched filopodia that extend from the spindle poles. When deprived

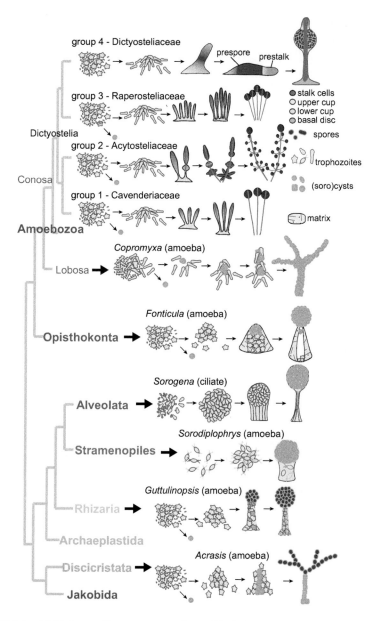

FIGURE 5.1 Multiple transitions to aggregative multicellularity in eukaryotes. During aggregative multicellularity cells feed and proliferate as single cells and collect into aggregates when stressed, usually by starvation. The aggregates then transform into a fruiting body (sorocarp) where the cells encapsulate and enter dormancy as either cysts or spores. This type of multicellularity evolved independently in six out of the eight major divisions of eukaryotes. Many organisms with aggregative multicellularity have retained the ability to encyst without aggregating, the survival strategy of their unicellular ancestors. Eukaryote phylogeny (grey lines) after (He et al., 2014). The Dictyostelia are the largest monophyletic group of organisms with aggregative multicellularity with the most extensive cell type specialization occurring in taxon group 4.

of food, the cells move in streams towards aggregation centers. A relatively firm basal core develops within the aggregate that consists of gelatinous matrix and dead and degenerating cells. Amoebas that surround and surmount this core develop into sorocytes; they retract the filopods, decrease in size and form a thin but distinctive cell wall. No cysts have been observed (Dykstra and Olive, 1975; Tice et al., 2016).

In Alveolata, the ciliate *Sorogena stoianovitchae* feeds underwater on smaller colpodid ciliates. Food depletion causes the cells to collect into a loose aggregate, which next compacts by cell adhesion. The cells secrete mucous material that forms a sheath around the aggregate and the structure now rises to the water surface. Continued matrix production and contraction of the sheath causes a stalk to form that pushes up the cells, which then encyst. Aggregation occurs only in the dark, but fruiting body formation requires a light period (Olive and Blanton, 1980; Sugimoto and Endoh, 2006).

In Opisthokonta, the nuclearid amoeba *Fonticula alba* displays long filopodia while feeding on bacteria. Upon starvation, the amoebas move together individually to form a mound that becomes surrounded by a slime sheath. The amoebas continue to secrete extracellular matrix that accumulates between the cell mass and the sheath. Starting from the top of the now volcano-shaped structure the amoebae encapsulate as spores, which are expelled through the top to form a globose sorus. The amoebas can also encyst individually without aggregating, with the cysts being morphologically identical to spores (Brown et al., 2009; Worley et al., 1979)

In Amoebozoa, sorocarpic multicellularity evolved in both the subphyla Lobosa and Conosa (Cavalier-Smith et al., 2015). In Lobosa, amoebae of the genus *Copromyxa* move by extending a single broad pseudopod. After depletion of their bacterial food, some amoebae encyst and become founder cells for the fruiting structure. The starving amoebas move towards an encysted founder cell, crawl on top and then also encapsulate to form sorocysts. This process continues until a tall sorocarp has formed, which is often branched. Amoebas can also encyst individually, and these microcysts are indistinguishable from the somewhat irregularly shaped sorocysts. Alternatively, they can form a round sphaerocyst after sexual fusion (Brown et al., 2011; Raper et al., 1978; Spiegel and Olive, 1978).

In Conosa, the Dictyostelia are the largest monophyletic group of organisms with sorocarpic multicellularity. The ˜150 known species of Dictyostelia can be subdivided into four major groups, recently classified as the families Cavenderiaceae (group 1), Acytosteliaceae (group 2), Raperosteliaceae (group 3), and Dictyosteliaceae (group 4) (Sheikh et al., 2018). Sorocarps in groups 1, 2, and 3 consist of at most two cell types, stalk cells and spores, and are often branched or clustered. Many species in these groups have retained the ability to encyst individually, and most use the dipeptide glorin as chemoattractant for aggregation. In contrast, group 4 species lost encystation and use 3'5'-adenosine monophosphate (cAMP) as chemoattractant. They generally form large, solitary, and unbranched sorocarps, which contain three novel cell types to support the stalk and spore mass. Many species throughout Dictyostelia have the additional ability to form sexual macrocysts. Here two starving cells fuse to form a zygote that subsequently attracts other starving cells and devours them to use their resources to construct a thick cell wall (Raper, 1984; Romeralo et al., 2013). The group 4 species *Dictyostelium discoideum* is a model system for research into the molecular mechanisms that control phagocytosis, motility, chemotaxis, morphogenesis, cell differentiation and evolution of sociality and is, apart from a few other Dictyostelia, the only sorocarpic or indeed only amoeboid organism with well-developed procedures for gene modification. We, therefore, concentrate in the remainder of this chapter on the evolution of multicellularity in the dictyostelid lineage.

5.3 PROXIMATE CAUSE OF *DICTYOSTELIUM* MULTICELLULARITY – HOW IT HAPPENED

5.3.1 REGULATION OF MULTICELLULAR DEVELOPMENT IN THE MODEL *D. DISCOIDEUM*

Over the past 40 years, the molecular mechanisms that control the developmental program of the model *D. discoideum* in group 4 have been thoroughly investigated. *D. discoideum* is one of the most morphologically complex Dictyostelia, and until recently there was only limited information on developmental control mechanisms in the other taxon groups. Research into the evolution of multicellularity in Dictyostelia, therefore, required a top-down approach. A well-resolved phylogeny of Dictyostelia (Schaap et al., 2006; Schilde et al., 2019), the availability of taxon group-representative genome sequences (Eichinger et al., 2005; Gloeckner et al., 2016; Heidel et al., 2011; Narita et al., 2020; Sucgang et al., 2011) and the development of genetic transformation for species outside of group 4 (Fey et al., 1995; Narita et al., 2020) were crucially important to address this problem.

A remarkable aspect of *D. discoideum* development is that so much of it is regulated by cAMP (Figure 5.2). As a secreted signal, cAMP controls cell movement and differentiation throughout development, while as an intracellular messenger, it mediates the effect of most other developmental signals. Intracellular cAMP is detected by cAMP-dependent protein kinase or PKA, while secreted cAMP targets G-protein coupled cAMP receptors (cARs). PKA activity is required for basal expression of aggregation genes (Schulkes and Schaap, 1995), for the expression of most prespore genes (Hopper et al., 1993), for the maturation of spores and stalk cells (Harwood et al., 1992; Hopper et al., 1993) and for keeping spores dormant in the spore head (Van Es et al., 1996). Upon starvation, PKA is upregulated by removal of the translational repressor PufA from the PKA 3' untranslated region (Souza et al., 1999). The adenylate cyclases AcgA and AcrA activate PKA in prespore and spore cells (Alvarez-Curto et al., 2007; Soderbom et al., 1999). AcgA is activated by solute stress caused by high levels of ammonium phosphate in the spore head, while AcrA appears to be constitutively active (Cotter et al., 1999; Van Es et al., 1996).

Other developmental signals act on the intracellular cAMP phosphodiesterase RegA to regulate PKA activity. RegA activity requires phosphorylation of its intrinsic response regulator domain, which in turn depends on the activity of sensor histidine kinases/phosphatases (Shaulsky et al., 1998; Thomason et al., 1998). These enzymes are the target for developmental signals, such as DhkA for the spore-inducing peptide SDF-2 (Wang et al., 1999), DhkC for the stalk inhibitor ammonia (Singleton et al., 1998), DhkB and DokA for respectively discadenine and high

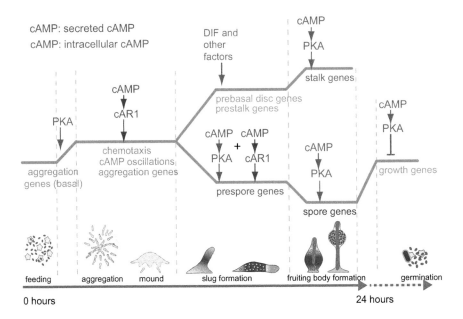

FIGURE 5.2 Roles of intracellular and secreted cAMP in *Dictyostelium discoideum*. Intracellular cAMP (in blue) acting on cAMP dependent protein kinase (PKA) and secreted cAMP (in red) acting on the cell surface cAMP receptor cAR1 regulate most aspects of the multicellular life cycle. See main text for a detailed description.

osmolarity (Schuster et al., 1996; Zinda and Singleton, 1998), which promote spore maturation and inhibit germination.

The cAMP receptor cAR1, AcaA and the extracellular cAMP phosphodiesterase PdsA and several other proteins, amongst which PKA and RegA, form an excitable network that can spontaneously generate cAMP oscillations (Laub and Loomis, 1998). Secreted cAMP acts as a chemoattractant, and the cAMP pulses coordinate both the aggregation of starving cells (Konijn et al., 1967) and the post-aggregative cell movements that result in sorogen migration and the upward projection of the fruiting body (Singer et al., 2019). For these post-aggregative roles, *acaA* becomes specifically expressed at the tip of the emerging sorogen, which thereby becomes the organizer of morphogenesis (Verkerke-van Wijk et al., 2001). AcaA hyperactivation by the stalk inducing factor c-di-GMP at the tip ensures that the *Dictyostelium* stalk is always formed at the morphogenetic organizer (Chen et al., 2017). The secreted cAMP pulses also act on cAR1 to further upregulate the expression of aggregation genes, such as *cAR1*, *acaA*, and *pdsA,* enabling rapid aggregation (Gerisch et al., 1975), while cAMP accumulates in aggregates, and acts on cAR1 to induce the expression of prespore genes (together with PKA) (Schaap and Van Driel, 1985).

Some developmental signaling events are thus far not known to require cAMP. Prespore cells secrete a chlorinated polyketide, DIF-1, that induces the differentiation of basal disc cells and contributes to robust stalk formation by increasing cytosolic Ca^{2+} (Kubohara et al., 2007; Poloz and O'Day, 2012; Schaap et al., 1996). Interaction between two highly polymorphic cell surface proteins, TgrC1 (a.k.a.

LagC) and TgrB1, mediates kin recognition in *D. discoideum,* allowing only closely related strains to contribute to the same sorogen. TgrC1 acts as ligand and TgrB1 as receptor in this interaction, which results in phosphorylation of TgrB1 and eventually in expression of a set of post-aggregative genes (Benabentos et al., 2009; Dynes et al., 1994; Hirose et al., 2017).

5.3.2 THE EVOLUTION OF CAMP SIGNALING IN THE DICTYOSTELIA

The dominant role of cAMP in all aspects of the *D. discoideum* life cycle is suggestive of a deeper ancestral role. Such a role was likely mediated by cAMP acting on PKA rather than cARs, since PKA is present in most eukaryotes, and cARs are thus far only found in Dictyostelia. Comparative genome analysis revealed that the PKA catalytic (PkaC) and regulatory (PkaR) subunits and the enzymes AcrA, AcgA, AcaA, and RegA are conserved throughout the four *Dictyostelium* taxon groups (Gloeckner et al., 2016) and that PKA, AcrA, and RegA are also present in the solitary Amoebozoa *Acanthamoeba castellani, Physarum polycephalum,* and *Protostelium aurantium* (Clarke et al., 2013; Hillmann et al., 2018; Schaap et al., 2015).

Deletion of *PkaC* in the group 2 Dictyostelid *Polysphondylium pallidum,* which can both encyst individually or aggregate to form fruiting bodies, blocked not only aggregation but also encystation. Deletion of *AcrA* and *AcgA* together also prevented encystation (Kawabe et al., 2015). Conversely, deletion of *RegA* in *P. pallidum* accelerated multicellular sporulation as it does in *D. discoideum*, and also caused precocious encystation during feeding. Pharmacological inactivation of RegA in *A. castellani* also caused amoebas to encyst precociously (Du et al., 2014). In conclusion, cAMP acting on PKA and regulated by at least RegA appears to have an ancestral role as intracellular messenger for the induction of encapsulation of starving solitary Amoebozoa into cysts (Figure 5.3). This role persisted in Dictyostelia and became elaborated to additionally control the initiation of aggregation and the encapsulation of starving amoebas in spores and stalk cells. The genomes of the solitary *A. castellani, Phy. polycephalum,* and *Pro. aurantium* Amoebozoa contain many sensor histidine kinases/phosphatases, and while no biological role for any of these enzymes has yet been demonstrated, it is likely that at least some of them regulate the activity of RegA.

While dictyostelids in groups 1, 2, and 3 do not use cAMP as an attractant to aggregate, cARs, and PdsA are conserved in these groups, suggesting at least some roles for extracellular cAMP. Loss of its two *cAR* genes or its *pdsA* gene had no effect on aggregation in *P. pallidum,* but disrupted fruiting body morphogenesis, with stalk cells differentiating in random clumps (Kawabe et al., 2009; 2012). Sp-cAMPS, a non-hydrolysable cAMP analog that disrupts oscillatory cAMP signaling, equally disrupted fruiting body morphogenesis in group 1, 2, and 3 species (Romeralo et al., 2013), indicating that cAMP oscillations ancestrally emerged to coordinate post-aggregative morphogenesis. Oscillatory cAMP signaling then moved forward in development in group 4 to organize aggregation as well. Addition of distal "early" promoters to proximate "late" promoters of *cAR1, pdsA,* and *acaA* genes likely enabled the use of cAMP as attractant for aggregation (Alvarez-Curto et al., 2005; Faure et al., 1990; Galardi-Castilla et al., 2010; Louis et al., 1993).

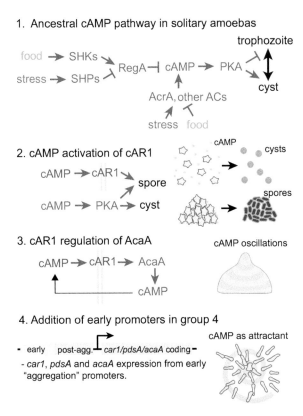

FIGURE 5.3   Evolution of cAMP signaling in Dictyostelia. Putative scenario for the evolution of intra- and extracellular cAMP signaling in Dictyostelia. See main text for description. SHK: sensor histidine kinase; SHP sensor histidine phosphatase. (Drawing revised from Kawabe et al., 2019.)

Organisation of morphogenesis may however not have been the earliest role for secreted cAMP. The *P. pallidum* cAR null cells made cysts instead of spores in their aberrant fruiting bodies and lost cAMP-induced prespore gene expression. Encystment only requires intracellular cAMP acting on PKA, while sporulation additionally requires extracellular cAMP acting on cARs. Without cARs, the *P. pallidum* presumptive spore cells reverted to the ancestral process of encystation (Kawabe et al., 2009), which points to what might be the most ancestral role for secreted cAMP. All starving Amoebazoa synthesize cAMP to activate PKA and encyst. Dictyostelia secrete most of their cAMP and may originally have used accumulation of secreted cAMP in aggregates to instruct the starving cells to form spores and not cysts.

The role of DIF-1 in basal disc and robust stalk cell differentiation is not conserved outside group 4, despite the fact that the genes encoding the enzymes, StlB, DmtA, and ChlA, which synthesize DIF-1, are present throughout Dictyostelia. However, a group 2 *dmtA* fails to restore a *D. discoideum dmtA* null mutant, indicating that it has another molecular function in group 2 (Mohri et al., 2014), while deletion of *stlB* in *Polyspondylium violaceum*, a sister species to group 4, causes more instead of fewer stalk cells to differentiate (Narita et al., 2020). Since basal disc cells are unique to group 4, it is likely that Dictyostelia co-evolved these cells in conjunction with the signal (DIF-1) that induces their differentiation.

GSK3 (glycogen synthase kinase 3) which, as component of the wingless/wnt pathway, determines cell fate decisions in animal development (Forde and Dale, 2007), was also reported to control cell fate choice in *D. discoideum*, by preventing the DIF-1 induced transition of prespore cells into basal disc cells (Harwood et al., 1995; Schilde et al., 2004). This role is not conserved in *P. pallidum*, where instead active GSK3 favors entry of starving cells into aggregation rather than encystation (Kawabe et al., 2018). While group 4 species such as *D. discoideum* lost encystation, remnants of this role in early development are still evident in a GSK3 requirement for expression of some early genes (Strmecki et al., 2007) and for efficient chemotaxis (Teo et al., 2010). Like the role of DIF-1, the role of GSK3 in basal disc differentiation appears to have co-evolved with the emergence of this novel cell type in group 4.

In conclusion, the dominant roles of secreted and intracellular cAMP in controlling *D. discoideum* multicellular development emerged from an ancestral role of intracellular cAMP as an intermediate for stress-induced encystation (Figure 5.3). A requirement for secreted cAMP acting on cARs to induce sporulation in aggregates (as opposed to encystation) may have been the first role of secreted cAMP. Next, incorporation of cARs, PdsA, AcaA, RegA, and PKA into an excitable network yielded the cAMP waves that enabled morphogenesis of well-structured fruiting bodies, while finally pre-aggregative expression of *cARs, pdsA,* and *acaA* enabled the use of cAMP as a chemoattractant in group 4.

5.4   ULTIMATE CAUSE OF *DICTYOSTELIUM* MULTICELLULARITY – WHY IT HAPPENED

Similar to many sorocarpic organisms, Dictyostelia evolved from a unicellular ancestor that survived starvation by differentiating into a cyst. The fact that sorocarpic multicellularity evolved so frequently across eukaryotes suggests an advantage to sporulating in fruiting bodies as opposed to unicellular encystment. An obvious benefit is improved spore dispersal from the aerially lifted spore head. This is particularly true for hydrophobic spores, such as those of fungi, which are carried by the wind. However, hydrophilic spores, such as those of Dictyostelia, depend for dispersal on sticking to soil invertebrates (Smith et al., 2014), or on being washed away by rain or snowmelt, the latter also causing dispersal of the soil-bound cysts. The aerial uplift of the spore agglomerate may further protect spores from decomposing influences in soil and predation by larger protists. Particularly for Dictyostelia with their small cells, predation may initially have encouraged starving amoebas to stick together, while going through encystation. Aggregation of solitary starving amoebas prior to encystation is reported for the holozoan amoeba *Capsaspora owczarzaki* (Sebe-Pedros et al., 2013), while the unicellular alga *Chlamydomonas reinhardtii* was shown to evolve multicellular structures when exposed to predation by ciliates for many generations (Herron et al., 2019).

For most sorocarpic protists, the spores only differ from cysts in being aloft, but in Dictyostelia, the two forms are also morphologically distinct (Figure 5.4A). Spores are predominantly elliptical with dense cytoplasm and a three-layered cell wall, while cysts are round and less dense with a two-layered wall (Hohl et al., 1970; Hohl and Hamamoto, 1969). This suggests that physiological differences between spores and cysts may have contributed to consolidation of the multicellular lifestyle of Dictyostelia. Long-term survival experiments of spores and cysts of representative species of each taxon group showed that spores and cysts, resuspended in water, survived temperatures above 20°C equally well, but that spores outperformed cysts at 4°C and -20°C. Cysts showed particularly poor survival when subjected to dry frost (Figure 5.4B). Among taxon groups, group 4 spores were the most cold-resistant. At the ultrastructural level, this was correlated with group 4 spores displaying a combination of dense cytoplasm and a thick spore wall. Global distribution data showed that group 4 species were frequently isolated from alpine and arctic zones, which was rarely the case for species in groups 1, 2, and 3. A fossil-calibrated phylogeny of Amoebozoa sets the split between the two major branches of Dictyostelia at 0.52 billion years ago, following the global glaciations known as "snowball earth" (Figure 5.4C). Combined, these observations suggest that *Dictyostelium* sporulation in multicellular fruiting bodies was an adaptation to cold climate (Lawal et al., 2020).

Whether climate change also played a role in the evolution of other sorocarpic protists is unknown. The fact that spores and cysts of *Fonticula alba*, *Copromyxa* spp., and *Acrasis* spp. are morphologically identical makes this unlikely. For these organisms, ultimate causes for their transitions to multicellularity may yet be found in protection from predation and/or improved spore dispersal and preservation. An important factor is the geological time at which the multicellular forms first appeared. Dictyostelia likely evolved on land since their aggregation and fruiting body formation require an air-water interface. They emerged when animal life was still bound to the oceans, invalidating the argument for improved spore dispersal by soil invertebrates. However, this may still have been a factor in the evolution of aggregative multicellularity in other organisms and in the later evolution of Dictyostelia (Smith et al., 2014).

The question remains why dictyostelid cysts did not simply evolve higher frost resistance on their own. The answer may lie in the stalk cells, which undergo extreme autolysis, not merely functioning to lift the spores, but also to provide them with metabolites for spore wall synthesis. This would be analogous to the sexual macrocysts of Dictyostelia cannabalizing their brethren to construct their very substantial cell walls. A flux of metabolites from stalk to spore cells has however yet to be demonstrated.

Inferring events that happened in the distant past is fraught with uncertainty. However, since the events that triggered multicellularity will also have contributed to shaping the mechanisms that control multicellular development, it is important to investigate ultimate cause, if only to have it questioned and analyzed further by other workers.

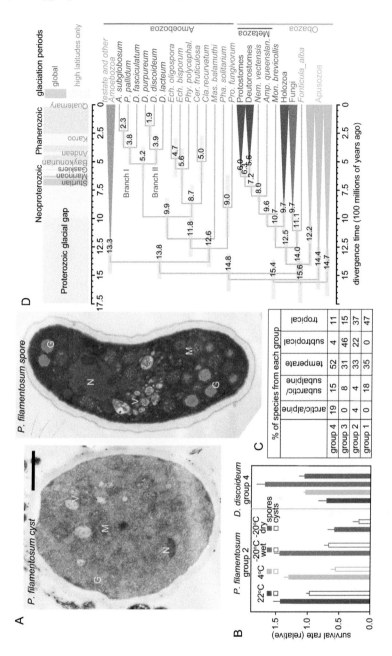

FIGURE 5.4 A cooling climate likely triggered multicellular sporulation in Dictyostelia. Spores are more compact than cysts and have a three-layered instead of maximally two-layered cell wall (A). Spores are more frost-resistant than cysts of the same species, with group 4 spores being most resistant (B), which is correlated with group 4 species being more frequently isolated from arctic and alpine habitats (C). The two major branches of Dictyostelia diverged 0.52 billion years ago, following several global glaciation periods (D). (Data taken from Lawal et al., 2020.). Bar in (A): 1 µm; N: nucleus; M: mitochondria; V: vesicles; G: granules.

ACKNOWLEDGEMENT

P.S. is funded by European Research Council grant 742288.

REFERENCES

Alvarez-Curto, E., Rozen, D.E., Ritchie, A.V., Fouquet, C., Baldauf, S.L., and Schaap, P. 2005. Evolutionary origin of cAMP-based chemoattraction in the social amoebae. *Proc Natl Acad Sci USA* 102: 6385–6390.

Alvarez-Curto, E., Saran, S., Meima, M., Zobel, J., Scott, C., and Schaap, P. 2007. cAMP production by adenylyl cyclase G induces prespore differentiation in *Dictyostelium* slugs. *Development* 134: 959–966.

Bass, D., Silberman, J.D., Brown, M.W. *et al.* 2016. Coprophilic amoebae and flagellates, including *Guttulinopsis*, *Rosculus* and *Helkesimastix*, characterise a divergent and diverse rhizarian radiation and contribute to a large diversity of faecal-associated protists. *Environmental Microbiology* 18: 1604–1619.

Benabentos, R., Hirose, S., Sucgang, R. *et al.* 2009. Polymorphic members of the *lag* gene family mediate kin discrimination in *Dictyostelium*. *Curr Biol* 19: 567–572.

Brown, Matthew W., Kolisko, M., Silberman, Jeffrey D., and Roger, Andrew J. 2012a. Aggregative multicellularity evolved independently in the eukaryotic supergroup Rhizaria. *Current Biology* 22: 1123–1127.

Brown, M.W., Silberman, J.D., and Spiegel, F.W. 2011. "Slime molds" among the Tubulinea (Amoebozoa): Molecular systematics and taxonomy of *Copromyxa*. *Protist* 162: 277–287.

Brown, M.W., Silberman, J.D., and Spiegel, F.W. 2012b. A contemporary evaluation of the acrasids (Acrasidae, Heterolobosea, Excavata). *Eur J Protistol* 48: 103–123.

Brown, M.W., Spiegel, F.W., and Silberman, J.D. 2009. Phylogeny of the "forgotten" cellular slime mold, *Fonticula alba*, reveals a key evolutionary branch within Opisthokonta. *Mol Biol Evol* 26: 2699–2709.

Cavalier-Smith, T., Fiore-Donno, A.M., Chao, E. *et al.* 2015. Multigene phylogeny resolves deep branching of Amoebozoa. *Mol Phylogenet Evol* 83: 293–304.

Chen, Z.H., Singh, R., Cole, C. *et al.* 2017. Adenylate cyclase A acting on PKA mediates induction of stalk formation by cyclic diguanylate at the *Dictyostelium* organizer. *Proc Natl Acad Sci U S A* 114: 516–521.

Clarke, M., Lohan, A.J., Liu, B. *et al.* 2013. Genome of *Acanthamoeba castellanii* highlights extensive lateral gene transfer and early evolution of tyrosine kinase signaling. *Genome Biol* 14: R11.

Cotter, D.A., Dunbar, A.J., Buconjic, S.D., and Wheldrake, J.F. 1999. Ammonium phosphate in sori of *Dictyostelium discoideum* promotes spore dormancy through stimulation of the osmosensor ACG. *Microbiology-Uk* 145: 1891–1901.

Du, Q., Schilde, C., Birgersson, E., Chen, Z.H., McElroy, S., and Schaap, P. 2014. The cyclic AMP phosphodiesterase RegA critically regulates encystation in social and pathogenic amoebas. *Cellular Signalling* 26: 453–459.

Dykstra, M.J., and Olive, L.S. 1975. *Sorodiplophrys*: an unusual sorocarp-producing protist. *Mycologia* 67: 873–879.

Dynes, J.L., Clark, A.M., Shaulsky, G., Kuspa, A., Loomis, W.F., and Firtel, R.A. 1994. LagC is required for cell-cell interactions that are essential for cell-type differentiation in *Dictyostelium*. *Genes Dev* 8: 948–958.

Eichinger, L., Pachebat, J.A., Glockner, G., *et al.* 2005. The genome of the social amoeba *Dictyostelium discoideum*. *Nature* 435: 43–57.

Faure, M., Franke, J., Hall, A.L., Podgorski, G.J., and Kessin, R.H. 1990. The cyclic nucleotide phosphodiesterase gene of *Dictyostelium discoideum* contains three promoters specific for growth, aggregation, and late development. *Mol Cell Biol* 10: 1921–1930.

Fey, P., Compton, K., and Cox, E.C. 1995. Green fluorescent protein production in the cellular slime molds *Polysphondylium pallidum* and *Dictyostelium discoideum*. *J Cell Biol* 165: 127–130.

Forde, J.E. and Dale, T.C. 2007. Glycogen synthase kinase 3: a key regulator of cellular fate. *Cell Molec Life Sci* 64: 1930–1944.

Galardi-Castilla, M., Garciandia, A., Suarez, T., and Sastre, L. 2010. The *Dictyostelium discoideum acaA* gene is transcribed from alternative promoters during aggregation and multicellular development. *PLoS One* 5: e13286.

Gerisch, G., Fromm, H., Huesgen, A., and Wick, U. 1975. Control of cell-contact sites by cyclic AMP pulses in differentiating *Dictyostelium* cells. *Nature* 255: 547–549.

Gloeckner, G., Lawal, H.M., Felder, M. *et al.* 2016. The multicellularity genes of dictyostelid social amoebas. *Nat Comm* 7: 12085.

Harwood, A.J., Hopper, N.A., Simon, M.N., Driscoll, D.M., Veron, M., and Williams, J.G. 1992. Culmination in *Dictyostelium* is regulated by the cAMP-dependent protein kinase. *Cell* 69: 615–624.

Harwood, A.J., Plyte, S.E., Woodgett, J., Strutt, H., and Kay, R.R. 1995. Glycogen synthase kinase 3 regulates cell fate in *Dictyostelium*. *Cell* 80: 139–148.

He, D., Fiz-Palacios, O., Fu, C.J., Fehling, J., Tsai, C.C., and Baldauf, S.L. 2014. An alternative root for the eukaryote tree of life. *Curr Biol* 24: 465–470.

Heidel, A., Lawal, H., Felder, M. *et al.* 2011. Phylogeny-wide analysis of social amoeba genomes highlights ancient origins for complex intercellular communication. *Genome Res* 21: 1882–1891.

Herron, M.D., Borin, J.M., Boswell, J.C. *et al.* 2019. De novo origins of multicellularity in response to predation. *Sci Rep* 9: 2328.

Hillmann, F., Forbes, G., Novohradska, S. *et al.* 2018. Multiple roots of fruiting body formation in Amoebozoa. *Genome Biol Evol* 10: 591–606.

Hirose, S., Chen, G., Kuspa, A., and Shaulsky, G. 2017. The polymorphic proteins TgrB1 and TgrC1 function as a ligand-receptor pair in *Dictyostelium* allorecognition. *J Cell Sci* 130: 4002–4012.

Hohl, H.R. and Hamamoto, S.T. 1969. Ultrastructure of spore differentiation in *Dictyostelium*: the prespore vacuole. *J Ultrastruct Res* 26: 442–453.

Hohl, H.R., Miura-Santo, L.Y., and Cotter, D.A. 1970. Ultrastructural changes during formation and germination of microcysts in *Polysphondylium pallidum*, a cellular slime mould. *J Cell Sci* 7: 285–306.

Hopper, N.A., Harwood, A.J., Bouzid, S., Véron, M., and Williams, J.G. 1993. Activation of the prespore and spore cell pathway of *Dictyostelium* differentiation by cAMP-dependent protein kinase and evidence for its upstream regulation by ammonia. *EMBO J* 12: 2459–2466.

Kaiser, D., Robinson, M., and Kroos, L. 2010. Myxobacteria, polarity, and multicellular morphogenesis. *Cold Spring Harb Perspect Biol* 2: a000380–a000380.

Kawabe, Y., Du, Q., Schilde, C., and Schaap, P. 2019. Evolution of multicellularity in Dictyostelia. *Int J Dev Biol* 63: 359–369.

Kawabe, Y., Morio, T., James, J.L., Prescott, A.R., Tanaka, Y., and Schaap, P. 2009. Activated cAMP receptors switch encystation into sporulation. *Proc Natl Acad Sci USA* 106: 7089–7094.

Kawabe, Y., Morio, T., Tanaka, Y., and Schaap, P. 2018. Glycogen synthase kinase 3 promotes multicellular development over unicellular encystation in encysting Dictyostelia. *Evodevo* 9: 12.

Kawabe, Y., Schilde, C., Du, Q., and Schaap, P. 2015. A conserved signalling pathway for amoebozoan encystation that was co-opted for multicellular development. *Sci Rep* 5: 9644.

Kawabe, Y., Weening, K.E., Marquay-Markiewicz, J., and Schaap, P. 2012. Evolution of self-organisation in Dictyostelia by adaptation of a non-selective phosphodiesterase and a matrix component for regulated cAMP degradation. *Development* 139: 1336–1345.

Konijn, T.M., Van De Meene, J.G., Bonner, J.T., and Barkley, D.S. 1967. The acrasin activity of adenosine-3',5'-cyclic phosphate. *Proc Natl Acad Sci USA* 58: 1152–1154.

Kubohara, Y., Arai, A., Gokan, N., and Hosaka, K. 2007. Pharmacological evidence that stalk cell differentiation involves increases in the intracellular Ca2+ and H+ concentrations in *Dictyostelium discoideum*. *Dev Growth Diff* 49: 253–264.

Laub, M.T. and Loomis, W.F. 1998. A molecular network that produces spontaneous oscillations in excitable cells of *Dictyostelium*. *Mol Biol Cell* 9: 3521–3532.

Lawal, H.M., Schilde, C., Kin, K. *et al.* 2020. Cold climate adaptation is a plausible cause for evolution of multicellular sporulation in Dictyostelia. *Sci Rep* 10, 8797.

Louis, J.M., Saxe III, C.L., and Kimmel, A.R. 1993. Two transmembrane signaling mechanisms control expression of the cAMP receptor gene *cAR1* during *Dictyostelium* development. *Proc Natl Acad Sci USA* 90: 5969–5973.

Mohri, K., Hata, T., Kikuchi, H., Oshima, Y., and Urushihara, H. 2014. Defects in the synthetic pathway prevent DIF-1 mediated stalk lineage specification cascade in the non-differentiating social amoeba, *Acytostelium subglobosum*. *Biol Open* 3: 553–560.

Narita, T.B., Kawabe, Y., Kin, K., Gibbs, R.A., Kuspa, A., Muzny, D.M., *et al.* 2020. Loss of the polyketide synthase StlB results in stalk cell overproduction in *Polysphondylium violaceum*. *Genome Biol Evol* 12: 674–683.

Olive, L.S. and Blanton, R.L. 1980. Aerial sorocarp development by the aggregative ciliate, *Sorogena-stoianovitchae*. *J Protozool* 27: 293–299.

Poloz, Y. and O'Day, D.H. 2012. Ca2+ signaling regulates *ecmB* expression, cell differentiation and slug regeneration in *Dictyostelium*. *Differentiation* 84: 163–175.

Raper, K.B. 1984. *The Dictyostelids*. Princeton New Jersey: Princeton University Press.

Raper, K.B., Worley, A.C., and Kessler, D. 1977. Observations on *Guttulinopsis vulgaris* and *Guttulinopsis nivea*. *Mycologia* 69: 1016–1030.

Raper, K.B., Worley, A.C., and Kurzynski, T.A. 1978. *Copromyxella*: a new genus of acrasidae. *Amer J Bot* 65: 111–1026.

Romeralo, M., Skiba, A., Gonzalez-Voyer, A. *et al.* 2013. Analysis of phenotypic evolution in Dictyostelia highlights developmental plasticity as a likely consequence of colonial multicellularity. *Proc Biol Sci* 280: 20130976.

Schaap, P., Barrantes, I., Minx, P. *et al.* 2015. The *Physarum polycephalum* genome reveals extensive use of prokaryotic two-component and metazoan-type tyrosine kinase signaling. *Genome Biol Evol* 8: 109–125.

Schaap, P., Nebl, T., and Fisher, P.R. 1996. A slow sustained increase in cytosolic Ca2+ levels mediates stalk gene induction by differentiation inducing factor in *Dictyostelium*. *EMBO J* 15: 5177–5183.

Schaap, P. and Van Driel, R. 1985. Induction of post-aggregative differentiation in *Dictyostelium discoideum* by cAMP. Evidence of involvement of the cell surface cAMP receptor. *Exp Cell Res* 159: 388–398.

Schaap, P., Winckler, T., Nelson, M. *et al.* 2006. Molecular phylogeny and evolution of morphology in the social amoebas. *Science* 314: 661–663.

Schilde, C., Araki, T., Williams, H., Harwood, A., and Williams, J.G. 2004. GSK3 is a multifunctional regulator of *Dictyostelium* development. *Development* 131: 4555–4565.

Schilde, C., Lawal, H.M., Kin, K., Shibano-Hayakawa, I., Inouye, K., and Schaap, P. 2019. A well supported multi gene phylogeny of 52 dictyostelia. *Mol Phylogenet Evol* 134: 66–73.

Schulkes, C. and Schaap, P. 1995. cAMP-dependent protein kinase activity is essential for preaggegative gene expression in *Dictyostelium*. *FEBS Lett* 368: 381–384.

Schuster, S.C., Noegel, A.A., Oehme, F., Gerisch, G., and Simon, M.I. 1996. The hybrid histidine kinase DokA is part of the osmotic response system of *Dictyostelium*. *EMBO J* 15: 3880–3889.

Sebe-Pedros, A., Irimia, M., Del Campo, J. *et al.* 2013. Regulated aggregative multicellularity in a close unicellular relative of metazoa. *eLife* 2: e01287.

Shaulsky, G., Fuller, D., and Loomis, W.F. 1998. A cAMP-phosphodiesterase controls PKA-dependent differentiation. *Development* 125: 691–699.

Sheikh, S., Thulin, M., Cavender, J.C. *et al.* 2018. A new classification of the Dictyostelids. *Protist* 169: 1–28.

Singer, G., Araki, T., and Weijer, C.J. 2019. Oscillatory cAMP cell-cell signalling persists during multicellular *Dictyostelium* development. *Comm Biol* 2: 139.

Singleton, C.K., Zinda, M.J., Mykytka, B., and Yang, P. 1998. The histidine kinase *dhkC* regulates the choice between migrating slugs and terminal differentiation in *Dictyostelium discoideum*. *Dev Biol* 203: 345–357.

Smith, J., Queller, D.C., and Strassmann, J.E. 2014. Fruiting bodies of the social amoeba *Dictyostelium discoideum* increase spore transport by *Drosophila*. *BMC Evol Biol* 14: 105.

Soderbom, F., Anjard, C., Iranfar, N., Fuller, D., and Loomis, W.F. 1999. An adenylyl cyclase that functions during late development of *Dictyostelium*. *Development* 126: 5463–5471.

Souza, G.M., daSilva, A.M., and Kuspa, A. 1999. Starvation promotes *Dictyostelium* development by relieving PufA inhibition of PKA translation through the YakA kinase pathway. *Development* 126: 3263–3274.

Spiegel, F.W. and Olive, L.S. 1978. New evidence for validity of *Copromyxa-protea*. *Mycologia* 70: 843–847.

Strmecki, L., Bloomfield, G., Araki, T. *et al.* 2007. Proteomic and microarray analyses of the *Dictyostelium* Zak1-GSK-3 signaling pathway reveal a role in early development. *Eukaryot Cell* 6: 245–252.

Sucgang, R., Kuo, A., Tian, X. *et al.* 2011. Comparative genomics of the social amoebae *Dictyostelium discoideum* and *Dictyostelium purpureum*. *Genome Biol* 12: R20.

Sugimoto, H. and Endoh, H. 2006. Analysis of fruiting body development in the aggregative ciliate *Sorogena stoianovitchae* (Ciliophora, Colpodea). *J Eukaryot Microbiol* 53: 96–102.

Teo, R., Lewis, K.J., Forde, J.E. *et al.* 2010. Glycogen synthase kinase-3 is required for efficient *Dictyostelium* chemotaxis. *Mol Biol Cell* 21: 2788–2796.

Thomason, P.A., Traynor, D., Cavet, G., Chang, W.-T., Harwood, A.J., and Kay, R.R. 1998. An intersection of the cAMP/PKA and two-component signal transduction systems in *Dictyostelium*. *EMBO J* 17: 2838–2845.

Tice, A.K., Silberman, J.D., Walthall, A.C., Le, K.N., Spiegel, F.W., and Brown, M.W. 2016. *Sorodiplophrys stercorea*: another novel lineage of sorocarpic multicellularity. *J Eukaryot Microbiol* 63: 623–628.

Van Es, S., Virdy, K.J., Pitt, G.S. *et al.* 1996. Adenylyl cyclase G, an osmosensor controlling germination of *Dictyostelium* spores. *J Biol Chem* 271: 23623–23625.

Verkerke-van Wijk, I., Fukuzawa, M., Devreotes, P.N., and Schaap, P. 2001. Adenylyl cyclase A expression is tip-specific in *Dictyostelium* slugs and directs StatA nuclear translocation and *CudA* gene expression. *Dev Biol* 234: 151–160.

Wang, N., Soderbom, F., Anjard, C., Shaulsky, G., and Loomis, W.F. 1999. SDF-2 induction of terminal differentiation in *Dictyostelium discoideum* is mediated by the membrane-spanning sensor kinase DhkA. *Mol Cell Biol* 19: 4750–4756.

Worley, A.C., Raper, K.B., and Hohl, M. 1979. *Fonticula alba*: A new cellular slime mold (Acrasiomycetes). *Mycologia* 71: 746–760.

Zinda, M.J. and Singleton, C.K. 1998. The hybrid histidine kinase *dhkB* regulates spore germination in *Dictyostelium discoideum*. *Dev Biol* 196: 171–183.

6 Group Formation
On the Evolution of
Aggregative Multicellularity

Marco La Fortezza
Institute for Integrative Biology
ETH Zürich, Zürich, Switzerland

Kaitlin A. Schaal
Institute for Integrative Biology
ETH Zürich, Zürich, Switzerland

Gregory J. Velicer
Institute for Integrative Biology
ETH Zürich, Zürich, Switzerland

CONTENTS

6.1 INTRODUCTION

Microbes have evolved many fascinating and complex behaviors. They were long thought of as self-sufficient single cells that compete with each other for resources, but in fact many of their interactions benefit each other. In particular, some microbes

DOI: 10.1201/9780429351907-8

can actively aggregate into cooperative multicellular groups. This ability has evolved independently multiple times throughout the tree of life, suggesting that it provides particular benefits to the organisms which possess it. Obligately multicellular organisms have also evolved aggregative behaviors (see Box 6.1), but here we focus on aggregation among reproductively autonomous cells.

Microbes can form a variety of multicellular associations, including filaments, biofilms, and fruiting bodies (Claessen et al. 2014; Grosberg and Strathmann 2007). Filaments form when newly divided cells adhere to one another rather than separating (Flärdh et al. 2012; Flores and Herrero 2010). Biofilms form when cells stick to each other and attach to a surface. This can involve clonal adhesion, as in filament formation, but the biofilm may also passively recruit external cells (passive aggregation) (Smith et al. 2015; Trunk et al. 2018), or these cells may actively join the biofilm using directed motility (active aggregation) (Houry et al. 2010). We do not cover microbial filaments or biofilms in this chapter because active aggregation is not the primary force driving the formation of either. Fruiting bodies, however, are raised or extended cell mounds generated primarily by active aggregation and within which cells often differentiate into stress-resistant spores.

Fruiting body formation has been documented in four Eukarya supergroups: Amoebozoa (*Dictyostelia* and *Copromyxa*), Discoba (*Acrasidae*), Holozoa (*Fonticula*), and SAR/Harosa (*Sorodiplophrys* in Stramenopiles, *Sorogena* in Alveolata, and

BOX 6.1   AGGREGATIVE BEHAVIORS

Here we use the terms *aggregative behavior* and *aggregation* to refer to those processes in which single units come together to form or join a group, which represents a higher-level unit of organization and potentially of selection. Aggregation occurs not only among autonomously reproducing cells (the focus of this chapter) but also among cells within animal bodies and among multicellular animals. Examples of cellular aggregation within animal bodies include aggregation among platelet cells during wound healing (Savage et al. 1998), germ-line and mesenchymal cells during embryogenesis (O'Shea 1987; Savage and Danilchik 1993), single sponge cells after dissociation (Padua et al. 2016; Wilson 1907), and cancer cells in the bloodstream during metastasis (Aceto et al. 2014). Animal groups such as herds, flocks, and schools also result from aggregative behavior. Many of the proposed advantages and disadvantages of aggregating into groups are similar for cells and animals, including protection from abiotic stresses and predation and enhanced migration (Krause and Ruxton 2002; West et al. 2015). Although very different from group formation, plants and fungi exhibit coming-together-like behavior in the act of fusing to form chimeric organisms composed of two or more distinct genotypes (Buss 1982). For example, hyphae of genetically distinct fungi can fuse and form chimeric fruiting bodies (Glass et al. 2004), and plant chimeras can emerge from genetically distinct individuals that merge during growth (Frank and Chitwood 2016).

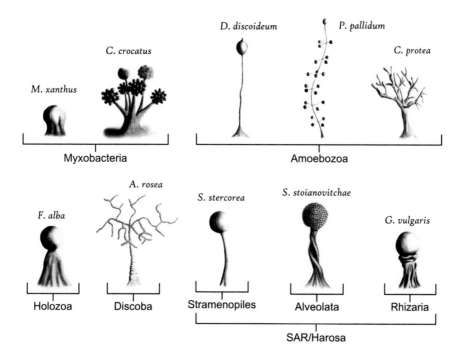

FIGURE 6.1 Representative examples of fruiting body morphologies across different taxa. Drawings are at different scales and are based on: *Myxococcus xanthus* and *Chondromyces crocatus* (direct observations), *Dictyostelium discoideum* (direct observation), *Polysphondylium pallidum* (Kawabe et al. 2015), *Copromyxa protea* (Brown and Silberman 2013), *Fonticula alba* (Brown and Silberman 2013), *Acrasis rosea* (Brown, Spiegel, and Silberman 2012), *Sorodiplophrys stercorea* (Dykstra and Olive 1975), *Sorogena stoianovitchae* (Olive and Blanton 1980), *Guttulinopsis vulgaris* (Brown and Silberman 2013) (drawings credit: Marco La Fortezza).

Guttulinopsis in Rhizaria) (Figure 6.1) (Brown et al. 2012; Brown et al. 2012b; Brown and Silberman 2013; Sugimoto & Endoh 2006). These organisms can be found in a variety of environments (Swanson et al. 1999) and have similar multicellular life cycles. Their process of aggregation often results in fruiting bodies (sometimes known as sorocarps), whose morphologies can differ greatly across species (see Chapters 5 and 8 and Figure 6.1). Of these taxa, *Dictyostelia* is the best described, thanks to the well-known model species *Dictyostelium discoideum* (Bonner 2009; Kessin 2001).

Some *Bacillus* species (Branda et al. 2001; Claessen et al. 2014) form fruiting bodies, but the most famous bacterial fruiting bodies are those formed by species in the order Myxococcales. The myxobacteria, or "slime bacteria," were first identified by Roland Thaxter (1892), who found and examined fruiting bodies on decaying wood. The best-known myxobacterium is *Myxococcus xanthus* (Yang and Higgs 2014). In addition to being models for studies of aggregative development, both *D. discoideum* and *M. xanthus* have become model experimental systems for studies of the ecology and evolution of microbial social behaviors (Medina et al. 2019; Velicer and Vos 2009).

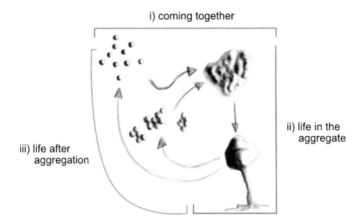

FIGURE 6.2 Stereotypical life cycle of aggregative multicellular organisms divided into three major stages (i) coming together: single cells and/or cell groups actively aggregate and adhere to one another to form a multicellular structure. (ii) life in the aggregate: cells interact within the aggregate while group-level behaviors and morphologies become more pronounced. In the drawing, an early multicellular aggregate develops into a mature fruiting body. In some cases, the aggregate may be dispersed as a whole to a new location by an external force. In other cases, the aggregate may not be dispersed before environmental conditions which previously favored aggregation change to again favor growth and disaggregation. iii) life after aggregation: once spores have been exposed to nutrients (e.g., prey) and germinated, cells may reduce local cell density by migrating away from the aggregate before living asocially (e.g., dictyostelids) or in cooperative groups (e.g., myxobacteria) until the next cycle of aggregation.

Active aggregation seems to be adaptive. Aggregative behaviors are energetically costly and are regulated by complex genetic pathways, so they presumably evolved via selection. However, other evolutionary processes may have played a role. In this chapter, we consider why aggregative multicellularity might be adaptive and which features may be nonadaptive. We discuss evolutionary hypotheses for the origin, maintenance, and complexification of several diverse forms of aggregative multicellularity, focusing on systems that use motility to aggregate. The chapter is organized according to three temporal phases of aggregate life cycles: first, the process of active aggregation, or coming together; second, life in the aggregate; and third, disaggregation and post-aggregate behavior (Figure 6.2). It ends with a discussion of chimeric aggregates (those composed of multiple genotypes) and an outlook on the broader relevance of studying this intriguing phenomenon.

6.2 COMING TOGETHER

Multicellular aggregates begin when cells seek each other out and stick together, increasing their local density. We consider two types of forces at work during this process: those responsible for increasing cell density and those responsible for maintaining it. Cell density can be increased by physical forces external to the aggregating cells, in which case this step of aggregation is passive (Arias Del Angel et al. 2020). For example, cells can encounter and stick to one another as they are transported by flowing liquid. Alternately, cells can aggregate by motility, an agent-like behavior (Arias Del Angel et al. 2020). In this case, cells call and search for each other by producing and responding to chemical cues or signals (Du et al. 2015; Kroos 2017). We will focus on forms of aggregative multicellularity in which cells use motility to actively increase their local density, such as fruiting body formation (Claessen et al. 2014; Du et al. 2015), rather than those generated more passively, such as flocculation of planktonic yeast cells (Verstrepen and Klis 2006; Trunk et al. 2018). Regardless of whether cell density increases passively or actively, cells will maintain high density under physical disturbance only if they establish stable physical contacts by producing cell-cell adhesins or by using chemoattraction and motility behavior to remain near one another, both of which constitute active behavior.

6.2.1 Aggregation in Myxobacteria and Dictyostelids

The myxobacteria and the dictyostelids are by far the most studied taxa in which motility-driven aggregation has evolved. In both, cells in sufficiently dense populations respond to nutrient deprivation by aggregating into cell mounds. These increase in size and, at least in some species, in morphological complexity as more cells join and development proceeds, culminating in the formation of a fruiting body. Some cells die during development, while others differentiate into stress-resistant spores (Kessin 2001; Lee et al. 2012; O'Connor and Zusman 1988; Stenhouse and Williams 1977; Wireman and Dworkin 1977). Other fruiting-body-forming eukaryotes aggregate in a similar fashion, with a few exceptions. For example, the *Copromyxa protea* fruiting body is composed entirely of living cells, which one by one move to the top of the structure and encyst (Spiegel and Olive 1978).

In the myxobacteria, mounds of aggregated cells gradually morph into elevated fruiting bodies with species-specific phenotypes (Grilione and Pangborn 1975; Spröer et al. 1999). Within fruiting bodies of *M. xanthus*, cells may die, convert to spherical spores, or remain rod-shaped. Development in *M. xanthus* involves complex regulatory pathways and multiple temporally regulated inter-cellular signals (Kroos 2017). The highly conserved bacterial stringent response, in which nutrient deprivation induces (p)ppGpp synthesis (Kroos 2017), has been evolutionarily integrated with a myxobacteria-specific quorum-sensing system, and the combination triggers aggregative motility behavior upon depletion of extracellular amino acids. In *M. xanthus,* the quorum signal that triggers aggregation is a mix of amino acids derived from degrading cell-surface peptides; this has been termed the A-signal (Kaplan and Plamann 1996). Simple aggregation of cells in response to A-signal may have been an early adaptation in the evolution of myxobacterial multicellularity prior to the evolution of more complex structures, though so far there are no studies that examine this hypothesis.

Aggregating and sporulating usually happen together. However, some *M. xanthus* genotypes sporulate extensively without aggregating (La Fortezza and Velicer 2021; Velicer et al. 1998), which raises the question of whether sporulation evolved prior to aggregation. Which of the genes necessary for spore production are also necessary for fruiting body formation remains to be fully determined. Many myxobacterial species actively aggregate prior to sporulating rather than simply sporulating in place in response to starvation cues, which could be done immediately and without expending energy on aggregating, suggesting that there are adaptive benefits to sporulating within a fruiting body. Additionally, most *M. xanthus* mutants that are defective at aggregation are also to some degree defective at spore production, indicating deep evolutionary integration of these processes.

In the dictyostelids, cells in nutrient-rich environments appear to interact with each other rarely and briefly (Kessin 2001). They usually feed on bacteria, which they find by following traces of folic acid that prey cells release into their surroundings (Pan et al. 1972). However, when prey are scarce, dictyostelid cells become social. This transition to multicellularity can be easily induced and observed in the lab by manipulating nutrient concentrations. *D. discoideum* cells secrete cAMP to actively attract each other, though the role of cAMP is not limited to chemotactic aggregation (Du et al. 2015, Kawabe et al. 2019; Meena and Kimmel 2017), and different species of amoebae use cAMP at different stages of aggregation (Kawabe et al. 2019). Like for the myxobacteria, there seems to be a dearth of dictyostelid studies testing for adaptive benefits that the earliest stages of aggregation may provide in their own right.

6.2.2   In Situ Aggregation Benefits: The Shelter Hypothesis

It seems plausible that the modern forms of aggregative multicellular structures evolved gradually. For example, in dictyostelids aggregation into fruiting bodies is hypothesized to have evolved through co-option of the process leading to macrocyst formation, a simpler form of aggregation present in many species (Kessin 2001). In general, aggregative structures presumably originated as slightly raised cell mounds which may have emerged in response to similar selective pressures. Simple biofilm associations have been reported to protect cells from abiotic stressors and biotic dangers (Oliveira et al. 2015; Trunk et al. 2018), and multicellularity has evolved in experimental populations of algae in response to predation pressure (Herron et al. 2019). Cells in the interior of an aggregate may be sheltered either by more distal cells (Kawaguchi et al. 2020; Smukalla et al. 2008) or by a layer of extracellular material produced at the group surface (Kirby et al. 2012; Vega and Gore 2014). The predatory nematode *Caenorhabditis elegans* has been observed to induce aggregation in *D. discoideum* (Kessin et al. 1996). This might be an indirect effect of increased competition for prey bacteria rather than an escape mechanism in direct response to *C. elegans*; regardless, these simple aggregates appear to protect cells from predation.

Larger groups may do a better job of protecting cells, and aggregation is a faster and more cost-effective way than cell division to increase group size (Pentz et al. 2020, Tarnita et al. 2013). Active aggregation into simple cell mounds may have increased protection from stress beyond what can be achieved within flat colonies or biofilms. Such simple shelter benefits are likely to have been among the forces selecting for the earliest evolutionary stages of active aggregation.

There are inherent challenges in attempting to test the shelter hypothesis. For one thing, we don't have access to the non-aggregating ancestors of aggregating lineages. One could use non-aggregating mutants of extant genotypes as an alternative, although there is always the possibility for the mutations to create confounding effects. One could compare these mutants with wild-type strains and with other mutants that arrest development after early aggregation in order to investigate whether there are shelter benefits realized at that point that are not contingent on later, more complex morphological development, sporulation, or aggregate dispersal.

6.2.3 INFLUENCE OF ECOLOGY ON AGGREGATIVE PROCESSES

Regardless of the selective forces that first favored primitive aggregation, the ecological factors involved in modern processes of aggregation differ across taxa. One natural isolate of *D. mucoroides* needs the presence of a specific fungus (*Mucor hiemalis*) in order to aggregate and develop into fruiting bodies (Ellison and Buss 1983). The protist *Acrasis rosea* generally requires a light:dark cycle in order to fruit (Reinhardt 1968). Aggregates of the myxobacterium *Stigmatella aurantiaca* form spore-bearing fruiting bodies only when exposed to light (Qualls et al. 1978). In *M. xanthus,* interaction with some other bacterial species can induce aggregation (Berleman and Kirby 2007), and the nutrient conditions triggering aggregation vary even among strains of the same species (Kraemer et al. 2010). These examples suggest that the evolution of aggregates from simple mounds into the range of structures seen today was likely influenced by varied and complex aspects of the specific ecologies of the organisms.

6.3   LIFE IN THE AGGREGATE

Primitive aggregation, once evolved, potentiates further adaptive innovations. For example, aggregating lineages might evolve cell-surface adhesins to stabilize the aggregate or changes in aggregate size and morphology that increase shelter benefits or provide aggregate-dispersal benefits. However, identifying adaptive versus non-adaptive features of aggregates and of cellular behavior within aggregates can be challenging.

6.3.1   ACTIVE AGGREGATE MIGRATION

While myxobacteria swarm cooperatively during vegetative growth (Muñoz-Dorado et al. 2016), aggregated *Dictyostelium* cells migrate together only during development in amorphous multicellular structures called slugs. Cells within a slug are not fully committed to continuing the developmental program until the slug begins to transform into a vertical fruiting body, at which point cells commit to terminal differentiation into either stalk cells or spores (Kessin 2001). Slugs exhibit photo- and thermotaxis, and slug migration has been hypothesized to be an adaptive group behavior that improves positioning for subsequent fruiting body formation and spore dispersal (Bonner 2009; Kessin 2001). Slugs can migrate across some terrain more easily than single cells, and larger slugs are able to migrate over a larger area than smaller slugs (Foster et al. 2002). As discussed in Chapter 8, motile slugs may provide dispersal-related benefits. However, not all dictyostelids form motile slugs (Bonner 2003, Nanjundiah 2016, see Chapter 8), and the evolutionary forces that led some genera to acquire this trait, whereas others did not, are worth investigating.

6.3.2 Cell Fates within Aggregates

One of the most evolutionarily fascinating features of many aggregative systems is that some cells die and some survive as spores. Joining an aggregate therefore can come with a high degree of fitness uncertainty. The extent of cell death may vary greatly across aggregative systems. In *Myxococcus,* only a minority of cells become heat-resistant spores under laboratory conditions, and extensive cell lysis occurs during aggregation, which has been suggested to be important for spore formation (Lee et al 2012; O'Connor & Zusman 1988; Wireman and Dworkin 1977). In dictyostelids, ~20–30% of the aggregated cells die and become part of a stalk while the remainder become viable spores elevated by the stalk (Kessin 2001). The mechanisms that determine whether aggregated dictyostelid cells die or survive as spores seem to be stochastic rather than deterministic, although developmental fate becomes more predictable if cells' physiological states are different at the onset of aggregation (Azhar et al. 2001; Leach et al. 1973; Nanjundiah and Sathe, 2011, 2013; Saran 1999). Among the *Copromyxa* amoebae and the Acrasids, all cells remain alive during aggregative development. An initial *Copromyxa* cell founds a fruiting body by sporulating, and subsequently arriving cells travel to the top of the rising fruiting body and sporulate in turn (Spiegel and Olive 1978).

Differing cell fates during development have been viewed as a social "division of labor" that harms some individuals and benefits others in order to serve a greater social good (Zhang et al. 2016). For social amoebae, it has been proposed that the extent of cell death in stalks may reflect an evolutionary tradeoff between the advantages of greater dispersal, which is hypothesized to be a result of increased cell death leading to taller stalks, versus a higher fraction of cells surviving to the next life cycle (Kaushik and Nanjundiah 2003; Votaw and Ostrowski 2017). For the myxobacteria, in addition to hypotheses regarding fruiting body height and dispersal, it has been suggested that components released by dying cells may be integrated into the spore coats of sporulating cells (Teintze et al. 1985), although this has not been clearly demonstrated. In all systems, there is still much to learn regarding the evolutionary forces that determine the extent of cell death during aggregative development.

6.3.3 Cells that Remain Outside the Aggregate

Once aggregation has begun, cells must cooperate for development to proceed. However, in both *M. xanthus* and *D. discoideum,* not all cells respond to developmental signals by entering aggregates. Such non-aggregating cells are referred to as "peripheral rods" in *M. xanthus* (O'Connor and Zusman 1991), whereas in *D. discoideum* they are called "loners" because individual cells remain isolated (Tarnita et al. 2015). *D. discoideum* loners avoid the prospect of dying in a stalk, and they may have an advantage if fresh nutrients become available while the rest of the population is involved in development because they can re-start vegetative growth and then divide more rapidly than aggregated cells, which eventually face an irreversible commitment to development (Dubravcic et al. 2014; Tarnita et al. 2015). It has been proposed that peripheral rods (O'Connor & Zusman 1991; Kroos 2017) and loners (Tarnita et al. 2015, Rainey 2015) may represent bet-hedging strategies that optimize long-term fitness in fluctuating environments. Consistent with this hypothesis, the frequency of *D. discoideum* loners in a population is a heritable trait and therefore potentially subject to selection (Dubravcic et al. 2014; Rossine et al. 2020). Evolution experiments which manipulate nutrient levels over time could be performed to test such bet-hedging hypotheses in both myxobacteria and social amoebae.

6.3.4   AGGREGATE DISPERSAL: THE DISPERSAL-BENEFIT HYPOTHESIS

It has long been assumed that fruiting bodies enhance dispersal of spores and that this was a driving force in the emergence of these structures (Bonner 1982, 2009; Kaiser 1993; 2001; Stanier 1942). The dispersal-benefit hypothesis is attractive in light of the observation that aggregates can be induced by harsh environmental conditions (Tarnita et al. 2013). Spores in taller fruiting bodies may have a greater likelihood of being picked up by an animal vector (smith et al. 2014) or translocated by water or air. Does aggregative structure formation in fact increase average dispersal relative to the absence of development? And if average dispersal is increased, does it actually increase average cell fitness – by making it more likely for spores to experience a resource-rich environment – relative to sheltering in place? Studies that address either of these questions experimentally are few; most treatments of these questions involve speculation regarding potential implications of fruiting body architecture and the evolutionary purpose of cell death during aggregative development. For example, it has been hypothesized that taller stalks better facilitate dispersal (Bonner 2009; Sathe et al. 2010), and smith et al. (2014) provide empirical evidence for this. It has also been proposed that utilizing dead cells to construct a stalk, as occurs in *Dictyostelium,* could be an adaptive innovation (Kaushik and Nanjundiah 2003) to produce taller, more robust stalks which might do more to facilitate dispersal than stalks built from living cells, as in the Acrasids (Brown et al. 2012) and *Copromyxa* (Spiegel and Olive 1978), or stalks built from secreted extracellular components, as found in the Holozoa (Brown 2009).

One difficulty with the dispersal-benefit hypothesis is that it does not explain the morphological diversity of fruiting bodies seen across taxa. Beyond the core commonality of elongation away from a surface, fruiting bodies vary dramatically in size, shape, color, and complexity, not only across major taxa divisions (see Chapters 5 and 8) but also within monophyletic groups. For example, some myxobacterial species generate highly elevated and elaborate morphologies whereas others merely aggregate into simple, slightly heightened cell mounds (Figure 6.1) (Dawid 2000). Such variation in fruiting body characteristics is hard to explain solely as a means to increase spore dispersal. Another difficulty is that mere aggregation into short, dense mounds may not suffice to provide dispersal-related benefits, and so the dispersal-benefit hypothesis may not explain the evolution of aggregation prior to the emergence of larger and more complex fruiting body morphologies.

6.3.5   NON-ADAPTIVE VS ADAPTIVE EXPLANATIONS OF MORPHOLOGY AND BEHAVIOR

Laboratory evolution experiments in myxobacteria have shown that fruiting body morphology can evolve rapidly (La Fortezza and Velicer 2021; Velicer et al. 1998). For example, one recent study has found that the morphological evolution of fruiting bodies can be strongly influenced by social interactions between genotypes during aggregative development (La Fortezza and Velicer 2021). However, different evolutionary forces may be responsible for the morphologies of aggregates in natural populations.

While fruiting body morphology may have been influenced by selection, historical contingency and chance have likely also played a role in determining which lineages evolved taller and more complex fruiting bodies and which evolved into smaller, simpler forms (Blount et al. 2018; Gould and Lewontin 1979; Travisano et al. 1995). Analogizing to organismal features, Gould and Lewontin (Gould and Lewontin 1979) famously remarked that the spandrels in Venice's San Marco Basilica can easily be viewed as having been designed to optimize display of the remarkable mosaics found on them, when in fact the spandrels are primary aspects of architectural design which support the cathedral domes. Their point was that non-adaptive explanations for origins of organismal features should not be excluded without sufficient cause.

The relevance of indirect evolution in aggregative systems is highlighted by outcomes from several evolution experiments with myxobacteria that have recently been dubbed MyxoEEs (Rendueles and Velicer 2020). Forces other than direct selection have been shown to drive evolution of a range of traits, including extreme cheating phenotypes (Velicer et al. 2000), social fitness inequalities between specific genotypes (referred to as cheating in *Dictyostelium* literature [e.g., Khare et al. 2009]), facultative social exploitation during development (Nair et al. 2018), kin-discrimination phenotypes (Rendueles et al. 2015), colony-level morphology (Rendueles et al. 2020), quality as phage host (Freund et al. 2021), and even susceptibility to cheating (Schaal et al. 2021). Disentangling which features of aggregative systems evolved as adaptations and which did not remains a major challenge for future research.

6.4   LIFE AFTER AGGREGATION

There are striking contrasts among organisms in how cells behave when aggregates disband. In particular, the myxobacteria appear to differ from amoebae in the degree to which germinating spores and vegetative cells continue to interact and to cooperate. Whereas amoebae may live comparatively solitary lives except when they undergo aggregative development, the myxobacteria remain social throughout their life cycle. For such systems that cooperate extensively during growth, aggregation may have density benefits that extend beyond the formation, survival, and dispersal of spores (Ramsey and Dworkin 1968).

Both dictyostelids and myxobacteria initiate spore germination once environmental conditions near the aggregate become favorable, but while high density inhibits spore germination in *D. discoideum* (Ceccarini and Cohen 1967; Cotter and Raper 1966; Kessin 2001), *M. xanthus* spores germinate better at higher density (Pande et al. 2020). Such positive density-dependence of spore germination may therefore have promoted the evolution and maintenance of fruiting body formation in the myxobacteria but appears unlikely to have done so in the dictyostelids.

Freshly germinated cells face many potential challenges, including abiotic stressors such as acidity, difficulties in finding and consuming prey, and encountering biotic enemies. Germinating in high-density aggregates appears to help myxobacteria meet such challenges. *M. xanthus* is sensitive to even mild acidity but grows much better under acid stress at higher density (Fiegna et al. 2021). The rate at which groups of *M. xanthus* cells swarm across surfaces depends greatly on local cell density (Kaiser and Crosby 1983), which suggests that spores germinating in aggregates may be better than single spores at moving toward prey and away from dangers. There is indirect evidence that myxobacteria may sometimes benefit from being at high density during predation (Rosenberg et al. 1977) and if this hypothesis is correct, spores germinating within fruiting bodies may have a predation advantage over non-aggregated spores. Additionally, like humans and other animals, bacteria frequently kill members of their own species (Granato et al. 2019); in the myxobacteria, higher density groups are more likely to survive these interactions (Rendueles et al. 2015). In pairwise competitions between *M. xanthus* natural isolates, the genotype in the majority tends to have an advantage at killing or antagonizing the opponent. Spores will fare better in such combat by having germinated together rather than alone.

Dictyostelids are not known to generally engage in social motility, group hunting, lethal frequency-dependent warfare, or cooperative growth under stress when nutrients are available (but see Rubin et al. 2019). Germinated *D. discoideum* cells are less social than germinated myxobacteria – moving, hunting, and reproducing as individuals before becoming cooperative only upon starvation. Advantages of aggregation in the dictyostelids may be limited to shelter and dispersal benefits, in contrast to the myxobacteria, whose cells appear to benefit from aggregate formation not only during development but also later during subsequent growth and collective migration. Such delayed indirect benefits of aggregation may have worked in concert with shelter and dispersal benefits in promoting its evolution in some systems.

6.5   GENETIC DIVERSITY WITHIN AGGREGATIVE SYSTEMS: THE CHALLENGE OF CHIMERISM

Because the types of aggregates examined here are not formed by a single dividing cell, they provide an opportunity for genetically distinct lineages to come together. While bacterial biofilms, for example, can contain many different genera (Nadell et al. 2009), even single-species aggregates often harbor genetic diversity (Buss 1982; Flowers et al. 2010; Filosa 1962; Kraemer and Velicer 2011; Nanjundiah 2016; Sathe et al 2010; Wielgoss et al. 2019). The potential for genetically distinct cells to cooperate, coalesce, and compete within cell aggregates creates the possibility for selection to operate at multiple levels – that of the single cell and that of the aggregate as a whole. In this way, whether aggregated groups are monoclonal or chimeric can determine the benefits and costs of actively aggregating with others (see Chapter 7).

Chimeric aggregates may be more frequent in nature than monoclonal ones. High local genetic diversity found in soil samples of both *Myxococcus* (Kraemer and Velicer 2011; Vos and Velicer 2006) and *Dictyostelium* (Fortunato et al. 2003;

Ostrowski et al. 2015) indicates that genetically distinct strains can be close enough to each other that co-aggregation is at least a possibility. In *M. xanthus*, chimeric aggregates have indeed been isolated (Wielgoss et al. 2019). In this case, genotypes from the same fruiting bodies were very closely related, suggesting that they recently diverged from a common ancestor and remained in close enough proximity to co-aggregate. The cellular slime mold *A. rosea* also forms chimeric fruiting bodies, although one strain tends to dominate, and the fruiting body shows features characteristic of those formed by that strain in monoculture (Reinhardt 1975). In dictyostelids, natural samples frequently yield distinct isolates capable of forming chimeric fruiting bodies (Buss 1982; Flowers et al. 2010; Filosa 1962; Nanjundiah 2016; Sathe et al 2010), although some *D. discoideum* fruiting bodies appear to be monoclonal (Gilbert et al. 2007, 2009). Additionally, distinct genotypes of *D. discoideum* (Gilbert et al. 2007), *D. giganteum* (Kaushik et al. 2006), and *D. mucoroides* (Filosa 1962) have been observed to readily co-aggregate in the lab. Future comparisons of whole-genome sequences among co-aggregated cells will provide a finer-scale understanding of the degree of chimerism in natural aggregative systems (Wielgoss et al. 2019).

Negative effects of chimerism often dominate discussions of this phenomenon. Co-aggregation of distinct genotypes can reduce aggregate fitness, a phenomenon known as "chimeric load" (Kraemer and Velicer 2011). Chimeric load can be observed even when all genotypes are equally proficient at development in monoculture. For example, chimeric *M. xanthus* fruiting bodies contain fewer spores when the mixed strains originated from different natural fruiting bodies (Mendes-Soares et al. 2014; Pande and Velicer 2018). Similarly, chimeric *D. discoideum* slugs migrate slower than clonal slugs of the same size (Foster et al. 2002). Such chimeric load might result from active antagonisms between genotypes or simply from dysfunctional divergence. A special form of chimeric load is "cheating load" (Fiegna and Velicer 2003; Travisano and Velicer 2004), in which the presence of aggregation-defective cheaters in groups of aggregation-proficient cells reduces group-level spore productivity. Such negative effects of chimerism might be expected to lead to the evolution of mechanisms that limit it, a prospect that has received much attention (e.g., Strassmann and Queller. 2011). However, intrinsic patterns of cell growth in spatially structured habitats may sometimes suffice to greatly restrict chimerism in aggregating groups (smith et al. 2016) even in the absence of kin-discriminatory traits.

In some cases, chimerism may increase the benefits of aggregation (Foster et al. 2002; Pande and Velicer 2018; Pineda-Krch and Lehtilä 2004). Strains with different phenotypes may be able to complement each other and increase overall group productivity (Bonner 1967). In *Myxococcus*, for example, natural isolates sampled from the same fruiting body produced fewer spores in monoclonal groups than when they underwent development together, suggesting synergistic interactions among these genotypes (Pande and Velicer 2018). In another example, in pairwise mixes of three strains of *D. giganteum,* there was a significant tendency for one strain to be favored over the other in spore formation. However, when all three strains were mixed together, such asymmetry was significantly reduced, suggesting that there are higher-order interactions stabilizing the coexistence of different genotypes (Kaushik et al. 2006). As with other aspects of aggregate formation, the effect of chimerism is influenced by ecological conditions and by divergence across taxa.

6.6 CONCLUSION

Modern evolutionary theory originated mainly from studies of plants and animals that develop from a single cell. But in many taxa, there are single-celled organisms that have a very different life cycle: living at lower density under good conditions and then actively aggregating and forming a multicellular structure when conditions are bad (Figure 6.2). The formation of these structures raises fascinating evolutionary questions. Why do cells aggregate even though this provides an opportunity for genetically different cells to take advantage? How have these complex processes emerged and changed over time? What is the relative strength of selection at different levels of these systems?

By organizing themselves into multicellular aggregates, these organisms challenge our idea of biological individuality (Ereshefsky and Pedroso 2012; Godfrey-Smith 2009; Lidgard and Nyhart 2017; Monte and Rainey 2014; West et al. 2015). In general, obligately multicellular organisms reproduce via a single-cell bottleneck, so the group as a whole is the unit of selection (Lewontin 1970). For aggregating organisms, however, the unit of selection is context-dependent. Developmental aggregates can be seen as a survival bottleneck, but they are composed of cells that retain reproductive autonomy for the majority of the organismal life cycle.

Group-level selection may explain the emergence of some aspects of aggregative multicellular processes. However, the adaptive character, if any, of many traits specific to multicellular aggregates has not been clearly demonstrated. Even the most basic adaptive roles of the fruiting body structure itself require further investigation, and counterpoints to the dispersal-benefit hypothesis should be explored more extensively. The evolutionary causes of variation in fruiting body morphologies are even less well explained, although study of more recently evolved aggregative organisms such as *Sorogena stoianovitchae* (Lasek-Nesselquist and Katz 2001) may provide fresh insights distinct from those gained by study of more ancient systems.

Studying aggregating organisms is difficult because they can be challenging to isolate and to observe under natural conditions. To date, for example, the cellular slime mold *Fonticula alba* seems to have only been isolated once (Brown et al. 2009). *Guttulinopsis vulgaris* (Brown et al. 2012b) and *Sorodiplophrys stercorea* (Tice 2016) are the only members of their supergroups (Rhizaria and Stramenopiles, respectively) known to show aggregative multicellularity. As a result, most studies of aggregative multicellularity have been conducted on dictyostelids and myxobacteria, and on very few species from those two groups. Development of new isolation methods may reveal a broader diversity of aggregative microbial systems in nature and so provide a more thorough understanding of the evolutionary forces that can shape aggregative multicellularity.

REFERENCES

Aceto, N., Aditya Bardia, David T. Miyamoto, Maria C. Donaldson, Ben S. Wittner, Joel A. Spencer, Min Yu, et al. 2014. "Circulating tumor cell clusters are oligoclonal precursors of breast cancer metastasis." *Cell* **158**: 1110–22. https://doi.org/10.1016/j.cell.2014.07.013.

Arias Del Angel, Juan A., Vidyanand Nanjundiah, Mariana Benítez, and Stuart A. Newman. 2020. "Interplay of mesoscale physics and agent-like behaviors in the parallel evolution of aggregative multicellularity." *EvoDevo* **11**: 21. https://doi.org/10.1186/s13227-020-00165-8

Azhar, M, P. K. Kennady, G. Pande, M. Espiritu, W. Holloman, D. Brazill, R. H. Gomer, and V. Nanjundiah. 2001. "Cell cycle phase, cellular Ca2+ and development in *Dictyostelium discoideum*." *The International Journal of Developmental Biology* **45**: 405–14. PMID: 11330860.

Berleman, James E., and John R. Kirby. 2007. "Multicellular development in *Myxococcus xanthus* is stimulated by predator-prey interactions." *Journal of Bacteriology* **189**: 5675–82. https://doi.org/10.1128/jb.00544-07.

Blount, Z. D., R. E. Lenski, and J. B. Losos. 2018. "Contingency and determinism in evolution: Replaying life's tape." *Science* **362**: eaam5979. https://doi.org/10.1126/science.aam5979.

Bonner, J. T. 1967. *The cellular slime molds.* (2nd. ed). Princeton, KJ: Princeton University Press.

Bonner, J. T. 1982. "Evolutionary strategies and developmental constraints in the cellular slime molds." *The American Naturalist* **119**, 530–52.

Bonner, J. T. 2003. "Evolution of development in the cellular slime molds." *Evolution & Development* **5**: 305–13. https://doi.org/10.1046/j.1525-142x.2003.03037.x.

Bonner, J. T. 2009. *The social amoebae: The biology of cellular slime molds.* Princeton; Oxford: Princeton University Press. https://doi.org/10.1515/9781400833283.129

Branda, S. S., J. E. Gonzalez-Pastor, S. Ben-Yehuda, R. Losick, and R. Kolter. 2001. "Fruiting body formation by *Bacillus subtilis*." *Proceedings of the National Academy of Sciences U.S.A.* **98**: 11621–26. https://doi.org/10.1073/pnas.191384198.

Brown, Matthew W., F. W., Spiegel, and Jeffrey D. Silberman. 2009. "Phylogeny of the "forgotten" cellular slime mold, *Fonticula alba*, reveals a key evolutionary branch within opisthokonta." *Molecular Biology & Evolution* **26**: 2699–2709. https://doi.org/10.1093/molbev/msp185

Brown, Matthew W., Jeffrey D. Silberman, and F. W., Spiegel. 2012. "A contemporary evaluation of the acrasids (Acrasidae, Heterolobosea, Excavata)." *European Journal of Protistology* **48**: 103–123. https://doi.org/10.1016/j.ejop.2011.10.001

Brown, Matthew W., Martin Kolisko, Jeffrey D. Silberman, and Andrew J. Roger. 2012b. "Aggregative multicellularity evolved independently in the eukaryotic supergroup Rhizaria." *Current Biology* **22**: 1123–27. https://doi.org/10.1016/j.cub.2012.04.021.

Brown, Matthew W., and Jeffrey D. Silberman. 2013. "The non-dictyostelid sorocarpic amoebae." *in Dictyostelids, Evolution, Genomics and Cell Biology.* 219–42. https://doi.org/10.1007/978-3-642-38487-5_12.

Buss, L. W. 1982. "Somatic cell parasitism and the evolution of somatic tissue compatibility." *Proceedings of the National Academy of Sciences U.S.A.* **79**: 5337–41. https://doi.org/10.1073/pnas.79.17.5337.

Ceccarini, Costante, and Arthur Cohen. 1967. "Germination inhibitor from the cellular slime mould *Dictyostelium discoideum*." *Nature* **214**: 1345–46. https://doi.org/10.1038/2141345a0.

Claessen, Dennis, Daniel E. Rozen, Oscar P. Kuipers, Lotte Søgaard-Andersen, and Gilles P. Van Wezel. 2014. "Bacterial solutions to multicellularity: A tale of biofilms, filaments and fruiting bodies." *Nature Reviews Microbiology* **12**: 115–24. https://doi.org/10.1038/nrmicro3178.

Cotter, D. A., and K. B. Raper. 1966. "Spore germination in *Dictyostelium discoideum*." *Proceedings of the National Academy of Sciences U.S.A.* **56**: 880–87. https://doi.org/10.1073/pnas.56.3.880.

Dawid, W. 2000. "Biology and global distribution of myxobacteria in soils." *FEMS Microbiology Reviews* **24**: 403–27. https://doi.org/10.1111/j.1574-6976.2000.tb00548.x.

Du, Qingyou, Yoshinori Kawabe, Christina Schilde, Zhi-hui Chen, and Pauline Schaap. 2015. "The evolution of aggregative multicellularity and cell–cell communication in the Dictyostelia." *Journal of Molecular Biology* **427**: 3722–33. https://doi.org/10.1016/j.jmb.2015.08.008.

Dubravcic, Darja, Minus Van Baalen, and Clément Nizak. 2014. "An evolutionarily significant unicellular strategy in response to starvation in *Dictyostelium* social amoebae." *F1000Research* **3**: 133. https://doi.org/10.12688/f1000research.4218.2.

Dykstra, Michael J., and Lindsay S. Olive. 1975. "*Sorodiplophrys*: An unusual sorocarp-producing protist." *Mycologia* **67**: 873. https://doi.org/10.2307/3758346.

Ellison, Aaron M., and Leo W. Buss. 1983. "A naturally occurring developmental synergism between the cellular slime mold, *Dictyostelium mucoroides* and the fungus, *Mucor hiemalis*." *American Journal of Botany* **70**: 298–302. https://doi.org/10.1002/j.1537-2197.1983.tb07870.x.

Ereshefsky, Marc, and Makmiller Pedroso. 2012. "Biological individuality: The case of biofilms." *Biology & Philosophy* **28**: 331–49. https://doi.org/10.1007/s10539-012-9340-4.

Fiegna, Francesca, and Gregory J. Velicer. 2003. "Competitive fates of bacterial social parasites: Persistence and self-induced extinction of *Myxococcus xanthus* cheaters." *Proceedings of the Royal Society B: Biological Sciences* **270**: 1527–34. https://doi.org/10.1098/rspb.2003.2387.

Fiegna, Francesca, and Gregory J. Velicer. 2005. "Exploitative and hierarchical antagonism in a cooperative bacterium." *PLoS Biology* **3**: e370. https://doi.org/10.1371/journal.pbio.0030370.

Fiegna, Francesca, Samay Pande, Hansrainer Peitz, and Gregory J. Velicer. 2021. "Widespread density dependence of bacterial growth under acid stress." *bioRxiv*. 2021.09.27.461844. https://doi.org/10.1101/2021.09.27.461844.

Filosa, M. F. 1962. "Heterocytosis in cellular slime molds." *The American Naturalist* **96**: 79–91. https://doi.org/10.1086/282209.

Flärdh, Klas, David M. Richards, Antje M. Hempel, Martin Howard, and Mark J Buttner. 2012. "Regulation of apical growth and hyphal branching in *Streptomyces*." *Current Opinion in Microbiology* **15**: 737–43. https://doi.org/10.1016/j.mib.2012.10.012.

Flores, Enrique, and Antonia Herrero. 2010. "Compartmentalized function through cell differentiation in filamentous *Cyanobacteria*." *Nature Reviews Microbiology* **8**: 39–50. https://doi.org/10.1038/nrmicro2242.

Flowers, Jonathan M., Si I. Li, Angela Stathos, Gerda Saxer, Elizabeth A. Ostrowski, David C. Queller, Joan E. Strassmann, and Michael D. Purugganan. 2010. "Variation, sex, and social cooperation: Molecular population genetics of the social amoeba *Dictyostelium discoideum*." *PLoS Genetics* **6**: e1001013. https://doi.org/10.1371/journal.pgen.1001013.

Fortunato, A., J. E. Strassmann, L. Santorelli, and D. C. Queller. 2003. "Co-occurrence in nature of different clones of the social amoeba, *Dictyostelium discoideum*." *Molecular Ecology* **12**: 1031–38. https://doi.org/10.1046/j.1365-294x.2003.01792.x.

Foster, Kevin R., Angelo Fortunato, Joan E. Strassmann, and David C. Queller. 2002. "The costs and benefits of being a chimera." *Proceedings of the Royal Society B: Biological Sciences* **269**: 2357–62. https://doi.org/10.1098/rspb.2002.2163.

Frank, Margaret H., and Daniel H. Chitwood. 2016. "Plant chimeras: The good, the bad, and the 'Bizzaria.'" *Developmental Biology* **419**: 41–53. https://doi.org/10.1016/j.ydbio.2016.07.003.

Freund L, Vasse M, Velicer GJ. 2021. Hidden paths to endless forms most wonderful: parasite-blind diversification of host quality. *Proceedings of the Royal Society B: Biological Sciences* 20210456. https://doi.org/10.1098/rspb.2021.0456.

Gilbert, Owen M., Kevin R. Foster, Natasha J. Mehdiabadi, Joan E. Strassmann, and David C. Queller. 2007. "High relatedness maintains multicellular cooperation in a social amoeba by controlling cheater mutants." *Proceedings of the National Academy of Sciences U.S.A.* **104**: 8913–17. https://doi.org/10.1073/pnas.0702723104.

Gilbert, Owen M., David C. Queller, and Joan E. Strassmann. 2009. "Discovery of a large clonal patch of a social amoeba: Implications for social evolution." *Molecular Ecology* **18**: 1273–81. https://doi.org/10.1111/j.1365-294x.2009.04108.x.

Glass, N. Louise, Carolyn Rasmussen, M. Gabriela Roca, and Nick D. Read. 2004. "Hyphal homing, fusion and mycelial interconnectedness." *Trends in Microbiology* **12**: 135–41. https://doi.org/10.1016/j.tim.2004.01.007.

Godfrey-Smith, Peter. 2009. *Darwinian populations and natural selection.* Oxford: Oxford University Press.

Gould, S. J., and R. C. Lewontin. 1979. "The spandrels of San Marco and the Panglossian paradigm: A critique of the adaptationist programme." *Proceedings of the Royal Society B: Biological Sciences* **205**: 581–98. https://doi.org/10.1098/rspb.1979.0086.

Granato, Elisa T., Thomas A. Meiller-Legrand, and Kevin R. Foster. 2019. "The evolution and ecology of bacterial warfare." *Current Biology* **29**: R521–37. https://doi.org/10.1016/j.cub.2019.04.024.

Grilione, P. L., and J. Pangborn. 1975. "Scanning electron microscopy of fruiting body formation by myxobacteria." *Journal of Bacteriology* **124**: 1558–65. https://doi.org/10.1128/jb.124.3.1558-1565.1975.

Grosberg, R. K., and R. R. Strathmann. 2007. "The evolution of multicellularity: A minor *major* transition?" *Annual Review of Ecology, Evolution & Systematics* 38, 621–654. https://doi.org/10.1146/annurev.ecolsys.36.102403.114735.

Herron, Matthew D., Joshua M. Borin, Jacob C. Boswell, Jillian Walker, I-Chen Kimberly Chen, Charles A. Knox, Margrethe Boyd, Frank Rosenzweig, and William C. Ratcliff. 2019. "De novo origins of multicellularity in response to predation." *Scientific Reports* **9**: 2328. https://doi.org/10.1038/s41598-019-39558-8.

Houry, A., R. Briandet, S. Aymerich, and M. Gohar. 2010. "Involvement of motility and flagella in *Bacillus cereus* biofilm formation." *Microbiology* **156**, 1009–1018.

Kaiser, Dale, and Cathy Crosby. 1983. "Cell movement and its coordination in swarms of *Myxococcus xanthus.*" *Cell Motility* **3**: 227–45. https://doi.org/10.1002/cm.970030304.

Kaiser, D. 1993. "Roland Thaxter's legacy and the origins of multicellular development." *Genetics* **135**: 249–54. https://doi.org/10.1093/genetics/135.2.249.

Kaiser, Dale. 2001. "Building a multicellular organism." *Genetics* **35**: 103–23. https://doi.org/10.1146/annurev.genet.35.102401.090145.

Kaplan, Heidi B., and Lynda Plamann. 1996. "A *Myxococcus xanthus* cell density-sensing system required for multicellular development." *FEMS Microbiology Letters* **139**: 89–95. https://doi.org/10.1111/j.1574-6968.1996.tb08185.x.

Kaushik, Sonia, and Vidyanand Nanjundiah. 2003. "Evolutionary questions raised by cellular slime mould development." *Proceedings of the Indian National Sciences Academy B* **69**: 825–852.

Kaushik, Sonia, Bandhana Katoch, and Vidyanand Nanjundiah. 2006. "Social behaviour in genetically heterogeneous groups of *Dictyostelium giganteum.*" *Behavioral Ecology and Sociobiology* **59**: 521–30. https://doi.org/10.1007/s00265-005-0077-9.

Kawabe, Yoshinori, Qingyou Du, Christina Schilde, and Pauline Schaap. 2019. "Evolution of multicellularity in Dictyostelia." *International Journal of Developmental Biology* **63**: 359–69. https://doi.org/10.1387/ijdb.190108ps.

Kawaguchi, Yuko, Mio Shibuya, Iori Kinoshita, Jun Yatabe, Issay Narumi, Hiromi Shibata, Risako Hayashi, et al. 2020. "DNA damage and survival time course of Deinococcal cell pellets during 3 years of exposure to outer space." *Frontiers in Microbiology* **11**: 2050. https://doi.org/10.3389/fmicb.2020.02050.

Kessin, R. H., G. G. Gundersen, V. Zaydfudim, and M. Grimson. 1996. "How cellular slime molds evade nematodes." *Proceedings of the National Academy of Sciences U.S.A.* **93**: 4857–61. https://doi.org/10.1073/pnas.93.10.4857.

Kessin, R. 2001. *Dictyostelium: Evolution, Cell Biology, and the Development of Multicellularity.* Cambridge: Cambridge University Press. doi:10.1017/CBO9780511525315

Khare, Anupama, Lorenzo A. Santorelli, Joan E. Strassmann, David C. Queller, Adam Kuspa, and Gad Shaulsky. 2009. "Cheater-resistance is not futile." *Nature* **461**: 980–82. https://doi.org/10.1038/nature08472.

Kirby, Amy E., Kimberly Garner, and Bruce R. Levin. 2012. "The relative contributions of physical structure and cell density to the antibiotic susceptibility of bacteria in biofilms." *Antimicrobial Agents and Chemotherapy* **56**: 2967–75. https://doi.org/10.1128/aac.06480-11.

Kraemer, Susanne A., Melissa A. Toups, and Gregory J. Velicer. 2010. "Natural variation in developmental life-history traits of the bacterium *Myxococcus xanthus.*" *FEMS Microbiology Ecology,* **73**: 226–233, https://doi.org/10.1111/j.1574-6941.2010.00888.x

Kraemer, Susanne A., and Gregory J. Velicer. 2011. "Endemic social diversity within natural kin groups of a cooperative bacterium." *Proceedings of the National Academy of Sciences of the U.S.A.* **108** (Supplement 2): 10823–30. https://doi.org/10.1073/pnas.1100307108.

Krause, Jens, and Graeme D. Ruxton. 2002. *Living in groups.* Oxford: Oxford University Press.

Kroos, Lee. 2017. "Highly signal-responsive gene regulatory network governing *Myxococcus* development." *Trends in Genetics* **33**: 3–15. https://doi.org/10.1016/j.tig.2016.10.006.

La Fortezza, M. and G. J. Velicer. 2021. "Social selection within aggregative multicellular development drives morphological evolution." *Proc Royal Soc B* **288**, 20211522 https://doi.org/10.1098/rspb.2021.1522

Lasek-Nesselquist, E., and L. A. Katz. 2001. "Phylogenetic position of *Sorogena stoianovitchae* and relationships within the class Colpodea (Ciliophora) based on SSU rDNA sequences." *Journal Eukaryotic Microbiology* **48**: 604–607.

Leach, C. K., J. M. Ashworth, and D. R. Garrod. 1973. "Cell sorting out during the differentiation of mixtures of metabolically distinct populations of *Dictyostelium discoideum.*" *Journal of Embryology and Experimental Morphology* **29**: 647–61.

Lee B., Holkenbrink C., Treuner-Lange A., Higgs P. I. 2012. *Myxococcus xanthus* developmental cell fate: heterogeneous accumulation of developmental regulatory proteins and reexamination of the role of MazF in developmental lysis. *Journal of Bacteriology* **194**: 3058–3068.

Lewontin, R. C. 1970. "The Units of Selection." *Annual Review of Ecology and Systematics* **1**: 1–18. https://doi.org/10.1146/annurev.es.01.110170.000245.

Li, Yinuo, Hong Sun, Xiaoyuan Ma, Ann Lu, Renate Lux, David Zusman, and Wenyuan Shi. 2003. "Extracellular polysaccharides mediate pilus retraction during social motility of *Myxococcus xanthus.*" *Proceedings of the National Academy of Sciences U.S.A.* **100**: 5443–48. https://doi.org/10.1073/pnas.0836639100.

Lidgard, Scott, and Lynn K. Nyhart. 2019. *"Biological individuality: Integrating scientific, philosophical, and historical perspectives."* https://doi.org/10.7208/9780226446592-002.

Manhes, Pauline, and Gregory J. Velicer. 2011. "Experimental evolution of selfish policing in social bacteria." *Proceedings of the National Academy of Sciences U.S.A.* **108**: 8357–62. https://doi.org/10.1073/pnas.1014695108.

Medina, James M., P.M. Shreenidhi, Tyler J. Larsen, David C. Queller, and Joan E. Strassmann. 2019. "Cooperation and conflict in the social amoeba *Dictyostelium discoideum.*" *International Journal of Developmental Biology* **63**: 371–82. https://doi.org/10.1387/ijdb.190158jm.

Meena, N. P. and Kimmel, A. R. 2017. "Chemotactic network responses to live bacteria show independence of phagocytosis from chemoreceptor sensing." *eLife* **6**, e24627. https://doi.org/10.7554/elife.24627.

Mendes-Soares, H., I. C. K. Chen, K. Fitzpatrick, and G. J. Velicer. 2014. "Chimaeric load among sympatric social bacteria increases with genotype richness." *Proceedings of the Royal Society B: Biological Sciences* **281**: 20140285–20140285. https://doi.org/10.1098/rspb.2014.0285.

Monte, Silvia De, and Paul B Rainey. 2014. "Nascent multicellular life and the emergence of individuality." *Journal of Biosciences* **39**: 237–48. https://doi.org/10.1007/s12038-014-9420-5.

Muñoz-Dorado, José, Francisco J. Marcos-Torres, Elena García-Bravo, Aurelio Moraleda-Muñoz, and Juana Pérez. 2016. "Myxobacteria: Moving, killing, feeding, and surviving together." *Frontiers in Microbiology* **7**: 2475–18. https://doi.org/10.3389/fmicb.2016.00781.

Nadell, Carey D., Joao B. Xavier, and Kevin R. Foster. 2009. "The sociobiology of biofilms." *FEMS Microbiology Reviews* **33**: 206–24. https://doi.org/10.1111/j.1574-6976.2008.00150.x.

Nair, Ramith R., Francesca Fiegna, and Gregory J. Velicer. 2018. "Indirect evolution of social fitness inequalities and facultative social exploitation." *Proceedings of the Royal Society B: Biological Sciences* **285**: 20180054. https://doi.org/10.1098/rspb.2018.0054.

Nanjundiah, Vidyanand, and Santosh Sathe. 2011. "Social selection and the evolution of cooperative groups: The example of the cellular slime moulds." *Integrative Biology* **3**: 329–42. https://doi.org/10.1039/c0ib00115e.

Nanjundiah, Vidyanand, and Santosh Sathe. 2013. "*Dictyostelids: Evolution, Genomics and Cell Biology,*" Berlin, Heidelberg: Springer Berlin Heidelberg 193–217. https://doi.org/10.1007/978-3-642-38487-5_11.

Nanjundiah, Vidyanand. 2016. "Cellular slime molds and aggregative multicellularity." in: *Multicellularity: Origins and evolution.* Cambridge, MA: The MIT Press.

O'Connor, K. A., and D. R. Zusman. 1988. "Reexamination of the role of autolysis in the development of *Myxococcus xanthus.*" *Journal of Bacteriology* **170**: 4103–12. https://doi.org/10.1128/jb.170.9.4103-4112.1988.

O'Connor, K. A., and D. R. Zusman. 1991. "Development in *Myxococcus xanthus* involves differentiation into two cell types, peripheral rods and spores." *Journal of Bacteriology* **173**: 3318–33. https://doi.org/10.1128/jb.173.11.3318-3333.1991.

O'Shea, K. S. 1987. "Differential deposition of basement membrane components during formation of the caudal neural tube in the mouse embryo." *Development* **99**: 509–19.

Olive, L. S., R. L. Blanton. 1980. "Aerial sorocarp development by the aggregative ciliate, *Sorogena stoianovitchae.*" *Journal of Protozoology* **27**: 293–9. https://doi.org/10.1111/j.1550-7408.1980.tb04260.x

Oliveira, Nuno M.,, Esteban Martinez-Garcia, Joao Xavier, William M Durham, Roberto Kolter, Wook Kim, and Kevin R. Foster. 2015. "Biofilm formation as a response to ecological competition." *PLoS Biology* **13**: e1002191. https://doi.org/10.1371/journal.pbio.1002191.

Ostrowski, Elizabeth A., Yufeng Shen, Xiangjun Tian, Richard Sucgang, Huaiyang Jiang, Jiaxin Qu, Mariko Katoh-Kurasawa, et al. 2015. "Genomic signatures of cooperation and conflict in the social amoeba." *Current Biology* **25**: 1661–65. https://doi.org/10.1016/j.cub.2015.04.059.

Padua, André, Pedro Leocorny, Márcio Reis Custódio, and Michelle Klautau. 2016. "Fragmentation, fusion, and genetic homogeneity in a calcareous sponge (Porifera, Calcarea)." *Journal of Experimental Zoology Part A: Ecological Genetics and Physiology* **325**: 294–303. https://doi.org/10.1002/jez.2017.

Pan, Pauline, E. M. Hall, and J. T. Bonner. 1972. "Folic acid as second chemotactic substance in the cellular slime moulds." *Nature New Biology* **237**: 181–82. https://doi.org/10.1038/newbio237181a0.

Pande, Samay, and Gregory J. Velicer. 2018. "Chimeric synergy in natural social groups of a cooperative microbe." *Current Biology* **28**: 262-267.e3. https://doi.org/10.1016/j.cub.2017.11.043.

Pande, Samay, Pau Pérez Escriva, Yuen-Tsu Nicco Yu, Uwe Sauer, and Gregory J. Velicer. 2020. "Cooperation and cheating among germinating spores." *Current Biology* **30**: 4745-4752.e4. https://doi.org/10.1016/j.cub.2020.08.091.

Pentz, Jennifer T., Pedro Márquez-Zacarías, G. Ozan Bozdag, Anthony Burnetti, Peter J. Yunker, Eric Libby, and William C. Ratcliff. 2020. "Ecological advantages and evolutionary limitations of aggregative multicellular development." *Current Biology*. https://doi.org/10.1016/j.cub.2020.08.006.

Pineda-Krch, M., and K. Lehtilä. 2004. "Costs and benefits of genetic heterogeneity within organisms." *Journal of Evolutionary Biology* **17**: 1167–77. https://doi.org/10.1111/j.1420-9101.2004.00808.x.

Qualls, G. T., K. Stephens, and D. White. 1978. "Light-stimulated morphogenesis in the fruiting myxobacterium *Stigmatella aurantiaca*." *Science* **201**: 444–45. https://doi.org/10.1126/science.96528.

Rainey, Paul B. 2015. "Precarious development: The uncertain social life of cellular slime molds." *Proceedings of the National Academy of Sciences U.S.A.* **112**: 2639–40. https://doi.org/10.1073/pnas.1500708112.

Ramsey, W. Scott, and Martin Dworkin. 1968. "Microcyst germination in *Myxococcus xanthus*." *Journal of Bacteriology* **95**: 2249–57. https://doi.org/10.1128/jb.95.6.2249-2257.1968.

Reinhardt, D. J. 1968. "The effects of light on the development of the cellular slime mold *Acrasis rosea*." *American Journal of Botany* **55**: 77–86. https://doi.org/10.1002/j.1537-2197.1968.tb06948.x

Reinhardt, D. J. 1975. Natural variants of the cellular slime mold *Acrasis rosea*. *The Journal of Protozoology* **22**: 309–317. https://doi.org/10.1111/j.1550-7408.1975.tb05176.x

Rendueles, Olaya, Michaela Amherd, and Gregory J. Velicer. 2015. "Positively frequency-dependent interference competition maintains diversity and pervades a natural population of cooperative microbes." *Current Biology* **25**: 1673–81. https://doi.org/10.1016/j.cub.2015.04.057.

Rendueles, Olaya, Peter C Zee, Iris Dinkelacker, Michaela Amherd, Sébastien Wielgoss, and Gregory J Velicer. 2015. "Rapid and widespread de novo evolution of kin discrimination." *Proceedings of the National Academy of Sciences U.S.A.* **112**: 9076–81. https://doi.org/10.1073/pnas.1502251112.

Rendueles, Olaya, and Gregory J. Velicer. 2020. "Hidden paths to endless forms most wonderful: Complexity of bacterial motility shapes diversification of latent phenotypes." *BMC Evolutionary Biology* **20**: 145. https://doi.org/10.1186/s12862-020-01707-3.

Rosenberg, E., K. H. Keller, and M Dworkin. 1977. "Cell density-dependent growth of *Myxococcus xanthus* on casein." *Journal of Bacteriology* **129**: 770–77. https://doi.org/10.1128/jb.129.2.770-777.1977.

Rossine, Fernando W., Ricardo Martinez-Garcia, Allyson E. Sgro, Thomas Gregor, and Corina E. Tarnita. 2020. "Eco-Evolutionary significance of 'Loners.'" *PLoS Biology* **18**: e3000642. https://doi.org/10.1371/journal.pbio.3000642.

Rubin M, Miller A. D, Katoh-Kurasawa M, Dinh C, Kuspa A, Shaulsky G. 2019. Cooperative predation in the social amoebae *Dictyostelium discoid*eum. *PLoS One* **14**: e0209438. https://doi.org/10.1371/journal.pone.0209438.

Saran, Shweta. 1999. "Calcium levels during cell cycle correlate with cell fate of *Dictyostelium discoideum*." *Cell Biology International* **23**: 399–405. https://doi.org/10.1006/cbir.1999.0379.

Sathe, Santosh, Sonia Kaushik, Albert Lalremruata, Ramesh K. Aggarwal, James C. Cavender, and Vidyanand Nanjundiah. 2010. "Genetic heterogeneity in wild isolates of cellular slime mold social groups." *Microbial Ecology* **60**: 137–48. https://doi.org/10.1007/s00248-010-9635-4.

Savage, Robert M., and Michael V. Danilchik. 1993. "Dynamics of germ plasm localization and its inhibition by ultraviolet irradiation in early cleavage *Xenopus* embryos." *Developmental Biology* **157**: 371–82. https://doi.org/10.1006/dbio.1993.1142.

Savage, Brian, Fanny Almus-Jacobs, and Zaverio M. Ruggeri. 1998. "Specific synergy of multiple substrate–receptor interactions in platelet thrombus formation under flow." *Cell* **94**: 657–66. https://doi.org/10.1016/s0092-8674(00)81607-4.

Schaal, K. A., Y. T. N. Yu, M. Vasse and G. J. Velicer. 2021. "Allopatric divergence limits cheating range and alters genetic requirements for a cooperative trait." *bioRxiv* 2021.01.07.425765. https://doi.org/10.1101/2021.01.07.425765.

Smith, Daniel R., Manuel Maestre-Reyna, Gloria Lee, Harry Gerard, Andrew H. J. Wang, and Paula I. Watnick. 2015. "In situ proteolysis of the *Vibrio cholerae* matrix protein RbmA promotes biofilm recruitment." *Proceedings of the National Academy of Sciences U.S.A.* **112**: 10491–96. https://doi.org/10.1073/pnas.1512424112.

smith, jeff, David C. Queller, and Joan E. Strassmann. 2014. "Fruiting bodies of the social amoeba *Dictyostelium discoideum* increase spore transport by *Drosophila*." *BMC Evolutionary Biology* **14**: 105. https://doi.org/10.1186/1471-2148-14-105.

Smith, Jeff, Joan E. Strassmann, and David C. Queller. 2016. "Fine-scale spatial ecology drives kin selection relatedness among cooperating amoebae." *Evolution* **70**: 848–59. https://doi.org/10.1111/evo.12895.

Smukalla, Scott, Marina Caldara, Nathalie Pochet, Anne Beauvais, Stephanie Guadagnini, Chen Yan, Marcelo D. Vinces, et al. 2008. "*FLO1* is a variable green beard gene that drives biofilm-like cooperation in budding yeast." *Cell* **135**: 726–37. https://doi.org/10.1016/j.cell.2008.09.037.

Spiegel, Frederick W., L. S. 1978. Olive. "New evidence for the validity of *Copromyxa protea*." *Mycologia* **70**: 843-847

Spröer, C., H. Reichenbach, and E. Stackebrandt. 1999. "The correlation between morphological and phylogenetic classification of myxobacteria." *International Journal of Systematic Bacteriology* **49**: 1255–62. https://doi.org/10.1099/00207713-49-3-1255.

Stanier, R. Y. 1942. "The Cytophaga group: a contribution to the biology of myxobacteria." *Bacteriological Reviews* **6**: 143–96.

Stenhouse, Fay O., and Keith L. Williams. 1977. "Patterning in *Dictyostelium discoideum*: The proportions of the three differentiated cell types (spore, stalk, and basal disk) in the fruiting body." *Developmental Biology* **59**: 140–52. https://doi.org/10.1016/0012-1606(77)90249-4.

Strassmann, Joan E., Yong Zhu, and David C. Queller. 2000. "Altruism and social cheating in the social amoeba *Dictyostelium discoideum*." *Nature* **408**: 965–67. https://doi.org/10.1038/35050087.

Strassmann, J. E., and D. C. Queller. 2011. "Evolution of cooperation and control of cheating in a social microbe." *Proceedings of the National Academy of Sciences U.S.A.* **108**: 10855–62. https://doi.org/10.1073/pnas.1102451108.

Sugimoto, Hiroki, and Hiroshi Endoh. 2006. "Analysis of fruiting body development in the aggregative ciliate *Sorogena stoianovitchae* (Ciliophora, Colpodea)." *The Journal of Eukaryotic Microbiology* **53**: 96–102. https://doi.org/10.1111/j.1550-7408.2005.00077.x.

Swanson, Andrew R., Eduardo M. Vadell, and James C. Cavender. 1999. "Global distribution of forest soil Dictyostelids." *Journal of Biogeography* **26**: 133–48. https://doi.org/10.1046/j.1365-2699.1999.00250.x.

Tarnita, Corina E, Clifford H Taubes, and Martin A Nowak. 2013. "Evolutionary construction by staying together and coming together." *Journal of Theoretical Biology* **320**: 10–22. https://doi.org/10.1016/j.jtbi.2012.11.022.

Tarnita, Corina E., Alex Washburne, Ricardo Martinez-Garcia, Allyson E. Sgro, and Simon A. Levin. 2015. "Fitness tradeoffs between spores and nonaggregating cells can explain the coexistence of diverse genotypes in cellular slime molds." *Proceedings of the National Academy of Sciences U.S.A.* **112**: 2776–81. https://doi.org/10.1073/pnas.1424242112.

Teintze, M., R. Thomas, T. Furuichi, M. Inouye, and S. Inouye. 1985. "Two homologous genes coding for spore-specific proteins are expressed at different times during development of *Myxococcus xanthus.*" *Journal of Bacteriology* **163**: 121–25. https://doi.org/10.1128/jb.163.1.121-125.1985.

Thaxter, R. 1892. "On the Myxobacteriaceæ, a New Order of Schizomycetes." *Botanical Gazette* **17**, 389–406. https://doi.org/10.1086/326866

Tice, Alexander K., Jeffrey D. Silberman, Austin C. Walthall, Khoa N. D. Le, Frederick W. Spiegel, Matthew W. Brown. 2016. "*Sorodiplophrys stercorea*: Another novel lineage of sorocarpic multicellularity." *Journal of Eukaryotic Microbiology* **63**: 623–628.

Travisano, M., J. Mongold, A. Bennett, and R. Lenski. 1995. "Experimental tests of the roles of adaptation, chance, and history in evolution." *Science* **267**: 87–90. https://doi.org/10.1126/science.7809610.

Travisano, M. and G. J. Velicer. 2004. "Strategies of microbial cheater control." *Trends in Microbiology* **12**: 72–78.

Trunk, Thomas, Hawzeen S. Khalil, Jack C. Leo.2018. "Bacterial autoaggregation." *AIMS Microbiology* **4**: 140–64. https://doi.org/10.3934/microbiol.2018.1.140.

Vega, Nicole M., and Jeff Gore. 2014. "Collective antibiotic resistance: Mechanisms and implications." *Current Opinion in Microbiology* **21**: 28–34. https://doi.org/10.1016/j.mib.2014.09.003.

Velicer, Gregory J., Lee Kroos, and Richard E. Lenski. 1998. "Loss of social behaviors by *Myxococcus xanthus* during evolution in an unstructured habitat." *Proceedings of the National Academy of Sciences U.S.A.* **95**: 12376–80. https://doi.org/10.1073/pnas.95.21.12376.

Velicer, G. J., L. Kroos, and R. E. Lenski. 2000. "Developmental cheating in the social bacterium *Myxococcus xanthus.*" *Nature* **404**: 598–601.

Velicer, Gregory J., and Michiel Vos. 2009. "Sociobiology of the myxobacteria." *Annual Review of Microbiology* **63**: 599–623. https://doi.org/10.1146/annurev.micro.091208.073158.

Verstrepen, Kevin J., and Frans M. Klis. 2006. "Flocculation, adhesion and biofilm formation in yeasts." *Molecular Microbiology* **60**: 5–15. https://doi.org/10.1111/j.1365-2958.2006.05072.x.

Vos, M., and G. J. Velicer. 2006. "Genetic population structure of the soil bacterium *Myxococcus xanthus* at the centimeter scale." *Applied and Environmental Microbiology* **72**: 3615–25. https://doi.org/10.1128/aem.72.5.3615-3625.2006.

Vos, Michiel, and Gregory J. Velicer. 2009. "Social conflict in centimeter- and global-scale populations of the bacterium *Myxococcus xanthus.*" *Current Biology* **19**: 1763–67. https://doi.org/10.1016/j.cub.2009.08.061.

Votaw, Heather R., and E. A. Ostrowski. 2017. "Stalk size and altruism investment within and among populations of the social amoeba." *Journal of Evolutionary Biology* **30**: 2017–30. https://doi.org/10.1111/jeb.13172.

Yang, Z. and P. I. Higgs. 2014. *Myxobacteria: genomics, cellular and molecular biology.* Caister Academic Press.

Wei, Xueming, Darshankumar T. Pathak, and Daniel Wall. 2011. "Heterologous protein transfer within structured myxobacteria biofilms." *Molecular Microbiology* **81**: 315–26. https://doi.org/10.1111/j.1365-2958.2011.07710.x.

West, Stuart A., Roberta M. Fisher, Andy Gardner, and E. Toby Kiers. 2015. "Major evolutionary transitions in individuality." *Proceedings of the National Academy of Sciences U.S.A.* **112**: 10112–19. https://doi.org/10.1073/pnas.1421402112.

Wielgoss, Sébastien, Rebekka Wolfensberger, Lei Sun, Francesca Fiegna, and Gregory J. Velicer. 2019. "Social genes are selection hotspots in kin groups of a soil microbe." *Science* **363**: 1342–1345. https://doi.org/10.1126/science.aar4416.

Wilson, H. V. 1907. "On some phenomena of coalescence and regeneration in sponges." *Journal of Experimental Zoology* **5**: 245–58. https://doi.org/10.1002/jez.1400050204.

Wireman, J. W., and M. Dworkin. 1977. "Developmentally induced autolysis during fruiting body formation by *Myxococcus xanthus*." *Journal of Bacteriology* **129**: 798–802.

Zhang, Zheren, Dennis Claessen, and Daniel E. Rozen. 2016. "Understanding microbial divisions of labor." *Frontiers in Microbiology* **7**: 2070. https://doi.org/10.3389/fmicb.2016.02070.

7 Group Maintenance in Aggregative Multicellularity

Israt Jahan
Department of Biology, Washington University
in St. Louis, St. Louis, MO, USA

Tyler Larsen
Department of Biology, Washington University
in St. Louis, St. Louis, MO, USA

Joan E. Strassmann
Department of Biology, Washington University
in St. Louis, St. Louis, MO, USA

David C. Queller
Department of Biology, Washington University
in St. Louis, St. Louis, MO, USA

CONTENTS

7.1 MULTICELLULARITY AND THE PROBLEM OF ALTRUISM

Life on earth has changed over its long history. The earliest organisms would have been too small to see, but over time organisms have diversified into endless forms, microscopic and macroscopic. The evolution of multicellularity – which happened not just once but more than 20 times – entailed individual cells banding together to

DOI: 10.1201/9780429351907-9

produce the large, often complex bodies of the organisms we see around us (Buss 1988; Bonner 1998; Kaiser 2001; Medina et al. 2003; Grosberg and Strathmann 2007; Ruiz-Trillo et al. 2007). The transition to multicellularity is regarded as one of the major evolutionary transitions (Szathmáry and Maynard Smith 1995), in which formerly independent units (cells) became so dependent upon one another that they thereafter replicated as a combined unit. Like the other major transitions, multicellularity represents a change in the level of organization upon which natural selection acts.

A multicellular organism can be thought of as a group of cells that cooperate, with little conflict, to perform functions that would be impossible for single cells (Queller and Strassmann 2009). Sometimes the benefits cells enjoy by working together can be as simple as advantages of being larger in size. Larger organisms can be better at both avoiding predators and being predators (Kessin et al. 1996; Kirk 2003; Müller et al. 2015; Pentz et al. 2015; Herron et al. 2019) even when their cells are not necessarily contiguous – for example, in the bacterium *Myxococcus xanthus*, which hunts other bacteria in large groups that cooperate to release high concentrations of bactericidal compounds to better lyse and digest their prey (Rosenberg, Keller, and Dworkin 1977; Daft, Burnham, and Yamamoto 1985; Dworkin and Kaiser 1985; Fraleigh and Burnham 1988; Berleman and Kirby 2009). Larger size can provide protective benefits as well – the cells towards the center of a multicellular group can be shielded by the cells on the periphery from chemical and environmental stress, as seen in flocculating yeasts (Smukalla et al. 2008).

A more complicated benefit of multicellularity is that it facilitates the division of labor (Grosberg and Strathmann 2007; Cooper et al. 2020). While single-celled organisms can only divide labor temporally or into different organellar compartments, multicellular organisms can use entire cells to specialize in different tasks. Thus, division of labor can also protect functions of an organism from other functions that would otherwise interfere. For example, in some cyanobacteria, like *Nostoc*, oxygen produced by photosynthesis interferes with the enzymes involved in nitrogen fixation, and so multicellular cyanobacteria split the tasks of photosynthesis and nitrogen fixation into separate cell types (Kumar, Mella-Herrera, and Golden 2010). Non-photosynthetic heterocyst cells specialize in fixing nitrogen by remaining anaerobic behind thick cell walls, and exporting nitrogen to photosynthetic neighbors. Division of labor is crucial for the function of very complex multicellular organisms like animals, whose bodies can require hundreds of separate cell types working in concert.

Cooperating within a multicellular group can also come with costs, ranging from the energetic costs of producing a public good that all the cells in the group use, like the bactericidal enzymes produced by *M. xanthus* cells in a swarm, to an extreme where some cells sacrifice their lives entirely to benefit other cells. The clearest example of the latter in the context of multicellularity is the cooperation between somatic and germline cells. Somatic cells – like the heterocysts in cyanobacteria – forgo reproduction entirely and thus have no individual fitness. In large multicellular organisms like animals, the vast majority of cells sacrifice themselves and only a minuscule minority are passed on to the next generation.

Costly acts invite exploitation. In the case of a costly public good, for example, it may be in the best interest of an individual cell to stop producing the public good, thus avoiding the cost but continuing to benefit from the public good's production by the rest of the group. An individual that does not cooperate or pay a full share towards the cost of cooperation but receives the benefits from a cooperative group is called a cheater (West, Griffin, and Gardner 2007). If allowed to proliferate, cheaters can result in the destruction of a cooperative trait and the subsequent failure of those dependent on it (Hardin 1968). At high frequency, cheaters can even drive entire populations to extinction (Fiegna and Velicer 2003). Cheaters have been observed in many cooperative systems in nature, and how such groups contend with the threat they pose has been a major question in evolutionary biology (West et al. 2006; 2007; Riehl and Frederickson 2016). Why should living things cooperate at all if it requires sacrificing their own fitness for others? Why does selection for cheating not preclude the evolution of costly cooperation? How do multicellular organisms persist?

Hamilton's theory of inclusive fitness gives us part of the answer – the altruistic act of sacrificing one's own fitness for others can be favored by selection when it benefits the altruist's relatives (Hamilton 1964a; 1964b). According to Hamilton's rule, costly cooperation should be selected for when $rb > c$, where b is the fitness benefit gained by the recipient of a cooperative trait, c is the cost incurred by the actor, and r is genetic relatedness between the recipient and the actor. Under Hamilton's rule, alleles underlying costly cooperation can be selected for because, while the costs may reduce a cooperator's own fitness (direct fitness), this cost can be compensated for if cooperation sufficiently benefits the fitness of other individuals likely to carry the same alleles (indirect fitness).

Hamilton's rule, with its emphasis on relatedness, goes a long way towards explaining the evolution and persistence of many multicellular organisms. Most familiar multicellular organisms like plants, animals, fungi, and red and brown algae are composed of cells that descend from a single cell and are therefore genetically identical, with relatedness between cells maximal at $r = 1$. Under Hamilton's Rule, altruism by somatic cells in a clonal organism can be favored by selection on the fitness of genetically identical germline cells so long as the total benefits of cooperation outweigh the total costs. Cells within a clonal organism have little to gain by conflict and so the single-celled bottlenecks they undergo can largely sidestep the risks of cheaters. Further, though new cheater mutations can arise within a clonal multicellular organism and reduce relatedness between cells, the single-celled bottleneck ensures that they would enjoy only a single generation of the benefits of cheating before producing progeny that are disadvantaged by consisting entirely of cheating cells (Queller 2000; Buss 2014).

7.2 AGGREGATIVE MULTICELLULARITY

Many lesser-known multicellular organisms, however, do not undergo an obligatory single-cell bottleneck and instead form by the aggregation of individual cells that may or may not be related. This path to multicellularity, called aggregative multicellularity, does not automatically ensure high relatedness among cells (Figure 7.1). Unrelated cells can join the same group, giving cheater genotypes opportunities to increase in frequency at cooperating genotypes' expense, even if doing so results in reduced fitness for the group as a whole. Such fitness reduction is particularly

Clonal and aggregative multicellularity

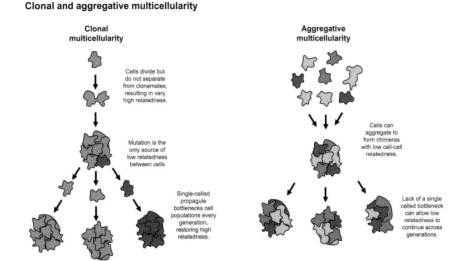

FIGURE 7.1   Clonal and aggregative multicellularity.

common with an increase of obligate cheaters, which cannot successfully cooperate on their own. Obligate cheaters readily appear and increase in laboratory populations of both *Myxococcus xanthus* and *Dictyostelium discoideum* (Velicer, Kroos, and Lenski 2000; Gilbert et al. 2007; Velicer and Vos 2009; Kuzdzal-Fick et al. 2011) and can even lead to the extinction of the population, as has been observed in *M. xanthus* (Fiegna and Velicer 2003). Aggregative multicellularity is nonetheless very common, with independently evolved examples in archaea, bacteria, and eukaryotes (Brown et al. 2012; Chimileski, Franklin, and Papke 2014; Chapter 5).

Many bacteria aggregate into cooperative single-species groups. One common form of aggregative multicellularity in bacteria is swarming, in which cells coordinate to move rapidly on solid or semi-solid surfaces via the production of public good surfactants (Harshey 2003; Butler, Wang, and Harshey 2010). Swarming motility has been described in *Bacillus subtilis* (Shapiro 1998; Aguilar et al. 2007), *Proteus mirabilis* (Shapiro 1998), *Pseudomonas aeruginosa* (Lai, Tremblay, and Déziel 2009), and Myxobacteria (Dale Kaiser, Robinson, and Kroos 2010; Velicer and Vos 2009). Cells within the swarm interact with each other and undergo morphological differentiation so that they can glide on surfaces (Julkowska et al. 2004).

Many bacteria – particularly Myxobacteria – also aggregate to form multicellular structures that facilitate survival in harsh conditions and during dispersal. Various Myxobacteria species produce aggregative fruiting bodies consisting of spores and sometimes a stalk that lifts the spores (Velicer and Vos 2009). Fruiting bodies can elevate spores, enhancing dispersal to new environments (White, Shropshire, and Stephens 1980; Huss 1989; Kessin et al. 1996; Kaiser 2001). The process of sporulation often entails death of most of the cells in the aggregate and may thus represent a form of germ/soma division of labor like that seen in conventional multicellular organisms.

Biofilm formation is another extremely common behavior in microbes that resembles aggregative multicellularity. In a biofilm, cells of one or more species adhere together on a surface, often producing a structure of secreted extracellular matrix that can protect the cells from antimicrobials, shear forces, and host immune systems. The close proximity of microbes within a biofilm can also facilitate the exchange of chemicals. Research over the past decade has revealed that biofilms can be quite complex and sometimes contain many interacting species (Webb, Givskov, and Kjelleberg 2003; Nadell, Xavier, and Foster 2009; Foster 2011; Xavier and Foster 2007; Claessen et al. 2014). When biofilms involve interspecific interactions where relatedness is necessarily zero, they may experience high levels of conflict and unresolved selfishness of members and would not usually qualify as organisms, and (Xavier and Foster 2007; Queller and Strassmann 2009).

Among eukaryotes, aggregative multicellularity has independent origins in Discicristata (Brown, Silberman, and Spiegel 2012), Stramenopiles (Tice et al. 2016), Alveolata (Sugimoto and Endoh 2006), Rhizaria (Brown et al. 2012), Holozoa (Brown, Spiegel, and Silberman 2009) and twice in Amoebozoa (Brown, Silberman, and Spiegel 2011). Most of these eukaryotes share similar lifestyles involving a unicellular stage wherein amoebas feed and divide independently by mitotic division. Upon starvation, amoebas aggregate and enter a multicellular stage that is concerned solely with development, and eventually morph into a fruiting body. There is considerable variation in fruiting body morphology and development among species (Kawabe et al. 2019).

The best-studied organism among the aggregating eukaryotes is the cellular slime mold *Dictyostelium discoideum* and its relatives. *D. discoideum* spends most of its time as a unicellular vegetative cell, hunting bacterial prey in temperate forest soils. Upon starvation, *D. discoideum* amoebas aggregate and undergo a series of developmental steps to form a multicellular motile slug and then a stalked fruiting body consisting of up to four different somatic cell types to support spore production and dispersal (Kessin 2001).

Despite its increased vulnerability to the risks posed by cheaters, aggregative multicellularity has nonetheless persisted in widely disparate taxa, which maintain their multicellularity with a series of mechanisms that limit cheaters from overtaking the cooperative group. In this chapter, we review some of the most important mechanisms, with a special focus on the two best studied aggregative multicellular taxa – the bacterium *Myxococcus xanthus* (Velicer and Vos 2009; Cao et al. 2015; Muñoz-Dorado et al. 2016) and the eukaryote *Dictyostelium discoideum* (Kessin 2001; Shaulsky and Kessin 2007; Li and Purugganan 2011; Medina et al. 2019; Ostrowski 2019). The social stages in the life cycles of *M. xanthus* and *D. discoideum* both involve thousands of individual cells aggregating to form multicellular structures in response to starvation and require the death of a large fraction of the aggregate (more than 90% in *M. xanthus* [Muñoz-Dorado et al. 2016] and approximately 20% in *D. discoideum* [Votaw and Ostrowski 2017]).

These two species thus face similar potential conflicts between cooperators that make the necessary sacrifices and cheaters that do not. In both species, cheater mutations are easy to find and cheaters can readily increase under experimental evolution (Ennis et al. 2000; Velicer, Kroos, and Lenski 2000; Santorelli et al. 2008; Kuzdzal-Fick et al. 2011). In both, cheating appears to be common between different clones

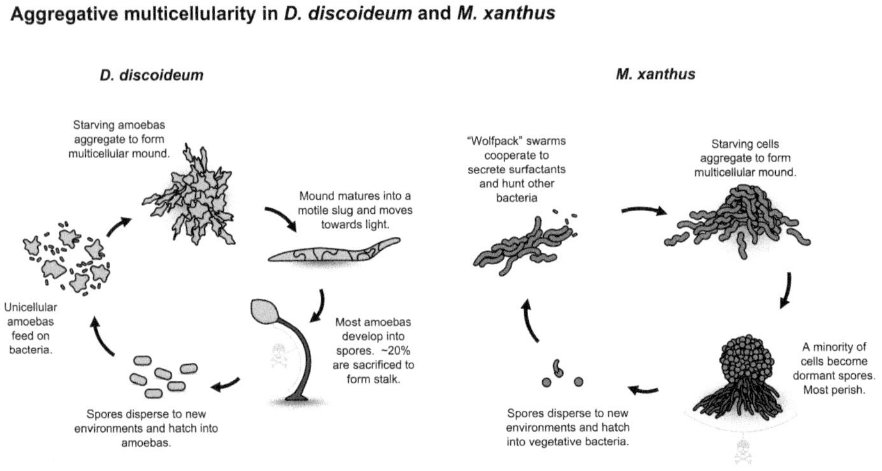

Aggregative multicellularity in *D. discoideum* and *M. xanthus*

D. discoideum   *M. xanthus*

FIGURE 7.2   Aggregative multicellularity in *Dictyostelium discoideum and Myxococcus xanthus.*

collected from nature (Strassmann, Zhu, and Queller 2000; Fiegna and Velicer 2005; Buttery et al. 2009; Vos and Velicer 2009) though alternative explanations might sometimes apply (Smith, Van Dyken, and Velicer 2014; Tarnita et al. 2015; Wolf et al. 2015). Despite their many differences, *M. xanthus* and *D. discoideum* employ some of the same kinds of mechanisms to prevent these conflicts from destabilizing coopera-tion (Figure 7.2).

7.3   MAINTAINING HIGH RELATEDNESS

Inclusive fitness theory's key insight is that even a very costly altruistic trait – like cells sacrificing themselves to help produce a *Dictyostelium* or *Myxococcus* fruiting body – can be selected for if the beneficiaries of that sacrifice are sufficiently related. Similarly, cheaters benefit only when they have someone to cheat, so when relat-edness is high and cheaters interact only with other cheaters, there is no incentive to cheat (Hamilton 1963; Frank 1995; 2003; Bourke 2013). As already discussed, relatedness is very high among cells in many conventional multicellular organisms because they undergo a single-cell bottleneck during development, while aggre-gative multicellular organisms without such a bottleneck can potentially include unrelated cells. Even without a strict single-cell bottleneck, however, relatedness between cells of these species may be kept high via other mechanisms. In *D. dis-coideum*, microsatellite markers showed relatedness between cells within the same fruiting bodies collected from nature to be very high – between 0.86 and 0.97 (Gilbert et al. 2007). It is also very high in fruiting bodies of *M. xanthus* isolated from the wild, which were composed of different clonal lineages even when the fruiting bodies were collected just micrometers apart (Kraemer et al. 2016). In a follow-up study on *M. xanthus,* the authors sequenced 120 clones from 6 naturally isolated fruiting bodies (Wielgoss et al. 2019). While each of the fruiting bodies included genetic variants at a few sites, these were distinguished by just a few mutational changes, which could be traced back to a recent ancestor. The accumulation of variation was due to recent *de novo* mutation in related cells rather than mixing of unrelated cells (Wielgoss et al. 2019).

7.3.1 POPULATION STRUCTURE

One very simple mechanism for maintaining high relatedness is via population structure, wherein cells tend to interact with close relatives as a passive consequence of how they grow and disperse. Though aggregative multicellular organisms like *Dictyostelium* and *Myxococcus* do not undergo an obligatory single-celled bottleneck, they do primarily reproduce via clonal division of a mother cell into two identical daughter cells. If movement is limited, microbial populations can grow up as patches of closely related cells radiating out from a single founder. When conditions arise that favor aggregation, cells interact with their nearest neighbors, which tend to be clonemates. In *D. discoideum* laboratory populations, only a couple of millimeters separation between genetically distinct founding spores is required to generate patches of high relatedness, despite the motility of amoebas (smith, Strassmann, and Queller 2016). When genetically different clones are plated in closer proximity to one another, there is increased mixing and relatedness in the resultant fruiting bodies is lower. Populations of *Dictyostelium* and *Myxococcus* have been observed to be structured the scale of millimeters to centimeters, which is consistent with the high relatedness observed between cells within fruiting bodies (Fortunato et al. 2003; Vos and Velicer 2008; Kraemer et al. 2016).

Further, even when multiple clones are mixed and relatedness is initially low, space constraints can result in a phenomenon called genetic demixing, wherein populations of cells separate into clonal sectors as they grow outwards (see Figure 7.3).

Structured growth and genetic demixing

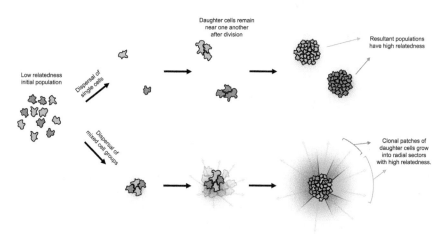

FIGURE 7.3 Structured growth and genetic demixing.

When space is limited, only a small fraction of cells along the edge of the group can divide. As these cells divide, their progeny push out radially to fill the available space, mostly crowding out the progeny of other cells and thus raising relatedness. This demixing was first shown to occur in bacterial colonies (Hallatschek et al. 2007; Nadell, Foster, and Xavier 2010). Buttery et al. (2012) showed the same phenomenon with *D. discoideum* grown on agar plates.

7.3.2 KIN DISCRIMINATION

While population structure can passively reduce opportunities for cheaters and cooperators to interact with one another, kin discrimination allows cooperators that do risk interacting with cheaters to selectively direct their cooperation towards those individuals most likely to share genes and avoid exploitation by those who do not (Tsutsui 2004). Kin discrimination is widespread in microbes (Strassmann, Gilbert, and Queller 2011; Strassmann 2016; Wall 2016; Medina et al. 2019). In *Dictyostelium* and *Myxococcus*, kin discrimination can help aggregating strains segregate from unrelated strains and preferentially develop in groups with high relatedness between cells.

In the social amoeba species *D. violaceum*, *D. purpureum*, and *D. giganteum*, clonemates sort into mostly separate aggregates in the laboratory (Mehdiabadi et al. 2006; 2009; Sathe et al. 2010; Kalla et al. 2011; Sathe, Khetan, and Nanjundiah 2014; Medina et al. 2019). Curiously, in the best studied *D. discoideum*, the degree of kin sorting in fruiting bodies is low (Ostrowski et al. 2008; Gilbert, Strassmann, and Queller 2012; Gruenheit et al. 2017). However, there are genes that seem to be involved in recognition in this species, whether or not they are used for sorting in nature.

Cell adhesion genes were suspected to be important in recognition from the earliest investigations in *Dictyostelium* (Bonner and Adams 1958; Sternfeld 1979). In *D. discoideum*, allorecognition is attributed to transmembrane proteins encoded by two sets of highly polymorphic essential genes, *tgrB1* and *tgrC1* (Benabentos et al. 2009; Hirose et al. 2011; 2015). Amoebas with compatible pairs of Tgr proteins bind and adhere to one another during aggregation, with a binding affinity that negatively correlates with the degree of segregation between *D. discoideum* genotypes (Gruenheit et al. 2017). Amoebas with sufficiently different *tgr* alleles segregate within a genetically mixed aggregate and form separate slugs (Hirose et al. 2011; Gruenheit et al. 2017), although these can later fuse and form chimeric fruiting bodies (Ho and Shaulsky 2015). Knocking out either *tgrB1* or *tgrC1* prevents development past the aggregate stage (Benabentos et al. 2009), but the presence of additional *tgrB1* and *tgrC1* alleles does not affect recognition as long as there is at least one compatible allele pair between the cells (Hirose et al. 2011).

The tgr recognition system appears to play some role in prevention of obligate cheaters even though it does not lead to complete sorting in *D. discoideum* fruiting bodies in the laboratory. Ho and colleagues found that an obligate social cheater lacking compatible Tgr proteins was excluded from the final fruiting body when *tgr* alleles were incompatible (Ho et al. 2013). They suggest that if obligate cheating occurs due to mutants that have effects early in development, the early segregation of cells within the aggregate and in the initial slugs can largely prevent cheating, even though the slugs later fuse (Ho and Shaulsky 2015). In addition to excluding cheaters, recognition may stimulate competition when chimeras do form. In one study (Noh et al. 2018), but not another (de Oliveira et al. 2019), genes that significantly change expression levels in chimeric mixtures showed rapid evolution, which is consistent with a history of evolutionary conflict between cell lineages.

Kin discrimination in *M. xanthus* is most readily observed in the formation of clear demarcation lines (zones of no or low-density cells) where two unrelated swarms meet on an agar plate. Swarms that are identical or closely related to each other readily merge and so produce no such demarcation zones (Senior 1977; Gibbs, Urbanowski, and Greenberg 2008; Patra et al. 2017).

One should be cautious in classifying a process that segregates clones as having necessarily evolved to discriminate kin from non-kin. In an experimental evolution study of natural isolates of *M. xanthus*, isolated clones rapidly evolved incompatibilities with one another or with their common ancestors (Rendueles et al. 2015), but because there were no foreign clones to interact with during this evolution, incompatibility seems unlikely to have evolved to recognize self or to exclude others. Instead, the result may simply reflect that genetically different clones may not function well together (Foster et al. 2002; Mendes-Soares et al. 2014) which can be due to social incompatibilities evolved in isolation (Rendueles et al. 2015; Ostrowski 2019). For example, if one clone changes its communication system for swarming, those changes might no longer work well with the systems in other clones. This is analogous to hybrid incompatibilities that evolve during allopatric speciation and is expected to occur in any context where complex interactions adapt in isolation from each other (Queller 2018). Even absent the threat of cheaters, these incompatibilities could result in a "chimeric load," such that chimeric aggregates function less well than clonal aggregates, and selection to reduce the likelihood of chimerism via kin-recognition-like mechanisms (Ostrowski 2019).

A less visible form of kin recognition in *M. xanthus* involves exchange of membrane components between adjacent cells (Pathak et al. 2013; Dey et al. 2016; Patra et al. 2017). This can serve both to allow related cells to coordinate swarming, development, and sharing of private goods like lipids (Pathak et al. 2012), lipopolysaccharides (Vassallo et al. 2015), and lipoproteins (Wall 2014) to assist in the repair of damaged cells, but also allow unrelated cells to harm one another by transferring toxic bacteriocins (Vassallo et al. 2017). Thus, outer membrane exchange acts as a kin discrimination system with dual purposes – its effects can be beneficial or harmful depending on relatedness between the cells (Wall 2016). Outer membrane exchange is facilitated by the two required genes *traA* and *traB* (Pathak et al. 2012). TraA is a highly polymorphic cell surface receptor that, similar to the tgr system in *D. discoideum,* facilitates adhesion to and interactions with cells that bear the same allele while TraB may involve the transport of TraA to the cell surface (Cao et al. 2015; 2019).

Recognition systems are not a certain defense against exploitation by cheaters. Cells do not always sort into perfectly related clonal patches; mixed strains of both *M. xanthus* and *D. discoideum* can readily form chimeric groups in the laboratory. In nature, the very high relatedness among cells in *M. xanthus* and *D. discoideum* fruiting bodies is likely to be achieved via a combination of passive population structure and active kin discrimination.

This high relatedness exerts stronger control over obligate cheaters than facultative ones. Facultative cheaters cheat when in mixtures but perform normally when they are alone. High relatedness simply reduces their opportunities to cheat. But for obligate cheaters, high relatedness adds an additional big cost by isolating them in groups of obligate cheaters that cannot perform well. Consistent with this difference, facultative cheating appears to be common among natural clones of *D. discoideum* (Strassmann, Zhu, and Queller 2000; Buttery et al. 2009) and *M. xanthus* (Fiegna and Velicer 2005; Vos and Velicer 2009). Yet, despite obligate cheaters readily evolving in the laboratory in both *D. discoideum* (Ennis et al. 2000; Gilbert et al. 2007; Kuzdzal-Fick et al. 2011) and *M. xanthus* (Velicer, Kroos, and Lenski 2000), they have never been isolated from nature in either species. Thus, high natural relatedness may be controlling the obligate mutations, which are the greatest threat to cooperation if they were to spread (Gilbert et al. 2007).

7.4 ENFORCEMENT OF COOPERATION

Mechanisms that increase relatedness between group members work by minimizing the opportunities for cheaters and cooperators to interact, but may not eliminate such opportunities entirely. A need thus remains to enforce cooperation in potentially exploitative group members that either could not be excluded by the mechanisms already discussed or that arose *de novo* via the accumulation of random mutations that lower relatedness and convert previously cooperative cells into cheaters. A particularly familiar example of the latter is the spontaneous appearance of cancerous cells in many multicellular organisms, which are analogous to cheaters selfishly exploiting the cooperation of the rest of the cells in the organism (Nunney 2013; Aktipis et al. 2015). As persistent a problem as cancer is for animals, however, cheaters may be even more destructive in aggregative multicellular organisms, because without the opportunity to reset each generation to maximal relatedness through a single-cell bottleneck, there is the potential for cheating genotypes to persist over multiple generations. Accordingly, just as animals have evolved genetic mechanisms to actively suppress cancerous cells, for instance by inducing apoptosis (Evan and Vousden 2001; Foster 2011; Singh and Boomsma 2015), aggregative multicellular organisms have evolved enforcement mechanisms to coerce cheaters within the social group to cooperate (Ågren, Davies, and Foster 2019).

The evolution of cheater resistance has been well documented in studies of both *M. xanthus* and *D. discoideum*. In two experimental evolution studies with *M. xanthus*, (Fiegna et al. 2006; Manhes and Velicer 2011), Velicer and colleagues evolved cheater suppression by repeated interactions between populations of a developmentally proficient strain and a developmentally defective cheater that had high fitness in chimeric mixtures with its ancestor, but could not sporulate on its own. In the first study, the cheater rapidly increased in frequency until it became so common that it no longer had enough developmentally competent cooperators to exploit, resulting in the entire population crashing and producing extremely few spores (Fiegna et al. 2006). Interestingly, at the fourth cycle of development, a new mutation occurred in the cheater that allowed for the rescue of the population. The resultant mutant strain (called phoenix because it "rose from the ashes" of the cheater) had a fitness advantage over both wild type and cheater. The phoenix strain can sporulate when it is common, an ability lost by the cheater strain, and rose to apparent fixation in the mixed population. In mixed culture, the phoenix mutant had an advantage over cheaters and wild type at all tested frequencies. The restoration of developmental proficiency by phoenix is not fully understood, but here it was not the wildtype that evolved cheater resistance, but a lineage of the cheater strain itself.

In a similar study (Manhes and Velicer 2011), a developmentally proficient strain was evolved in the presence of a *csgA*— cheater mutant (Velicer, Kroos, and Lenski 2000). By repeatedly replacing the cheater each generation, it was held evolutionarily static, and the non-cheater thus could evolve resistance without evolutionary retaliation. After 20 cycles of repeated development, the wild-type cooperators evolved to have higher fitness than cheaters in mixed populations. Interestingly, the evolved populations not only suppressed cheating by cheaters but increased the absolute fitness of ancestral wild type in a three-party mixture. This benefit to wild type was seen only in the presence of cheater. This benefit of cheater suppression to a third party was likened to policing behavior in some social animals (Manhes and Velicer 2011; Zanette et al. 2012).

Similar studies of cheater resistance have been performed in *D. discoideum*. In a study by Khare et al., the authors evolved the wild type AX4 in the presence of *chtC* mutants which were prevented from evolving. They selected over four rounds of development and isolated a "noble resister" mutant (*rccA*) that was resistant to being cheated by *chtC* but did not cheat AX4 or the original cheater *chtC* (Khare et al. 2009). Hollis performed an evolution experiment using the strains NC4 and AX2, the former of which is a strong cheater of the latter (Hollis 2012). NC4 and AX2-GFP were allowed to grow and develop together but the cheating NC4 was prevented from coevolving. After only 10 generations of social development, AX2-GFP developed defenses against cheating in the social stage along with differences in competitive growth in the vegetative stage. Levin and colleagues employed an experimental design to look for the evolution of cheater resistance in a situation where the cheater was free to evolve in response and found that non-cheaters evolved to resist the obligate social cheaters without themselves cheating (Levin et al. 2015). This study showed that resistance can evolve on a rapid timescale before the cheater becomes fixed.

7.5   PLEIOTROPIC COSTS OF CHEATING

Cheating can be costly. If the costs are large enough, the benefits gained by being able to exploit cooperators may be too small to compensate and cheating does not spread. Such "intrinsic defector inferiority" (Travisano and Velicer 2004) becomes more interesting when the cheating phenotype itself is beneficial but the allele causing this phenotype also has pleiotropic costs (Foster et al. 2004). To put it the other way around, if the cooperator allele has pleiotropic benefits, it may be protected against cheating. Alleles, or sets of tightly linked alleles, that encode for cooperative traits coupled to other essential functions may thus be naturally resistant to cheaters.

Those cooperator alleles that have persisted in aggregative multicellular organisms may therefore be a biased set with pleiotropic advantages (Foster et al. 2004).

There are plausible examples in both *D. discoideum* (Medina et al. 2019), and *M. xanthus* (Velicer and Vos 2009). For example, the *D. discoideum* gene *dimA* is required to receive the signaling molecule DIF-1 that causes cells to differentiate into sterile prestalk cells during development. Cells with *dimA* knocked out are therefore expected to cheat by becoming spores instead of stalk. They do move to the prespore region of the slug, and yet these cells are largely excluded from maturing into spores by an unknown pleiotropic effect (Foster et al. 2004). For this reason, cheating on prestalk cell production leads to a reduction in spores and thus is unlikely to be favored by natural selection (Foster et al. 2004). A similar example can be seen in the *csA* – mutants. These mutants lack functional gp80 adhesion proteins and, on agar substrate, cheat their ancestor with functional gp80, perhaps because their impaired cell adhesion causes them to slide to the prespore region at the back of the slug (Queller et al. 2003). Interestingly, these mutants do not succeed when grown on the more realistic substrate of soil because their impaired adhesion prevents them from getting into an aggregation in the first place (Queller et al. 2003).

Pleiotropic effects can sometimes occur in different individuals, as in sexually antagonistic pleiotropy or pleiotropy underlying local adaptation (Paaby and Rockman 2013). In this light, obligate cheaters can be viewed as carrying a pleiotropic cost (Medina et al. 2019). Though cheaters in mixtures gain a cheating benefit, cheaters that occur in all-cheater groups experience a large cost and this can prevent cheaters from increasing.

It has been recently argued (dos Santos, Ghoul, and West 2018) that pleiotropy cannot explain the maintenance of cooperation because mutations will eventually occur that separate the pleiotropic traits, freeing cheaters from their pleiotropic handicap and enabling their spread. This is a good point for many pleiotropic effects. For example, many bacterial genes are upregulated by quorum sensing and can thus be regarded as pleiotropic effects of the alleles underlying quorum sensing. Cheater strains that did not participate in quorum sensing would suffer costs due to misregulating so many other genes, but it seems likely that these effects can be easily separated by mutations in regulatory regions that remove individual genes from quorum sensing control (dos Santos, Ghoul, and West 2018). The effects of *D. discoideum*'s *csaA* knockout allele might also be separable provided a mutant could arise that allowed the gene to be fully expressed during aggregation but suppressed in the slug stage. Some of the obligate cheaters seem less clear. Can one always repair a gene that is developmentally defective on its own? But at least one case is known from *M. xanthus* of an obligate cheater recovering developmental function (Fiegna et al. 2006).

Yet there still appears to be some role for pleiotropy. In general, whether pleiotropy can suppress cheaters relates to the larger unanswered evolutionary question of how much pleiotropy constrains evolution. An example that weighs against the argument that pleiotropic effects can always be separated is senescence. Senescence is thought to result, at least in part, from selection for alleles that pleiotropically cause fitness gains early in life but fitness losses later in life (Williams 1957). If mutation could effectively separate these effects, the later deleterious effects would not have accumulated. Returning to social traits, models suggest the importance of pleiotropy in limiting cancer in large organisms, where cell lineages persist for many generations (Ågren, Davies, and Foster 2019). Finally, pleiotropy may be crucial for certain kinds of frequency-dependent cooperation. Some cooperative genes act synergistically such that cooperation is favored only if it can become sufficiently common. Here, pleiotropic effects could allow cooperation to initially increase and if this happens before a mutation separates the pleiotropic effects, cooperation may have reached a frequency where it can be favored on its own (Queller 2019). Understanding the role of pleiotropy in the evolution and maintenance of cooperation will require additional theoretical and empirical work.

7.6 CONCLUSIONS

In this chapter, we discussed how cheaters are controlled in aggregative multicellular organisms, in particular in the best studied systems – *D. discoideum* and *M. xanthus*. We categorized the mechanisms of group maintenance into three main categories – maintaining high relatedness, enforcement of cooperation, and pleiotropic costs of cheating.

High relatedness among group members makes cooperation easier to evolve and maintain because cooperators tend to benefit cooperators and cheaters tend to harm cheaters. At sufficiently high relatedness, cheaters can only hurt themselves. *D. discoideum* and *M. xanthus* can maintain high relatedness passively through structured growth and dispersal that results in structured populations with patches of closely related cells. Relatedness over short spatial scales thus remains high, and it is only at the interfaces between patches that there is opportunity for exploitation. Over longer spatial scales, relatedness will be lower, but the crucial measure is relatedness at the scale of aggregations in which interactions are occurring. Questions remain about the mechanisms of structured growth and dispersal in *Dictyostelium* and *Myxococcus*. What are the vectors for transportation of spores? How well do spores disperse together? What are the distances over which spores can be dispersed? How large a role does demixing play in maintaining relatedness in nature?

A more active way to increase relatedness is kin discrimination – preferentially cooperating with relatives and avoiding harm from non-kin. Kin discrimination has been observed in diverse organisms and may be especially important to aggregative multicellular organisms, in which it can be observed playing a role in clonal segregation (again driving high relatedness within groups) and, for *M. xanthus,* transfer of private goods or toxins via membrane exchange. The potential importance of kin discrimination in *M. xanthus* is underscored by the finding that *M. xanthus* readily evolves social incompatibilities in the laboratory.

The apparent importance of high relatedness in aggregative multicellular organisms suggests that they are not really so different from the more familiar organisms that develop from a single cell. In both cases, high relatedness is maintained. In both cases, this can be due to daughter cells remaining in proximity to each other, although they are detached in one and attached in the other. Kin-recognition or self-recognition systems reinforce high relatedness. High relatedness may be easier to maintain with an obligatory single-cell bottleneck, but aggregative organisms can also achieve it.

When relatedness is imperfect and cheaters and cooperators can interact, a second mechanism – enforcement – comes into play. In *D. discoideum* and *M. xanthus*, cooperators can evolve to resist cheaters in the laboratory, increasing group productivity in the presence of cheaters. These policing mechanisms are somewhat similar to the suppression of cheaters in eusocial insect groups via aggression. Studies of enforcement are relatively new in the field of microbial social evolution and there are multiple unanswered questions. What genes underlie cheater resistance? Do power hierarchies exist between wild cheaters and cooperators, such that strains differ in which other strains they can exploit? Are interactions between cheaters and cheater-resistant cooperators frequency dependent? Do they result in evolutionary arms races?

Finally, cheating may sometimes be inherently costly due to pleiotropic costs of cheating on other important traits. Examples are known from both *D. discoideum* and *M. xanthus*, arguably including obligate cheaters. Sometimes the costly pleiotropic trait may be evolutionarily separable from the cheating trait, allowing cheating to increase. But sometimes, it may not be separable, or it may separate too late, allowing the frequency dependent cooperative trait to reach a sustainable level. Separability is thus an important topic for future studies.

Multicellularity is an ancient, diverse, and highly successful strategy, partly because multicellular organisms have evolved mechanisms by which they control the threat of cheating. In the most familiar multicellular organisms, this is accomplished via the single-cell bottleneck, but there are many examples of aggregative multicellular organisms without single-cell bottlenecks, for which cheating is predicted to be an especially key threat. Though they may seem primitive, aggregative multicellular organisms, too, have ways of maintaining their groups against cheaters, as exemplified by the diverse mechanisms seen in *Dictyostelium* and *Myxococcus*.

ACKNOWLEDGEMENTS

We thank Trey Scott of the Queller/Strassmann group; Kaitlin Schaal, Marco La Fortezza, and Greg Velicer of ETH Zurich; and two anonymous reviewers for helpful comments. Our own research is supported by the National Science Foundation Grant Numbers IOS 1656756 and DEB 1753743. Israt Jahan is supported by a fellowship from McDonnell International Scholars Academy.

REFERENCES

Ågren, J. Arvid, Nicholas G. Davies, and Kevin R. Foster. 2019. "Enforcement Is Central to the Evolution of Cooperation." *Nature Ecology and Evolution* 3 (July): 1018–1029. doi:10.1038/s41559-019-0907-1.

Aguilar, Claudio, Hera Vlamakis, Richard Losick, and Roberto Kolter. 2007. "Thinking about Bacillus Subtilis as a Multicellular Organism." *Current Opinion in Microbiology* 10 (6): 638–643. doi: 10.1016/j.mib.2007.09.006.

Aktipis, C. A., Amy M. Boddy, Gunther Jansen, Urszula Hibner, Michael E. Hochberg, Carlo C. Maley, and Gerald S. Wilkinson. 2015. "Cancer across the Tree of Life: Cooperation and Cheating in Multicellularity." *Philosophical Transactions of the Royal Society B: Biological Sciences* 370 (1673): 20140219–20140219. doi:10.1098/rstb.2014.0219.

Aktipis, C. Athena, Amy M. Boddy, Gunther Jansen, Urszula Hibner, Michael E. Hochberg, Carlo C. Maley, and Gerald S. Wilkinson. 2015. "Cancer across the Tree of Life: Cooperation and Cheating in Multicellularity." *Philosophical Transactions of the Royal Society B: Biological Sciences* 370 (1673): 20140219–20140219. doi:10.1098/rstb.2014.0219.

Benabentos, Rocio, Shigenori Hirose, Richard Sucgang, Tomaz Curk, Mariko Katoh, Elizabeth A. Ostrowski, Joan E. Strassmann, et al. 2009. "Polymorphic Members of the Lag Gene Family Mediate Kin Discrimination in Dictyostelium." *Current Biology* 19 (7): 567–572. doi: 10.1016/j.cub.2009.02.037.

Berleman, James E., and John R. Kirby. 2009. "Deciphering the Hunting Strategy of a Bacterial Wolfpack." *FEMS Microbiology Reviews* 33 (5): 942–957. doi: 10.1111/j.1574-6976.2009.00185.x.

Bonner, J. T., and M. S. Adams. 1958. "Cell Mixtures of Different Species and Strains of Cellular Slime Moulds." *Journal of Embryology and Experimental Morphology* 6 (2): 346–356. http://www.ncbi.nlm.nih.gov/pubmed/13575646.

Bonner, John Tyler. 1998. "The Origins of Multicellularity." *Integrative Biology: Issues, News, and Reviews* 1 (1): 27–36. doi: 10.1002/(sici)1520-6602(1998)1:1<27::aid-inbi4>3.3.co;2-y.

Bourke, Andrew F.G. 2013. *Principles of Social Evolution. Principles of Social Evolution.* Oxford University Press. doi: 10.1093/acprof:oso/9780199231157.001.0001.

Brown, Matthew W., Martin Kolisko, Jeffrey D. Silberman, and Andrew J. Roger. 2012. "Aggregative Multicellularity Evolved Independently in the Eukaryotic Supergroup Rhizaria." *Current Biology* 22 (12): 1123–1127. doi: 10.1016/j.cub.2012.04.021.

Brown, Matthew W., Jeffrey D. Silberman, and Frederick W. Spiegel. 2011. "'Slime Molds' among the Tubulinea (Amoebozoa): Molecular Systematics and Taxonomy of Copromyxa." *Protist* 162 (2): 277–287. doi: 10.1016/j.protis.2010.09.003.

Brown, Matthew W., Jeffrey D. Silberman, and Frederick W. Spiegel. 2012. "A Contemporary Evaluation of the Acrasids (Acrasidae, Heterolobosea, Excavata)." *European Journal of Protistology* 48 (2): 103–123. doi: 10.1016/j.ejop.2011.10.001.

Brown, Matthew W., Frederick W. Spiegel, and Jeffrey D. Silberman. 2009. "Phylogeny of the 'Forgotten' Cellular Slime Mold, Fonticula alba, Reveals a Key Evolutionary Branch within Opisthokonta." *Molecular Biology and Evolution* 26 (12): 2699–2709. doi:10.1093/molbev/msp185.

Buss, Leo W. 1988. *The Evolution of Individuality.* Princeton University Press. doi:10.1515/9781400858712.

Buss, Leo W. 2014. "The Evolution of Development." In *The Evolution of Individuality.* doi: 10.1515/9781400858712.27.

Butler, Mitchell T., Qingfeng Wang, and Rasika M. Harshey. 2010. "Cell Density and Mobility Protect Swarming Bacteria against Antibiotics." *Proceedings of the National Academy of Sciences of the United States of America* 107 (8): 3776–3781. doi: 10.1073/pnas.0910934107.

Buttery, Neil J., Chandra N. Jack, Boahemaa Adu-Oppong, Kate T. Snyder, Christopher R.L. Thompson, David C. Queller, and Joan E. Strassmann. 2012. "Structured Growth and Genetic Drift Raise Relatedness in the Social Amoeba Dictyostelium discoideum." *Biology Letters* 8 (5): 794–797. doi: 10.1098/rsbl.2012.0421.

Buttery, Neil J., Daniel E. Rozen, Jason B. Wolf, and Christopher R.L. Thompson. 2009. "Quantification of Social Behavior in D. discoideum Reveals Complex Fixed and Facultative Strategies." *Current Biology* 19 (16): 1373–1377. doi: 10.1016/j.cub.2009.06.058.

Cao, Pengbo, Arup Dey, Christopher N. Vassallo, and Daniel Wall. 2015. "How Myxobacteria Cooperate." *Journal of Molecular Biology* 427 (23): 3709–3721. doi: 10.1016/j.jmb.2015.07.022.

Cao, Pengbo, Xueming Wei, Ram Prasad Awal, Rolf Müller, and Daniel Wall. 2019. "A Highly Polymorphic Receptor Governs Many Distinct Self- Recognition Types within the Myxococcales Order." Edited by Bonnie Bassler. *MBio* 10 (1): 1–15. doi: 10.1128/mBio.02751-18.

Chimileski, Scott, Michael J. Franklin, and R. T. Papke. 2014. "Biofilms Formed by the Archaeon Haloferax volcanii Exhibit Cellular Differentiation and Social Motility, and Facilitate Horizontal Gene Transfer." *BMC Biology* 12 (1): 65–65. doi: 10.1186/s12915-014-0065-5.

Claessen, Dennis, Daniel E. Rozen, Oscar P. Kuipers, Lotte Søgaard-Andersen, and Gilles P. Van Wezel. 2014. "Bacterial Solutions to Multicellularity: A Tale of Biofilms, Filaments and Fruiting Bodies." *Nature Reviews Microbiology* 12 (2): 115–124. doi: 10.1038/nrmicro3178.

Cooper, Guy Alexander, Jorge Peña, Ming Liu, and Stuart Andrew West. 2020. "Mechanisms to Divide Labour in Social Microorganisms." *BioRxiv.* doi: 10.1101/2020.02.25.964643.

Daft, M. J., J. C. Burnham, and Yoko Yamamoto. 1985. "Lysis of Phormidium luridum by Myxococcus fulvus in Continuous Flow Cultures." *Journal of Applied Bacteriology* 59 (1): 73–80. doi: 10.1111/j.1365-2672.1985.tb01778.x.

de Oliveira, Janaina Lima, Atahualpa Castillo Morales, Balint Stewart, Nicole Gruenheit, Jennifer Engelmoer, Suzanne Battom Brown, Reinaldo A. de Brito, et al. 2019. "Conditional Expression Explains Molecular Evolution of Social Genes in a Microbe." *Nature Communications* 10 (1): 3284–3284. doi: 10.1038/s41467-019-11237-2.

Dey, Arup, Christopher N. Vassallo, Austin C. Conklin, Darshankumar T. Pathak, Vera Troselj, and Daniel Wall. 2016. "Sibling Rivalry in Myxococcus Xanthus Is Mediated by Kin Recognition and a Polyploid Prophage." *Journal of Bacteriology* 198 (6): 994–1004. doi: 10.1128/JB.00964-15.

dos Santos, Miguel, Melanie Ghoul, and Stuart A. West. 2018. "Pleiotropy, Cooperation, and the Social Evolution of Genetic Architecture." *PLoS Biology* 16 (10): 1–25. doi: 10.1371/journal.pbio.2006671.

Dworkin, Martin, and Dale Kaiser. 1985. "Cell Interactions in Myxobacterial Growth and Development." *Science* 230 (4721): 18–24. doi: 10.1126/science.3929384.

Ennis, Herbert L., Dee N. Dao, Stefan U. Pukatzki, and Richard H. Kessin. 2000. "Dictyostelium Amoebae Lacking an F-Box Protein Form Spores Rather than Stalk in Chimeras with Wild Type." *Proceedings of the National Academy of Sciences of the United States of America* 97 (7): 3292–3297. doi: 10.1073/pnas.97.7.3292.

Evan, Gerard I., and Karen H. Vousden. 2001. "Proliferation, Cell Cycle and Apoptosis in Cancer." *Nature.* doi: 10.1038/35077213.

Fiegna, Francesca, and Gregory J. Velicer. 2003. "Competitive Fates of Bacterial Social Parasites: Persistence and Self-Induced Extinction of Myxococcus xanthus Cheaters." *Proceedings of the Royal Society B: Biological Sciences* 270 (1523): 1527–1534. doi: 10.1098/rspb.2003.2387.

Fiegna, Francesca, and Gregory J. Velicer. 2005. "Exploitative and Hierarchical Antagonism in a Cooperative Bacterium." *PLoS Biology.* doi: 10.1371/journal.pbio.0030370.

Fiegna, Francesca, Yuen Tsu N. Yu, Supriya V. Kadam, and Gregory J. Velicer. 2006. "Evolution of an Obligate Social Cheater to a Superior Cooperator." *Nature* 441 (7091): 310–314. doi: 10.1038/nature04677.

Fortunato, A., J. E. Strassmann, L. Santorelli, and D. C. Queller. 2003. "Co-Occurrence in Nature of Different Clones of the Social Amoeba, Dictyostelium discoideum." *Molecular Ecology* 12 (4): 1031–1038. doi: 10.1046/j.1365-294X.2003.01792.x.

Foster, Kevin R. 2011. "The Sociobiology of Molecular Systems." *Nature Reviews Genetics.* doi: 10.1038/nrg2903.

Foster, Kevin R., Angelo Fortunato, Joan E. Strassmann, and David C. Queller. 2002. "The Costs and Benefits of Being a Chimera." *Proceedings of the Royal Society B: Biological Sciences* 269 (1507): 2357–2362. doi: 10.1098/rspb.2002.2163.

Foster, Kevin R., Gad Shaulsky, Joan E. Strassmann, David C. Queller, and Chris R.L. Thompson. 2004. "Pleiotropy as a Mechanism to Stabilize Cooperation." *Nature* 431 (7009): 693–696. doi: 10.1038/nature02894.

Fraleigh, Peter C., and Jeffrey C. Burnham. 1988. "Myxococcal Predation on Cyanobacterial Populations: Nutrient Effects." *Limnology and Oceanography* 33 (3): 476–483. doi: 10.4319/lo.1988.33.3.0476.

Frank, Steven A. 1995. "Mutual Policing and Repression of Competition in the Evolution of Cooperative Groups." *Nature* 377: 520–522. doi: 10.1038/377520a0.

Frank, Steven A. 2003. "Perspective: Repression of Competition and the Evolution of Cooperation." *Evolution* 57 (4): 693–705. doi: 10.1111/j.0014-3820.2003.tb00283.x.

Gibbs, Karine A., Mark L. Urbanowski, and E. Peter Greenberg. 2008. "Genetic Determinants of Self Identity and Social Recognition in Bacteria." *Science* 321 (5886): 256–259. doi: 10.1126/science.1160033.

Gilbert, Owen M., Kevin R. Foster, Natasha J. Mehdiabadi, Joan E. Strassmann, and David C. Queller. 2007. "High Relatedness Maintains Multicellular Cooperation in a Social Amoeba by Controlling Cheater Mutants." *Proceedings of the National Academy of Sciences of the United States of America* 104 (21): 8913–8917. doi: 10.1073/pnas.0702723104.

Gilbert, Owen M., Joan E. Strassmann, and David C. Queller. 2012. "High Relatedness in a Social Amoeba: The Role of Kin-Discriminatory Segregation." *Proceedings of the Royal Society B: Biological Sciences* 279 (1738): 2619–2624. doi: 10.1098/rspb.2011.2514.

Grosberg, Richard K., and Richard R. Strathmann. 2007. "The Evolution of Multicellularity: A Minor Major Transition?" *Annual Review of Ecology, Evolution, and Systematics* 38 (1): 621–654. doi: 10.1146/annurev.ecolsys.36.102403.114735.

Gruenheit, Nicole, Katie Parkinson, Balint Stewart, Jennifer A. Howie, Jason B. Wolf, and Christopher R.L. Thompson. 2017. "A Polychromatic 'greenbeard' Locus Determines Patterns of Cooperation in a Social Amoeba." *Nature Communications* 8: 14171. doi: 10.1038/ncomms14171.

Hallatschek, Oskar, Pascal Hersen, Sharad Ramanathan, and David R. Nelson. 2007. "Genetic Drift at Expanding Frontiers Promotes Gene Segregation." *Proceedings of the National Academy of Sciences of the United States of America* 104 (50): 19926–19930. doi: 10.1073/pnas.0710150104.

Hamilton, W. D. 1963. "The Evolution of Altruistic Behavior." *The American Naturalist* 97 (896): 354–356. doi: 10.1086/497114.

Hamilton, W.D. 1964a. "The Genetical Evolution of Social Behaviour. I." *Journal of Theoretical Biology* 7 (1): 17–52. doi: 10.1016/0022-5193(64)90039-6.

Hamilton, W. D. 1964b. "The Genetical Evolution of Social Behaviour. II." *Journal of Theoretical Biology* 7 (1): 17–52. doi: 10.1016/0022-5193(64)90039-6.

Hardin, Garrett. 1968. "The Tragedy of the Commons." *Science* 162 (3859): 1243–1248. doi: 10.1126/science.162.3859.1243.

Harshey, Rasika M. 2003. "Bacterial Motility on a Surface: Many Ways to a Common Goal." *Annual Review of Microbiology* 57 (1): 249–273. doi: 10.1146/annurev.micro.57.030502.091014.

Herron, Matthew D., Joshua M. Borin, Jacob C. Boswell, Jillian Walker, I. Chen Kimberly Chen, Charles A. Knox, Margrethe Boyd, Frank Rosenzweig, and William C. Ratcliff. 2019. "De Novo Origins of Multicellularity in Response to Predation." *Scientific Reports* 9 (1): 2328–2328. doi: 10.1038/s41598-019-39558-8.

Hirose, Shigenori, Rocio Benabentos, Hsing I. Ho, Adam Kuspa, and Gad Shaulsky. 2011. "Self-Recognition in Social Amoebae Is Mediated by Allelic Pairs of Tiger Genes." *Science* 333 (6041): 467–470. doi: 10.1126/science.1203903.

Hirose, Shigenori, Balaji Santhanam, Mariko Katoh-Kurosawa, Gad Shaulsky, and Adam Kuspa. 2015. "Allorecognition, via Tgrb1 and Tgrc1, Mediates the Transition from Unicellularity to Multicellularity in the Social Amoeba Dictyostelium discoideum." *Development (Cambridge)* 142 (20): 3561–3570. doi: 10.1242/dev.123281.

Ho, Hsing I., Shigenori Hirose, Adam Kuspa, and Gad Shaulsky. 2013. "Kin Recognition Protects Cooperators against Cheaters." *Current Biology* 23 (16): 1590–1595. doi: 10.1016/j.cub.2013.06.049.

Ho, Hsing I., and Gad Shaulsky. 2015. "Temporal Regulation of Kin Recognition Maintains Recognition-Cue Diversity and Suppresses Cheating." *Nature Communications* 6 (1): 7144–7144. doi: 10.1038/ncomms8144.

Hollis, Brian. 2012. "Rapid Antagonistic Coevolution between Strains of the Social Amoeba Dictyostelium discoideum." *Proceedings of the Royal Society B: Biological Sciences* 279 (1742): 3565–3571. doi: 10.1098/rspb.2012.0975.

Huss, M. J. 1989. "Dispersal of Cellular Slime Molds by Two Soil Invertebrates." *Mycologia* 81 (5): 677–682. doi: 10.2307/3759871.

Julkowska, Daria, Michal Obuchowski, I. Barry Holland, and Simone J. Séror. 2004. "Branched Swarming Patterns on a Synthetic Medium Formed by Wild-Type Bacillus subtilis Strain 3610: Detection of Different Cellular Morphologies and Constellations of Cells as the Complex Architecture Develops." *Microbiology* 150 (6): 1839–1849. doi: 10.1099/mic.0.27061-0.

Kaiser, D. 2001. "Building a Multicellular Organism." *Annual Review of Genetics* 35 (1): 103–123. doi: 10.1146/annurev.genet.35.102401.090145.

Kaiser, Dale, Mark Robinson, and Lee Kroos. 2010. "Myxobacteria, Polarity, and Multicellular Morphogenesis." *Cold Spring Harbor Perspectives in Biology* 2 (8): a000380–a000380. doi: 10.1101/cshperspect.a000380.

Kalla, Sara E., David C. Queller, Andrea Lasagni, and Joan E. Strassmann. 2011. "Kin Discrimination and Possible Cryptic Species in the Social Amoeba Polysphondylium violaceum." *BMC Evolutionary Biology* 11 (1): 31–31. doi: 10.1186/1471-2148-11-31.

Kawabe, Yoshinori, Qingyou Du, Christina Schilde, and Pauline Schaap. 2019. "Evolution of Multicellularity in Dictyostelia." *International Journal of Developmental Biology* 63 (9–10): 359–369. doi: 10.1387/ijdb.190108ps.

Kessin, Richard H. 2001. *Dictyostelium: Evolution, Cell Biology, and the Development.* Cambridge University Press. Vol. 33. Cambridge University Press. doi: 10.1017/CBO9780511525315.

Kessin, Richard H., Gregg G. Gundersen, Victor Zaydfudim, Mark Grimson, and R. Lawrence Blanton. 1996. "How Cellular Slime Molds Evade Nematodes." *Proceedings of the National Academy of Sciences of the United States of America* 93 (10): 4857–4861. doi: 10.1073/pnas.93.10.4857.

Khare, Anupama, Lorenzo A. Santorelli, Joan E. Strassmann, David C. Queller, Adam Kuspa, and Gad Shaulsky. 2009. "Cheater-Resistance Is Not Futile." *Nature* 461 (7266): 980–982. doi: 10.1038/nature08472.

Kirk, David L. 2003. "Seeking the Ultimate and Proximate Causes of Volvox Multicellularity and Cellular Differentiation." *Integrative and Comparative Biology* 43 (2): 247–253. doi:10.1093/icb/43.2.247.

Kraemer, Susanne A., Sébastien Wielgoss, Francesca Fiegna, and Gregory J. Velicer. 2016. "The Biogeography of Kin Discrimination across Microbial Neighbourhoods." *Molecular Ecology* 25 (19): 4875–4888. doi: 10.1111/mec.13803.

Kumar, Krithika, Rodrigo A. Mella-Herrera, and James W. Golden. 2010. "Cyanobacterial Heterocysts." *Cold Spring Harbor Perspectives in Biology.* doi: 10.1101/cshperspect.a000315.

Kuzdzal-Fick, Jennie J., Sara A. Fox, Joan E. Strassmann, and David C. Queller. 2011. "High Relatedness Is Necessary and Sufficient to Maintain Multicellularity in Dictyostelium." *Science* 334 (6062): 1548–1551. doi: 10.1126/science.1213272.

Lai, Sandra, Julien Tremblay, and Eric Déziel. 2009. "Swarming Motility: A Multicellular Behaviour Conferring Antimicrobial Resistance." *Environmental Microbiology* 11 (1): 126–136. doi: 10.1111/j.1462-2920.2008.01747.x.

Levin, S. R., D. A. Brock, D. C. Queller, and J. E. Strassmann. 2015. "Concurrent Coevolution of Intra-Organismal Cheaters and Resisters." *Journal of Evolutionary Biology* 28 (4): 756–765. doi: 10.1111/jeb.12618.

Li, Si I., and Michael D. Purugganan. 2011. "The Cooperative Amoeba: Dictyostelium as a Model for Social Evolution." *Trends in Genetics* 27 (2): 48–54. doi: 10.1016/j.tig.2010.11.003.

Manhes, Pauline, and Gregory J. Velicer. 2011. "Experimental Evolution of Selfish Policing in Social Bacteria." *Proceedings of the National Academy of Sciences of the United States of America* 108 (20): 8357–8362. doi: 10.1073/pnas.1014695108.

Medina, James M., P. M. Shreenidhi, Tyler J. Larsen, David C. Queller, and Joan E. Strassmann. 2019. "Cooperation and Conflict in the Social Amoeba Dictyostelium discoideum." *International Journal of Developmental Biology* 63 (9–10): 371–382. doi: 10.1387/ijdb.190158jm.

Medina, Mónica, Allen G. Collins, John W. Taylor, James W. Valentine, Jere H. Lipps, Linda Amaral-Zettler, and Mitchell L. Sogin. 2003. "Phylogeny of Opisthokonta and the Evolution of Multicellularity and Complexity in Fungi and Metazoa." *International Journal of Astrobiology* 2 (3): 203–211. doi: 10.1017/S1473550403001551.

Mehdiabadi, Natasha J., Chandra N. Jack, Tiffany Talley Farnham, Thomas G. Platt, Sara E. Kalla, Gad Shaulsky, David C. Queller, and Joan E. Strassmann. 2006. "Kin Preference in a Social Microbe." *Nature* 442 (7105): 881–882. doi: 10.1038/442881a.

Mehdiabadi, Natasha J., Marcus R. Kronforst, David C. Queller, and Joan E. Strassmann. 2009. "Phylogeny, Reproductive Isolation and Kin Recognition in the Social Amoeba Dictyostelium purpureum." *Evolution* 63 (2): 542–548. doi: 10.1111/j.1558-5646.2008.00574.x.

Mendes-Soares, Helena, I. Chen Kimberly Chen, Kara Fitzpatrick, and Gregory J. Velicer. 2014. "Chimaeric Load among Sympatric Social Bacteria Increases with Genotype Richness." *Proceedings of the Royal Society B: Biological Sciences* 281: 20140285. doi: 10.1098/rspb.2014.0285.

Müller, Susanne, Sarah N. Strack, Sarah E. Ryan, Daniel B. Kearns, and John R. Kirby. 2015. "Predation by Myxococcus xanthus Induces Bacillus subtilis to Form Spore-Filled Megastructures." *Applied and Environmental Microbiology* 81 (1): 203–210. doi: 10.1128/AEM.02448-14.

Muñoz-Dorado, José, Francisco J. Marcos-Torres, Elena García-Bravo, Aurelio Moraleda-Muñoz, and Juana Pérez. 2016. "Myxobacteria: Moving, Killing, Feeding, and Surviving Together." *Frontiers in Microbiology* 7 (MAY): 1–18. doi: 10.3389/fmicb.2016.00781.

Nadell, Carey D., Kevin R. Foster, and João B. Xavier. 2010. "Emergence of Spatial Structure in Cell Groups and the Evolution of Cooperation." *PLoS Computational Biology* 6 (3): e1000716–e1000716. doi: 10.1371/journal.pcbi.1000716.

Nadell, Carey D., Joao B. Xavier, and Kevin R. Foster. 2009. "The Sociobiology of Biofilms." *FEMS Microbiology Reviews* 33 (1): 206–224. doi: 10.1111/j.1574-6976.2008.00150.x.

Tsutsui, Neil D. 2004. "Scents of Self: The Expression Component of Self/Nonself Recognition Systems." *Annales Zoologici Fennici* 41 (6): 713–727. https://www.jstor.org/stable/pdf/23736139.pdf?refreqid=excelsior%3A226f5010468ee9e0f9c15731a922d942.

Noh, Suegene, Katherine S. Geist, Xiangjun Tian, Joan E. Strassmann, and David C. Queller. 2018. "Genetic Signatures of Microbial Altruism and Cheating in Social Amoebas in the Wild." *Proceedings of the National Academy of Sciences of the United States of America* 115 (12): 3096–3101. doi: 10.1073/pnas.1720324115.

Nunney, Leonard. 2013. "The Real War on Cancer: The Evolutionary Dynamics of Cancer Suppression." *Evolutionary Applications* 6 (1): 11–19. doi: 10.1111/eva.12018.

Ostrowski, Elizabeth A. 2019. "Enforcing Cooperation in the Social Amoebae." *Current Biology* 29 (11): R474–R484. doi: 10.1016/j.cub.2019.04.022.

Ostrowski, Elizabeth A., Mariko Katoh, Gad Shaulsky, David C. Queller, and Joan E. Strassmann. 2008. "Kin Discrimination Increases with Genetic Distance in a Social Amoeba." *PLoS Biology* 6 (11): 2376–2382. doi: 10.1371/journal.pbio.0060287.

Paaby, Annalise B., and Matthew V. Rockman. 2013. "The Many Faces of Pleiotropy." *Trends in Genetics* 29 (2): 66–73. doi: 10.1016/j.tig.2012.10.010.

Pathak, Darshankumar T., Xueming Wei, Alex Bucuvalas, Daniel H. Haft, Dietlind L. Gerloff, and Daniel Wall. 2012. "Cell Contact-Dependent Outer Membrane Exchange in Myxobacteria: Genetic Determinants and Mechanism." *PLoS Genetics* 8 (4): e1002626–e1002626. doi: 10.1371/journal.pgen.1002626.

Pathak, Darshankumar T., Xueming Wei, Arup Dey, and Daniel Wall. 2013. "Molecular Recognition by a Polymorphic Cell Surface Receptor Governs Cooperative Behaviors in Bacteria." *PLoS Genetics* 9 (11): e1003891–e1003891. doi: 10.1371/journal. pgen.1003891.

Patra, Pintu, Christopher N. Vassallo, Daniel Wall, and Oleg A. Igoshin. 2017. "Mechanism of Kin-Discriminatory Demarcation Line Formation between Colonies of Swarming Bacteria." *Biophysical Journal* 113 (11): 2477–2486. doi: 10.1016/j.bpj.2017.09.020.

Pentz, Jennifer T., Tami Limberg, Nicholas Beermann, and William C. Ratcliff. 2015. "Predator Escape: An Ecologically Realistic Scenario for the Evolutionary Origins of Multicellularity." *Evolution: Education and Outreach* 8 (1): 13. doi: 10.1186/ s12052-015-0041-8.

Queller, David C. 2018. *Isolation Mismatch".207–210. This Idea Is Brilliant: Lost, Overlooked, and Underappreciated Scientific Concepts Everyone Should Know.* Edited by Brockman, John. Harper Perennial.

Queller, D. C. 2000. "Relatedness and the Fraternal Major Transitions." *Philosophical Transactions of the Royal Society B: Biological Sciences* 355 (1403): 1647–1655. doi: 10.1098/rstb.2000.0727.

Queller, David C. 2019. "Pleiotropy and Synergistic Cooperation." *PLoS Biology* 17 (6): 3–5. doi: 10.1371/journal.pbio.3000320.

Queller, David C., Eleonora Ponte, Salvatore Bozzaro, and Joan E. Strassmann. 2003. "Single-Gene Greenbeard Effects in the Social Amoeba Dictyostelium discoideum." *Science* 299 (5603): 105–106. doi: 10.1126/science.1077742.

Queller, David C., and Joan E. Strassmann. 2009. "Beyond Society: The Evolution of Organismality." *Philosophical Transactions of the Royal Society B: Biological Sciences* 364 (1533): 3143–3155. doi: 10.1098/rstb.2009.0095.

Radzvilavicius, Arunas L., and Neil W. Blackstone. 2018. "The Evolution of Individuality Revisited." *Biological Reviews* 93 (3): 1620–1633. doi: 10.1111/brv.12412.

Rendueles, Olaya, Peter C. Zee, Iris Dinkelacker, Michaela Amherd, Sébastien Wielgoss, and Gregory J. Velicer. 2015. "Rapid and Widespread de Novo Evolution of Kin Discrimination." *Proceedings of the National Academy of Sciences of the United States of America* 112 (29): 9076–9081. doi: 10.1073/pnas.1502251112.

Riehl, Christina, and Megan E. Frederickson. 2016. "Cheating and Punishment in Cooperative Animal Societies." *Philosophical Transactions of the Royal Society B: Biological Sciences* 371 (1687): 20150090. doi: 10.1098/rstb.2015.0090.

Rosenberg, E., K. H. Keller, and M. Dworkin. 1977. "Cell Density Dependent Growth of Myxococcus xanthus on Casein." *Journal of Bacteriology* 129 (2): 770–777. doi: 10.1128/jb.129.2.770-777.1977.

Ruiz-Trillo, Iñaki, Gertraud Burger, Peter W.H. Holland, Nicole King, B. Franz Lang, Andrew J. Roger, and Michael W. Gray. 2007. "The Origins of Multicellularity: A Multi-Taxon Genome Initiative." *Trends in Genetics* 23 (3): 113–118. doi: 10.1016/j. tig.2007.01.005.

Santorelli, Lorenzo A., Christopher R.L. Thompson, Elizabeth Villegas, Jessica Svetz, Christopher Dinh, Anup Parikh, Richard Sucgang, et al. 2008. "Facultative Cheater Mutants Reveal the Genetic Complexity of Cooperation in Social Amoebae." *Nature* 451 (7182): 1107–1110. doi: 10.1038/nature06558.

Sathe, S., N. Khetan, and V. Nanjundiah. 2014. "Interspecies and Intraspecies Interactions in Social Amoebae." *Journal of Evolutionary Biology* 27 (2): 349–362. doi: 10.1111/jeb.12298.

Sathe, Santosh, Sonia Kaushik, Albert Lalremruata, Ramesh K. Aggarwal, James C. Cavender, and Vidyanand Nanjundiah. 2010. "Genetic Heterogeneity in Wild Isolates of Cellular Slime Mold Social Groups." *Microbial Ecology* 60 (1): 137–148. doi: 10.1007/s00248-010-9635-4.

Senior, B. W. 1977. "The Dienes Phenomenon: Identification of the Determinants of Compatibility." *Journal of General Microbiology* 102 (2): 235–244. doi: 10.1099/00221287-102-2-235.

Shapiro, James A. 1998. "Thinking about Bacterial Populations as Multicellular Organisms." *Annual Review of Microbiology* 52 (1): 81–104. doi: 10.1146/annurev. micro.52.1.81.

Shaulsky, Gad, and Richard H. Kessin. 2007. "The Cold War of the Social Amoebae." *Current Biology* 17 (16): R684–R692. doi: 10.1016/j.cub.2007.06.024.

Singh, Manvir, and Jacobus J. Boomsma. 2015. "Policing and Punishment across the Domains of Social Evolution." *Oikos* 124 (8): 971–982. doi: 10.1111/oik.02064.

Smith, J., Joan E. Strassmann, and David C. Queller. 2016. "Fine-Scale Spatial Ecology Drives Kin Selection Relatedness among Cooperating Amoebae." *Evolution* 70 (4): 848–859. doi: 10.1111/evo.12895.

smith, Jeff, J. David Van Dyken, and Gregory J. Velicer. 2014. "Nonadaptive Processes Can Create the Appearance of Facultative Cheating in Microbes." *Evolution* 68 (3): 816–826. doi: 10.1111/evo.12306.

Smukalla, Scott, Marina Caldara, Nathalie Pochet, Anne Beauvais, Stephanie Guadagnini, Chen Yan, Marcelo D. Vinces, et al. 2008. "FLO1 Is a Variable Green Beard Gene That Drives Biofilm-like Cooperation in Budding Yeast." *Cell* 135 (4): 726–737. doi: 10.1016/j.cell.2008.09.037.

Sternfeld, J. 1979. "Evidence for Differential Cellular Adhesion as the Mechanism of Sorting-out of Various Cellular Slime Mold Species." *Journal of Embryology and Experimental Morphology* 53 (1902): 163–178.

Strassmann, J. E., Y. Zhu, and D. C. Queller. 2000. "Altruism and Social Cheating in the Social Amoeba Dictyostelium discoideum." *Nature* 408 (6815): 965–967. doi:10.1038/35050087.

Strassmann, Joan E. 2016. "Kin Discrimination in Dictyostelium Social Amoebae." *Journal of Eukaryotic Microbiology* 63 (3): 378–383. doi: 10.1111/jeu.12307.

Strassmann, Joan E., Owen M. Gilbert, and David C. Queller. 2011. "Kin Discrimination and Cooperation in Microbes." *Annual Review of Microbiology* 65 (1): 349–367. doi: 10.1146/annurev.micro.112408.134109.

Sugimoto, Hiroki, and Hiroshi Endoh. 2006. "Analysis of Fruiting Body Development in the Aggregative Ciliate Sorogena stoianovitchae (Ciliophora, Colpodea)." *Journal of Eukaryotic Microbiology* 53 (2): 96–102. doi: 10.1111/j.1550-7408.2005.00077.x.

Szathmáry, Eörs, and John Maynard Smith. 1995. "The Major Evolutionary Transitions." *Nature* 374 (6519): 227–232. doi: 10.1038/374227a0.

Tarnita, Corina E., Alex Washburne, Ricardo Martinez-Garcia, Allyson E. Sgro, and Simon A. Levin. 2015. "Fitness Tradeoffs between Spores and Nonaggregating Cells Can Explain the Coexistence of Diverse Genotypes in Cellular Slime Molds." *Proceedings of the National Academy of Sciences of the United States of America* 112 (9): 2776–2781. doi: 10.1073/pnas.1424242112.

Tice, Alexander K., Jeffrey D. Silberman, Austin C. Walthall, Khoa N.D. Le, Frederick W. Spiegel, and Matthew W. Brown. 2016. "Sorodiplophrys stercorea: Another Novel Lineage of Sorocarpic Multicellularity." *The Journal of Eukaryotic Microbiology* 63 (5): 623–628. doi: 10.1111/jeu.12311.

Travisano, Michael, and Gregory J. Velicer. 2004. "Strategies of Microbial Cheater Control." *Trends in Microbiology* 12 (2): 72–78. doi: 10.1016/j.tim.2003.12.009.

Vassallo, Christopher N., Pengbo Cao, Austin Conklin, Hayley Finkelstein, Christopher S. Hayes, and Daniel Wall. 2017. "Infectious Polymorphic Toxins Delivered by Outer Membrane Exchange Discriminate Kin in Myxobacteria." *eLife* 6: 1–24. doi: 10.7554/eLife.29397.

Vassallo, Christopher, Darshankumar T. Pathak, Pengbo Cao, David M. Zuckerman, Egbert Hoiczyk, and Daniel Wall. 2015. "Cell Rejuvenation and Social Behaviors Promoted by LPS Exchange in Myxobacteria." *Proceedings of the National Academy of Sciences of the United States of America* 112 (22): E2939–E2946. doi: 10.1073/pnas.1503553112.

Velicer, Gregory J., Lee Kroos, and Richard E. Lenski. 2000. "Developmental Cheating in the Social Bacterium Myxococcus xanthus." *Nature* 404 (6778): 598–601. doi: 10.1038/35007066.

Velicer, Gregory J., and Michiel Vos. 2009. "Sociobiology of the Myxobacteria." *Annual Review of Microbiology* 63 (1): 599–623. doi: 10.1146/annurev.micro.091208.073158.

Vos, Michiel, and Gregory J. Velicer. 2008. "Isolation by Distance in the Spore-Forming Soil Bacterium Myxococcus xanthus." *Current Biology* 18 (5): 386–391. doi: 10.1016/j.cub.2008.02.050.

Vos, Michiel, and Gregory J. Velicer. 2009. "Social Conflict in Centimeter-and Global-Scale Populations of the Bacterium Myxococcus xanthus." *Current Biology* 19 (20): 1763–1767. doi: 10.1016/j.cub.2009.08.061.

Votaw, H. R., and E. A. Ostrowski. 2017. "Stalk Size and Altruism Investment within and among Populations of the Social Amoeba." *Journal of Evolutionary Biology* 30 (11): 2017–2030. doi: 10.1111/jeb.13172.

Wall, Daniel. 2014. "Molecular Recognition in Myxobacterial Outer Membrane Exchange: Functional, Social and Evolutionary Implications." *Molecular Microbiology* 91 (2): 209–220. doi: 10.1111/mmi.12450.

Wall, Daniel. 2016. "Kin Recognition in Bacteria." *Annual Review of Microbiology* 70 (1): 143–160. doi:10.1146/annurev-micro-102215-095325.

Webb, Jeremy S., Michael Givskov, and Staffan Kjelleberg. 2003. "Bacterial Biofilms: Prokaryotic Adventures in Multicellularity." *Current Opinion in Microbiology* 6 (6): 578–585. doi: 10.1016/j.mib.2003.10.014.

West, S. A., A. S. Griffin, and A. Gardner. 2007. "Social Semantics: Altruism, Cooperation, Mutualism, Strong Reciprocity and Group Selection." *Journal of Evolutionary Biology* 20 (2): 415–432. doi: 10.1111/j.1420-9101.2006.01258.x.

West, Stuart A., Stephen P. Diggle, Angus Buckling, Andy Gardner, and Ashleigh S. Griffin. 2007. "The Social Lives of Microbes." *Annual Review of Ecology, Evolution, and Systematics* 38: 53–77. doi: 10.1146/annurev.ecolsys.38.091206.095740.

West, Stuart A., Ashleigh S. Griffin, Andy Gardner, and Stephen P. Diggle. 2006. "Social Evolution Theory for Microorganisms." *Nature Reviews Microbiology* 4 (8): 597–607. doi: 10.1038/nrmicro1461.

White, D., W. Shropshire, and K. Stephens. 1980. "Photocontrol of Development by Stigmatella aurantiaca." *Journal of Bacteriology* 142 (3): 1023–1024. doi: 10.1128/jb.142.3.1023-1024.1980.

Wielgoss, Sébastien, Rebekka Wolfensberger, Lei Sun, Francesca Fiegna, and Gregory J. Velicer. 2019. "Social Genes Are Selection Hotspots in Kin Groups of a Soil Microbe." *Science* 363 (6433): 1342–1345. doi: 10.1126/science.aar4416.

Williams, George C. 1957. "Pleiotropy, Natural Selection, and the Evolution of Senescence." *Evolution* 11 (4): 398–411. doi: 10.1111/j.1558-5646.1957.tb02911.x.

Wolf, Jason B., Jennifer A. Howie, Katie Parkinson, Nicole Gruenheit, Diogo Melo, Daniel Rozen, and Christopher R.L. Thompson. 2015. "Fitness Trade-Offs Result in the Illusion of Social Success." *Current Biology* 25 (8): 1086–1090. doi: 10.1016/j.cub.2015.02.061.

Xavier, Joao B., and Kevin R. Foster. 2007. "Cooperation and Conflict in Microbial Biofilms." *Proceedings of the National Academy of Sciences of the United States of America* 104 (3): 876–881. doi: 10.1073/pnas.0607651104.

Zanette, Lorenzo R.S., Sophie D.L. Miller, Christiana M.A. Faria, Edd J. Almond, Tim J. Huggins, William C. Jordan, and Andrew F.G. Bourke. 2012. "Reproductive Conflict in Bumblebees and the Evolution of Worker Policing." *Evolution* 66 (12): 3765–3777. doi: 10.1111/j.1558-5646.2012.01709.x.

8 Group Transformation
Fruiting Body and Stalk Formation

Cathleen Broersma
School of Natural Sciences,
Massey University, Auckland, New Zealand

Elizabeth A. Ostrowski
School of Natural Sciences,
Massey University, Auckland, New Zealand

CONTENTS

8.1 INTRODUCTION

Dispersal is essential to the life-history of many organisms (Bowler and Benton 2005; Kokko and López-Sepulcre 2006). It can enhance survival when environmental conditions deteriorate, promote outbreeding, and broaden a species' geographic

DOI: 10.1201/9780429351907-10

range. Dispersal often occurs in response to environmental cues, is regulated during ontogeny, and may involve specialized structures or cells (propagules) that promote long-distance travel and survival under adverse conditions.

Dispersal can potentially be accomplished by locomotion, and many single-celled organisms have mechanisms of active locomotion that enable travel over short distances. For example, some prokaryotic and eukaryotic microorganisms can swim using their flagella. Amoebae crawl across surfaces by extending and retracting pseudopods. Some organisms, such as myxobacteria, can glide over surfaces alone or in groups. However, microscopic organisms typically rely on passive mechanisms for dispersal over greater distances. Passive mechanisms include dispersal by wind or water or by hitching a ride on a larger organism. It can be facilitated by a resting state, such as a spore or cyst.

In addition to forming hardy cysts or spores that protect cells in harsh environments, many organisms also produce fruiting bodies that promote dispersal by lifting spores up into the air. Fruiting bodies are produced by diverse taxa, ranging from fungi to bacteria to amoebae. Most have a similar morphology, consisting of some sort of stalk that lifts and supports a spore-producing head, resulting in a lollipop or umbrella morphology. Fruiting bodies can consist of only one or a few spores produced by cell division ("sporocarps") or they can be multicellular, usually produced through aggregation of cells ("sorocarps") (Kang et al. 2017; Spiegel et al. 2017). Formation of single-celled vs multicellular fruiting bodies likely entails different costs and benefits, as do the different ways of achieving these structures, through aggregation or cell division, referred to as "coming together" versus "staying together" (Tarnita, Taubes, and Nowak 2013).

In this chapter, we discuss what is known about the evolution and function of stalked fruiting bodies in taxa that exhibit aggregative multicellularity. We begin by discussing amoeboid organisms throughout the tree of life that form these structures. The convergent evolution of similar morphologies, accomplished through diverse means, suggests that they are adaptations. Nevertheless, while it seems likely that stalked fruiting bodies confer a benefit, exactly what they are an adaptation for, and why they evolved and persist, remains subject to debate. In the later sections, we turn our attention to a large clade of sorocarpic amoebae—the Dictyostelia—that are well-studied for their aggregative multicellularity from both a developmental and an evolutionary perspective.

8.2 AGGREGATIVE FRUITING IS FOUND THROUGHOUT THE EUKARYOTIC TREE OF LIFE

Many taxa that undergo aggregative multicellularity to form fruiting bodies have an amoeboid single-cell state. Thus, these organisms are called "sorocarpic amoebae" or "cellular slime molds." The first sorocarpic amoeba to be discovered, *Dictyostelium mucoroides*, was isolated by Brefeld in 1869. The Dictyostelia, which comprises a large clade within the Amoebozoa, is still the most well-studied of the taxa that undergo this morphological transformation (Shadwick et al. 2009). However, sorocarpic amoebae can be found in five of the six eukaryotic supergroups: Amoebozoa, Opisthokonta, Excavata, Stramenopiles and Rhizaria (Figure 8.1, Chapter 5).

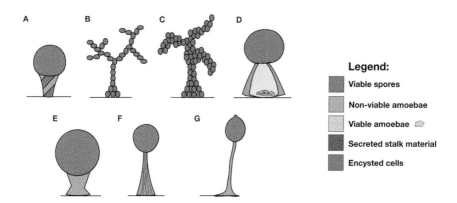

FIGURE 8.1 Fruiting body morphologies throughout the eukaryotic tree of life. (A) *Guttulinops is vulgaris*—Rhizaria, (B) *Acrasis rosea*—Excavata, (C) *Copromyxa protea*—Amoebozoa, (D) *Fonticula alba*—Opisthokonta, (E) *Sorodiplophrys stercorea*—Stramenopiles, (F) *Sorogena stoianovitchae*—Alveolata, (G) *Dictyostelium discoideum*—Amoebozoa. Fruiting bodies are not drawn to scale.

Molecular reconstruction indicates that sorocarpy evolved independently in most of these taxa, as summarized in Brown and Silberman (2013).

Sorocarpic amoebae have a characteristic life cycle, consisting of separate unicellular and multicellular stages, the latter of which is achieved through aggregation. In all cases, the transition to multicellularity involves a switch from a stage of feeding and cell division to one of development and differentiation. When nutrients are abundant, the single-celled amoebae feed on soil microorganisms and increase in number by cell division. Upon starvation, anywhere from a few to hundreds of thousands of amoebae aggregate through chemotaxis and cooperatively form a fruiting body structure, consisting of dispersal propagules and a stalk that supports them. Despite these commonalities, however, there are numerous differences among different types of sorocarpic amoebae in the formation and morphology of their fruiting bodies. Below, we emphasize some notable differences among the different taxa and how these different structures, achieved through different routes, nevertheless result in a stalked fruiting body structure (Figure 8.1).

8.2.1 Guttulinopsidae (Rhizaria)

Species from the genus *Guttulinopsis* are the only known aggregative fruiters in the supergroup Rhizaria, which contains mostly unicellular eukaryotes (Brown et al. 2012). In the most common species, *Guttulinopsis vulgaris,* the stalk is composed of multiple different compartments, some of which contain dead cells, which can hold aloft one or several sori (Figure 8.1A) (Raper, Worley, and Kessler 1977). *Guttulinopsis* thus represents an example of reproductive division of labor; only some cells undertake reproduction and others instead provide non-reproductive, structural support.

8.2.2 Acrasidae (Excavata)

The species *Acrasis rosea* was the first non-dictyostelid sorocarpic amoeba to be discovered (Olive and Stoianovitch 1960). Following aggregation into a mound, a single cell at the base of the mound differentiates into an encysted stalk cell. Cells on top of the stalk cell subsequently encyst as well, resulting in a stalk that extends upward and holds aloft the rest of the population. After stalk formation is complete, the remaining cells in the aggregate form chains and encyst to become spore cells. This results in a fruiting body that resembles a tree-like structure, consisting of a main branch and smaller offshoots (Figure 8.1B). Although all cells in the fruiting body are viable, it is unknown whether there are fitness costs or benefits associated with adopting different positions in the fruiting body, for example, higher up from the ground (Kaushik and Nanjundiah 2003).

8.2.3 Copromyxa (Amoebozoa)

The genus *Copromyxa* consists of only two species, only one of which, *Copromyxa protea,* exhibits aggregative fruiting (Brown, Silberman, and Spiegel 2011). In *C. protea*, fruiting body formation involves aggregation of cells, with those at the apex becoming encysted (Figure 8.1C). Similar to *A. rosea*, the fruiting body consists of a column of cells, and the existence of any position-dependent fitness difference is unknown. Kaushik and Nanjundiah (2003) describe this process as "coming together and sticking to each other," a more primitive evolutionary form of multicellularity compared to other species of sorocarpic amoebae that show complex division of labor, where cells adopt distinct and irreversible cell fates.

8.2.4 Fonticula (Opisthokonta)

Multicellularity evolved multiple times within the Opisthokonta, manifested by different forms: metazoans (animals) and fungi, in which multicellularity arises by cell division from a single starting cell, and aggregative fruiting forms (Brown, Spiegel, and Silberman 2009; Fisher, Shik, and Boomsma 2020; Ruiz-Trillo et al. 2007). Among the latter is the taxon *Fonticula,* which currently consists of only a single species, *Fonticula alba. F. alba* was isolated only once by Olive and Stoianovitch (1960) and never rediscovered, although the original isolate has been retained. Following aggregation into a mound, the cells secrete a Golgi-derived extracellular matrix, forming a hollow volcano-shaped stalk tube. With the exception of a small number of cells that continue to produce stalk material, the cells in the stalk are mechanically forced out of the top of the structure. These cells then encyst, forming a ball of spores suspended in a thin slime sheet, which collapses a few days after maturation (Figure 8.1D) (Brown, Spiegel, and Silberman 2009; Worley, Raper, and Hohl 1979).

8.2.5 Sorodiplophrys (Stramenopiles)

Little has been published about *Sorodiplophrys stercorea* (Dykstra and Olive 1975; Tice et al. 2016). Similar to other sorocarpic amoebae, however, single-celled amoebae aggregate and form a fruiting body. The final fruiting body structure consists of a small, thick stalk composed of secreted material and dead cells that supports a spherical, golden, mucoid sorus (Figure 8.1E). The presence of dead stalk cells indicates that this organism shows reproductive division of labor as well.

Finally, while many examples of aggregative multicellular fruiting body formation involve sorocarpic amoebae, it is worth noting that there are additional non-amoeboid taxa that undergo cooperative fruiting. Myxobacteria are a clade of prokaryotic organisms that undergo aggregative fruiting in response to starvation. Their fruiting bodies can vary dramatically among different species. For example, some species form stalks that lift up the spores, whereas in others, the stalk is reduced or absent (Velicer and Vos 2009). The best-known species of myxobacteria is *Myxococcus xanthus*, which has been used as a model system for studies on cooperation and conflict (see Chapter 6). The life cycle of *M. xanthus* is similar to that of social amoebae: the soil-dwelling bacteria prey upon other microbes, sometimes cooperatively as a swarm (Mauriello et al. 2010). Upon starvation, the bacteria aggregate into mounds and form fruiting bodies, where only a fraction of the cells differentiate into spores, and others remain as rod-shaped cells or undergo autolysis (Figure 8.1F) (Nariya and Inouye 2008; Varon, Cohen, and Rosenberg 1984). The percentage of cells that become viable spores is much lower than in the eukaryotic species discussed so far, with a non-spore percentage of up to 90% in *M. xanthus*, at least under laboratory conditions (Velicer and Vos 2009). The reason for the variation in fruiting body morphology among these species is not well understood (Velicer and Vos 2009).

Sorogena stoianovitchae (eukaryotic supergroup Alveolata) is unique among ciliates in undergoing aggregative fruiting (Olive and Blanton 1980; Sugimoto and Endoh 2006). In its unicellular stage, it feeds on other ciliate species. Upon food shortage, however, it aggregates beneath the water surface and forms an aerial fruiting body. The stalk is produced via collective secretion of a mucous material by the entire population. The stalk lifts the population out of the water, after which each of the cells become encysted and together form a sorus (Sugimoto and Endoh 2006).

8.3   FRUITING BODY FORMATION IN THE AMOEBOZOA

Within the Amoebozoa, the group historically known as the Eumycetozoa (true slime molds) consists of three major groups of organisms: the protostelids, myxogastrids, and dictyostelids. The latter two groups are monophyletic, whereas molecular analyses indicate that protostelids are not. For this reason, these organisms are now sometimes referred to as "protostelioid amoebae" rather than "protostelids," to emphasize common elements of their morphology in lieu of a phylogenetic classification (Shadwick et al. 2009). Protostelioid amoebae undergo sporocarpic development (Spiegel et al. 2017). The amoeba secretes an extracellular matrix, which forms a stalk that lifts the amoeba up. The amoeba then differentiates into a spore, sometimes following cell division (Furtado and Olive 1971; Lahr et al. 2011; Spiegel et al. 2017). Thus, sporocarpy in protostelids results in the production of microscopic fruiting bodies that contain only one or a few cells.

The plasmodial slime molds belong to the monophyletic group the Myxogastrea, another taxon of sporocarpic amoeba. Plasmodial slime molds are named for the slimy structure they produce—called a plasmodium—which forms when amoeboid cells undergo repeated rounds of mitosis without cytokinesis. This process results in a single, massive, multi-nucleated cell with a continuous cytoplasm, which can reach many meters in size. When conditions turn bad, the plasmodium produces masses of stalked fruiting bodies, a process that occurs not through growth, but through rearrangement of the existing biomass (Stephenson and Schnittler 2017). Thus, they also form fruiting bodies, albeit not through aggregation. The spores are mostly wind-dispersed and germinate to form the plasmodium again. Plasmodial slime molds were the likely inspiration for the 1950s horror film *The Blob*. Like protostelids, the stalks produced by plasmodial slime molds are acellular (i.e., secreted). Unlike the protostelids, however, their fruiting bodies are macroscopic.

Finally, the dictyostelids consist of more than 160 species (Romeralo et al. 2011). Their phylogenetic tree contains many long, unbroken branches, which suggests that they have been undersampled and that the true diversity of the group is even greater (Romeralo et al. 2011). Although the phylogeny has been revised over the years, recent phylogenies based on SSU rRNA and alpha-tubulin sequences group dictyostelids into two major clades, the Dictyosteliales and the Acytosteliales, each of which is composed of two groups—resulting in groups 1–4, referred to below. These groups are then further subdivided (e.g., into groups 2A and 2B) (Romeralo et al. 2011). The model organism *Dictyostelium discoideum*, discussed in detail below, belongs to group 4.

Comparative analyses indicate that the formation of a stalked fruiting body is conserved within dictyostelids (Heidel et al. 2011; Romeralo et al. 2011, 2013; Schaap et al. 2006; Sucgang et al. 2011). Romeralo et al. (2013) combined genetic data from 99 species with phenotypic data based on 24 traits. This work suggests that the last common ancestor of dictyostelids (~0.6–1.0 billion years ago) formed fruiting body structures that lift spores up in the air. However, as we emphasize below, dictyostelids show substantial variation in the formation and appearance of their fruiting bodies. The evolutionary drivers of such diversity in fruiting structures are still being investigated. Nevertheless, this diversity makes this group suitable to study the function of a stalk, its associated costs and benefits, and the possible functional constraints on this structure.

8.4 MORPHOLOGICAL VARIATION AMONG DICTYOSTELIDS, WITH EMPHASIS ON STALK FORMATION

In this section, we discuss fruiting body formation and function from the perspective of the model organism *Dictyostelium discoideum*. We start by describing the life cycle and stalk formation in *D. discoideum*. We then describe some of the morphological variation within dictyostelids, particularly in when and how they form their stalked fruiting bodies.

D. discoideum is a soil-dwelling amoeba frequently isolated from the upper layer of the soil of deciduous forests in the temperate zone (Landolt, Stephenson, and

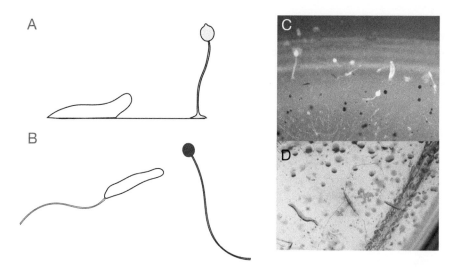

FIGURE 8.2 Stalked versus stalkless migration among dictyostelid species. (A) *D. discoideum* and (B) *D. purpureum,* both group 4 species. *D. discoideum* slugs lack a stalk and lie flat on the substrate, with their tips in the air. The slugs later transform into yellow fruiting bodies that have an upright, sturdy stalk and a flattened disk of cells at the base. *D. purpureum* slugs have a stalk at the rear of the slug that is lengthened throughout the migration period. The fruiting bodies consist of purple sori atop a slender stalk, and they lack a basal disk. (C) *D. discoideum* and *D. purpureum* co-occurring on a soil isolation plate. The pale yellow fruiting bodies are *D. discoideum*, whereas the dark fruiting bodies are *D. purpureum*. (D) A different soil isolation plate, showing an unidentified dictyostelid species with stalked migration.

Cavender 2006; Swanson, Vadell, and Cavender 1999). Its life cycle is broadly similar to many of the sorocarpic amoebae described in previous sections. However, its developmental cycle is more complex, as it involves division of labor, extensive cell-cell communication and coordinated cell death. *D. discoideum* initiates aggregation in response to deteriorating environmental conditions. Starving cells cease phagocytosis, secrete extracellular cyclic adenosine monophosphate (cAMP), and respond chemotactically to sources of cAMP, which results in cell streaming and aggregation to form a mound. In the mound, cells initially differentiate into either pre-stalk or pre-spore cells, indicative of their eventual cell fates in the later fruiting body (see Chapter 5). The pre-spore cells move to the top of the mound, forming a tip that elongates into a finger-like structure that falls to the surface. The resulting worm-like structure, called a slug, migrates away from the site of aggregation (Figure 8.2). Following migration and upon detecting cues such as overhead light, the slug transforms into a multicellular fruiting body. Cells at the anterior of the slug undergo apoptosis, having vacuolized and hardened to form a rigid, dead cellular stalk. The remaining cells move to the top of the stalk and differentiate into viable spores, which disperse and germinate to release single-celled amoebae (Figure 8.1G). In addition to the fruiting body stage, which results in asexually produced spores, *D. discoideum* also has a multicellular sexual stage. The sexual stage also involves aggregation and cell sacrifice. The final result is a durable structure, called a macrocyst, that is not stalked (Bloomfield 2010, 2011; see Chapter 5). Meiosis takes place during the formation of the macrocyst, and the amoebae that later emerge are recombinants.

D. discoideum is a model system for cell biology, developmental biology, chemotaxis, and host-pathogen interactions (reviewed in Bozzaro 2019; Williams 2010). It is genetically tractable, has a precise 24-hour development cycle, and differentiation results in a small number of distinct cell types. *D. discoideum* is also notable for its relatively stable cell-type proportions: approximately 80% of cells in the posterior of the slug form the spore-containing sorus, whereas ~20% of cells in the anterior die and form the stalk. These cell-type proportions can partially re-establish following perturbations, for example, by ablation of either the anterior (prestalk) or posterior (prespore) sections of the slug (Ràfols et al. 2001; Raper 1940). The robustness of its spore-stalk cell proportions is of interest to both developmental biologists who focus on how multicellular organisms achieve and maintain specific cell-type proportions, as well as evolutionary biologists interested in how and why some cells die to promote the fitness of other cells.

While *D. discoideum* is by far the best studied of the social amoebae species, it exhibits a variety of traits that are somewhat uncommon among dictyostelids. Below, we focus on three morphological traits, namely stalk composition, slug morphology, and fruiting body morphology, comparing *D. discoideum* to other dictyostelid species. Although it is difficult to ascertain the adaptive significance of this variation in morphology, we discuss functional implications of the different structures and some of their potential costs and benefits.

8.4.1 Cellular vs Acellular Stalks

Aggregative multicellularity potentially allows unrelated cells to collaborate to form a multicellular individual. It therefore presents opportunities for conflict, especially if there are fitness costs associated with adoption of a particular cell fate. This problem is particularly severe in social amoebae, because the cells that form the stalk die, whereas the remainder live, thus providing a potentially large fitness advantage to strains that avoid the stalk fate. Conflict is thus thought to emerge over which cells will adopt the dead-cell fate and which will survive into the next generation. Because different genotypes can co-aggregate, selection has the opportunity to favor genotypes that behave selfishly by avoiding the stalk fate (Ostrowski 2019).

Stalk formation, however, does not necessarily require self-sacrifice. For example, the acytostelids (group 2A) have acellular stalks, consisting of a hollow tube that is made from secreted cellulose, with all cells subsequently forming viable spores atop the stalk (Mohri et al. 2013). In contrast, cellular stalks are made from an inner layer of hardened vacuolized stalk cells and an outer layer of cellulose (Gezelius 1959). Why some species evolved cellular stalks, the formation of which depends on the death of a fraction of the population, while others form stalked fruiting bodies without cell sacrifice, is not known.

At present, we can only speculate about why these differences might have evolved. One possibility is that cellular stalks provide stronger support, allowing larger aggregate sizes and taller stalks that support more spores. For example, acytostelids have smaller aggregates and smaller structures compared to species with cellular stalk formation (170–1200 mm versus 1200–8200 mm), which might be consistent with a weaker stalk in the former (Raper 1956b; Schaap et al. 2006). Additionally, Kaushik and Nanjundiah (2003) point out that production of an acellular stalk might be energetically costly to the cells and therefore detract from their ability to survive for long periods. The division of labor achieved through formation of a cellular stalk might thus entail benefits for the survival of the spores.

One possible benefit of an acellular stalk is the ability to produce a fruiting body with a smaller population. This potential benefit is apparent in *Dictyostelium lacteum*, the only species known to be capable of producing both cellular and acellular stalks. When food availability is low, the species forms a small, acellular stalk, similar to that produced by acytostelids; only at higher cell numbers is a larger, cellular stalk formed (Bonner and Dodd 1962; Bonner 1982). Currently, *D. lacteum* is the only species known to be plastic for acellular versus cellular stalk formation, but it is possible that other dictyostelids possess similar plasticity but remain to be discovered, or their plasticity has simply not been noticed. Finally, although some have speculated that acytostelids could represent an intermediate stage in the evolutionary transition from simple (e.g., single cell type, no division of labor) to complex multicellularity (multiple cell types with division of labor) (Bonner 2003; Olive 1975), a molecular study by Schaap et al. (2006) concluded that the most recent common ancestor of the dictyostelids likely already displayed cellular stalk formation, suggesting that acellular stalk formation is a derived trait.

8.4.2 Stalked Migration

In some dictyostelid species, aggregation of cells leads to the formation of a slug that migrates away from the point of aggregation (Figure 8.2). For example, in *D. discoideum*, the slug forms approximately 12 hours after the onset of starvation, and it can travel long distances, up to 6 cm in a week (Jack et al. 2011, 2015). The slugs are strongly phototactic, moving towards a directional light source. Migration ceases once light is overhead, which triggers culmination to form a fruiting body. The combination of attractants (light and heat) and repellents (high ammonia levels) is thought to direct slugs upwards through the soil, into an open area suitable for fruiting body formation (Bonner and Lamont 2005; Raper 1984).

Slug migration is thought to have evolved several times in the major groups of the dictyostelids (Romeralo et al. 2013). In *D. discoideum*, the stalk is not formed until after slug migration, during the final stages of development (Figure 8.2A). However, in the majority of dictyostelid species, a stalk is continuously produced from the rear of the slug during its migration (Figure 8.2B). In these species, the fraction of cells that will die and form the stalk is not constant, but depends on the distance traveled by the slug. Ancestral trait reconstruction suggests that the last common ancestor of the dictyostelids was likely a stalked migrator, with only a few species (*D. discoideum, D. polycephalum, D. citrinium, D. intermedium* and *D. dimigraformum*) having evolved stalkless migration (Romeralo et al. 2013; Schaap et al. 2006, Schaap 2007).

Several studies have addressed the potential costs and benefits of stalked versus stalkless migration (Bonner 1982; Gadagkar and Bonner 1994; Jack et al. 2011).

Jack et al. (2011) quantified the costs of slug migration in *D. purpureum* (a stalked migrating species) and *D. discoideum* (which does not undergo stalked migration). Both species showed a tradeoff between migration and sporulation, in that they show a similar decrease in sporulation after accounting for differences in migration distance. The authors suggest that the two species could have adapted their behavior to different stages of the life cycle. As *D. purpureum* produces much taller fruiting bodies than *D. discoideum* (>7 mm versus 3–7 mm), the authors speculate that *D. purpureum* invests less in migration (active dispersal) and more in fruiting body size (passive dispersal). In contrast, *D. discoideum* invests relatively more in active dispersal so it can reach suitable fruiting locations further away, but produces a shorter fruiting body structure.

That the stalk might serve a different or additional purpose in different dictyostelids was proposed by Bonner (1982) and later tested by Gilbert et al. (2012). They showed that the stalked migrator *D. giganteum* could use its stalk as a bridge to traverse small gaps in the substratum. In contrast, slugs of the stalkless migrator *D. discoideum* were not able to traverse these same gaps, suggesting one potential advantage of stalked migration. Regardless of some potential benefits associated with stalked migration, it is surprising that cells co-aggregate to form a slug with no guarantee as to how large of a stalk—and thus, how big of a cell sacrifice—will be made.

Because the costs of stalk formation are likely to be greater for stalked migrators like *D. purpureum*, one might expect that levels of relatedness within cooperative fruiting bodies would likewise need to be greater. Indeed, while both *D. discoideum* and *D. purpureum* possess mechanisms of kin discrimination, whereby cells of different strains separate out during chimeric development (Mehdiabadi et al. 2006; Ostrowski et al. 2008), *D. purpureum* shows stronger kin discrimination than *D. discoideum*. It would be interesting to know whether *D. purpureum* has a stronger history of selection on the genes that underpin its kin discrimination. These genes have been identified in *D. discoideum* (Hirose et al. 2011), but are not yet known in *D. purpureum*.

8.4.3 CLUSTERING AND BRANCHING PATTERNS

In *D. discoideum,* each aggregate gives rise to a fruiting body, which consists of a relatively thick, non-branched stalk that holds aloft a single sorus. This morphology is common among species in group 4, but outside of this group, there is a large variety of structures that differ in their degree of clustering and branching (Figure 8.3). In some species, secondary tips form after aggregation, giving rise to multiple, closely spaced fruiting bodies (referred to as "gregarious" development) or fruiting body structures with multiple sori emanating from one stalk ("like flowers in a vase"; Raper 1956). Fruiting bodies can also be branched and/or consist of whorls (Baldauf and Strassmann 2017; Schaap et al. 2006). The extent of the branching and clustering is also plastic, as it can depend on cell density (Bonner and Dodd 1962; Romeralo et al. 2013). The general pattern was that larger structures—i.e., branched and clustered—tend to form in response to high cell density, whereas unbranched and solitary structures emerge at low cell density.

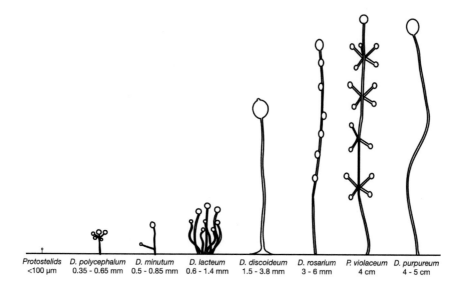

Protostelids	D. polycephalum	D. minutum	D. lacteum	D. discoideum	D. rosarium	P. violaceum	D. purpureum
<100 μm	0.35 - 0.65 mm	0.5 - 0.85 mm	0.6 - 1.4 mm	1.5 - 3.8 mm	3 - 6 mm	4 cm	4 - 5 cm

FIGURE 8.3 Examples of clustering and branching patterns in the fruiting bodies of proto-stelids and dictyostelids. Approximate fruiting body height is indicated below each species, based on descriptions in Raper (1984), except for *P. violaceum*, where the value listed is the mean stalk length from Romeralo et al. (2013). Note that fruiting body size can vary substantially depending on plating conditions, so only an approximate range is provided.

8.5 ALTRUISM, STALK FORMATION, AND THE MAINTENANCE OF MULTICELLULARITY IN *DICTYOSTELIUM DISCOIDEUM*

In previous sections, we discussed numerous examples of aggregative multicellularity that result in the formation of a stalked fruiting body. We emphasized that, in many cases, many or all the cells in the fruiting body remain viable, although there could be fitness costs associated with exactly which role is adopted by a given cell. We also emphasized that some species form fruiting bodies with secreted, non-cellular stalks, and others form extensive cellular stalks throughout migration, resulting in a potentially large and unpredictable degree of cell sacrifice.

The formation of cellular stalks by dictyostelids is of special interest to evolutionary biologists. Stalk formation is likely altruistic, in that some cells give up opportunities for direct fitness in order to form a structure that appears to benefit the rest. This differentiation into spore and stalk cells is analogous to the differentiation into soma and germline that is seen in complex multicellularity. Stalk formation is a clear example of reproductive division of labor, where some cells specialize in reproduction, and others specialize in non-reproductive functions.

Certain features of the *Dictyostelium* life cycle mean that its reproductive division of labor might be evolutionarily fragile. Aggregative multicellularity potentially permits multiple different genotypes to co-aggregate and form chimeric multicellular structures. This genetic diversity, in combination with strong fitness consequences for becoming stalk versus spore, means that natural selection can operate during this stage of the life cycle. Thus, all else being equal (and it may not be), natural selection should favor genetic variants that can avoid the costly role of the stalk and disproportionately adopt the high-fitness spore fate. The problems posed by chimerism in *Dictyostelium* and its potential consequences for the evolution of multicellularity are discussed more fully in Chapter 7. Given these opportunities for selection to favor stalk-avoiders, one long-standing question is the extent to which opportunities for conflict may have influenced how multicellularity evolves (e.g., whether aggregative multicellularity is successful) and whether it is evolutionarily stable. In *Dictyostelium*, this possibility has led to interest in whether stalkless forms might evolve in nature.

8.5.1 HUNTING FOR STALKLESS STRAINS IN NATURE

There have been limited attempts to find stalkless strains in nature. Buss (1982) reported the existence of a stalkless *Dictyostelium* strain. While growing and isolating *Dictyostelium* fruiting bodies from soil samples, he observed two distinct fruiting body morphologies. One strain showed a standard stalked morphology, whereas the other produced a ball of spores directly on the substratum, without a stalk. Buss described the latter strain as a "somatic cell parasite," because it was capable of "reproducing itself and spreading infectiously" at the expense of the rest of the population (Buss 1982). To study whether such a stalkless strain could be maintained in the population over multiple generations, Buss co-cultured the stalked and stalkless strain over a number of generations. The stalkless strain increased in frequency and, if it started at a sufficiently high frequency in the population, it could take over. However, when the stalkless strain was mixed with a different set of strains isolated from the same soil sample, no chimeric fruiting bodies were observed, indicative of some sort of recognition mechanism present in these strains to distinguish self from non-self. Unfortunately, the stalkless strain was subsequently lost (Buss, pers. communication), making it impossible to study it further. This means that it is not possible to carry out molecular analyses to establish which dictyostelid species it was or to identify the genetic changes responsible for its unusual morphology.

More recently, Gilbert et al. (2007) attempted to isolate stalkless strains from natural populations. Unfortunately, dictyostelid fruiting bodies are too small and infrequent to find simply by directly examining soil samples. Moreover, finding stalkless strains—given that they do not form the recognizable macroscopic fruiting body—is even more challenging. However, previous work showed that fruiting bodies can sometimes be observed on animal dung pellets (Raper 1984; Stephenson and Landolt 1992). Gilbert and colleagues collected dung pellets, brought them back to the lab, and incubated them with or without additional bacteria for food until fruiting bodies formed (Gilbert et al. 2007). They then collected the spores from 95 of these fruiting bodies and plated them at low density on a lawn of bacteria. Under these conditions, well-spaced spores germinate and divide to produce circular plaques—clearings in the bacterial lawn where the amoebae have devoured the prey. At the center of each plaque, where the amoeba cell density is high and food has been depleted, fruiting bodies will form if the strain is capable of multicellularity. Thus, dilution plating of spores to see whether they give rise to fruiting bodies is a way to screen for the presence of strains that have lost multicellularity, despite previously joining a stalked fruiting body with others, or that exhibit other morphological alterations. However, despite screening >3,300 plaques, Gilbert and colleagues observed no stalkless morphologies.

To our knowledge, the works by Buss (1982) and Gilbert et al. (2007) are the only studies that have attempted to identify and/or quantify the frequency of stalkless strains in nature. In the future, as new methods for single-cell genomics improve, it might be possible to use culture-independent methods to isolate, sequence, and identify each amoeba cell in a soil sample, enabling identification of natural isolates that have lost multicellular development or evolved novel morphologies not currently recognized. For now, however, whether *D. discoideum*'s aggregative fruiting can lead to selfishness that threatens the maintenance of the stalk in nature remains unknown.

8.5.2 LABORATORY AND THEORETICAL STUDIES OF STALKLESSNESS

The observation by Buss (1982) of a stalkless morphology motivated laboratory studies of stalk-avoiding mutants as well as theory to address the circumstances under which stalklessness might evolve and persist. For example, two studies examined an insertion mutant (*fbxA*⁻ mutant) that contributes less to the stalk when in chimerae with the wild-type strain (Ennis et al. 2000; Nelson et al. 2000). When developed clonally, the mutant forms aberrant fruiting bodies that contain few to no spores or fails to initiate stalk production altogether (Ennis et al. 2000; Gilbert et al. 2007). However, when co-developed with the wild-type strain, it produces a disproportionate fraction of the spores. Ennis and colleagues (2000) speculated that the *fbxA* gene takes part in the regulation of a complex involved in the differentiation into spore or stalk cells, where deletion of the gene causes the stalk cell differentiation pathway to be halted.

Gilbert and colleagues (2007) subsequently used the *fbxA*⁻ mutant to examine the extent to which such a strain that does not contribute fairly to the stalk might increase in frequency in a population, owing to its advantage in spore production in chimerae. Such a strain may face a disadvantage at high frequency, if it has displaced the very strain it relies on to sporulate. The impacts of these different frequencies can be quantified as relatedness, which encompasses the degree to which the mutant interacts with self ($r = 1$) or with the stalk-proficient wild-type ($r = 0$) to build a fruiting body. The authors found that when $r > 0.75$, the *fbxA*⁻ mutant decreased in frequency, indicating net selection against the mutant. This work demonstrated that sufficiently high relatedness could be essential for preventing the invasion and takeover of populations by non-stalk-forming strains.

The above empirical examples suggest that stalk-avoiding strains might be selected against when relatedness is high—but is relatedness in natural populations sufficiently high to accomplish this? Relatively little is known about relatedness in nature, especially over the small spatial scales in soil where different strains might encounter one another and co-develop to form chimeric fruiting bodies. Fortunato and colleagues genotyped natural isolates from minute soil samples collected using a plastic straw with a diameter of 6 mm (Fortunato et al. 2003). Of 26 soil samples that contained *Dictyostelium*, 63% yielded more than one genotype, with as many as nine distinct genotypes present in a single soil sample. These results yielded an estimate of average genetic relatedness of 0.52. In addition, Gilbert collected and genotyped 88 individual fruiting bodies that emerged from 25 dung piles incubated in the lab (Gilbert et al. 2007). Seventy-seven percent of the fruiting bodies contained only a single genotype, which yielded a minimum relatedness of $r = 0.86$, supporting their hypothesis that relatedness might be high enough to select against the *fbxA*⁻ mutant in nature (i.e., 0.86>0.75).

The hypothesis that high genetic relatedness would be sufficient to stop the spread of non-stalked mutants was further supported by Kuzdzal-Fick et al. (2011). They carried out laboratory evolution experiments that involved multicellular development under either low or high relatedness conditions (i.e., in genetically diverse or clonal groups, respectively). Approximately one-third of the populations evolved at low relatedness harbored strains that could not form a stalked fruiting body when developed clonally, but were nonetheless disproportionately represented among the spores when co-developed with their ancestor. Conversely, no losses of multicellularity occurred in the populations evolved under high relatedness. Taken together, these experiments demonstrate that high relatedness may be an important condition for the maintenance of the stalked fruiting body in this organism.

Hudson et al. (2002) developed a mathematical model to address the evolutionary stability of stalk formation. The model made two assumptions about how stalks influence fitness: (1) the fitness of a completely stalkless strain would be low and (2) dispersal success increases with stalk size. They showed that stalk formation could still be maintained in populations founded by genetically unrelated individuals. However, they also showed that selfish genotypes, those that contribute less to the stalk, could drive the evolution of suboptimal stalk sizes—that is, reductions in the allocation to the stalk relative to what would be optimal, assuming that dispersal is an important fitness component. Similarly, a model by Brännström and Dieckmann (2005) showed that it was possible to obtain coexistence of multiple genotypes (e.g., stalked and stalkless) within a single population. Taken together, these experiments suggest that one way to identify stalkless strains in nature might be to look for them under natural conditions where relatedness is low (i.e., genetic diversity is high) or the importance of dispersal is low. Moreover, while it remains unknown how essential stalk formation is for dispersal (as opposed to simply protecting the spores by lifting them in the air), one study did show that spores from an intact, stalked fruiting body were more likely to be acquired by an insect vector than those that were placed directly on the substrate; the spores also survived passage in the insect gut when ingested (smith, Queller, and Strassmann 2014). This finding thus provides some support for the role of the stalk in facilitating spore dispersal.

8.5.3 NATURALLY OCCURRING VARIATION IN STALK SIZE

The model by Hudson et al. (2002) suggested that the size of the stalk in *D. discoideum* could reflect a tradeoff between its cost to the cells that die and its benefit to the surviving spores, and thus that its size may be sub-optimal, at least for dispersal. Similarly, the study by Brännström and Dieckmann (2005) supported the idea that within-population polymorphisms could arise, and thus we might expect to observe natural variation in the degree of stalk investment. To what extent is there evidence that this variation exists?

Within *D. discoideum*, only a few studies have looked at natural variation in overall stalk or fruiting body size, as well as equity of spore-stalk allocation between strains (Buttery et al. 2009; Votaw and Ostrowski 2017). Buttery et al. (2009) estimated spore-stalk allocation in six strains from a site in North Carolina. Spore allocation was estimated in two ways: from images of individual fruiting bodies, followed by estimation of relative sorus and stalk volume, and by comparing the spore production of strains clonally and following development in pairwise chimerae, under the assumption that more spores indicate less allocation to stalk. Using the latter approach, they observed variation in spore allocation (up to 2.8–fold difference) among strains (Buttery et al. 2009).

Votaw and Ostrowski (2017) looked at variation among strains, among sites, and in clonal versus pairwise chimerae using strains from two geographically distant sites, one in North Carolina and one in Texas. In addition to imaging and measuring stalk height and spore number, GFP reporter strains were used to estimate spore-stalk allocation. They observed variation in stalk allocation within both populations, but also larger fruiting bodies for Texas strains compared to strains from North Carolina. These results underscore that larger stalks can be accomplished through two routes: either by allocating more cells to the stalk or by making larger fruiting bodies altogether. In addition, variation in relative stalk size was observed in both populations, similar to the findings of Buttery et al. (2009). These studies together indicate some polymorphism in clonal and chimeric spore-stalk allocation among strains within a site, but also that the morphology of the fruiting bodies can evolve divergently among sites.

Overall, the studies of altruistic stalk formation by *D. discoideum*, clonally and in chimerae, illustrate the potential for conflict in organisms that undergo aggregative multicellularity with division of labor. Aggregation can lead to genetic diversity within the multicellular organism, providing fuel for natural selection, and division of labor generates a strong competitive advantage to cells that avoid the dead stalk fate. Studies in this organism thus help to validate the predictions of evolutionary theory about the problems of aggregative multicellularity.

8.6   CONCLUSIONS

Aggregative fruiting has independently evolved in five of the six supergroups of the Eukarya, suggesting that the formation of a stalk is a morphological adaptation. Nevertheless, there is a substantial diversity in how these structures are formed— whereas some stalks are composed of dead cells (and thus involve cell sacrifice), others form by secretion, such that all cells potentially survive as spores. Differences in how stalks are formed in organisms that undergo aggregative multicellularity has important implications for the evolutionary maintenance of this trait.

Substantial variation in how stalked fruiting bodies form exists within dictyostelids alone, a large clade consisting of more than 150 species within the Amoebozoa. Here we described variation among species in the composition of the stalk (acellular versus cellular), the timing of its production (during or after migration), and its branching morphology (branched or unbranched, whorled or not). Unfortunately, the explanations for the variation in these features are not known, although several studies provide information about some of the potential functions of the stalk. For example, spores that sat atop stalks were more likely to be picked up by an insect vector (smith, Queller, and Strassman 2014), and stalks emanating from the rear of the slug can help in traversing gaps in the soil (Gilbert et al. 2012)—yet, whether and how these features are used in nature remains to be seen. In addition, while some stalk features may provide a fitness advantage, there may also be functional constraints imposed by development or physics. For example, acellular stalks have the benefit of not necessitating cell sacrifice, but these structures may not support as many spores, or the spores may be of lower quality (Kaushik and Nanjundiah 2003).

In those species that form cellular stalks, the death of the cells that form the stalk presents an opportunity for conflict, as strains that avoid forming the stalk and preferentially form spores should have a fitness advantage. In the model organism *D. discoideum*, studies of stalk-avoiding mutants and the behaviors of natural isolates, as well as mathematical models, have all contributed to our understanding of the evolutionary maintenance of altruistic stalk formation. These studies confirm an essential role for relatedness, but whether relatedness is high enough in nature to prevent takeover by stalk-avoiding strains remains uncertain. Future studies would benefit from considering how multicellular morphologies have been impacted by variation in relatedness across populations, as well as of other factors that might also promote the evolutionary maintenance of cooperative multicellularity.

REFERENCES

Baldauf, S. L. & Strassmann, J. E. Dictyostelia. 2017. *Handbook of the Protists*, edited by John M. Archibald, Alastair Simpson and Claudio Slamovits. Springer Press.

Bloomfield, Gareth. 2011. "Genetics of Sex Determination in the Social Amoebae." *Development, Growth & Differentiation* 53 (4): 608–16.

Bloomfield, Gareth, Jason Skelton, Alasdair Ivens, Yoshimasa Tanaka, and Robert R. Kay. 2010. "Sex Determination in the Social Amoeba *Dictyostelium discoideum*." *Science* 330 (6010): 1533–36.

Bonner, John Tyler, and Marya R. Dodd. 1962. "Aggregation Territories in the Cellular Slime Molds." *The Biological Bulletin* 122 (1).

Bonner, J. T. 1982. "Evolutionary Strategies and Developmental Constraints in the Cellular Slime Molds." *The American Naturalist* 119 (4): 530–52.

Bonner, John Tyler. 2003. "Evolution of Development in the Cellular Slime Molds." *Evolution and Development* 5 (3): 305–13.

Bonner, J. T., and D. S. Lamont. 2005. "Behavior of Cellular Slime Molds in the Soil." *Mycologia* 97 (1): 178–84.

Bowler, Diana E., and Tim G. Benton. 2005. "Causes and Consequences of Animal Dispersal Strategies: Relating Individual Behaviour to Spatial Dynamics." *Biological Reviews of the Cambridge Philosophical Society* 80 (2): 205–25.

Bozzaro, Salvatore. 2019. "The Past, Present and Future of *Dictyostelium* as a Model System." *The International Journal of Developmental Biology* 63 (8–9–10): 321–31.

Brännström, A., and U. Dieckmann. 2005. "Evolutionary Dynamics of Altruism and Cheating among Social Amoebas." Proceedings of the Royal Society B 272 (1572): 1609–16.

Brefeld, O. 1869. "Ein Neuer Organismus Aus Der Verwandtschaft Der Myxomyceten." *Abhandlungen Der Senckenbergischen Naturforschenden Gesellschaft Frankfurt* 7: 85–107.

Brown, Matthew W., Martin Kolisko, Jeffrey D. Silberman, and Andrew J. Roger. 2012. "Aggregative Multicellularity Evolved Independently in the Eukaryotic Supergroup Rhizaria." *Current Biology* 22 (12): 1123–27.

Brown, Matthew W., and Jeffrey D. Silberman. 2013. "The Non-Dictyostelid Sorocarpic Amoebae." In *Dictyostelids: Evolution, Genomics and Cell Biology*, edited by Maria Romeralo, Sandra Baldauf, and Ricardo Escalante, 219–42. Berlin, Heidelberg: Springer Berlin Heidelberg.

Brown, Matthew W., Jeffrey D. Silberman, and Frederick W. Spiegel. 2011. "'Slime Molds' among the Tubulinea (Amoebozoa): Molecular Systematics and Taxonomy of *Copromyxa*." *Protist* 162 (2): 277–87.

Brown, Matthew W., Frederick W. Spiegel, and Jeffrey D. Silberman. 2009. "Phylogeny of the 'Forgotten' Cellular Slime Mold, *Fonticula alba*, Reveals a Key Evolutionary Branch within Opisthokonta." *Molecular Biology and Evolution* 26 (12): 2699–2709.

Buss, L. W. 1982. "Somatic Cell Parasitism and the Evolution of Somatic Tissue Compatibility." *Proceedings of the National Academy of Sciences of the United States of America* 79 (17): 5337–41.

Buttery, Neil J., Daniel E. Rozen, Jason B. Wolf, and Christopher R. L. Thompson. 2009. "Quantification of Social Behavior in *D. discoideum* Reveals Complex Fixed and Facultative Strategies." *Current Biology* 19 (16): 1373–77.

Dykstra, Michael J., and Lindsay S. Olive. 1975. "*Sorodiplophrys*: An Unusual Sorocarp-Producing Protist." *Mycologia* 67 (4): 873–79.

Ennis, H. L., D. N. Dao, S. U. Pukatzki, and R. H. Kessin. 2000. "*Dictyostelium* Amoebae Lacking an F-Box Protein Form Spores rather than Stalk in Chimeras with Wild Type." *Proceedings of the National Academy of Sciences of the United States of America* 97 (7): 3292–97.

Fisher, R. M., J. Z. Shik, and J. J. Boomsma. 2020. "The Evolution of Multicellular Complexity: The Role of Relatedness and Environmental Constraints." Proceedings of the Royal Society B 287 (1931): 20192963.

Fortunato, A., J. E. Strassmann, L. Santorelli, and D. C. Queller. 2003. "Co-Occurrence in Nature of Different Clones of the Social Amoeba, *Dictyostelium discoideum*." *Molecular Ecology* 12 (4): 1031–38.

Furtado, J. S., and L. S. Olive. 1971. "Ultrastructural Evidence of Meiosis in *Ceratiomyxa fruticulosa*." *Mycologia* 63 (2): 413–16.

Gadagkar, Raghavendra, and J. Bonner. 1994. "Social Insects and Social Amoebae." *Journal of Biosciences* 19: 219–245.

Gezelius, K. 1959. "The Ultrastructure of Cells and Cellulose Membranes in Acrasiae." *Experimental Cell Research* 18 (November): 425–53.

Gilbert, Owen M., Kevin R. Foster, Natasha J. Mehdiabadi, Joan E. Strassmann, and David C. Queller. 2007. "High Relatedness Maintains Multicellular Cooperation in a Social Amoeba by Controlling Cheater Mutants." *Proceedings of the National Academy of Sciences of the United States of America* 104 (21): 8913–17.

Gilbert, Owen M., Jennie J. Kuzdzal-Fick, David C. Queller, and Joan E. Strassmann. 2012. "Mind the Gap: A Comparative Study of Migratory Behavior in Social Amoebae." *Behavioral Ecology and Sociobiology* 66 (9): 1291–96.

Heidel, Andrew J., Hajara M. Lawal, Marius Felder, Christina Schilde, Nicholas R. Helps, Budi Tunggal, Francisco Rivero, et al. 2011. "Phylogeny-Wide Analysis of Social Amoeba Genomes Highlights Ancient Origins for Complex Intercellular Communication." *Genome Research* 21 (11): 1882–91.

Hirose, Shigenori, Rocio Benabentos, Hsing-I Ho, Adam Kuspa, and Gad Shaulsky. 2011. "Self-Recognition in Social Amoebae Is Mediated by Allelic Pairs of *Tiger* Genes." *Science* 333 (6041): 467–70.

Hudson, Richard Ellis, Juliann Eve Aukema, Claude Rispe, and Denis Roze. 2002. "Altruism, Cheating, and Anticheater Adaptations in Cellular Slime Molds." *The American Naturalist* 160 (1): 31–43.

Jack, Chandra N., Neil Buttery, Boahemaa Adu-Oppong, Michael Powers, Christopher R. L. Thompson, David C. Queller, and Joan E. Strassmann. 2015. "Migration in the Social Stage of *Dictyostelium discoideum* Amoebae Impacts Competition." *PeerJ* 3 (October): e1352.

Jack, C. N., B. Adu-Oppong, M. Powers, D. C. Queller, and J. E. Strassmann. 2011. "Cost of Movement in the Multicellular Stage of the Social Amoebae *Dictyostelium discoideum* and *D. purpureum.*" *Ethology Ecology & Evolution* 23 (4): 358–67.

Kang, Seungho, Alexander K. Tice, Frederick W. Spiegel, Jeffrey D. Silberman, Tomáš Pánek, Ivan Cepicka, Martin Kostka, et al. 2017. "Between a Pod and a Hard Test: The Deep Evolution of Amoebae." *Molecular Biology and Evolution* 34 (9): 2258–70.

Kaushik, Sonia, and Vidyanand Nanjundiah. 2003. "Evolutionary Questions Raised by Cellular Slime Mould Development." Proceedings of the Indian National Science Academy-Part B 69: 825–52.

Kokko, Hanna, and Andrés López-Sepulcre. 2006. "From Individual Dispersal to Species Ranges: Perspectives for a Changing World." *Science* 313 (5788): 789–91.

Kuzdzal-Fick, Jennie J., Sara A. Fox, Joan E. Strassmann, and David C. Queller. 2011. "High Relatedness Is Necessary and Sufficient to Maintain Multicellularity in *Dictyostelium.*" *Science* 334 (6062): 1548–51.

Lahr, Daniel J. G., Laura Wegener Parfrey, Edward A. D. Mitchell, Laura A. Katz, and Enrique Lara. 2011. "The Chastity of Amoebae: Re-Evaluating Evidence for Sex in Amoeboid Organisms." Proceedings of the Royal Society B 278 (1715): 2081–90.

Landolt, John C., Steven L. Stephenson, and James C. Cavender. 2006. "Distribution and Ecology of Dictyostelid Cellular Slime Molds in Great Smoky Mountains National Park." *Mycologia* 98 (4): 541–49.

Mauriello, Emilia M. F., Tâm Mignot, Zhaomin Yang, and David R. Zusman. 2010. "Gliding Motility Revisited: How Do the Myxobacteria Move without Flagella?" *Microbiology and Molecular Biology Reviews* 74 (2): 229–49.

Mehdiabadi, Natasha J., Chandra N. Jack, Tiffany Talley Farnham, Thomas G. Platt, Sara E. Kalla, Gad Shaulsky, David C. Queller, and Joan E. Strassmann. 2006. "Social Evolution: Kin Preference in a Social Microbe." *Nature* 442 (7105): 881–82.

Mohri, Kurato, Yu Kiyota, Hidekazu Kuwayama, and Hideko Urushihara. 2013. "Temporal and Non-Permanent Division of Labor during Sorocarp Formation in the Social Amoeba *Acytostelium subglobosum.*" *Developmental Biology* 375 (2): 202–9.

Nariya, Hirofumi, and Masayori Inouye. 2008. "MazF, an mRNA Interferase, Mediates Programmed Cell Death during Multicellular *Myxococcus* Development." *Cell* 132 (1): 55–66.

Nelson, M. K., A. Clark, T. Abe, A. Nomura, N. Yadava, C. J. Funair, K. A. Jermyn, S. Mohanty, R. A. Firtel, and J. G. Williams. 2000. "An F-Box/WD40 Repeat-Containing Protein Important for *Dictyostelium* Cell-Type Proportioning, Slug Behaviour, and Culmination." *Developmental Biology* 224 (1): 42–59.

Olive, L. S. 1975. *The Mycetozoans.* New York: Academic Press.

Olive, Lindsay S., and Carmen Stoianovitch. 1960. "Two New Members of the Acrasiales." *Bulletin of the Torrey Botanical Club* 87 (1): 1–20.

Olive, L. S., and R. L. Blanton. 1980. "Aerial Sorocarp Development by the Aggregative Ciliate, *Sorogena stoianovitchae*." *The Journal of Protozoology* 27 (3): 293–99.

Ostrowski, Elizabeth A. 2019. "Enforcing Cooperation in the Social Amoebae." *Current Biology* 29 (11): R474–84.

Ostrowski, Elizabeth A., Mariko Katoh, Gad Shaulsky, David C. Queller, and Joan E. Strassmann. 2008. "Kin Discrimination Increases with Genetic Distance in a Social Amoeba." *PLoS Biology* 6 (11): e287.

Ràfols, I., A. Amagai, Y. Maeda, H. K. MacWilliams, and Y. Sawada. 2001. "Cell Type Proportioning in *Dictyostelium* Slugs: Lack of Regulation within a 2.5-Fold Tolerance Range." *Differentiation; Research in Biological Diversity* 67 (4–5): 107–16.

Raper, Kenneth B. 1940. "Pseudoplasmodium Formation and Organization in *Dictyostelium discoideum*." *Journal of the Elisha Mitchell Scientific Society* 56 (2): 241–82.

Raper, K. B. 1956. "*Dictyostelium polycephalum* n. sp.: A New Cellular Slime Mould with Coremiform Fructifications." *Journal of General Microbiology* 14: 716–32.

Raper, K. B. 1956b. "Factors Affecting Growth and Differentiation in Simple Slime Molds." *Mycologia* 48: 169.

Raper, Kenneth Bryan. 1984. *The Dictyostelids*. Princeton, NJ: Princeton University Press.

Raper, Kenneth B., Ann C. Worley, and Dietrich Kessler. 1977. "Observations on *Guttulinopsis vulgaris* and *Guttulinopsis nivea*." *Mycologia* 69 (5): 1016–30.

Romeralo, Maria, James C. Cavender, John C. Landolt, Steven L. Stephenson, and Sandra L. Baldauf. 2011. "An Expanded Phylogeny of Social Amoebas (Dictyostelia) Shows Increasing Diversity and New Morphological Patterns." *BMC Evolutionary Biology* 11 (March): 84.

Romeralo, Maria, Anna Skiba, Alejandro Gonzalez-Voyer, Christina Schilde, Hajara Lawal, Sylwia Kedziora, Jim C. Cavender, Gernot Glöckner, Hideko Urushihara, and Pauline Schaap. 2013. "Analysis of Phenotypic Evolution in Dictyostelia Highlights Developmental Plasticity as a Likely Consequence of Colonial Multicellularity." Proceedings of the Royal Society B 280 (1764): 20130976.

Ruiz-Trillo, Iñaki, Gertraud Burger, Peter W. H. Holland, Nicole King, B. Franz Lang, Andrew J. Roger, and Michael W. Gray. 2007. "The Origins of Multicellularity: A Multi-Taxon Genome Initiative." *Trends in Genetics* 23 (3): 113–18.

Schaap, Pauline. 2007. "Evolution of Size and Pattern in the Social Amoebas." *Bioessays* 29 (7): 635–44.

Schaap, Pauline, Thomas Winckler, Michaela Nelson, Elisa Alvarez-Curto, Barrie Elgie, Hiromitsu Hagiwara, James Cavender, et al. 2006. "Molecular Phylogeny and Evolution of Morphology in the Social Amoebas." *Science* 314 (5799): 661–63.

Shadwick, Lora L., Frederick W. Spiegel, John D. L. Shadwick, Matthew W. Brown, and Jeffrey D. Silberman. 2009. "Eumycetozoa = Amoebozoa?: SSUrDNA Phylogeny of Protosteloid Slime Molds and Its Significance for the Amoebozoan Supergroup." *PLoS One* 4 (8): e6754.

Smith, Jeff, David C. Queller, and Joan E. Strassmann. 2014. "Fruiting Bodies of the Social Amoeba *Dictyostelium discoideum* Increase Spore Transport by *Drosophila*." *BMC Evolutionary Biology* 14 (May): 105.

Spiegel, Frederick W., Lora L. Shadwick, George G. Ndiritu, Matthew W. Brown, Maria Aguilar, and John D. Shadwick. 2017. "Protosteloid Amoebae (Protosteliida, Protosporangiida, Cavosteliida, Schizoplasmodiida, Fractoviteliida, and Sporocarpic Members of Vannellida, Centramoebida, and Pellitida)." In *Handbook of the Protists*, edited by John M. Archibald, Alastair G. B. Simpson, Claudio H. Slamovits, Lynn Margulis, Michael Melkonian, David J. Chapman, and John O. Corliss, 1–38. Cham: Springer International Publishing.

Stephenson, Steven L., and John C. Landolt. 1992. "Vertebrates as Vectors of Cellular Slime Moulds in Temperate Forests." *Mycological Research* 96 (8): 670–72.

Stephenson, Steven L., and Martin Schnittler. 2017. "Myxomycetes." In *Handbook of the Protists*, edited by John M. Archibald, Alastair G. B. Simpson, Claudio H. Slamovits, Lynn Margulis, Michael Melkonian, David J. Chapman, and John O. Corliss, 1–27. Cham: Springer International Publishing.

Sucgang, Richard, Alan Kuo, Xiangjun Tian, William Salerno, Anup Parikh, Christa L. Feasley, Eileen Dalin, et al. 2011. "Comparative Genomics of the Social Amoebae *Dictyostelium discoideum* and *Dictyostelium purpureum*." *Genome Biology* 12 (2): R20.

Sugimoto, Hiroki, and Hiroshi Endoh. 2006. "Analysis of Fruiting Body Development in the Aggregative Ciliate *Sorogena stoianovitchae* (Ciliophora, Colpodea)." *The Journal of Eukaryotic Microbiology* 53 (2): 96–102.

Swanson, Andrew R., Eduardo M. Vadell, and James C. Cavender. 1999. "Global Distribution of Forest Soil Dictyostelids." *Journal of Biogeography* 26 (1): 133–48.

Tarnita, Corina E., Clifford H. Taubes, and Martin A. Nowak. 2013. "Evolutionary Construction by Staying Together and Coming Together." *Journal of Theoretical Biology* 320 (March): 10–22.

Tice, Alexander K., Jeffrey D. Silberman, Austin C. Walthall, Khoa N. D. Le, Frederick W. Spiegel, and Matthew W. Brown. 2016. "*Sorodiplophrys stercorea*: Another Novel Lineage of Sorocarpic Multicellularity." *The Journal of Eukaryotic Microbiology* 63 (5): 623–28.

Varon, Mazal, Shulamit Cohen, and Eugene Rosenberg. 1984. "Autocides Produced by *Myxococcus xanthus*." *Journal of Bacteriology* 160 (3): 1146–50.

Velicer, Gregory J., and Michiel Vos. 2009. "Sociobiology of the Myxobacteria." *Annual Review of Microbiology* 63: 599–623.

Votaw, H. R., and E. A. Ostrowski. 2017. "Stalk Size and Altruism Investment within and among Populations of the Social Amoeba." *Journal of Evolutionary Biology* 30 (11): 2017–30.

Williams, Jeffrey G. 2010. "*Dictyostelium* Finds New Roles to Model." *Genetics* 185 (3): 717–26.

Worley, Ann C., Kenneth B. Raper, and Marianne Hohl. 1979. "*Fonticula alba*: A New Cellular Slime Mold (Acrasiomycetes)." *Mycologia* 71 (4): 746–60.

Section 3

Clonal Multicellularity

9 Phylogenetics of Clonal Multicellularity

Michelle M. Leger
Institute of Evolutionary Biology (CSIC-Universitat
Pompeu Fabra), Barcelona, Spain

Iñaki Ruiz-Trillo
Institute of Evolutionary Biology (CSIC-Universitat
Pompeu Fabra), Barcelona, Spain

CONTENTS

DOI: 10.1201/9780429351907-12

9.1   INTRODUCTION

Clonal multicellularity involves the formation of a multicellular entity in which all cells are the offspring of a single progenitor cell; the term covers both the **complex multicellularity** most familiar from animals, plants and fungi, and the **simple multicellularity** that is displayed by a wide range of microbial organisms (Macario and de Macario 2001; Knoll 2011; Herron et al. 2013; Lyons and Kolter 2015).

Complex multicellular organisms are multicellular in three dimensions and have extensive tissue differentiation, *i.e.* spatial division of labor between individual cells, and a complex body plan arising from a centralized developmental program (Knoll 2011). As a consequence of this differentiation, they possess sophisticated systems of cellular communication, nutrient and oxygen transport, and coordinated developmental regulation. These features allow them to overcome limitations associated with their greater size and three-dimensional organization, including individual cells' lack of access to external sources of nutrients and gas exchange, and the need for coordination between different cells and tissues (Knoll 2011). Complex multicellular organisms may reproduce sexually by means of unicellular gametes, or clonally by means of unicellular spores or multicellular propagules. The distinction between complex and simple multicellularity is somewhat blurry, and the presence of eight cell types has been proposed as an arbitrary rule of thumb (Cock and Collén 2015). This type of multicellularity has only been described in eukaryotic lineages: the red algae (Rhodophyceae), land plants (Embryophyta), brown algae (Phaeophyceae), two groups of fungi (Ascomycota and Basidiomycota), and animals (Metazoa) (Figure 9.1).

In contrast, simple multicellularity involves the formation of simple two- or three-dimensional structures (clusters, balls, filaments or sheets) consisting of cells that adhere to one another, and may engage in simple cell-cell communication, but without the complex spatial differentiation or coordinated developmental regulation found in complex multicellularity (Knoll 2011). This type of multicellularity may represent a single stage of a complex lifestyle involving one or more unicellular amoeboid, flagellate, amoeboflagellate or cystic stages. Alternatively, it may arise in response to environmental conditions, such as the presence of prey organisms or environmental stressors. Simple multicellularity has been described many times in both eukaryotic and prokaryotic groups (Figure 9.1), some (but not all) of which are close relatives of complex multicellular organisms, making them attractive subjects of study to understand the evolution of complex multicellularity (Kirk 2000; Dayel et al. 2011; Suga and Ruiz-Trillo 2013; Umen 2014; Yamagishi et al. 2014;

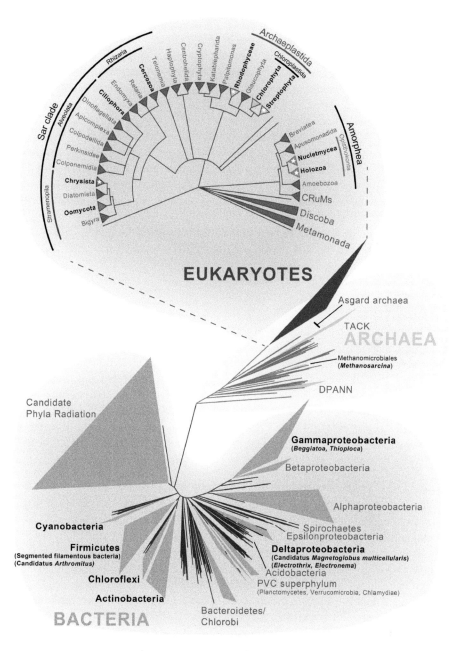

FIGURE 9.1 Simplified schematic tree of life showing lineages (in bold) in which clonal multicellularity has been described (note that the endosymbiotic events that gave rise to mitochondria and plastids are not depicted). White stars indicate lineages in which complex clonal multicellularity has been described. For the sake of simplicity, some lineages of uncertain placement have been left unlabeled. Tree of life modified from (Hug et al. 2016); eukaryote tree topology based on (Burki et al. 2020). Parts of the eukaryote tree highlighted in green, blue and orange are shown in more detail in Figures 9.2, 9.3 and 9.5.

Sebé-Pedrós et al. 2016; Booth et al. 2018; Parra-Acero et al. 2018; Bringloe et al. 2020; Faktorová et al. 2020; Parra-Acero et al. 2020; Levin et al.) (see Chapter 13).

This chapter provides an overview of groups in which simple and/or complex clonal multicellularity has been described, loosely organized by taxonomic affiliation (Figure 9.1). A wider range of eukaryotes (e.g. *Physarum* and Xenophyophorea) are unicellular but multinucleate (reviewed in Adl et al. [2019]), in some cases reaching macroscopic sizes (Haindl and Holler 2005; Lecroq et al. 2009). These organisms are generally out of the scope of this chapter, but we mention cases that are closely related to truly multicellular species, or in which true multicellularity has been reported in some stages.

Living organisms have traditionally been classified into a small number of kingdoms (e.g. Cavalier-Smith 1998; Margulis and Chapman 2009); but while that system is still referenced in much of the literature focusing on specific groups, it has over the past few decades been superseded in eukaryotes by a classification scheme of nameless ranked systematics based on group monophyly (Adl et al. 2005). This chapter uses the most recently revised version of this scheme (Adl et al. 2019).

9.2   EUKARYOTA

9.2.1   ARCHAEPLASTIDA

The Archaeplastida are a eukaryotic supergroup characterized by the ancestral presence of primary plastids. It includes three major groups, the relative branching order of which remains unclear (Leebens-Mack et al. 2019; Burki et al. 2020): the Rhodophyceae (red algae), the Glaucophyta (glaucophyte algae), and the Chloroplastida, a clade that includes the land plants and their closest algal relatives (Streptophyta), the green algae (Chlorophyta) (Adl et al. 2019), and the Prasinodermophyta (Li et al. 2020). Multicellularity is present in the Prasinodermophyta and at least eight lineages of Chloroplastida, while complex, embryonic multicellularity emerged only in the land plants (Embryophyta) and in the red algae (Herron et al. 2013) (Figure 9.2A).

9.2.1.1   Streptophyta (Chloroplastida, Archaeplastida, Eukaryota)

Land plants form a clade that emerges from within the paraphyletic group of algae known as charophytes; together, they comprise the Streptophyta (Wickett et al. 2014; Leebens-Mack et al. 2019) (Figures 9.2A–C). Other than the earliest-branching Mesostigmatophyceae, at least some members of all of the charophyte groups form simple clonal multicellular structures, including **sarcinoids** (three-dimensional clusters) in the Chlorokybophyceae; unbranched filaments in the Klebsormidiophyceae and Zygnematophyceae; three-dimensional branched **thalli** (undifferentiated vegetative tissue) in the Charophyceae; and elaborate two-dimensional structures in the Coleochaetophyceae (Umen 2014) (Figure 9.2A, D). Of these, the most recent analyses place Zygnematophyceae as a sister group to land plants (Wickett et al. 2014; Gitzendanner et al. 2018; Cheng et al. 2019; Leebens-Mack et al. 2019). Despite the relatively simple nature of filaments in the immediate sister group to land plants (Figure 9.2A), key features underpinning land plant multicellularity are shared with several or all charophytes, in some cases with apparent secondary loss in the Zygnematophyceae (Donoghue and Paps 2020). These features include the formation of new cell walls on a scaffold known as the phragmoplast, microscopic channels

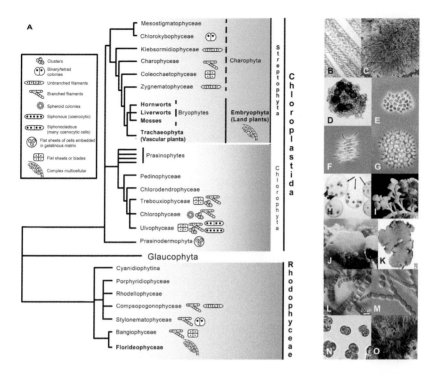

FIGURE 9.2   Clonal multicellularity in Archaeplastida. **A**, Schematic phylogeny showing the major groups within Archaeplastida, indicating types of multicellular or multinucleate structures found in each one. Larger groups in which multicellularity has been reported are shown in coloured boxes; lineages in which complex multicellularity has been reported are shown in bold. **B**, *Spirogyra* sp. (Zygnematophyceae, Streptophyta, Chloroplastida), © Bob Blaylock 2009, distributed under a *CC BY–SA 3.0 license*. **C**, Hornwort (Embryophyta, Streptophyta, Chloroplastida) © Jason Hollinger 2007, distributed under a *CC BY 2.0 license*. **D**, *Coleochaete orbicularis* strain CFD5b1 (EZ) (Coleochaetophyceae, Charophyta, Streptophyta, Chloroplastida) isolated from Lake Tomohawk, Oneida County, WI USA. A ring of seven mature zygotes is visible, as are the sheaths and hairs for which *Coleochaete* is named. Image courtesy Charles Delwiche, all rights reserved. **E**, *Crucigeniella* sp. (Trebouxiophyceae, Chlorophyta, Chloroplastida), **F**, *Scenedesmus dimorphus* (Chlorophyceae, Chlorophyta, Chloroplastida), **G**, *Pediastrum duplex* var. gracillimum (Chlorophyceae), *Chlorella* sp. (Chlorophyceae); **E – G,** modified from © Andrei Savitsky 2019, distributed under a *CC BY-SA 4.0 license*. **H**, *Volvox* sp. (Chlorophyta, Chloroplastida) showing reproductive cells (thick arrows) and somatic cells (thin arrows), modified from © Frank Fox 2011, distributed under a *CC BY-SA 3.0 license*. **I**, *Caulerpa racemosa* (Ulvophyceae, Chlorophyta, Chloroplastida), © Nick Hobgood 2007, distributed under a *CC BY-SA 3.0 license*. **J**, *Dictyosphaeria cavernosa* (Ulvophyceae, Chlorophyta, Chloroplastida), modified from © Frederik Leliaert 2013, distributed under a *CC BY-SA 3.0 license*. **K**, *Verdigellas peltata* (Palmophyllophyceae, Prasinodermophyta) reprinted from (Zechman et al. 2010), © 2010 Phycological Society of America, with permission from John Wiley and Sons. **L – M,** *Boldia erythrosiphon* and *Rhodochaete parvula* (Compsopogonophyceae, Rhodophyceae), reprinted from (Muñoz-Gómez et al. 2017), © 2017 Elsevier Ltd., with permission from Elsevier. **N**, *Bangiopsis franklynottii* (Stylonematophyceae, Rhodophyceae), modified from (West et al. 2014) (distributed under a *CC BY–NC 3.0 license*). **O**, Red algal seaweed (Rhodophyceae), © John Martin Davies 2011, distributed under a *CC BY–SA 3.0 license*.

through the cell wall known as plasmodesmata that allow communication between cells, three-dimensional cell division, cell wall biosynthesis, and orthologs of key plant developmental hormones (reviewed in McCourt et al. 2004; Mikkelsen et al. 2014; Umen 2014; Ishizaki 2017; Ohtaka et al. 2017; de Vries and Archibald 2018). Because of these shared features, and the widespread nature of simple clonal multicellularity in charophyte algal groups (Figure 9.2A), the ancestors of land plants most probably formed similar simple multicellular structures. Charophytes are being developed as model organisms for research into basic biological processes such as plant development and stress physiology, and to yield insight into the evolution of land plants (Domozych et al. 2016).

Complex multicellularity arose subsequently in the immediate ancestors of land plants as they adapted to terrestrialization (Figure 9.2A) (see Chapter 17). Charophyte algae are **haplontic** (possessing a multicellular haploid lifestage and a diploid unicellular stage, with the zygote undergoing meiosis immediately after karyogamy); one of the main changes undergone during the emergence of land plants was the switch to a **haplodiplontic** lifestyle, with alternation of generations between haploid (**gametophyte**) and diploid (**sporophyte**) stages that were both multicellular (Ishizaki 2017). Based on fossil evidence, this switch may have been preceded by the emergence of spores with dessication-resistant coating (Brown and Lemmon 2011). Another major change was the emergence of differentiated three-dimensional structures: while the xylem and phloem emerged only later, in the vascular plants (Trachaeophyta), differentiation is present in even the earliest-branching plants (Harrison 2017).

The earliest branching groups within Embryophyta are the hornworts, liverworts, and mosses (Figure 9.2A). Together, these are informally known as the bryophytes. Early phylogenetic analyses differed with respect to the monophyly or paraphyly of these three lineages, and if the latter, which lineage(s) might be sister to vascular plants (reviewed in Puttick et al. [2018]). The most recent large phylogenomic analyses either weakly or strongly support bryophyte monophyly (Puttick et al. 2018; Leebens-Mack et al. 2019; de Sousa et al. 2019; Harris et al. 2020), with the implication that stomata originated once on the plant stem lineage, and were secondarily lost in liverworts and many mosses (Harris et al. 2020).

9.2.1.2   Chlorophyta (Chloroplastida, Archaeplastida, Eukaryota)

The Chlorophyta are the sister group to Streptophyta, comprising the core Chlorophyta (the Chlorophyceae, Ulvophyceae, Trebouxiophyceae, Pedinophyceae and Chlorodendrophyceae [Fučíková et al. 2014]) and a paraphyletic assemblage known as the prasinophytes (Fang et al. 2018; Adl et al. 2019) (Figure 9.2A). While the monophyly of Chlorophyta is well supported (Cocquyt et al. 2010; Fang et al. 2018; Gitzendanner et al. 2018; Leebens-Mack et al. 2019), their internal relationships have mainly been inferred from 18S rDNA, plastid or mitochondrial genes, and remain inconsistently resolved, with doubt surrounding the monophyly of the Trebouxiophyceae in some, and the Ulvophyceae in most, analyses (Leliaert et al. 2012a; Fučíková et al. 2014; Leliaert and Lopez-Bautista 2015; Lemieux et al. 2015; Turmel et al. 2017; Fang et al. 2018) (Figure 9.2A).

While the Pedinophyceae and Chlorodendrophyceae are single celled (Leliaert et al. 2012a), the Chlorophyceae, Ulvophyceae and Trebouxiophyceae contain lineages with a dazzling variety of multicellular structures (Herron et al. 2013; Umen 2014) (Figures 9.2E–J), many of which are familiar as seaweeds in marine environments (Cortona et al. 2020) (Figure 9.2I), or as terrestrial biofilms or components of lichens (Gustavs et al. 2016). These include undifferentiated clusters, branched filaments, or macroscopic sheets in the Trebouxiophyceae (Figure 9.2E); undifferentiated sheets or chains, or branched filaments with a differentiated **rhizoid** (rootlike anchoring structure) in the Chlorophyceae (Figures 9.2F, G); chains of cells, or spheroid colonies of cells with the flagella turned outwards, and which may reach over 500 individuals in some species (Yamashita et al. 2016) in the Volvocine green algae (Chlorophyceae) (Figure 9.2H), which are models for the development and evolution of multicellularity (Kirk 2000; Nedelcu and Michod 2004). The most striking morphologies are found in the Ulvophyceae (Leliaert et al. 2012a) (Figure 9.2I, J). In addition to undifferentiated clusters and differentiated branching filaments, these include members of the Dasycladales (e.g. *Acetabularia*) that are not multicellular, but in which uninucleated single cells nevertheless reach up to 10 cm in length and possess complex morphology. Other members of the Dasycladales and of the Bryopsidales are **siphonous**, with a thallus consisting of a single tubular or even bladed multinucleate cell reaching up to several meters in length, and containing up to millions of nuclei (Figure 9.2I). In contrast, the Cladophorales and some Ulotrichales are **siphonocladous**, with a multicellular thallus in which each cell is multinucleate (Leliaert et al. 2012b) (Figure 9.2J). Finally, members of the Ulvales and Ulotrichales form large multicellular sheets of uninucleate cells, with a differentiated rhizome (Herron et al. 2013).

9.2.1.3   Prasinodermophyta (Chloroplastida, Archaeplastida, Eukaryota)

The Prasinodermophyta had previously been characterized as the earliest branching class within Chlorophyta, named Palmophyllophyceae (Leliaert et al. 2016), but were recently established as a third phylum within Chloroplastida (Li et al. 2020). Some members of this group, the Palmophyllales, form unique macroscopic thalli consisting of separated, undifferentiated coccoid cells embedded in a gelatinous matrix to form flattened sheets (Figure 9.2K); some members attach to the substrate by means of a holdfast disc (Zechman et al. 2010).

9.2.1.4   Red Algae (Rhodophyceae, Archaeplastida, Eukaryota)

The oldest fossil evidence of complex multicellularity, independent from that in land plants, takes the form of 1200 million-year-old fossils attributed to a red alga; as in plants, simple multicellularity is much more widespread in the group.

Traditionally, red algae were divided into the Bangiophyceae and the Florideophyceae; recent molecular studies have substituted this classification scheme with one based on seven lineages that are consistently monophyletic: the Cyanidiophyceae, Porphyridiophyceae, Stylonematophyceae, Compsopogonophyceae, Rhodellophyceae, Bangiophyceae, and Florideophyceae (Yoon et al. 2006; Verbruggen et al. 2010; Yang et al. 2016; Muñoz-Gómez et al. 2017). A recent phylogenomic analysis has clarified the relationships between these lineages, grouping them into three subphyla: Cyanidiophytina (Cyanidiophyceae), Proteorhodophytina (Porphyridiophyceae, Stylonematophyceae, Compsopogonophyceae) and Rhodellophyceae), and Eurhodophytina (Bangiophyceae, and Florideophyceae) (Muñoz-Gómez et al. 2017) (Figure 9.2A).

The Cyanidiophytina, Porphyridiophyceae and Rhodellophyceae are exclusively unicellular, while the Compsopogonophyceae and some members of the Stylonematophyceae form filaments (Cock and Collén 2015) (Figures 9.2A, L–N), suggesting either multiple acquisitions or multiple losses of simple multicellularity within Proteorhodophytina. Eurhodophytina appears to be ancestrally multicellular. Bangiophyceae species form filaments or leaf-shaped thalli (Yoon et al. 2017), with some differentiation (e.g. rhizomes). Meanwhile, the Florideophyceae have diverse complex multicellular body plans that include tissue differentiation within the crust-like, discoid or leaf-shaped thallus, and a distinctive tripartite haplodiplontic life-style comprising one haploid (gametophytic) and two diploid (carposporophytic and tetrasporophytic) lifestages (Yoon et al. 2017) (Figure 9.2O). Plasmodesmata equivalents (pit plugs) are present in both Bangiophyceae and Florideophyceae, but also in Compsopogonophyceae (Scott et al. 1988), showing that they pre-date Eurhodophytina.

9.2.2 SAR CLADE

The Sar clade (Adl et al. 2019) is formed by three lineages: the Stramenopiles (or heterokonts), the Alveolates, and the Rhizaria. Multicellularity has only been reported in the stramenopiles, the alveolates, and most recently in one rhizarian species (Figure 9.1).

9.2.2.1 Brown Algae (Phaeophyceae, Stramenopila, Sar, Eukaryota)

Despite their outward similarity to green and red algae, brown algae are only distantly related to these taxa. Instead, they represent an independent origin of complex multicellularity within stramenopiles (also known as heterokonts) – a diverse group that also includes water molds, single-celled algae such as diatoms, and commensals or parasites of plants and animals. Brown algae are found primarily in marine habitats, though up to seven transitions to freshwater have occurred (Bringloe et al. 2020). They represent important food sources, carbon sinks, and habitats for other aquatic organisms (Kawai and Henry 2016; Bringloe et al. 2020). These habitats include large undersea forests made up of kelp (Figures 9.3B, C) that may reach more than 50m in height (Vergés and Campbell 2020), and floating mats of *Sargassum* (Figure 9.3D) that can stretch 8850km in length and weigh >20 million tons (Wang et al. 2019).

Traditionally, brown algal classification schemes have relied on morphology, life-cycles, and types of fertilization. As in red and green algae, the multicellular stage is morphologically variable between species, taking the form of microscopic or macroscopic filamentous tufts, flattened crusts, or large thalli that may be simple or branched, occasionally flattened, and sometimes differentiated into blades (flattened structures), stipes (morphologically similar to stems, and providing support to the blades) and holdfasts (anchoring structures) (Kawai and Henry 2016; Bringloe et al. 2020); thallus architecture may be **haplostichous** (filamentous, with filaments growing by transverse cleavage), **polystichous** (with filaments growing by transverse or longitudinal cleavage) or **parenchymatous** (with cells dividing in all directions from the central filament in an undifferentiated manner, leading to a loss of filamentous shape) (Silberfeld et al. 2010).

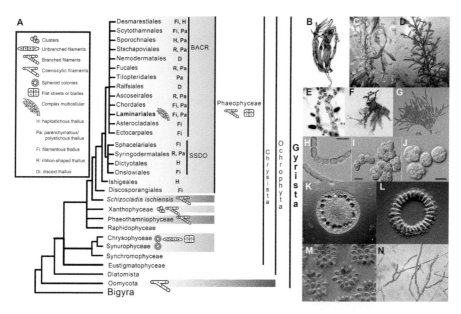

FIGURE 9.3 Clonal multicellularity in Stramenopila. **A**, Schematic phylogeny showing the major groups within Stramenopila, indicating clonal multicellular features found in each one (Bringloe et al. 2020). Larger groups in which multicellularity has been reported are shown in colored boxes; lineages in which complex multicellularity has been reported are shown in bold. **B**, *Laminaria digitata* (Laminariales, Phaeophyceae, Ochrophyta) © Stemonitis 2007 (modified by Thiotrix), distributed under a *CC BY 2.5 license*. **C**, Kelp (Laminariales) © Eric Kilby 2015, distributed under a *CC BY 2.0 license*. **D**, *Sargassum muticum* (Fucales, Phaeophyceae, Ochrophyta) © Lamiot 2015, distributed under a *CC BY–SA 4.0 license*. **E**, *Ectocarpus* sp. (Ectocarpales, Phaeophyceae, Ochrophyta), © Curtis Clark 2012, distributed under a *CC BY–SA 3.0 license*. **F**, *Ectocarpus siliculosus* (Ectocarpales) thallus, © Mike Wilcox 2014, distributed under a *CC BY–SA 4.0 license*. **G**, *Schizocladia ischiensis* (Schizocladiophyceae, Ochrophyta), from (Rizouli et al. 2020) (distributed under a *CC BY–SA 4.0 license*). **H**, Unbranched filaments of *Xanthonema bristolianum* (Xanthophyceae, Ochrophyta). **I**, sarcinoid pseudofilaments of *Heterococcus leptosiroides* (Xanthophyceae). **J**, autosporangia (left), and single cells (right) of *Pleurochloris meiringensis* (Xanthophyceae). **H – J**, reprinted by permission of the publisher Taylor &Francis Ltd (Rybalka et al. 2020), scale bars 10mm. **K**, *Chrysostephanosphaera globulifera* (Chrysophyceae, Ochrophyta), **L**, *Cyclonexis* sp. (Chrysophyceae, Ochrophyta), **M**, *Synura petersenii* (Synurophyceae, Ochrophyta). **K – M**, © Steffen Clauß and Eckhard Völcker (modified from http://www.penard.de/), with permission to the authors. **N**, *Pythium debaryanum* (Oomycota) aseptate hyphae © Tashkoskip 2015, distributed under a *CC BY–SA 4.0 license*.

Brown algal lifecycles are similarly variable. Species may be **diplontic** (possessing a diploid multicellular stage and a haploid unicellular stage), or haplodiplontic (Liu et al. 2017; Bringloe et al. 2020). In haplodiplontic species, both stages may be morphologically similar (**isodiplontic**) or dissimilar (**heterodiplontic**) (Liu et al. 2017); in heterodiplontic species, the diploid stage is usually, but not always, larger (Bringloe et al. 2020).

More recently, phylogenetic analyses using molecular data have superseded morphology- and life history-based classification schemes, and revealed the variable distribution of each type of trait among brown algal species (Silberfeld et al. 2010; Bringloe et al. 2020) (Figure 9.3A). Most phylogenetic analyses of brown algae have been carried out using a small number of markers or focusing on specific subgroups (reviewed in (Bringloe et al. [2020]). However, large-scale multimarker analyses have recovered Discosporangiales and Ishigeales as the earliest branching lineages, followed by a grouping of Sphacelariales, Syringodermatales, Dictyotales, Onslowiales (the SSDO clade) (Silberfeld et al. 2014) (Figure 9.3A), and resolved relationships within the remaining brown algae (a previously poorly resolved grouping known as the brown algal crown radiation (BACR) (Silberfeld et al. 2010). The same study suggested that filamentous thallous architecture was likely secondarily derived convergently in several brown algal lineages.

Within Phaeophyceae, only the Laminariales (kelp) (Figures 9.3 B, C) show significant tissue differentiation. This includes distinct differences between the blades, stipe and holdfast, and specialized phloem-like structures that transport nutrients (Bringloe et al. 2020).

The small, filamented genus *Ectocarpus* (Figures 9.3E, F) has been developed as a model system to understand multicellular evolution and development in brown algae (see Chapter 15), as well as basic biological questions such as the control and evolution of lifecycles and reproductive systems (Coelho et al. 2020).

While Phaeophyceae contains exclusively complex multicellular lineages, simple multicellularity is found in members of its three closely related lineages, the Schizocladiophyceae (consisting of the single species *Schizocladia ischiensis*; Figure 9.3G), Xanthophyceae (Figures 9.3H–J) and Phaeothamniophyceae (Derelle et al. 2016; Ševčíková et al. 2016; Ševčíková et al. 2019). All three groups include at least some species that form filamentous thalli without spatial cell differentiation (Yamagishi et al. 2014; Nicholls and Wujek 2015); of these, at least some members of the Xanthophyceae are known to be haplontic (Pereira and Neto 2014; Sahoo and Kumar 2015). This suggests that simple multicellular thalli were present in at least the last common ancestor of Phaeophyceae and Schizocladiophyceae (Bringloe et al. 2020), and possibly earlier. However, both spatial cell differentiation and cell wall polysaccharide metabolism pathways appear to be innovations of Phaeophyceae (Yamagishi et al. 2014).

9.2.2.2 Chrysophyceae and Synurophyceae (Stramenopila, Sar, Eukaryota)

Simple multicellularity appears to have evolved independently in two other closely related groups of stramenopiles, the Chrysophyceae (commonly, golden-brown algae or golden algae; Figures 9.3K–L) and the Synurophyceae (or synurids; Figure 9.3M) (Figures 9.3A, 3C, 3D, 3E). Many species form clonal colonies with a wide variety of morphologies between species (Nicholls and Wujek 2015; Kristiansen and Škaloud 2017), and in at least some cases consisting of exclusively male or female cells, some of which detach from the main colony to become gametes (Kristiansen and Škaloud 2017). Colonial species are found in all groups of the Chrysophyceae except for the earliest-branching Paraphysomonadida (Kristiansen and Škaloud 2017). One species, *Hydrurus foetida*, forms macroscopic (10–30cm), gelatinous, undifferentiated thalli (Nicholls and Wujek 2015).

9.2.2.3 Oomycetes (Oomycota/Peronosporomycetes, Stramenopila, Sar, Eukaryota)

A different form of multicellularity is found within the Oomycetes (Figure 9.3A), a group that includes parasitic pathogens of plants, and free-living saprophytic molds. Oomycetes form **aseptate hyphae** (Figure 9.3N): structures consisting of branching, clonally multinucleate cells (**coenocytes**) through successive rounds of nuclear division without cytokinesis, superficially similar to those found in fungi (Latijnhouwers et al. 2003).

9.2.2.4 Sessile Peritrich Ciliates (Peritricha, Ciliophora, Alveolata, Sar, Eukaryota)

Ciliates are large (typically >50μm) single-celled eukaryotes named for the characteristic cilia (short, hairlike organelles) arranged on their surface. They are found in a variety of marine, freshwater and soil environments, and as parasites and commensals of animals (reviewed in Lynn [2016]). While the vast majority of described ciliates are unicellular (whether motile or sessile), clonal colonies have been reported in members of subclass Peritricha (Figure 9.1; Figure 9.4).

Within peritrichs, the colonial lifestyle has been reported in members of the families Usconophryidae, Ophrydiidae, Zoothamnidae (Figures 9.4A, B), Vorticellidae (Figures 9.4C–E), Operculariidae, Scyphidiidae, and Epistylididae (Lynn 2016), all of which are placed (together with a number of noncolonial species) in the order Sessilida. Colonies vary widely between species, containing between two cells and thousands (Clamp and Williams 2006; Lynn 2016; Volland et al. 2018), and may measure up to 1.5 cm in height (Clamp and Williams 2006; Volland et al. 2018).

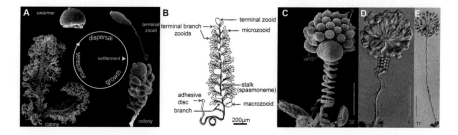

FIGURE 9.4 Clonal multicellular ciliates. **A**, Life cycle of *Zoothamnium niveum* (Peritricha, Ciliophora) © Bright, Espada-Hinojosa, Lagkouvardos and Volland 2014 (Bright et al. 2014) (distributed under a *CC BY–SA 4.0 license*). **B**, Drawing of *Zoothamnium niveum* colony showing different cell types, modified from © Bright, Espada-Hinojosa, Lagkouvardos and Volland 2014 (Bright et al. 2014) (distributed under a *CC BY–SA 4.0 license*). **C–E**, *Apocarchesium arndti* (Peritricha, Ciliophora), reprinted from (Norf and Foissner 2010), © Norf and Foissner 2010, with permission from John Wiley and Sons; **C** contracted colony with two macrozooids (MAZ), 27 visible ordinary zooids (Z) and stalk (ST), **D**, extended colony, **E**, partially contracted colony.

They have best been described in species of *Zoothamnium*. Here, motile single cells (known as **telotrochs** or **swarmers**) settle and divide by binary fission (Figures 9.4A, B), giving rise to several individual cells within the larger colony (**zooids**) that branch off from a single contractile stalk (**spasmoneme**; Figures 9.4C–E) attached to the substrate. Overall colony morphology varies widely between species, with varying spasmoneme structure and branching patterns that have emerged convergently in several species (Clamp and Williams 2006).

Spatial differentiation has been described in two genera (Sun et al. 2011; Herron et al. 2013): *Zoothamnium* (Fauré-Fremiet 1930) and *Apocarchesium* (Ji and Kusuoka 2009; Norf and Foissner 2010), with smaller feeding **microzooids**, larger **macrozooids** that detach from the colony as new telotrochs, and actively proliferating **terminal zooids** at the tip of each colony (Fauré-Fremiet 1930; Bauer-Nebelsick et al. 1996b; Bauer-Nebelsick et al. 1996a) (Figure 9.4B).

Uncertainty surrounds the phylogenetic placement and relationships of colonial ciliates. Based on morphological characteristics, the order Sessilida was originally placed sister to order Mobilida, within subclass Peritricha. This classification scheme was subsequently challenged by molecular data that cast doubt on the sister relationship between Sessilida and Mobilida (Gong et al. 2006), and as a result, members of the order Mobilida were formally removed from Peritricha to the new subclass Mobilia (Zhan et al. 2009). However, debate has persisted (Gong et al. 2010; Sun et al. 2011; Zhan et al. 2013; Gao et al. 2016); most recently, a robust phylogenomic analysis based on a much larger multigene dataset (Gentekaki et al. 2017) recovered a monophyletic relationship between Sessilida and Mobilida that reflects the original classification scheme. Traditional clades within sessilid peritrichs have similarly been revisited based on molecular data: most notably, recent molecular data suggest that genus *Zoothamnium* is in fact paraphyletic (Zhuang et al. 2018).

9.2.2.5 *Viridiuvalis adhaerens* (Chlorarachnea, Cercozoa, Rhizaria, Sar, Eukaryota)

Chlorarachniophytes are a small yet morphologically diverse group of coccoid, flagellated or amoeboid single-celled algae that fall within the larger group Cercozoa. Recently, the first instance of clonal colonies was reported in a chloroarachniophyte, *Viridiuvalis adhaerens* (Shiratori et al. 2017) (Figure 9.1, Cercozoa). Here, daughter cells are believed to be retained within the mother cell wall, resulting in sarcinoid clusters of two to many cells, and occasionally compound clusters.

9.2.3 Amorphea

Amorphea is a clade comprised of the Holozoa (animals and their closest relatives), the Nucletmycea (or Holomycota – fungi and their closest relatives), the Amoebozoa, and two poorly understood lineages of single-celled organisms, the Breviatea and the Apusomonadida (Figure 9.1). Within this group, the Holozoa and Nucletmycea are most closely related to one another and form the subgroup Opisthokonta. Clonal multicellularity has been reported only in the Opisthokonta, though aggregative multicellularity is present in the Amoebozoa (see Chapter 5).

9.2.3.1 Fungi (Opisthokonta, Amorphea, Eukaryota)

The fungi are generally understood as the major clade within Nucletmycea, with complex multicellularity present in two of its largest lineages, the Ascomycota and Basidiomycota (Figure 9.5A). However, the group is unusual in that there is a lack of consensus

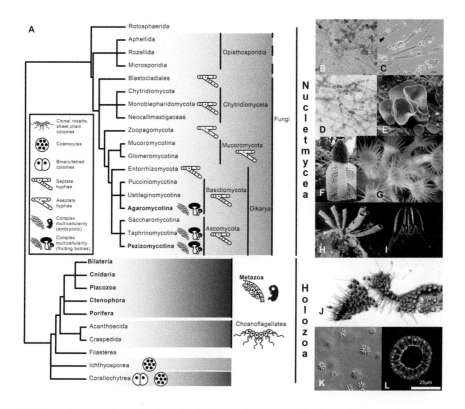

FIGURE 9.5 Clonal multicellularity in Opisthokonta. **A**, Schematic phylogeny showing the major groups within Opisthokonta, indicating types of multicellular or multinucleate structures found in each one. Larger groups in which multicellularity has been reported are shown in coloured boxes; lineages in which complex multicellularity has been reported are shown in bold. **B**, *Allomyces* (Blastocladiales, Fungi) strain WJD103 mature thallus, © TelosCricket 2011, distributed under a *CC BY–SA 4.0 license*. **C**, *Mucor racemosus* (Mucoromycotina, Fungi), © Medmyco 2011, distributed under a *CC BY–SA 4.0 license*. **D**, *Alternia* sp. (Pezizomycotina, Ascomycota, Fungi) septate hyphae © Tashkkoskip 2008, distributed under a *CC BY–SA 4.0 license*. **E**, *Aleuria aurantia* (Pezizomycotina, Ascomycota, Fungi) © Holger Krisp 2009, distributed under a *CC BY–SA 3.0 license*. **F**, *Phallus indusiatus* (Agaromycotina, Basidiomycota, Fungi) fruiting body (https://www.pxfuel.com). **G – I** © Joan J. Soto Àngel, with permission to the authors: **G**, *Parazoanthus axinellae* (Cnidaria, Metazoa, Holozoa), **H**, Sponge (Porifera, Metazoa, Holozoa), **I**, *Bolinopsis infundibulum* (Ctenophora, Metazoa, Holozoa). **J**, Cross-section through *Choanoeca flexa* cup-shaped colony, © Thibaut Brunet, with permission to the authors. **K**, *Salpingoeca rosetta* (Choanoflagellatea, Holozoa) rosette colonies © Mark J. Dayel 2009, distributed under a *CC BY–SA 3.0 license*. **L**, *Sphaeroforma arctica* (Ichthyosporea, Holozoa) polarized layer of cells, © Omaya Dudin, with permission to the authors.

surrounding the groups that should be included within Fungi itself (and thus, the very definition of "Fungi"), and those that should be counted merely as sister lineages.

Briefly, Nucletmycea comprises two clades: the Rotosphaerida (or Nucleariida), containing amoeboid protists, and a second comprising seven major groups: the Opisthosporidia, Chytridiomyceta (including Chytridiomycota, Monoblephari-diomycota and Neocallimastigaceae), Blastocladiales, Zoopagomycota, Mucoromycota (including Glomeromycotina and Mucoromycotina), Ascomycota and Basidiomycota (Adl et al. 2019; James et al. 2020) (Figure 9.5A). Of these, there is broad consensus on the inclusion of the latter six within Fungi. Meanwhile, Opisthosporidia comprises three lineages of unicellular, mainly parasitic organisms (Aphellida, Rozellida, Microsporidia); these lineages are excluded from the Fungi in some schemes (Karpov et al. 2014; Torruella et al. 2018), but included in most others (Ahrendt et al. 2018; Hibbett et al. 2018; Tedersoo et al. 2018; Naranjo-Ortiz and Gabaldón 2019; James et al. 2020) because the monophyly of Opisthosporidia is not always recovered. Of the remaining six lineages, Blastocladiales, Chytridiomyceta and Zoopagomycota are usually recovered as the earliest branching three clades; studies have differed regarding the branching order of the Blastocladiales and Chytridiomyceta (and the monophyly of taxa previously considered to form Chytridiomycota) (reviewed in Naranjo-Ortiz and Gabaldón [2019] and James et al. [2020]), though recent phylogenies place Blastocladiales as the earlier branching of the two (Ahrendt et al. 2018; Kiss et al. 2019). Ascomycota and Basidiomycota have traditionally been placed together in the larger group Dikarya; recent analyses have also placed the small, parasitic group Entorrhizomycota within this group, either as the earliest branching lineage, or sister to Basidiomycota (reviewed in Hibbett et al. [2018]); Mucoromycota form the sister group to this clade (Spatafora et al. 2016). Earlier taxonomic schemes included a further major group, the Zygomycota; this grouping has been confirmed to be paraphyletic, and its members are now placed within the Mucoromycota and Zoopagomycota (Liu et al. 2009; Spatafora et al. 2016).

Many fungi form **mycelia** (vegetative networks of hyphae; Figures 9.5B–D), which may be either septate (with cell walls between cells) or aseptate (*i.e.*, coenocytic) (Stajich et al. 2009). True hyphae are present in the Zoopagomycota, Mucoromycota, Ascomycota and Basidiomycota, where they have undergone many secondary losses (Harris 2011), and in some members of the Chytridiomyceta and Blastocladiales, where they may have originated independently (Dee et al. 2015). In addition to hyphae, members of two groups of Ascomycota (Pezizomycotina and Taphrinomycotina) and one group of Basidiomycota (Agaricomycotina) form complex multicellular fruiting bodies (reviewed in Stajich et al. [2009] and Naranjo-Ortiz and Gabaldón [2019]; see Chapter 14; Figures 9.5E,F).

9.2.3.2   Holozoa (Opisthokonta, Amorphea, Eukaryota)

Holozoa is the clade that includes animals (Metazoa) and their closest unicellular relatives (Figure 9.5A). Within this clade, Metazoa (Figures 9.5G–I) represents the best understood and most morphologically diverse group, with by far the largest described number of taxa, including several well-established model organisms. All animals form a clade – that is, they descend from a single common ancestor that already displayed complex multicellularity arising through embryonic development, and that possessed the key signaling, adhesion and genetic regulation systems ubiquitous in modern animals (Sebé-Pedrós et al. 2017; Paps and Holland 2018; Richter et al. 2018; Fernández and Gabaldón 2020).

The phylogeny of the earliest-branching animals remains the subject of intense debate (Figure 9.5A), involving sophisticated large-scale phylogenomic analyses as well as morphological evidence, with rival camps favoring sponges (Porifera; Figure 9.5H) (e.g. Feuda et al. 2017; Simion et al. 2017) or comb jellies (Ctenophora; Figure 9.5I) (e.g., Shen et al. 2017; Whelan et al. 2017; Laumer et al. 2019), respectively, as the first lineage to diverge from other animals (reviewed in King and Rokas 2017) (Figure 9.5). This question has profound implications for our understanding of the timing of animal origins (Dohrmann and Wörheide 2017) and of the key events in animal evolution. Comb jellies appear to possess more features in common with animals than do sponges, including neurons, specialized gametes and striated muscle (Ryan et al. 2013; Moroz et al. 2014; Dunn et al. 2015). Uncertainty surrounding the branching order of these taxa, therefore, affects our assessment of the complexity of the last animal common ancestor, and whether these traits were acquired only once in animals, after sponges diverged from the lineage that gave rise to the rest of animals (in the Porifera-first scenario) (Simion et al. 2017); whether they were ancestrally present, but secondarily lost in sponges (in the Ctenophora-first scenario) (Ryan and Chiodin 2015); or whether some features evolved convergently in ctenophores and in other animals (in the Ctenophora-first scenario) (Moroz 2015).

In recent years, improved sequencing technologies and discoveries of new organisms have improved our understanding of the closest unicellular relatives of animals. Broadly, four major lineages of these organisms are recognized: Choanoflagellatea, Filasterea, Ichthyosporea, and Corallochytrea (or Pluriformea) (Figure 9.5A). Each lineage includes at least some members with a complex lifecycle that includes a simple multicellular stage: in the case of Filasterea, this stage is an aggregative one (Sebé-Pedrós et al. 2013), while the other three lineages each display a clonal multicellular stage.

The closest unicellular lineage to animals, **Choanoflagellatea** (Figures 9.5J, K), is also the best understood. Choanoflagellates are single-celled bacterivorous flagellates, ubiquitous in marine and freshwater environments (reviewed in Leadbeater 2015; Richter and Nitsche 2017).

Some choanoflagellate species have long been known to form transient simple clonal multicellular structures (James-Clark 1866), arising through clonal division, with between two to hundreds of cells remaining connected to one another by their collar microvilli, by cytoplasmic bridges, and/or by a shared extracellular matrix (Hibberd 1975; Karpov and Coupe 1998; Dayel et al. 2011; Brunet et al. 2019; Levin et al.). Different colony morphologies exist, several of which may be present in a single species (Dayel et al. 2011; Leadbeater 2015; Brunet and King 2017; Carr et al. 2017): chains or sheets of cells, clusters (sometimes present on stalks), or "rosettes" – spherical clusters oriented around a hollow center containing secreted extracellular matrix that controls colony shape (Dayel et al. 2011; Larson et al. 2020) (Figure 9.5K). The latter structure has made *Salpingoeca rosetta* a model for understanding the genetic basis of simple clonal multicellularity, and the subject of investigation into commonalities between its simple multicellularity and the complex multicellularity of animals (Dayel et al. 2011; Fairclough et al. 2013; Levin et al.). Recent 3-D reconstruction of colony shapes from electron micrographs showed apparent morphological differences between individual cells within single colonies, raising the possibility that these colonies exhibit some degree of spatial cell differentiation (Laundon et al. 2019; Naumann and Burkhardt 2019).

Uncertainty remains regarding relationship between the two deepest lineages within Holozoa, **Ichthyosporea** and **Corallochytrea** (or **Pluriformea**): the two lineages may be sister to one another (Grau-Bové et al. 2017; López-Escardó et al. 2019), or Ichthyosporea may be the earliest branching of the two, with Pluriformea branching sister to the clade comprising Filasterea, Choanoflagellatea and Metazoa (Hehenberger et al. 2017). Known species of ichthyosporeans are mainly parasites of fish, amphibians or arthropods, with a smaller number of free-living species. They form large, rounded or elongated coenocytes (Marshall and Berbee 2011; Mendoza and Vilela 2013), which may grow from two to up to hundreds of nuclei within a single species (Ondracka et al. 2018); cell walls subsequently form around the individual nuclei to produce daughter cells. Interestingly, in at least one species of ichthyosporean, newly cellularized daughter cells form a layer that superficially resembles the epithelium that forms during early animal development (Dudin et al. 2019) (Figure 9.5L). Coenocytes have also been reported in Corallochytrea/Pluriformea, as have binary colonies arising through clonal division (Raghu-kumar 1987; Torruella et al. 2015; Kożyczkowska et al. 2020; Tikhonenkov et al. 2020). Because both lineages possess homologs of animal cell adhesion genes and transcription factors (de Mendoza et al. 2015; Torruella et al. 2015; Grau-Bové et al. 2017), representative taxa make attractive targets to understand the functions of these genes outside animals, and their possible role in simple multicellularity (Suga and Ruiz-Trillo 2013; Waller et al. 2018; Faktorová et al. 2020).

9.3 ARCHAEA

9.3.1 *METHANOSARCINA* SPP. (METHANOSARCINALES, EURYARCHAEOTA, ARCHAEA)

Multicellularity has been described in only one group of Archaea: the genus *Methanosarcina*. Members of this group are metabolically diverse, capable of producing methane through three different metabolic pathways and using nine different substrates, the most among any group of methanogens (Galagan et al. 2002). They are found in a wide range of environments, including marine and freshwater sediments, soil, decaying organic matter, anaerobic sewage digesters, and ungulate rumens (Galagan et al. 2002).

Methanosarcina species form simple multicellular structures in response to a range of environmental stressors: temperature, pH, mechanical, osmotic, and antibiotic-induced (Macario and de Macario 2001). Two main types of structure have been described: **packets** (cubelike clusters arising through cell division, coated in a fibrous extracellular matrix; Figure 9.6A) (Robinson 1986) and **lamina** (flat sheets or meshes reaching several cm in length, consisting of actively dividing cells of various shapes and sizes held together by intercellular connective material (Mayerhofer et al. 1992)). It has been suggested that cell size and morphological heterogeneity

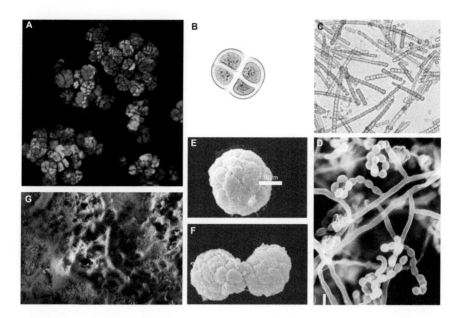

FIGURE 9.6 Clonal multicellular structures in prokaryotes. **A**, *Methanosarcina barkeri* (Methanomicrobiales, Archaea) packets, reprinted from (Lambie et al. 2015) (distributed under a *CC BY–SA 4.0 license*). **B**, *Chroococcus turgidus* (Cyanobacteria, Bacteria) © Burn12121212 2020, distributed under a *CC BY–SA 4.0 license*. **C**, *Cylindrospermum* sp. (Cyanobacteria, Bacteria) © Willem van Aken, CSIRO, 2007, distributed under a *CC BY–SA 3.0 license*. **D**, *Streptomyces* sp. SF1293 (Actinomycetes, Bacteria) mycelium, © 2002–2020 The Society for Actinomycetes Japan (http://atlas.actino.jp/), Contributors T. Shomura, S. Amano & T. Niida, used with permission. **E – F**, Single (**E**) and dividing (**F**) Multicellular magnetotactic prokaryote (MMP) *Ca*. Magnetoglobus multicellularis (Deltaproteobacteria, Bacteria) multicellular individuals, reprinted from (Keim et al. 2004), © 2004 Federation of European Microbiological Societies, with permission from Elsevier. **G**, Beggiatoaceae (Gammaproteobacteria, Bacteria) mat in Guaymas Basin (individual orange filaments approx. 40μm), from (Buckley et al. 2019) (distributed under a *CC BY–SA 4.0 license*).

within these structures, and possible localized differences in protein and extracellular matrix secretion, may point to some degree of cell differentiation (Macario and de Macario 2001), although this remains to be studied in detail.

9.4 BACTERIA

9.4.1 CYANOBACTERIA (CYANOBACTERIA, BACTERIA)

The oldest evidence for multicellular structures comes from cyanobacterial fossils. Formerly known as blue-green algae, Cyanobacteria are gram-negative, photosynthetic bacteria found ubiquitously in aquatic and terrestrial habitats, including polar environments, deserts and thermal springs (Herrero et al. 2016), and in symbiosis with fungi to form lichens (Rikkinen 2013). While some species exist only as single cells, others form multicellular filaments consisting of chains of cells covered by a mucilaginous sheath, with a structurally and functionally continuous periplasm that may allow communication and coordination between individual cells within the filament (Herrero et al. 2016) (Figures 9.6B, C). Fossilized cyanobacterial filaments have been preserved as part of stromatolites (layered sedimentary formations of mineral compounds produced by cyanobacteria) (Golubic and Seong-Joo 1999), and in sediment layers at least 1900 million years old (Hofmann 1976) though bacterial filaments that may be cyanobacterial have been found in 3465-million-year old rock formations (Schopf and Kudryavtsev 2009).

Cyanobacteria have traditionally been classified into five subsections based on their ability to form filaments: subsections I and II, which do not form filaments; subsection III (e.g., Oscillatoriales), whose filaments are undifferentiated, comprising only vegetative cells; subsection IV (e. g. Nostocales), in which vegetative cells may terminally differentiate into specialized **heterocysts** (nitrogen-fixing cells), and/or **akinocysts** (resistant resting cells), with the filaments branching in a single plane (**false branching**); subsection V (e. g. Stigonematales), in which cells may likewise undergo terminal differentiation, but with the filaments branching in all planes, giving rise to complex **true branching** patterns (Castenholz et al. 2001). Of these, only subsections IV and V are supported by molecular data (Tomitani et al. 2006; Schirrmeister et al. 2015), while each of the other subsections is polyphyletic. Interestingly, IV and V together form a clade in the same analyses. Overall, this lends support to multiple origins of the undifferentiated filamentous lifestyle in Cyanobacteria, and a single origin of differentiated filaments, likely from within a clade of cyanobacteria possessing undifferentiated filaments (Schirrmeister et al. 2015). Tomitani et al. (Tomitani et al. 2006) have suggested that this filament differentiation emerged in response to the Great Oxygenation Event, which likely compromised the reducing conditions necessary for nitrogen fixation in existing cyanobacteria, leading to selection for specialized heterocysts capable of fixing nitrogen under oxidative conditions.

9.4.2 ACTINOBACTERIA (ACTINOBACTERIA, BACTERIA)

A different type of multicellular structure is present in members of the Actinobacteria. These are gram-positive bacteria that are cosmopolitan, but best known for their importance in soil ecosystems, their use as sources of antibiotics, and pathogenic members such as the causative agents of tuberculosis, diphtheria and leprosy (Barka et al. 2015). While the earliest-branching Actinobacteria are unicellular rod- or coccoid-shaped organisms, most species form transient or permanent multicellular mycelia in solid or liquid media (Chandra and Chater 2014; Barka et al. 2015) (Figure 9.6D). On surfaces, cells in some species differentiate by growing upwards to form aerial hyphae, which produce asexual spores (Barka et al. 2015); in others, spores may form directly on the mycelium, or individual hyphae may fragment as a means of vegetative reproduction. The morphology (chains, loops, spirals, vertices or bags) and length of spore chains vary considerably between species (Barka et al. 2015). *Streptomyces* has been the most extensively studied as a model for mycelium development and aerial hyphae formation; it appears that homologs of genes involved in mycelium development in other actinobacteria have become co-opted for spore formation in this genus (Chandra and Chater 2014). Interestingly, nutrient depletion in *Streptomyces* leads to autoproteolytic degradation of the mycelium to provide the raw materials for aerial hyphae (and thus, spore) formation, in a bacterial analog of autophagy (Barka et al. 2015).

9.4.3 Multicellular Magnetotactic Prokaryotes (*Candidatus* Magnetoglobus multicellularis, Deltaproteobacteria, Bacteria)

Multicellular Magnetotactic Prokaryotes (MMPs) are an intriguing case within multicellular bacteria, in that they appear to lack a unicellular stage altogether. Instead, they exist as spherical clonal colonies of 15–45 pyramidal cells organized around a central compartment, with the flagella of each cell arranged outward (Figure 9.6E); cells are capable of swimming only as part of the colony. Division of the colony begins with growth and then coordinated division of all cells within the colony, resulting in doubling in size and cell number of the whole colony. This is followed by elongation of the colony, which progresses further at the midpoint than in the rest of the colony. This creates a constriction at the midpoint (Figure 9.6F), which progresses until the two daughter colonies separate. MMPs do not appear to possess different cell types, yet their cells undergo a high level of coordination, and some degree of division of labor based on positioning. MMPs appear to form a clade together with uncultured sequences in Deltaproteobacteria (Abreu et al. 2007).

9.4.4 Other Bacterial Taxa (Bacteria)

Multicellular filaments have also been reported in other diverse, distantly related bacteria species (Lyons and Kolter 2015). While these filaments are known to be clonal, little is known about their formation in most species. The examples below provide an exciting glimpse into how much still remains to be understood of their cell biology, evolutionary origins and environmental interactions.

The candidate genera ***Electrothrix* and *Electronema*** (**Desulfobulbacea, Deltaproteobacteria**) (Trojan et al. 2016), known as cable bacteria, form multicellular filaments with ridges running along their surfaces. Incredibly, these filaments act as living wires, conducting electrons over distances of several centimeters through marine or freshwater sediments, allowing them to couple oxygen reduction at the sediment surface with sulfide oxidation in the deeper, anaerobic layers of sediment (Pfeffer et al. 2012; Bjerg et al. 2018).

The segmented filamentous bacteria (SFB; ***Candidatus* Savagella, Clostridiales** [Jonsson et al. 2020]) are host-specific intestinal symbionts of vertebrates. In recent years, they have drawn attention for their ability to attach to the epithelial cells in the small intestine and stimulate innate and adaptive immune cell activation, particularly T helper 17 activation (Flannigan and Denning 2018). Although their effects on organismal health are not yet fully understood, they are likely to play a critical role in post-natal maturation of the gut immune system. Filaments form from a single infectious particle that adheres to epithelial cells, forming an anchoring segment known as a holdfast. Once anchored, this segment elongates and then divides by transverse septum formation. Filaments may reach up to 90 segments and 1000μm in length. The free ends of filaments of >50μm undergo differentiation and asymmetric division leading to new infectious particle formation (Schnupf et al. 2013).

Another group of bacteria with similar filamentous morphology (**Candidatus Arthromitus, Lachnospiracea, Clostridiales**) is found in arthropod guts. Filaments in individual species were 1.5μm or 0.8 – 1μm in diameter, with oblong, rounded or oval endospores, respectively. Although initially classified as members of the SFB based on morphology, these were later found to place separately from the SFB, in a more distantly related group of Clostridiales known as the Lachnospiraceae (Thompson et al. 2012).

Filamentous forms with variable size and morphology belonging to the **Chloroflexi** have been reported in activated sludge wastewater treatment plants, where they form an important structural support for organic compound degradation and fermentation (Speirs et al. 2019). Molecular data have only recently begun to supersede morphology as a means of identification, and little is understood of their physiology.

Two genera of **Gammaproteobacteria** form conspicuously large filaments and mats in sulfide-rich environments (coastal sediments, salt marshes, hydrothermal vents, cold seeps). **Beggiatoa** species form individual filaments 1 – 200 μm in diameter, and up to 10 cm in length (Figure 9.6G); **Thioploca** species form bundles of multiple filaments enveloped in a single gelatinous sheath; each bundle may be 2.5 – 80 μm in diameter, and 2 – 5 cm in length (Teske and Nelson 2006). Despite these differences, 16S rDNA phylogenies suggest that *Beggiatoa* is in fact paraphyletic, with *Thioploca* forming either one or two clades branching from within it (Teske and Nelson 2006).

9.5 CONCLUSIONS

Complex multicellularity has arisen only six times, and then only in eukaryotes. However, recent studies of a growing number of lineages have made it clear that simple multicellularity is more widespread than previously believed – more than is appreciated even by many researchers working on questions relating to multicellularity in specific systems. With greater efforts to sample new organisms from a wider range of environments, and with cheaper and higher quality genome sequencing, it is likely that simple clonal multicellularity will be discovered in an ever-larger number of lineages, and this represents an exciting area for future research.

It is apparent that most complex clonal multicellular organisms have at least some close relatives with simple multicellular stages, and that some of them share key genes underpinning the multicellular lifestyle of their complex multicellular relatives. This suggests that clonal multicellularity progresses in sequential steps, with the early emergence of genes implicated in simple multicellular processes potentiating its retention (or emergence) in several closely related lineages, followed by the complexification of multicellularity in some of those lineages, involving increased spatial division of labor and cellular coordination. Efforts to study simple multicellularity in close relatives of complex multicellular organisms will allow us to understand how this complexification occurred in different lineages.

ACKNOWLEDGEMENTS

This work was supported by Juan de la Cierva-Incorporación fellowship IJC2018-036657-I to M.M.L and grants (BFU2017-90114-P and PID2020-120609GB-I00) from Ministerio de Economía y Competitividad (MINECO), Agencia Estatal de Investigación (AEI), and Fondo Europeo de Desarrollo Regional (FEDER) to I.R.-T. We thank Joan J. Soto Àngel, Thibaut Brunet, Charles F. Delwiche, Omaya Dudin, Nataliya Rybalka, T. Shomura, S. Amano, T. Niida and the Society for Actinomycetes Japan, Steffen Clauß and Eckhard Völcker for generously sharing images, Núria Ros-Rocher, Yana Eglit and two anonymous reviewers for helpful feedback, and Alexandra A. Elbakyan for creating Sci-Hub, which made this review possible.

BIBLIOGRAPHY

Abreu, F., J. L. Martins, T. S. Silveira, et al. 2007. *Candidatus* Magnetoglobus multicellularis, a multicellular, magnetotactic prokaryote from a hypersaline environment. *International Journal of Systematic and Evolutionary Microbiology* 57:1318–1322.

Adl, S. M., A. G. B. Simpson, M. Farmer, et al. 2005. The new higher level classification of Eukaryotes with emphasis on the taxonomy of protists. *Journal of Eukaryotic Microbiology* 52:399–451.

Adl, S. M., D. Bass, C. E. Lane, et al. 2019. Revisions to the classification, nomenclature, and diversity of Eukaryotes. *Journal of Eukaryotic Microbiology* 66:4–119.

Ahrendt, S. R., C. A. Quandt, D. Ciobanu, et al. 2018. Leveraging single-cell genomics to expand the fungal tree of life. *Nature Microbiology* 3:1417–1428.

Barka, E. A., P. Vatsa, L. Sanchez, et al. 2015. Taxonomy, physiology, and natural products of Actinobacteria. *Microbiology and Molecular Biology Reviews: MMBR* 80:1–43.

Bauer-Nebelsick, M., C. F. Bardele, and J. A. Ott. 1996a. Electron microscopic studies on *Zoothamnium niveum* (Hemprich & Ehrenberg, 1831) Ehrenberg 1838 (Oligohymenophora, Peritrichida), a ciliate with ectosymbiotic, chemoautotrophic bacteria. *European Journal of Protistology* 32:202–215.

Bauer-Nebelsick, M., C. F. Bardele, and J. A. Ott. 1996b. Redescription of *Zoothamnium niveum* (Hemprich & Ehrenberg, 1831) Ehrenberg, 1838 (Oligohymenophora, Peritrichida), a ciliate with ectosymbiotic, chemoautotrophic bacteria. *European Journal of Protistology* 32:18–30.

Bjerg, J. T., H. T. S. Boschker, S. Larsen, et al. 2018. Long-distance electron transport in individual, living cable bacteria. *Proceedings of the National Academy of Sciences* 115:5786–5791.

Booth, D. S., H. Szmidt-Middleton, and N. King. 2018. Choanoflagellate transfection illuminates their cell biology and the ancestry of animal septins. *Molecular Biology of the Cell* 29:3026–3038.

Bright, M., S. Espada-Hinojosa, I. Lagkouvardos, and J.-M. Volland. 2014. The giant ciliate *Zoothamnium niveum* and its thiotrophic epibiont *Candidatus* Thiobios zoothamnicoli: a model system to study interspecies cooperation. *Frontiers in Microbiology* 5:145.

Bringloe, T. T., S. Starko, R. M. Wade, et al. 2020. Phylogeny and evolution of the brown algae. *Critical Reviews in Plant Sciences* 39:281–321.

Brown, R. C., and B. E. Lemmon. 2011. Spores before sporophytes: hypothesizing the origin of sporogenesis at the algal–plant transition. *New Phytologist* 190:875–881.

Brunet, T., and N. King. 2017. The origin of animal multicellularity and cell differentiation. *Developmental Cell* 43:124–140.

Brunet, T., B. T. Larson, T. A. Linden, et al. 2019. Light-regulated collective contractility in a multicellular choanoflagellate. *Science* 366:326–334.

Buckley, A., B. MacGregor, and A. Teske. 2019. Identification, expression and activity of candidate nitrite reductases from orange Beggiatoaceae, Guaymas Basin. *Frontiers in Microbiology* 10:644.

Burki, F., A. J. Roger, M. W. Brown, and A. G. B. Simpson. 2020. The new tree of Eukaryotes. *Trends in Ecology & Evolution* 35:43–55.

Carr, M., D. J. Richter, P. Fozouni, et al. 2017. A six-gene phylogeny provides new insights into choanoflagellate evolution. *Molecular Phylogenetics and Evolution* 107:166–178.

Castenholz, R. W., A. Wilmotte, M. Herdman, et al. 2001. Phylum BX. Cyanobacteria. In *Bergey's Manual® of Systematic Bacteriology*, ed. D. R. Boone, R. W. Castenholz, and G. M. Garrity, 473–599. New York, NY: Springer New York.

Cavalier-Smith, T. 1998. A revised six-kingdom system of life. *Biological Reviews* 73:203–266.

Chandra, G., and K. F. Chater. 2014. Developmental biology of Streptomyces from the perspective of 100 actinobacterial genome sequences. *FEMS Microbiology Reviews* 38:345–379.

Cheng, S., W. Xian, Y. Fu, et al. 2019. Genomes of subaerial Zygnematophyceae provide insights into land plant evolution. *Cell* 179:1057-1067.e14.

Clamp, J. C., and D. Williams. 2006. A molecular phylogenetic investigation of *Zoothamnium* (Ciliophora, Peritrichia, Sessilida)1. *Journal of Eukaryotic Microbiology* 53:494–498.

Cock, J. M., and J. Collén. 2015. Independent emergence of complex multicellularity in the brown and red algae. In *Evolutionary Transitions to Multicellular Life*, ed. I. Ruiz-Trillo and A. M. Nedelcu, 2:335–361. Advances in Marine Genomics. Dordrecht: Springer Netherlands.

Cocquyt, E., G. H. Gile, F. Leliaert, et al. 2010. Complex phylogenetic distribution of a non-canonical genetic code in green algae. *BMC Evolutionary Biology* 10:327.

Coelho, S. M., A. F. Peters, D. Müller, and J. M. Cock. 2020. *Ectocarpus*: an evo-devo model for the brown algae. *EvoDevo* 11:19.

Cortona, A. D., C. J. Jackson, F. Bucchini, et al. 2020. Neoproterozoic origin and multiple transitions to macroscopic growth in green seaweeds. *Proceedings of the National Academy of Sciences* 117:2551–2559.

Dayel, M. J., R. A. Alegado, S. R. Fairclough, et al. 2011. Cell differentiation and morphogenesis in the colony-forming choanoflagellate *Salpingoeca rosetta*. *Developmental Biology* 357:73–82.

Dee, J. M., M. Mollicone, J. E. Longcore, R. W. Roberson, and M. L. Berbee. 2015. Cytology and molecular phylogenetics of Monoblepharidomycetes provide evidence for multiple independent origins of the hyphal habit in the Fungi. *Mycologia* 107:710–728.

Derelle, R., P. López-García, H. Timpano, and D. Moreira. 2016. A phylogenomic framework to study the diversity and evolution of stramenopiles (=heterokonts). *Molecular Biology and Evolution* 33:2890–2898.

Dohrmann, M., and G. Wörheide. 2017. Dating early animal evolution using phylogenomic data. *Scientific Reports* 7:3599.

Domozych, D., Z. A. Popper, and I. Sorensen. 2016. Charophytes: evolutionary giants and emerging model organisms. *Frontiers in Plant Science* 7:1470.

Donoghue, P., and J. Paps. 2020. Plant evolution: assembling land plants. *Current Biology* 30:R81–R83.

Dudin, O., A. Ondracka, X. Grau-Bové, et al. 2019. A unicellular relative of animals generates a layer of polarized cells by actomyosin-dependent cellularization. *eLife* 8:e49801.

Dunn, C. W., S. P. Leys, and S. H. D. Haddock. 2015. The hidden biology of sponges and ctenophores. *Trends in Ecology & Evolution* 30:282–291.

Fairclough, S. R., Z. Chen, E. Kramer, et al. 2013. Premetazoan genome evolution and the regulation of cell differentiation in the choanoflagellate *Salpingoeca rosetta*. *Genome Biology* 14:R15.

Faktorová, D., E. R. R. Nisbet, J. A. Fernández Robledo, et al. 2020. Genetic tool development in marine protists: emerging model organisms for experimental cell biology. *Nature Methods* 17:481–494.

Fang, L., F. Leliaert, P. M. Novis, et al. 2018. Improving phylogenetic inference of core Chlorophyta using chloroplast sequences with strong phylogenetic signals and heterogeneous models. *Molecular Phylogenetics and Evolution* 127:248–255.

Fauré-Fremiet, E. 1930. Growth and differentiation of the colonies of *Zoothamnium alternans* (clap. and lachm.). *The Biological Bulletin* 58:28–51.

Fernández, R., and T. Gabaldón. 2020. Gene gain and loss across the metazoan tree of life. *Nature Ecology & Evolution* 4:524–533.

Feuda, R., M. Dohrmann, W. Pett, et al. 2017. Improved modeling of compositional heterogeneity supports sponges as sister to all other animals. *Current Biology* 27:3864–3870.e4.

Flannigan, K. L., and T. L. Denning. 2018. Segmented filamentous bacteria-induced immune responses: a balancing act between host protection and autoimmunity. *Immunology* 154:537–546.

Fučíková, K., F. Leliaert, E. D. Cooper, et al. 2014. New phylogenetic hypotheses for the core Chlorophyta based on chloroplast sequence data. *Frontiers in Ecology and Evolution* 2:63.

Galagan, J. E., C. Nusbaum, A. Roy, et al. 2002. The genome of *M. acetivorans* reveals extensive metabolic and physiological diversity. *Genome Research* 12:532–542.

Gao, F., A. Warren, Q. Zhang, et al. 2016. The all-data-based evolutionary hypothesis of ciliated protists with a revised classification of the phylum Ciliophora (Eukaryota, Alveolata). *Scientific Reports* 6:24874.

Gentekaki, E., M. Kolisko, Y. Gong, and D. Lynn. 2017. Phylogenomics solves a long-standing evolutionary puzzle in the ciliate world: the subclass Peritrichia is monophyletic. *Molecular Phylogenetics and Evolution* 106:1–5.

Gitzendanner, M. A., P. S. Soltis, G. K. S. Wong, B. R. Ruhfel, and D. E. Soltis. 2018. Plastid phylogenomic analysis of green plants: a billion years of evolutionary history. *American Journal of Botany* 105:291–301.

Golubic, S., and L. Seong-Joo. 1999. Early cyanobacterial fossil record: preservation, palaeoenvironments and identification. *European Journal of Phycology* 34:339–348.

Gong, Y., K. Xu, Z. Zhan, et al. 2010. Alpha-tubulin and small subunit rRNA phylogenies of peritrichs are congruent and do not support the clustering of mobilids and sessilids (Ciliophora, Oligohymenophorea). *Journal of Eukaryotic Microbiology* 57:265–272.

Gong, Y. C., Y. H. Yu, E. Villalobo, F. Y. Zhu, and W. Miao. 2006. Reevaluation of the phylogenetic relationship between mobilid and sessilid Peritrichs (Ciliophora, Oligohymenophorea) based on small subunit rRNA genes sequences. *Journal of Eukaryotic Microbiology* 53:397–403.

Grau-Bové, X., G. Torruella, S. Donachie, et al. 2017. Dynamics of genomic innovation in the unicellular ancestry of animals. *eLife* 6:e26036.

Gustavs, L., R. Schumann, U. Karsten, and M. Lorenz. 2016. Mixotrophy in the terrestrial green alga *Apatococcus lobatus* (Trebouxiophyceae, Chlorophyta). *Journal of Phycology* 52:311–314.

Haindl, M., and E. Holler. 2005. Use of the giant multinucleate plasmodium of *Physarum polycephalum* to study RNA interference in the myxomycete. *Analytical Biochemistry* 342:194–199.

Harris, B. J., C. J. Harrison, A. M. Hetherington, and T. A. Williams. 2020. Phylogenomic evidence for the monophyly of bryophytes and the reductive evolution of stomata. *Current Biology* 30:2001–2012.e2.

Harris, S. D. 2011. Hyphal morphogenesis: an evolutionary perspective. *Fungal Biology* 115:475–484.

Harrison, C. J. 2017. Development and genetics in the evolution of land plant body plans. *Philosophical Transactions of the Royal Society B: Biological Sciences* 372:20150490.

Hehenberger, E., D. V. Tikhonenkov, M. Kolísko, et al. 2017. Novel predators reshape Holozoan phylogeny and reveal the presence of a two-component signaling system in the ancestor of animals. *Current Biology* 27:2043–2050 e6.

Herrero, A., J. Stavans, and E. Flores. 2016. The multicellular nature of filamentous heterocyst-forming cyanobacteria. *FEMS Microbiology Reviews* 40:831–854.

Herron, M. D., A. Rashidi, D. E. Shelton, and W. W. Driscoll. 2013. Cellular differentiation and individuality in the "minor" multicellular taxa. *Biological Reviews* 88:844–861.

Hibberd, D. J. 1975. Observations on the ultrastructure of the choanoflagellate *Codosiga botrytis* (Ehr.) Saville-Kent with special reference to the flagellar apparatus. *Journal of Cell Science* 17:191–219.

Hibbett, D. S., M. Blackwell, T. Y. James, et al. 2018. Phylogenetic taxon definitions for Fungi, Dikarya, Ascomycota and Basidiomycota. *IMA Fungus* 9:291–298.

Hofmann, H. J. 1976. Precambrian microflora, Belcher Islands, Canada: significance and systematics. *Journal of Paleontology* 50:1040–1073.

Hug, L. A., B. J. Baker, K. Anantharaman, et al. 2016. A new view of the tree of life. *Nature Microbiology* 1:1–6.

Ishizaki, K. 2017. Evolution of land plants: insights from molecular studies on basal lineages. *Bioscience, Biotechnology, and Biochemistry* 81:73–80.

James, T. Y., J. E. Stajich, C. T. Hittinger, and A. Rokas. 2020. Toward a fully resolved fungal tree of life. *Annual Review of Microbiology* 74:291–313.

James-Clark, H. 1866. Conclusive proofs of the animality of the ciliate sponges, and of their affinities with the Infusoria flagellata. *American Journal of Science* Series 2 Vol. 42:320–324.

Ji, D., and Y. Kusuoka. 2009. A description of *Apocarchesium rosettum* n. gen., n. sp. and a redescription of *Ophrydium eichornii* Ehrenberg, 1838, two freshwater peritrichous ciliates from Japan. *European Journal of Protistology* 45:21–28.

Jonsson, H., L. W. Hugerth, J. Sundh, E. Lundin, and A. F. Andersson. 2020. Genome sequence of segmented filamentous bacteria present in the human intestine. *Communications Biology* 3:1–9.

Karpov, S., M. A. Mamkaeva, V. Aleoshin, et al. 2014. Morphology, phylogeny, and ecology of the aphelids (Aphelidea, Opisthokonta) and proposal for the new superphylum Opisthosporidia. *Frontiers in Microbiology* 5:112.

Karpov, S. A., and S. J. Coupe. 1998. A revision of choanoflagellate genera *Kentrosiga* Schiller, 1953 and *Desmarella* Kent, 1880. *Acta Protozoologica* 37:23–27.

Kawai, H., and E. C. Henry. 2016. Phaeophyta. In *Handbook of the Protists*, ed. J. M. Archibald, A. G. B. Simpson, C. H. Slamovits, et al., 1–38. Cham: Springer International Publishing.

Keim, C. N., J. L. Martins, F. Abreu, et al. 2004. Multicellular life cycle of magnetotactic prokaryotes. *FEMS Microbiology Letters* 240:203–208.

King, N., and A. Rokas. 2017. Embracing uncertainty in reconstructing early animal evolution. *Current Biology* 27:R1081–R1088.

Kirk, D. L. 2000. *Volvox* as a model system for studying the ontogeny and phylogeny of multicellularity and cellular differentiation. *Journal of Plant Growth Regulation* 3:265–274.

Kiss, E., B. Hegedüs, M. Virágh, et al. 2019. Comparative genomics reveals the origin of fungal hyphae and multicellularity. *Nature Communications* 10:4080.

Knoll, A. H. 2011. The multiple origins of complex multicellularity. *Annual Review of Earth and Planetary Sciences* 39:217–239.

Kożyczkowska, A., S. R. Najle, E. Ocaña-Pallarès, et al. 2020. Stable transfection in the protist *Corallochytrium limacisporum* allows identification of novel cellular features among unicellular relatives of animals. *bioRxiv*:2020.11.12.379420.

Kristiansen, J., and P. Škaloud. 2017. Chrysophyta. In *Handbook of the Protists*, ed. J. M. Archibald, A. G. B. Simpson, and C. H. Slamovits, 331–366. Cham: Springer International Publishing.

Lambie, S. C., W. J. Kelly, S. C. Leahy, et al. 2015. The complete genome sequence of the rumen methanogen *Methanosarcina barkeri* CM1. *Standards in Genomic Sciences* 10:57.

Larson, B. T., T. Ruiz-Herrero, S. Lee, et al. 2020. Biophysical principles of choanoflagellate self-organization. *Proceedings of the National Academy of Sciences* 117:1303–1311.

Latijnhouwers, M., P. J. G. M. De Wit, and F. Govers. 2003. Oomycetes and fungi: similar weaponry to attack plants. *Trends in Microbiology* 11:462–469.

Laumer, C. E., R. Fernández, S. Lemer, et al. 2019. Revisiting metazoan phylogeny with genomic sampling of all phyla. *Proceedings of the Royal Society B: Biological Sciences* 286:20190831.

Laundon, D., B. T. Larson, K. McDonald, N. King, and P. Burkhardt. 2019. The architecture of cell differentiation in choanoflagellates and sponge choanocytes. *PLoS Biol* 17:e3000226.

Leadbeater, B. S. C. 2015. *The choanoflagellates: evolution, biology, and ecology*. Cambridge, United Kingdom: Cambridge University Press.

Lecroq, B., A. J. Gooday, M. Tsuchiya, and J. Pawlowski. 2009. A new genus of xenophyophores (Foraminifera) from Japan Trench: morphological description, molecular phylogeny and elemental analysis. *Zoological Journal of the Linnean Society* 156:455–464.

Leebens-Mack, J. H., M. S. Barker, E. J. Carpenter, et al. 2019. One thousand plant transcriptomes and the phylogenomics of green plants. *Nature* 574:679–685.

Leliaert, F., and J. M. Lopez-Bautista. 2015. The chloroplast genomes of *Bryopsis plumosa* and *Tydemania expeditiones* (Bryopsidales, Chlorophyta): compact genomes and genes of bacterial origin. *BMC Genomics* 16:204.

Leliaert, F., D. R. Smith, H. Moreau, et al. 2012a. Phylogeny and Molecular Evolution of the Green Algae. *Critical Reviews in Plant Sciences* 31. Taylor & Francis:1–46.

Leliaert, F., D. R. Smith, H. Moreau, et al. 2012b. Phylogeny and molecular evolution of the green algae. *Critical Reviews in Plant Sciences* 31:1–46.

Leliaert, F., A. Tronholm, C. Lemieux, et al. 2016. Chloroplast phylogenomic analyses reveal the deepest-branching lineage of the Chlorophyta, Palmophyllophyceae class. nov. *Scientific Reports* 6:25367.

Lemieux, C., A. T. Vincent, A. Labarre, C. Otis, and M. Turmel. 2015. Chloroplast phylogenomic analysis of chlorophyte green algae identifies a novel lineage sister to the Sphaeropleales (Chlorophyceae). *BMC Evolutionary Biology* 15:264.

Levin, T. C., A. J. Greaney, L. Wetzel, and N. King. The rosetteless gene controls development in the choanoflagellate *S. rosetta*. *eLife* 3:e04070.

Li, L., S. Wang, H. Wang, et al. 2020. The genome of *Prasinoderma coloniale* unveils the existence of a third phylum within green plants. *Nature Ecology & Evolution* 4:1220–1231.

Liu, X., K. Bogaert, A. H. Engelen, et al. 2017. Seaweed reproductive biology: environmental and genetic controls. *Botanica Marina* 60:89–108.

Liu, Y., E. T. Steenkamp, H. Brinkmann, et al. 2009. Phylogenomic analyses predict sistergroup relationship of nucleariids and Fungi and paraphyly of zygomycetes with significant support. *BMC Evolutionary Biology* 9:272.

López-Escardó, D., X. Grau-Bové, A. Guillaumet-Adkins, et al. 2019. Reconstruction of protein domain evolution using single-cell amplified genomes of uncultured choanoflagellates sheds light on the origin of animals. *Philosophical Transactions of the Royal Society B: Biological Sciences* 374:20190088.

Lynn, D. H. 2016. Ciliophora. In *Handbook of the Protists*, ed. J. M. Archibald, A. G. B. Simpson, C. H. Slamovits, et al., 1–52. Cham: Springer International Publishing.

Lyons, N. A., and R. Kolter. 2015. On the evolution of bacterial multicellularity. *Current Opinion in Microbiology* 24:21–28.

Macario, A. J. L., and E. C. de Macario. 2001. The molecular chaperone system and other anti-stress mechanisms in Archaea. *Frontiers in Bioscience* 6:d262-283.

Margulis, L., and M. J. Chapman. 2009. *Kingdoms and domains: an illustrated guide to the phyla of life on earth.* Amsterdam: Academic Press.

Marshall, W. L., and M. L. Berbee. 2011. Facing unknowns: living cultures (*Pirum gemmata* gen. nov., sp. nov., and *Abeoforma whisleri*, gen. nov., sp. nov.) from invertebrate digestive tracts represent an undescribed clade within the unicellular Opisthokont lineage Ichthyosporea (Mesomycetozoea). *Protist* 162:33–57.

Mayerhofer, L. E., A. J. Macario, and E. C. de Macario. 1992. Lamina, a novel multicellular form of *Methanosarcina mazei* S-6. *Journal of Bacteriology* 174:309–314.

McCourt, R. M., C. F. Delwiche, and K. G. Karol. 2004. Charophyte algae and land plant origins. *Trends in Ecology & Evolution* 19:661–666.

de Mendoza, A., H. Suga, J. Permanyer, M. Irimia, and I. Ruiz-Trillo. 2015. Complex transcriptional regulation and independent evolution of fungal-like traits in a relative of animals. *eLife* 4:e08904.

Mendoza, L., and R. Vilela. 2013. Presumptive synchronized nuclear divisions without cytokinesis in the *Rhinosporidium seeberi* parasitic life cycle. *Microbiology* 159:1545–51.

Mikkelsen, M. D., J. Harholt, P. Ulvskov, et al. 2014. Evidence for land plant cell wall biosynthetic mechanisms in charophyte green algae. *Annals of Botany* 114:1217–1236.

Moroz, L. L. 2015. Convergent evolution of neural systems in ctenophores. *Journal of Experimental Biology* 218:598–611.

Moroz, L. L., K. M. Kocot, M. R. Citarella, et al. 2014. The ctenophore genome and the evolutionary origins of neural systems. *Nature* 510:109–114.

Muñoz-Gómez, S. A., F. G. Mejía-Franco, K. Durnin, et al. 2017. The new red algal subphylum Proteorhodophytina comprises the largest and most divergent plastid genomes known. *Current Biology* 27:1677–1684.e4.

Naranjo-Ortiz, M. A., and T. Gabaldón. 2019. Fungal evolution: diversity, taxonomy and phylogeny of the Fungi. *Biological Reviews* 94:2101–2137.

Naumann, B., and P. Burkhardt. 2019. Spatial cell disparity in the colonial choanoflagellate *Salpingoeca rosetta*. *Frontiers in Cell and Developmental Biology* 7:231.

Nedelcu, A. M., and R. E. Michod. 2004. Evolvability, modularity, and individuality during the transition to multicellularity in volvocalean green algae. In *Modularity in Development and Evolution*, Schlosser, G. and Wagner, G Eds., 31. Chicago, IL: Univ. Chicago Press.

Nicholls, K. H., and D. E. Wujek. 2015. Chrysophyceae and Phaeothamniophyceae. In *Freshwater Algae of North America*, ed. J. D. Wehr, R. G. Sheath, and J. P. Kociolek, 537–586. Amsterdam: Academic Press.

Norf, H., and W. Foissner. 2010. A new flagship peritrich (Ciliophora, Peritrichida) from the River Rhine, Germany: *Apocarchesium arndti* n. sp. *Journal of Eukaryotic Microbiology* 57:250–264.

Ohtaka, K., K. Hori, Y. Kanno, M. Seo, and H. Ohta. 2017. Primitive auxin response without TIR1 and Aux/IAA in the charophyte alga *Klebsormidium nitens*. *Plant Physiology* 174:1621–1632.

Ondracka, A., O. Dudin, and I. Ruiz-Trillo. 2018. Decoupling of nuclear division cycles and cell size during the coenocytic growth of the Ichthyosporean *Sphaeroforma arctica*. *Current Biology* 28:1964–1969.

Paps, J., and P. W. H. Holland. 2018. Reconstruction of the ancestral metazoan genome reveals an increase in genomic novelty. *Nature Communications* 9:1730.

Parra-Acero, H., N. Ros-Rocher, A. Perez-Posada, et al. 2018. Transfection of *Capsaspora owczarzaki*, a close unicellular relative of animals. *Development* 145:dev162107.

Parra-Acero, H., M. Harcet, N. Sánchez-Pons, et al. 2020. Integrin-mediated adhesion in the unicellular holozoan *Capsaspora owczarzaki*. *Current Biology* 30:4270–4275.e4.

Pereira, L., and J. M. Neto. 2014. *Marine algae: biodiversity, taxonomy, environmental assessment, and biotechnology.* Boca Raton, FL: CRC Press.

Pfeffer, C., S. Larsen, J. Song, et al. 2012. Filamentous bacteria transport electrons over centimeter distances. *Nature* 491:218–221.

Puttick, M. N., J. L. Morris, T. A. Williams, et al. 2018. The interrelationships of land plants and the nature of the ancestral embryophyte. *Current Biology* 28:733–745.e2.

Raghu-kumar, S. 1987. Occurrence of the thraustochytrid, *Corallochytrium limacisporum* gen. et sp. nov. in the coral reef lagoons of the lakshadweep islands in the Arabian Sea. *Botanica Marina* 30:83–90.

Richter, D. J., and F. Nitsche. 2017. Choanoflagellatea. In *Handbook of the Protists*, 1479–1496. New York, NY: Springer Berlin Heidelberg.

Richter, D. J., P. Fozouni, M. Eisen, and N. King. 2018. Gene family innovation, conservation and loss on the animal stem lineage. *eLife* 7:e34226.

Rikkinen, J. 2013. Molecular studies on cyanobacterial diversity in lichen symbioses. *MycoKeys* 6:3–32.

Rizouli, A., F. C. Küpper, P. Louizidou, et al. 2020. The minute alga *Schizocladia ischiensis* (Schizocladiophyceae, Ochrophyta) isolated by germling emergence from 24 m depth off Rhodes (Greece). *Diversity* 12:102.

Robinson, R. W. 1986. Life cycles in the methanogenic Archaebacterium *Methanosarcina mazei. Applied and Environmental Microbiology* 52:17–27.

Ryan, J. F., and M. Chiodin. 2015. Where is my mind? How sponges and placozoans may have lost neural cell types. *Philosophical Transactions of the Royal Society B: Biological Sciences* 370. Royal Society:20150059.

Ryan, J. F., K. Pang, C. E. Schnitzler, et al. 2013. The genome of the ctenophore *Mnemiopsis leidyi* and Its implications for cell type evolution. *Science* 342:1242592.

Rybalka, N., T. Mikhailyuk, T. Darienko, et al. 2020. Genotypic and phylogenetic diversity of new isolates of terrestrial Xanthophyceae (Stramenopiles) from maritime sandy habitats. *Phycologia* 0:1–9.

Sahoo, D., and S. Kumar. 2015. Xanthophyceae, Euglenophyceae and Dinophyceae. In *The Algae World*, ed. D. Sahoo and J. Seckbach, 259–305. Cellular Origin, Life in Extreme Habitats and Astrobiology. Dordrecht: Springer Netherlands.

Schirrmeister, B. E., M. Gugger, and P. C. J. Donoghue. 2015. Cyanobacteria and the great oxidation event: evidence from genes and fossils. *Palaeontology* 58:769–785.

Schnupf, P., V. Gaboriau-Routhiau, and N. Cerf-Bensussan. 2013. Host interactions with segmented filamentous bacteria: an unusual trade-off that drives the post-natal maturation of the gut immune system. *Seminars in Immunology* 25. Microbiota and the Immune System, an Amazing Mutualism Forged by Co-Evolution:342–351.

Schopf, J. W., and A. B. Kudryavtsev. 2009. Confocal laser scanning microscopy and Raman imagery of ancient microscopic fossils. *Precambrian Research* 173. World Summit on Ancient Microscopic Fossils:39–49.

Scott, J., J. Thomas, and B. Saunders. 1988. Primary pit connections in *Compsopogon coeruleus* (Balbis) Montagne (Compsopogonales, Rhodophyta). *Phycologia* 27: 327–333.

Sebé-Pedrós, A., M. Irimia, J. Del Campo, et al. 2013. Regulated aggregative multicellularity in a close unicellular relative of metazoa. *eLife* 2:e01287.

Sebé-Pedrós, A., C. Ballare, H. Parra-Acero, et al. 2016. The dynamic regulatory genome of *Capsaspora* and the origin of animal multicellularity. *Cell* 165:1224–1237.

Sebé-Pedrós, A., B. M. Degnan, and I. Ruiz-Trillo. 2017. The origin of Metazoa: a unicellular perspective. *Nature Reviews Genetics* 18:498–512.

Ševčíková, T., V. Klimeš, V. Zbránková, et al. 2016. A comparative analysis of mitochondrial genomes in Eustigmatophyte algae. *Genome Biology and Evolution* 8:705–722.

Ševčíková, T., T. Yurchenko, K. P. Fawley, et al. 2019. Plastid genomes and proteins illuminate the evolution of Eustigmatophyte algae and their bacterial endosymbionts. *Genome Biology and Evolution* 11:362–379.

Shen, X. X., C. T. Hittinger, and A. Rokas. 2017. Contentious relationships in phylogenomic studies can be driven by a handful of genes. *Nature Ecology & Evolution* 1:1–10.

Shiratori, T., S. Fujita, T. Shimizu, T. Nakayama, and K. Ishida. 2017. *Viridiuvalis adhaerens* gen. et sp. nov., a novel colony-forming chlorarachniophyte. *Journal of Plant Research* 130:999–1012.

Silberfeld, T., J. W. Leigh, H. Verbruggen, et al. 2010. A multi-locus time-calibrated phylogeny of the brown algae (Heterokonta, Ochrophyta, Phaeophyceae): investigating the evolutionary nature of the "brown algal crown radiation." *Molecular Phylogenetics and Evolution* 56:659–674.

Silberfeld, T., F. Rousseau, and B. de Reviers. 2014. An updated classification of brown algae (Ochrophyta, Phaeophyceae). *Cryptogamie, Algologie* 35:117–156.

Simion, P., H. Philippe, D. Baurain, et al. 2017. A large and consistent phylogenomic dataset supports sponges as the sister group to all other animals. *Current Biology* 27:958–967.

de Sousa, F., P. G. Foster, P. C. J. Donoghue, H. Schneider, and C. J. Cox. 2019. Nuclear protein phylogenies support the monophyly of the three bryophyte groups (Bryophyta Schimp.). *New Phytologist* 222:565–575.

Spatafora, J. W., Y. Chang, G. L. Benny, et al. 2016. A phylum-level phylogenetic classification of zygomycete fungi based on genome-scale data. *Mycologia* 108:1028–1046.

Speirs, L. B. M., D. T. F. Rice, S. Petrovski, and R. J. Seviour. 2019. The phylogeny, biodiversity, and ecology of the Chloroflexi in activated sludge. *Frontiers in Microbiology* 10:2015.

Stajich, J. E., M. L. Berbee, M. Blackwell, et al. 2009. Primer – The Fungi. *Current Biology* 19: R840–R845.

Suga, H., and I. Ruiz-Trillo. 2013. Development of ichthyosporeans sheds light on the origin of metazoan multicellularity. *Developmental Biology* 377:284–92.

Sun, P., J. C. Clamp, D. Xu, Y. Kusuoka, and M. Hori. 2011. Molecular phylogeny of the family Vorticellidae (Ciliophora, Peritrichia) using combined datasets with a special emphasis on the three morphologically similar genera *Carchesium*, *Epicarchesium* and *Apocarchesium*. *International Journal of Systematic and Evolutionary Microbiology*, 61:1001–1010.

Tedersoo, L., S. Sánchez-Ramírez, U. Kõljalg, et al. 2018. High-level classification of the Fungi and a tool for evolutionary ecological analyses. *Fungal Diversity* 90:135–159.

Teske, A., and D. C. Nelson. 2006. The genera *Beggiatoa* and *Thioploca*. In *The Prokaryotes*, ed. M. Dworkin, S. Falkow, E. Rosenberg, K. H. Schleifer, and E. Stackebrandt, 784–810. Springer New York.

Thompson, C. L., R. Vier, A. Mikaelyan, T. Wienemann, and A. Brune. 2012. *Candidatus* Arthromitus revised: segmented filamentous bacteria in arthropod guts are members of Lachnospiraceae. *Environmental Microbiology* 14:1454–1465.

Tikhonenkov, D. V., E. Hehenberger, A. S. Esaulov, et al. 2020. Insights into the origin of metazoan multicellularity from predatory unicellular relatives of animals. *BMC Biology* 18:39.

Tomitani, A., A. H. Knoll, C. M. Cavanaugh, and T. Ohno. 2006. The evolutionary diversification of cyanobacteria: molecular–phylogenetic and paleontological perspectives. *Proceedings of the National Academy of Sciences* 103:5442–5447.

Torruella, G., A. de Mendoza, X. Grau-Bové, et al. 2015. Phylogenomics reveals convergent evolution of lifestyles in close relatives of animals and fungi. *Current Biology* 25:2404–10.

Torruella, G., X. Grau-Bové, D. Moreira, et al. 2018. Global transcriptome analysis of the aphelid *Paraphelidium tribonemae* supports the phagotrophic origin of fungi. *Communications Biology* 1:1–11.

Trojan, D., L. Schreiber, J. T. Bjerg, et al. 2016. A taxonomic framework for cable bacteria and proposal of the candidate genera *Electrothrix* and *Electronema*. *Systematic and Applied Microbiology* 39:297–306.

Turmel, M., C. Otis, and C. Lemieux. 2017. Divergent copies of the large inverted repeat in the chloroplast genomes of ulvophycean green algae. *Scientific Reports* 7:994.

Umen, J. G. 2014. Green algae and the origins of multicellularity in the plant kingdom. *Cold Spring Harbor Perspectives in Biology* 6:a016170.

Verbruggen, H., C. A. Maggs, G. W. Saunders, et al. 2010. Data mining approach identifies research priorities and data requirements for resolving the red algal tree of life. *BMC Evolutionary Biology* 10:16.

Vergés, A., and A. H. Campbell. 2020. Kelp forests. *Current Biology* 30:R919–R920.

Volland, J. M., A. Schintlmeister, H. Zambalos, et al. 2018. NanoSIMS and tissue autoradiography reveal symbiont carbon fixation and organic carbon transfer to giant ciliate host. *The ISME Journal* 12:714–727.

de Vries, J., and J. M. Archibald. 2018. Plant evolution: landmarks on the path to terrestrial life. *New Phytologist* 217:1428–1434.

Waller, R. F., P. A. Cleves, M. Rubio-Brotons, et al. 2018. Strength in numbers: collaborative science for new experimental model systems. *PLoS Biol* 16:e2006333.

Wang, M., C. Hu, B. B. Barnes, et al. 2019. The great Atlantic *Sargassum* belt. *Science* 365:83–87.

West, J. A., S. L. de Goer, and G. C. Zuccarello. 2014. A new species of *Bangiopsis*: *B. franklynottii* sp. nov. (Stylonematophyceae, Rhodophyta) from Australia and India and comments on the genus. *Algae* 29:101–109.

Whelan, N. V., K. M. Kocot, T. P. Moroz, et al. 2017. Ctenophore relationships and their placement as the sister group to all other animals. *Nature Ecology & Evolution* 1:1737–1746.

Wickett, N. J., S. Mirarab, N. Nguyen, et al. 2014. Phylotranscriptomic analysis of the origin and early diversification of land plants. *Proceedings of the National Academy of Sciences* 111:E4859–E4868.

Yamagishi, T., D. G. Müller, and H. Kawai. 2014. Comparative transcriptome analysis of *Discosporangium mesarthrocarpum* (Phaeophyceae), *Schizocladia ischiensis* (Schizocladiophyceae), and *Phaeothamnion confervicola* (Phaeothamniophyceae), with special reference to cell wall-related genes. *Journal of Phycology* 50:543–551.

Yamashita, S., Y. Arakaki, H. Kawai-Toyooka, et al. 2016. Alternative evolution of a spheroidal colony in volvocine algae: developmental analysis of embryogenesis in *Astrephomene* (Volvocales, Chlorophyta). *BMC Evolutionary Biology* 16:243.

Yang, E. C., S. M. Boo, D. Bhattacharya, et al. 2016. Divergence time estimates and the evolution of major lineages in the florideophyte red algae. *Scientific Reports* 6:21361.

Yoon, H. S., K. M. Müller, R. G. Sheath, F. D. Ott, and D. Bhattacharya. 2006. Defining the major lineages of red algae (rhodophyta). *Journal of Phycology* 42:482–492.

Yoon, H. S., W. Nelson, S. C. Lindstrom, et al. 2017. Rhodophyta. In *Handbook of the Protists*, ed. J. M. Archibald, A. G. B. Simpson, and C. H. Slamovits, 89–133. Cham: Springer International Publishing.

Zechman, F. W., H. Verbruggen, F. Leliaert, et al. 2010. An unrecognized ancient lineage of green plants persists in deep marine waters. *Journal of Phycology* 46:1288–1295.

Zhan, Z., K. Xu, A. Warren, and Y. Gong. 2009. Reconsideration of phylogenetic relationships of the subclass Peritrichia (Ciliophora, Oligohymenophorea) based on small subunit ribosomal RNA gene sequences, with the establishment of a new subclass Mobilia Kahl, 1933. *Journal of Eukaryotic Microbiology* 56:552–558.

Zhan, Z., K. Xu, and M. Dunthorn. 2013. Evaluating molecular support for and against the monophyly of the Peritrichia and phylogenetic relationships within the Mobilida (Ciliophora, Oligohymenophorea). *Zoologica Scripta* 42:213–226.

Zhuang, Y., J. C. Clamp, Z. Yi, and D. Ji. 2018. Phylogeny of the families Zoothamniidae and Epistylididae (Protozoa: Ciliophora: Peritrichia) based on analyses of three rRNA-coding regions. *Molecular Phylogenetics and Evolution* 118:99–107.

10 Group Formation
Hypotheses for the Evolution of Clonal Multicellularity

Stefania E. Kapsetaki
Biodesign Institute, Arizona State University, Tempe, AZ, USA

Roberta M. Fisher
Department of Biology, University of Copenhagen, Copenhagen, Denmark

CONTENTS

10.1 INTRODUCTION

Multicellularity occurs all over the tree of life in myriad different forms and, depending on the estimate, has evolved independently between 8 and 25 times (Fisher et al., 2013; Grosberg & Strathmann, 2007; Knoll, 2011; Lyons & Kolter, 2015; Niklas, 2014; Niklas & Newman, 2013). Multicellularity underpins much of the complex life that we can see, but increasingly we are becoming aware of the plethora of microbial species that are multicellular or have cooperative multicellular behaviors. In Chapters 5–8, we heard about the myxobacteria and about *Dictyostelium* and the cellular slime molds.

DOI: 10.1201/9780429351907-13

However, regardless of the taxa in which multicellularity has evolved, the way in which multicellular groups form has important ramifications for cooperative behavior, multicellular complexity, and the potential to evolve obligate multicellularity (which we will come to later). In this chapter, we focus on clonal multicellularity, where multicellular groups form through daughter cells sticking to mother cells after cell division. This type of multicellularity has evolved in at least 12 different lineages and has led to some of the most complex and diverse multicellular species (Fisher et al., 2013).

First, we briefly review the different lineages where multicellularity is found and ways in which multicellular groups can form, before focusing on clonal multicellularity. Next, we give a primer in social evolution theory to allow us to explore the reasons why clonality has allowed the evolution of extreme cooperative behavior, such as altruistic somatic cells. We then give some examples of clonal multicellular taxa to explore the advantages this type of group formation can pose in the natural world. Finally, we ask whether clonality is the only route to complex, obligate multicellularity, like we see in animals and plants, and finish by reflecting on the major evolutionary transitions in individuality.

10.2   WHAT DO WE MEAN BY MULTICELLULARITY?

What do we mean by multicellularity? On a basic level, we define multicellularity as when cells stick together (hence 'multicell') and have been selected to do so. This definition captures a huge variety of multicellular phenotypes, including Bacteria and Archaea that form cooperative groups (e.g. Bonner, 1998, 2009; LaPaglia & Hartzell, 1997; Mayerhofer et al., 1992), fungal hyphae, plant multicellularity, fruiting bodies in ciliates, a range of algal phenotypes, and of course, metazoans. However, it is clear that multicellularity can either be facultative, where cells are able to survive and reproduce independently or be part of a group, or obligate, where cells are permanently part of a group and cannot survive and reproduce independently of that group. As an example, humans are obligately multicellular. Our somatic cells (e.g. skin cells, neurons, muscle cells) cannot separate themselves from our multicellular bodies and live an independent existence. They exist as part of a multicellular whole and are terminally differentiated into their respective somatic phenotypes so that they have no (sexually) reproductive function. Our gametes may exist transiently as unicells, but this is an existence which either ends in death or a new, obligately multicellular individual. This is in stark contrast to species with facultative multicellularity, where individual cells need not join a multicellular group and may do so only under specific conditions. An example of this is the cellular slime mold *Dictyostelium*. In this species, single cells can survive and proliferate in the environment quite happily (Hashimura et al., 2019; Strassmann & Queller, 2011), only coming together to form a multicellular fruiting body during times of hardship where food supplies run low (Bonner, 2009; Strassmann et al., 2000). The fruiting body phase is in no way required for *Dictyostelium* amoebae to feed, move, communicate and divide (Kessin, 2001; Strassmann, 2010).

Multicellular species also form multicellular groups in distinctly different ways. These can be broadly classified as non-clonal (aggregative) multicellularity and clonal multicellularity. A possible third category, hyphal multicellularity as described below, displays similarities to both types of group formation but seems (at least in practice) to more closely resemble clonal multicellularity.

10.2.1   Non-clonal (Aggregative) Multicellularity

We will only describe aggregative multicellularity very briefly here, as it has been covered extensively in Chapters 5–8. However, it is worth stating that non-clonal aggregative multicellularity has evolved independently in five different lineages and is, therefore, more common than clonal aggregative multicellularity, which has evolved independently in three different lineages, as a way of forming multicellular groups (Fisher et al., 2013). Succinctly put, aggregative multicellularity occurs when many individual cells in the environment come together to form a multicellular group. They do this not through mitotic cell division, but through either active movement or chance. This results in a group made of often genetically distinct cells (which is why it is also referred to as 'non-clonal' multicellularity). Some classic examples of species with non-clonal group formation include *Dictyostelium* (mentioned here before, and in detail in Chapter 7), *Myxobacteria* (Chapter 6), and many species of green algae in the Chlorophyta.

10.2.2   Hyphal Multicellularity

Hyphal multicellularity is specific to Fungi, a lineage where multicellularity is somewhat distinct from all others. Multicellular fungi are composed of many hyphae (tubular structures that form by apical extension), and these hyphae form a mycelium that spreads out through a substrate to forage for nutrients (Olsson et al., 2002). This is a unique setup for a few reasons. Firstly, hyphae are not compartmentalized so do not need to solve conflicts between cells in the same way as non-clonal species (Scott et al., 2019). Secondly, there can also be multiple nuclei in one hypha. This makes hyphal multicellularity similar to the non-clonality that occurs in aggregative multicellularity. However, the way hyphae form is through cell division and does often result in an alignment of fitness interests across hyphae, which more closely resembles clonal group formation (Kuhn et al., 2001; Marleau et al., 2011). Hyphal multicellularity in the fungi probably didn't evolve in response to predators, like we think might have happened in other lineages, but instead helped with foraging for nutrients (Heaton et al., 2020; Olsson et al., 2002; Richards et al., 2017).

10.2.3   Clonal Multicellularity

Clonal multicellularity is defined as when multicellular groups form through daughter cells remaining attached to their mother cells after division (Figure 10.1). This can be very simple, as displayed in *Chlamydomonas reinhardtii* (see later section), where several rounds of divisions take place leading to multiple cells attached with an extracellular matrix. This simple clonal group formation is found in, for example, *Scenedesmus*, *Candida albicans* and *Physarum polycephalum* (Baldauf & Doolittle, 1997; Berman, 2006; Engelberg et al., 1998; Everhart & Keller, 2008; Kapsetaki et al., 2017; Lürling, 2001; Whiteway & Bachewich, 2007). However, in the metazoan and plant lineages, large and complex multicellular bodies are formed through much the same process – a zygote divides many thousands of times to form

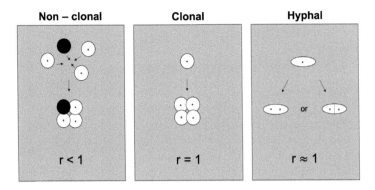

FIGURE 10.1 Multicellular group formation. Sketches showing non-clonal, clonal and hyphal multicellular group formation and the consequences of each for relatedness between cells.

a body with perhaps up to 100 quadrillion individual cells (Zhang et al., 2005). The fundamental similarity between all multicellular species that form this way is that (given everything else being equal) all cells in the multicellular body are clonally related. In other words, they are all copies of each other formed through mitosis (hence, the name). This mode of group formation results in special relatedness conditions between cells, which we deal with in more depth in our 'Social evolution primer' (below). This is in stark contrast to non-clonal group formation, where cells with different genetic backgrounds form a multicellular group.

Whilst aggregative and hyphal multicellularity can result in myriad of multicellular forms and, in some cases, impressively complex and coordinated behavior (as described in Chapters 5–8), only clonal multicellularity has led to the levels of complexity demonstrated in the lineages of animals and plants. Only clonal multicellularity, where all cells in the body result from one (or very few) initial cells, has led to permanent division of labor and >100 different cell types (Figure 10.2) (Fisher et al., 2013; Nagy et al., 2020). And whilst multicellularity has evolved >24 times, obligate multicellular organisms have in fact only evolved nine times, and only in

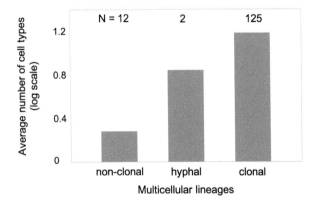

FIGURE 10.2 Cell-type complexity in non-clonal, hyphal and clonal multicellular lineages. N indicates the number of lineages in each category. (Reproduced from Fisher et al., 2020.)

species with clonal multicellularity (Fisher et al., 2013). It is clear there is something unique about clonal multicellularity. To understand what, it is necessary to go back to social evolution theory.

10.3   SOCIAL EVOLUTION PRIMER

At a fundamental level, multicellularity is a cooperative behavior between cells. Here, we will just briefly touch on how cooperative behaviors can evolve between cells, but for an in-depth explanation of the evolution of cooperative behavior, we direct the reader to Bourke (2011), Davies et al. (2012) and West et al. (2007b). Simply put, in order for cooperation to evolve the cooperative behavior needs to be beneficial and the benefits of cooperating need to be felt by the individual cells themselves or the relatives of those cells. For example, if clumping together in a group means that each individual cell has a lower risk of being eaten by a predator (Kapsetaki & West, 2019; Lürling, 1999b, 2020; Lürling et al., 1996; Van Donk et al., 1993), then cooperation can be favored – there is a clear survival benefit to cooperating (not being eaten) and that benefit is felt by the individual cells in the group. In the above example, the cells experience *direct* benefits – the cooperating cells are the cells that receive the benefit of cooperation. However, cooperation can also evolve when cooperating cells experience *indirect* benefits – where it is the *relatives* of the cooperating cells that receive the benefits of cooperation. This requires that the cells are genetically related so that the benefits of cooperation are directed towards other individuals that carry the same genes as the cooperator (for a more thorough explanation, see Fisher et al., 2013 and West et al., 2007b).

When cells are clonally related, i.e. when relatedness (r) = 1, this creates specific conditions that allow altruism to evolve. This is because altruistic behaviors (Foster et al., 2006; Hamilton, 1964a, 1964b; West et al., 2007a) do not provide direct benefits to the cooperating cell, but instead provide indirect benefits to related cells. When the condition of $r = 1$ is consistent in time (throughout the lifespan of the multicellular body) and space (between all cells in the body), unconditionally altruistic cell types can evolve, that will always cooperate despite receiving no direct benefits. If $r < 1$ between cells, this opens up the possibility for natural selection to favor cells not cooperating.

Clonal group formation guarantees that all cells in the group will be genetically identical ($r = 1$), as they have all arisen through mitotic divisions from a single (or very few) cell(s). It is for this reason that multicellular species that develop through clonal group formation have been able to evolve an unconditional division of labor leading to many different cell types in obligately multicellular bodies. This has not been achieved by species developing through non-clonal group formation because relatedness between cells has not been high enough (and/or consistently high enough in time and space) to allow altruistic cooperation to evolve between cells (Fisher et al., 2013). For these species, there has always been the (at least theoretical) possibility that cells could 'do better' by not cooperating, and hence natural selection has not favored the evolution of unconditional altruistic behavior, such as permanent division of labor (for a detailed explanation, see Cooper & West, 2018).

As an interesting side note, this observation is reflected in other cooperative systems, most strikingly in animal societies. Just like multicellularity, animals from a variety of taxonomic groups live in cooperative groups (to varying degrees), including spiders, aphids, fish, meerkats, lions, bees, birds, monkeys – the list goes on. However, consistent with social evolution theory, the only lineages where unconditionally altruistic behaviors have evolved and cooperative groups have developed into obligate associations, is where groups begin with a strictly monogamous sexual pair (e.g. in ants) or where groups develop clonally (e.g. aphids) (Boomsma, 2007, 2009; Boomsma & Gawne, 2018; West et al., 2015).

It is worth noting that not all clonally developing species have become obligately multicellular. Clonal group formation does not guarantee that a species will evolve obligate multicellularity. This is because relatedness alone does not determine whether cooperation can evolve. It is a crucial point that clonal genetic relatedness is a necessary condition of obligate multicellularity, but it is not sufficient. The benefits (and costs) of multicellular cooperation still need to be high (and low) enough over time and through space to allow the evolution of unconditional altruism and obligate multicellularity (Foster et al., 2006; West et al., 2015). That is why there are many examples (see below) of multicellular species that develop clonally, but that have remained facultatively multicellular.

10.4   BENEFITS AND COSTS OF CLONAL MULTICELLULARITY

In order to understand what the benefits and costs of clonal multicellularity might be, we cannot just focus on obligate, complex multicellular species. This is because it can be difficult to disentangle the benefits and costs of multicellularity in species with hundreds of cell types, and even more so in species that are obligately multicellular where we cannot compare the free-living cells with their multicellular equivalents. It is also difficult to disentangle the current benefits of multicellularity (e.g. the benefits of having specialized organs) with the initial benefits of clonality. For clonal group formation to have been favored, there must have been consistent benefits of clonal group formation versus unicellularity, and this can be especially hard to evaluate in highly complex species such as metazoans.

Ideally, we want to be able to 'look back in time' to see what the initial benefits may have been for forming multicellular groups through clonal group formation. Here, we focus on four examples of multicellular species that develop clonally and allow us to explore what the benefits and costs of clonal multicellularity might be.

10.5   CYANOBACTERIA

Cyanobacteria are a large and diverse group of bacteria with an estimated 2698 species (Nabout et al., 2013). They include species such as *Microcystis aeruginosa* and *Nostoc thermotolerans*. For our purposes, what is particularly interesting about the cyanobacteria is that they include some of the simplest but oldest examples of clonal multicellularity (and, in fact, multicellularity in general). Evidence of multicellular filaments of cyanobacteria have been found in rocks from three billion years ago (Schirrmeister et al., 2011, 2013).

Certain species, notably *Oscillatoria sancta, Anabaena* sp., and *Cylindrospermum* sp., form long filaments of individual cells that originate from one precursor cell

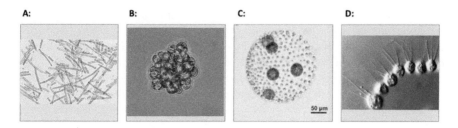

FIGURE 10.3   Examples of clonal multicellularity. A: Cyanobacteria (*Cylindrospermum* sp). Image courtesy of Adelaide laboratories of CSIRO Land and Water (1993). B: *Chlamydomonas reinhardtii*. Image courtesy of Josh Ming Borin (Herron et al., 2019). C: *Volvox carteri*. Image courtesy of Shelton & Michod, 2010. D: Choanoflagellate (*Desmarella moniliformis*). Image by Sergey Karpov, CC BY-SA 3.0 via Wikimedia Commons.

through clonal group formation (see Figure 10.3) (Schirrmeister et al., 2011). These filaments are obligately multicellular in the sense that the individual cells cannot survive and reproduce on their own (i.e. distinct from the group). Several species also have a division of labor between cells in the filament. For example, *Nostoc* species have cells called 'heterocysts' which fix nitrogen at night and vegetative cells that photosynthesize during the day (Bonner, 2003; Golden & Yoon, 2003; Kaiser, 2001; Schirrmeister et al., 2011). Nitrogen from the atmosphere is converted into ammonia via the enzyme nitrogenase, and this enzyme is inactivated by oxygen (Gallon, 1981; Rossetti et al., 2010). Such division of labor evolves when different partners (vegetative and heterocyst cells) receive accelerating fitness benefits from their environment when their tasks do not mix well or become more efficient (Cooper & West, 2018; Rossetti et al., 2010; West et al., 2015; West & Cooper, 2016). Accelerating fitness benefits means that the fitness of the individual performing the tasks increases over time (West et al., 2015). The fixation of nitrogen and photosynthesis are both beneficial to the cyanobacteria.

Cyanobacteria provide an example of how clonal group formation has allowed extreme morphological, functional, and spatiotemporal division of labor even in a relatively simple species (Cooper & West, 2018; Flores & Herrero, 2010; Schirrmeister et al., 2011). In fact, Rossetti et al. 2010 show that spatial structure, specifically the organization of cells in compartments, is important in preventing the spread of cheating cells, allowing the evolution of cooperation and division of labor in this species.

10.6   *CHLAMYDOMONAS REINHARDTII*

Chlamydomonas reinhardtii is a single-celled eukaryotic alga. When these cells are exposed to predators, they form multicellular groups through clonal group formation (Lurling & Beekman, 2006). One cell divides to gradually form a palmelloid cluster of many cells – a behavior that increases their survival (Lurling & Beekman, 2006). Cellular clumping in response to predation is selectively advantageous for the prey and has been observed in many more species of green algae (Kapsetaki & West, 2019; Lüring et al., 1997; Lürling, 1999a; Lürling et al., 1996; Van Donk et al., 1993).

However, these groups are not obligately multicellular (Figure 10.3). When removing the predators, algal groups break apart into single cells (Fisher et al., 2016;

Kapsetaki et al., 2016). One of the explanations for this phenomenon is that being in a group can be costly for individual cells (Kirk, 1994;; Kapsetaki & West, 2019; Lürling, 1999a; Ploug et al., 1999; Reynolds, 1984; Tollrian & Dodson, 1999; Trainor, 1998) because cells in the group may have less access to light due to higher sinking rates, and may pay a cost of producing extracellular adhesive molecules (Francis Rice Trainor, 1998; Kirk, 1994; Lürling, 1999a; Ploug et al., 1999; Reynolds, 1984; Tollrian & Dodson, 1999). *Chlamydomonas* is, therefore, a good example of a species that can form multicellular groups through clonal group formation, but does so facultatively.

An interesting consideration is the experimental evolution of multicellularity in *C. reinhardtii* by Herron et al. (2019) in response to predation pressure by *Paramecium tetraurelia*. Multicellular life cycles had a fitness advantage over the unicellular ancestors, showing that multicellularity conferred a significant benefit to *C. reinhardtii* cells. They were also able to show that multicellular groups of *C. reinhardtii* were stable over multiple generations, suggesting that obligate multicellularity could possibly evolve in the lab given the right conditions (Herron et al., 2019).

10.7   *VOLVOX CARTERI*

C. reinhardtii is a species in the order Chlamydomonadales also known as Volvocales (Buss, 1987; Grosberg & Strathmann, 2007; Koufopanou, 1994). This group contains more than 1760 species, many of which display multicellular phenotypes to varying degrees (Fritsch & West, 1927; Herron, 2009; Umen, 2020). Here, we focus on *Volvox carteri* (Figure 10.3), arguably one of the most extensively studied multicellular species in the group (Herron et al., 2009), but a detailed account of multicellularity in the Volvocine algae is given in Chapter 11.

Volvox carteri forms spherical colonies composed of up to 6000 cells (Kochert, 1968; Smith, 1944), each of which bears a striking resemblance to *Chlamydomonas*. *V. carteri* is obligately multicellular, and its cells are functionally and reproductively differentiated (Buss, 1987; Grosberg & Strathmann, 2007; Koufopanou, 1994). Every colony contains around 16 larger reproductive cells that are enclosed by an outer layer of many more somatic cells that are terminally differentiated and contribute only to motility and photosynthesis (Matt & Umen, 2016; Shelton et al., 2012).

Division of labor between reproductive and somatic cells is achieved because some cells, during the first five rounds of cell division, express the gene regA whose protein inhibits cell growth (Herron, 2016; Meissner et al., 1999). These cells become smaller than 8μm, they keep their flagella, and become somatic non-reproductive cells. Other cells do not express regA and grow much larger, lose their flagella and become reproductive germ cells (Hallmann, 2011; Kirk et al., 1993; Koufopanou, 1994).

Clonality plays a major role in *V. carteri*. The somatic and germ cells of *V. carteri* are clonal. In other words, their fitness interests are aligned. If a cell was to 'choose' to maximize its inclusive fitness by dividing (direct fitness) or sacrificing its reproduction to help a neighboring cell divide (indirect fitness), either 'choice' would be similar in terms of fitness, since the neighboring cell here is a clone (Foster et al., 2006). This is the case in *V. carteri*. Somatic cells have altruistically sacrificed their ability to reproduce, they are sterile, and only the germ cells reproduce. In this case of clonality, the maximal fitness of an individual cell is equal to the maximum fitness of the group (Davies et al., 2012; Michod et al., 2003, 2006).

10.8   CHOANOFLAGELLATES

In marine and freshwater environments, we find some single-celled eukaryotes, the choanoflagellates. They share a common ancestor with multicellular animals (e.g. Brunet & King, 2017; Fairclough et al., 2010; King, 2005; Koehl, 2020; Leadbeater, 2015; Mikhailov et al., 2009; Salvini-Plawen, 1978; Valentine & Marshall, 2015), and are therefore widely used as model organisms in the study of multicellular origins (Brunet et al., 2019; Brunet & King, 2017; Fairclough et al., 2013; King, 2004; King et al., 2008; Richter & King, 2013), including the benefits and costs of group formation. Groups of choanoflagellates have an evolutionary advantage over single cells in terms of their ability to forage (Kirkegaard & Goldstein, 2016; Nichols et al., 2009; Roper et al., 2013) and avoid predation (Koehl, 2020). Prey bacteria *Algoriphagus machipongonensis* release a sulfonolipid, RIF-1, in the extracellular environment. This localized concentration of RIF-1 can trigger facultative multicellular group formation in their predator *Salpingoeca rosetta* (Alegado et al., 2012). Group formation can provide a selective advantage to the choanoflagellate cells since they are better able to capture their prey than single cells (Kreft, 2010). The groups are formed by cells remaining attached after division (Dayel et al., 2011; Fairclough et al., 2010; Koehl, 2020; Laundon et al., 2019). Such attachment allows high relatedness that minimizes conflict between cells (Buss, 1987; Grosberg & Strathmann, 2007), despite the potential costs that come with living in a multicellular group such as production and secretion of an extracellular matrix (Cavalier-Smith, 2017).

10.9   THE ROLE OF THE ENVIRONMENT

All species are shaped by their environment, and multicellular species are of course no exception to this (Bonner, 1998; Darwin, 1859). In the section above, we have considered how the ecological benefits and costs of cooperation influence multicellularity in clonal species, and these can vary depending on the species and ecological conditions. However, it is becoming clear that the environment could play a larger role in determining the course of multicellular evolution and that the environment may impact non-clonal and clonal multicellular species in different ways.

A recent comparative study across 14 independent transitions to multicellularity has shown that in aquatic environments a higher proportion of lineages that originated there form groups clonally than non-clonally (Fisher et al., 2020). This includes multicellular lineages such as animals, plants, green algae, red algae and brown algae, who all originated in the sea and develop through clonal group formation. This result suggests (although is by no means definitive) that aquatic environment, in general, has an impact on which mode of group formation works best. One hypothesis, first suggested by Bonner (1998), is that in order to reap the benefits of being in a group, cells in the water need to stick together directly after division to avoid being dispersed by water currents. The study also showed that there have been more transitions to obligate multicellularity in aquatic environments compared to on land, which supports the previous observation that obligate multicellularity has only ever evolved in groups with clonal group formation (Fisher et al., 2013).

Temperature, calcium concentration, resource availability and artificial selection can also affect multicellular group formation. At low temperatures, the facultatively multicellular algae *Scenedesmus* form multicellular groups (Egan & Trainor, 1989; Trainor, 1992; 1993). High calcium concentrations induce aggregation in the cyanobacteria *Microcystis*, but when calcium concentrations decrease, they form colonies clonally by cell division (Chen & Lürling, 2020). In terms of resources, when they are abundant, we see monoclonal multicellular groups being prevalent, whereas when resources are scarce, polyclonal groups are more common (Hamant et al., 2019). In experimental conditions of artificial selection for fast-settling multicellular yeast, unicellular yeast can evolve into clonal multicellular clusters with division of labor (Ratcliff et al., 2012).

10.10   IS CLONAL DEVELOPMENT THE ONLY WAY?

In this chapter, we have focused on the observation that all instances of obligate multicellularity have evolved in species with clonal group formation. But is clonality always necessary in the formation of an obligate multicellular organism? Here, we give several examples of how obligate multicellular organisms can 'break the rules' of clonality and challenge our idea that clonality is essential for obligateness.

10.11   PLACOZOA

Placozoa are an obligately multicellular phylum of animals consisting of three species: *Trichoplax adhaerens*, *Hoilungia hongkongensis*, and *Polyplacotoma mediterranea* (Bernd Schierwater & DeSalle, 2018; Eitel et al., 2013; Schierwater & Eitel, 2019). These microscopic marine organisms (Eitel et al., 2013; Grell & Ludwig, 1971; Pearse & Voigt, 2007; Signorovitch et al., 2006; Voigt et al., 2004) break up into single cells upon exposure to certain chemicals (colchicine, vinblastine, and seawater without divalent ions). Removal of these chemicals makes the cells reaggregate (Ruthmann & Terwelp, 1979). This phenomenon raises several questions. First, can cells of the disaggregated individual grow independently, as in a facultatively multicellular organism? Second, are non-clonal cells able to join the group during reaggregation? Third, do the reaggregated cells form an obligate multicellular organism (West et al., 2015)? Future experiments would need to confirm Ruthmann & Terwelp's (1979) findings of disaggregation and reaggregation in Placozoa, and further assess whether addition of non-clonal cells in the disaggregated clonal cellular culture leads to chimeric obligately multicellular individuals. This will help us answer whether obligate multicellularity can arise non-clonally in placozoa.

10.12   MICROCHIMERISM

Microchimerism refers to the presence of non-clonal cells inside the brain, thyroid, breast tissue and bloodstream of mammals, including cattle, dogs, Rhesus monkeys and even humans (Axiak-Bechtel et al., 2013; Bakkour et al., 2014; Barinaga, 2002; Boddy et al., 2015; Muehlenbachs et al., 2015; Owen, 1945; Turin et al., 2007; Van

Dijk et al., 1996; Youssoufian & Pyeritz, 2002). These cells are exchanged between mother and fetus during pregnancy for many generations and can survive for over 27 years inside the mother as seen in humans (Barinaga, 2002; Boddy et al., 2015; Chan et al., 2012; Kinder et al., 2017; O'Donoghue et al., 2004; Yan et al., 2005). According to social evolution theory, non-clonal cells can be a source of conflict inside the multicellular organism, as the genetic fitness interests of all cells in the multicellular group are not aligned (Burt & Trivers, 2006; Gardner & Grafen, 2009; Queller & Strassmann, 2009; Roze & Michod, 2001; West & Gardner, 2013). Even though the presence of microchimerism cells is a fact, they do not outnumber the majority of the cells in our body which are clonal. Furthermore, centuries of transplantation experiments have shown that despite some observations of donor transplant cells moving to the semen of a recipient patient (Long & Chilton, 2019), multicellular organisms are known to reject foreign tissue via rapid immune responses (Brent, 1996; Medawar, 1948). The immune system tightly regulates such sources of conflict within the multicellular organism after obligateness has evolved.

10.13   CANCER

Another source of conflict within the obligate multicellular organism that challenges the notion of clonality being necessary in obligate multicellularity, are nucleotide substitutions among clonal cells. If these substitutions lead to cellular over proliferation and movement to other tissues, these non-clonal cells now can cause disease inside the multicellular body in the form of cancer. Cancer is found within the body of multicellular organisms across the tree of life (Aktipis et al., 2015), even between different bodies as in transmissible cancers (Metzger et al., 2016; Murchison, 2009; Murchison et al., 2014). However, this does not mean that the multicellular organism is non-clonal and obligate at the same time. The first steps in its development are clonal, with single-cell bottlenecks minimizing the potential for conflict among cells (Buss, 1987; Fisher et al., 2013; Grosberg & Strathmann, 2007; Hurst et al., 1996; Michod, 1999; 2003; Michod et al., 1997; Niklas, 2014). If the first few divisions of multicellular life are non-clonal, with substitutions accumulating among cells, the embryo usually dies and is naturally aborted (Pandey et al., 2005; Tur-Torres et al., 2017).

Slime molds reveal that even if cells in a multicellular group are highly related, they are not necessarily on the 'road towards obligate multicellularity'. Slime molds form multicellular aggregates upon starvation (Bourke, 2011; Fisher et al., 2013; Smith et al., 2014). High relatedness ($r = 0.97$) prevents cheating cells from spreading in the population and allows high levels of altruism (Gilbert et al., 2007; Strassmann et al., 2000). Cells in the stalk altruistically sacrifice their ability to reproduce, and can only pass their genes to the next generation by helping the spores that reproduce. Still, these multicellular groups are facultatively multicellular (Gilbert et al., 2007). There is conflict among cells, as there has been selection for within-group kin discrimination (Mehdiabadi et al., 2006; Strassmann et al., 2000). For altruistic behavior to become unconditional (permanent), any deviation from $r = 1$ means that ($rb < c$) the cost of cooperating will be larger, even if only slightly larger, than the benefit of cooperating and this will lead to a breakdown of the evolution of unconditional altruism (Bourke, 2011; Hamilton, 1964a, 1964b; West et al., 2007b, 2015).

10.14   CONCLUSION

In this chapter, we have tried to describe the importance and consequences of clonal group formation in multicellular evolution. Our purpose has been to provide an overview of the topic and we have tried to direct the reader to more detailed accounts where necessary.

The question, however, remains: why do some species form multicellular groups clonally, rather than by other means? We have described the consequences and advantages of clonal group formation *once it has evolved*, but what about the initial benefit? This is a much more difficult question to answer. We would argue that a combination of the physical environment and a species' biology imposes constraints on what type of group formation is possible – as always with evolution, there is some luck and chance at its core! It is possible that the physical environment of living in water created conditions that made clonal group formation much more likely than non-clonal group formation. It is also possible that certain species were predisposed to form groups clonally due to almost random facts of their biology, e.g. those with cell walls or particular proteins in their extracellular matrix. Whatever the reason, it is a fact that clonal group formation, under the influence of many ecological, intercellular, and intracellular, known and unknown factors, has led to some of the most impressive and complex multicellular radiations in the living world.

A crucial point that we have only briefly touched upon is the necessity of clonal group formation for the major evolutionary transition to multicellularity. In other words – the evolution of a multicellular individual. This is not a chapter devoted to the major evolutionary transitions in individuality, but it is worth noting that multicellularity has been included amongst the 8 or so transitions from the outset. Maynard-Smith & Szathmary recognized that the transition from unicells to multicells represented a 'key event in the history of life' (Maynard Smith & Szathmary, 1995) but it has only been more recently (Bourke, 2011; Michod, 2007; West et al., 2015) that the underlying evolutionary theory behind the major evolutionary transitions in individuality has been refined.

It is now clear that in order to transition from one level of individuality, single cells, to a new level of individuality, the obligate multicellular organism, isn't just made more likely by clonal group formation, but *requires* it. To use Andrew Bourke's terminology, clonal group formation allows not only group formation and group maintenance but allows the final step in the evolution of multicellularity, group transformation (which could also be called the major evolutionary transition to individuality). This point leads us nicely to the next chapter in this book, which deals with multicellular group maintenance (Chapter 11).

REFERENCES

Aktipis, C. A., Boddy, A. M., Jansen, G., Hibner, U., Hochberg, M. E., Maley, C. C., & Wilkinson, G. S. (2015). Cancer across the tree of life: cooperation and cheating in multicellularity. *Philosophical Transactions of the Royal Society B: Biological Sciences*, *370*(1673), 20140219–20140219. https://doi.org/10.1098/rstb.2014.0219

Alegado, R. A., Brown, L. W., Cao, S., Dermenjian, R. K., Zuzow, R., Fairclough, S. R., Clardy, J., & King, N. (2012). A bacterial sulfonolipid triggers multicellular development in the closest living relatives of animals. *eLife 1*, e00013. https://doi.org/10.7554/eLife.00013

Axiak-Bechtel, S. M., Kumar, S. R., Hansen, S. A., & Bryan, J. N. (2013). Y-chromosome DNA is present in the blood of female dogs suggesting the presence of fetal microchimerism. *PLoS ONE*, *8*(7), e68114. https://doi.org/10.1371/journal.pone.0068114

Bakkour, S., Baker, C. A., Tarantal, A. F., Wen, L., Busch, M. P., Lee, T. H., & McCune, J. M. (2014). Analysis of maternal microchimerism in rhesus monkeys (*Macaca mulatta*) using real-time quantitative PCR amplification of MHC polymorphisms. *Chimerism*, *5*(1), 6–15. https://doi.org/10.4161/chim.27778

Baldauf, S. L., & Doolittle, W. F. (1997). Origin and evolution of the slime molds (Mycetozoa). *Proceedings of the National Academy of Sciences*, *94*(22), 12007–12012.

Barinaga, M. (2002). Cells exchanged during pregnancy live on. *Science*, *296*(5576), 2169–2172. https://www.science.org/doi/abs/10.1126/science.296.5576.2169

Berman, J. (2006). Morphogenesis and cell cycle progression in *Candida albicans*. In *Current Opinion in Microbiology* (Vol. 9, Issue 6, pp. 595–601). https://doi.org/10.1016/j.mib.2006.10.007

Boddy, A. M., Fortunato, A., Wilson Sayres, M., & Aktipis, A. (2015). Fetal microchimerism and maternal health: a review and evolutionary analysis of cooperation and conflict beyond the womb. *BioEssays*, *37*(10), 1106–1118. https://doi.org/10.1002/bies.201500059

Bonner, J. (2009). *First signals: the evolution of multicellular development*. http://books.google.com/books?hl=en&lr=&id=RgB-NxcpTrEC&oi=fnd&pg=PP10&dq=evolution+of+multicellularity&ots=mDZMffuL2w&sig=vSgKAoowA_CS4U8738D0MXqs_oQ

Bonner, J. T. (1998). The origins of multicellularity. *Integrative Biology: Issues, News, and Reviews*, *1*(1), 27–36. https://doi.org/10.1002/(SICI)1520-6602(1998)1:1<27:AID-INBI4>3.3.CO;2-Y

Bonner, J. T. (2003). Evolution of development in the cellular slime molds. *Evolution & Development*, *5*(3), 305–313.

Boomsma, J. J. (2007). Kin selection versus sexual selection: why the ends do not meet. *Current Biology*, *17*(16), R673–R683.

Boomsma, J. J. (2009). Lifetime monogamy and the evolution of eusociality. *Philosophical Transactions of the Royal Society B: Biological Sciences*, *364*(1533), 3191–3207. https://doi.org/10.1098/rstb.2009.0101

Boomsma, J. J., & Gawne, R. (2018). Superorganismality and caste differentiation as points of no return: how the major evolutionary transitions were lost in translation. *Biological Reviews*, *93*(1), 28–54.

Bourke, A. F. G. (2011). *Principles of social evolution*. Oxford University Press, Oxford.

Brent, L. (1996). *A history of transplantation immunology*. Elsevier Science, UK.

Brunet, T., & King, N. (2017). The origin of animal multicellularity and cell differentiation. *Developmental Cell* (43, 124–140). https://doi.org/10.1016/j.devcel.2017.09.016

Brunet, T., Larson, B. T., Linden, T. A., Vermeij, M. J. A., McDonald, K., & King, N. (2019). Light-regulated collective contractility in a multicellular choanoflagellate. *Science*, *366*(6463), 326–334. https://doi.org/10.1126/science.aay2346

Burt, A., & Trivers, R. L. (2006). *Genes in Conflict*. Belknap, Cambridge, MA.

Buss, L. W. (1987). *The evolution of individuality*. Princeton Univ. Press, Princeton, NJ.

Cavalier-Smith, T. (2017). Origin of animal multicellularity: precursors, causes, consequences—the choanoflagellate/sponge transition, neurogenesis and the Cambrian explosion. *Philosophical Transactions of the Royal Society B: Biological Sciences*, *372*(1713), 20150476.

Chan, W. F. N., Gurnot, C., Montine, T. J., Sonnen, J. A., Guthrie, K. A., & Nelson, J. L. (2012). Male microchimerism in the human female brain. *PLoS One*, *7*(9), e45592.

Chen, H., & Lürling, M. (2020). Calcium promotes formation of large colonies of the cyanobacterium *Microcystis* by enhancing cell-adhesion. *Harmful Algae*, *92*, 101768.

Cooper, G. A., & West, S. A. (2018). Division of labour and the evolution of extreme specialization. *Nature Ecology & Evolution*, *2*(7), 1161–1167.

Darwin, C. (1859). *The origin of species by means of natural election, or the preservation of favored races in the struggle for life*. J. Murray, London.

Davies, N. B., Krebs, J. R., & West, S. A. (2012). *An introduction to behavioural ecology*. John Wiley & Sons, UK.

Dayel, M. J., Alegado, R. A., Fairclough, S. R., Levin, T. C., Nichols, S. A., McDonald, K., & King, N. (2011). Cell differentiation and morphogenesis in the colony-forming choanoflagellate *Salpingoeca rosetta*. *Developmental Biology*, *357*(1), 73–82. https://doi.org/10.1016/j.ydbio.2011.06.003

Egan, P. F., & Trainor, F. R. (1989). Low cell density: the unifying principle for unicell development in *Scenedesmus* (Chlorophyceae). *British Phycological Journal*, *24*(3), 271–283. https://doi.org/10.1080/00071618900650291

Eitel, M., Osigus, H. J., DeSalle, R., & Schierwater, B. (2013). Global diversity of the Placozoa. *PloS One*, *8*(4), e57131.

Engelberg, D., Mimran, A., Martinetto, H., Otto, J., Simchen, G., Karin, M., & Fink, G. R. (1998). Multicellular stalk-like structures in *Saccharomyces cerevisiae*. *Journal of Bacteriology*, *180*(15), 3992–3996.

Everhart, S. E., & Keller, H. W. (2008). Life history strategies of corticolous myxomycetes: the life cycle, plasmodial types, fruiting bodies, and taxonomic orders. *Fungal Diversity*, *29*, 1–16.

Fairclough, S. R., Chen, Z., Kramer, E., Zeng, Q., Young, S., Robertson, H. M., Begovic, E., Richter, D. J., Russ, C., Westbrook, M. J., Manning, G., Lang, B. F., Haas, B., Nusbaum, C., & King, N. (2013). Premetazoan genome evolution and the regulation of cell differentiation in the choanoflagellate *Salpingoeca rosetta*. *Genome Biology*, *14*(2), 1–15. https://doi.org/10.1186/gb-2013-14-2-r15

Fairclough, S. R., Dayel, M. J., & King, N. (2010). Multicellular development in a choanoflagellate. *Current Biology*, *20*(20), R875–R876.

Fisher, R. M., Shik, J. Z., & Boomsma, J. J. (2020). The evolution of multicellular complexity: the role of relatedness and environmental constraints. *Proceedings of the Royal Society B*, *287*(1931), 20192963. https://doi.org/10.1098/rspb.2019.2963

Fisher, Roberta M., Cornwallis, C. K., & West, S. A. (2013). Group formation, relatedness, and the evolution of multicellularity. *Current Biology*, *23*(12), 1120–1125. https://doi.org/10.1016/j.cub.2013.05.004

Fisher, Roberta May, Bell, T., & West, S. A. (2016). Multicellular group formation in response to predators in the alga *Chlorella vulgaris*. *Journal of Evolutionary Biology*, *29*(3), 551–559.

Flores, E., & Herrero, A. (2010). Compartmentalized function through cell differentiation in filamentous cyanobacteria. *Nature Reviews. Microbiology*, *8*(1), 39–50. https://doi.org/10.1038/nrmicro2242

Foster, K. R., Wenseleers, T., & Ratnieks, F. L. W. (2006). Kin selection is the key to altruism. *Trends in Ecology and Evolution*. 21(2), 57–60 https://doi.org/10.1016/j.tree.2005.11.020

Fritsch, F. E., & West, G. S. (1927). *A treatise on the British Freshwater Algae, in which are included all the pigmented Protophyta hitherto found in British freshwaters*. New and revised edition, in great part rewritten by F. E. Fritsch. The University Press, Cambridge.

Gallon, J. R. (1981). The oxygen sensitivity of nitrogenase: a problem for biochemists and micro-organisms. *Trends in Biochemical Sciences*, *6*, 19–23.

Gardner, A., & Grafen, A. (2009). Capturing the superorganism: a formal theory of group adaptation. *Journal of Evolutionary Biology*, *22*(4), 659–671.

Gilbert, O. M., Foster, K. R., Mehdiabadi, N. J., Strassmann, J. E., & Queller, D. C. (2007). High relatedness maintains multicellular cooperation in a social amoeba by controlling cheater mutants. *Proceedings of the National Academy of Sciences*, 104(21), 8913–8917. www.pnas.org/cgi/content/full/

Golden, J. W., & Yoon, H. S. (2003). Heterocyst development in *Anabaena*. *Current Opinion in Microbiology*, 6(6), 557–563. https://doi.org/10.1016/j.mib.2003.10.004

Grell, K. G., & Ludwig, C. (1971). *Trichoplax Adhaerens (Placozoa): Bewegung und Organisation*. Inst. fd wiss. Film, Göttingen.

Grosberg, R. K., & Strathmann, R. R. (2007). The evolution of multicellularity: a minor major transition? *Annual Review of Ecology, Evolution, and Systematics*, 38(1), 621–654.

Hallmann, A. (2011). Evolution of reproductive development in the volvocine algae. *Sexual Plant Reproduction* (24(2), 97–112). https://doi.org/10.1007/s00497-010-0158-4

Hamant, O., Bhat, R., Nanjundiah, V., & Newman, S. A. (2019). Does resource availability help determine the evolutionary route to multicellularity? *Evolution and Development*, 21(3), 115–119. https://doi.org/10.1111/ede.12287

Hamilton. (1964a). The genetical evolution of social behaviour. I. *Journal of Theoretical Biology*, 7(1), 1–16. https://doi.org/10.1016/0022-5193(64)90038-4

Hamilton, W. D. (1964b). The genetical evolution of social behaviour. II. *Journal of Theoretical Biology*, 7(1), 17–52. https://doi.org/10.1016/0022-5193(64)90039-6

Hashimura, H., Morimoto, Y. V, Yasui, M., & Ueda, M. (2019). Collective cell migration of *Dictyostelium* without cAMP oscillations at multicellular stages. *Communications Biology*, 2(1), 1–15.

Heaton, L. L. M., Jones, N. S., & Fricker, M. D. (2020). A mechanistic explanation of the transition to simple multicellularity in fungi. *Nature Communications*, 11(1). https://doi.org/10.1038/s41467-020-16072-4

Herron, M. (2009). Many from one: lessons from the volvocine algae on the evolution of multicellularity. *Communicative & Integrative Biology*, 2(4), 368–370.

Herron, M. D. (2016). Origins of multicellular complexity: *Volvox* and the volvocine algae. *Molecular Ecology*, n/a-n/a. https://doi.org/10.1111/mec.13551

Herron, M. D, Borin, J. M., Boswell, J. C., Walker, J., Chen, I.-C. K., Knox, C. A., Boyd, M., Rosenzweig, F., & Ratcliff, W. C. (2019). De novo origins of multicellularity in response to predation. *Scientific Reports*, 9(1), 1–9.

Herron, M. D, Hackett, J. D., Aylward, F. O., & Michod, R. E. (2009). Triassic origin and early radiation of multicellular volvocine algae. *Proceedings of the National Academy of Sciences*, 106(9), 3254–3258.

Hurst, L. D., Atlan, A., & Bengtsson, B. O. (1996). Genetic conflicts. *The Quarterly Review of Biology*, 71(3), 317–364. https://doi.org/10.1086/419442

Kaiser, D. (2001). Building a multicellular organism. *Annual Review of Genetics*, 35(1), 103–123.

Kapsetaki, S. E., Fisher, R. M., & West, S. A. (2016). Predation and the formation of multicellular groups in algae. *Evolutionary Ecology Research*, 17(5), 651–669.

Kapsetaki, S. E., Tep, A., & West, S. A. (2017). How do algae form multicellular groups? *Evolutionary Ecology Research*, 18, 663–675.

Kapsetaki, S. E., & West, S. A. S. A. (2019). The costs and benefits of multicellular group formation in algae*. *Evolution*, 73(6), 1296–1308. https://doi.org/10.1111/evo.13712

Kessin, R. H. (2001). *Dictyostelium: evolution, cell biology, and the development of multicellularity*. No. 38. Cambridge University Press, Cambridge, UK.

Kinder, J. M., Stelzer, I. A., Arck, P. C., & Way, S. S. (2017). Immunological implications of pregnancy-induced microchimerism. *Nature Reviews Immunology*, 17(8), 483.

King, N. (2004). The unicellular ancestry of animal development. *Developmental Cell*, 7(3), 313–325.

King, N. (2005). Choanoflagellates. *Current Biology*, 15(4), R113–R114.

King, N., Westbrook, M. J., Young, S. L., Kuo, A., Abedin, M., Chapman, J., Fairclough, S., Hellsten, U., Isogai, Y., Letunic, I., Marr, M., Pincus, D., Putnam, N., Rokas, A., Wright, K. J., Zuzow, R., Dirks, W., Good, M., Goodstein, D., … Rokhsar, D. (2008). The genome of the choanoflagellate *Monosiga brevicollis* and the origin of metazoans. *Nature*, *451*(7180), 783–788. https://doi.org/10.1038/nature06617

Kirk, J. T. O. (1994). *Light and photosynthesis in aquatic ecosystems.* Cambridge University Press, Cambridge, UK.

Kirk, M. M., Ransick, A., McRae, S. E., & Kirk, D. L. (1993). The relationship between cell size and cell fate in *Volvox carteri. The Journal of Cell Biology*, *123*(1), 191–208. https://doi.org/10.1083/jcb.123.1.191

Kirkegaard, J. B., & Goldstein, R. E. (2016). Filter-feeding, near-field flows, and the morphologies of colonial choanoflagellates. *Physical Review E*, *94*(5), 052401.

Knoll, A. H. (2011). The multiple origins of complex multicellularity. *Annual Review of Earth and Planetary Sciences*, *39*, 217–239.

Kochert, G. (1968). Differentiation of reproductive cells in *Volvox carteri. The Journal of Protozoology*, *15*(3), 438–452.

Koehl, M. A. R. (2020). Selective factors in the evolution of multicellularity in choanoflagellates. *Journal of Experimental Zoology Part B: Molecular and Developmental Evolution*, *336*(3), 315–326. https://doi.org/10.1002/jez.b.22941

Koufopanou, V. (1994). The evolution of soma in the volvocales. *The American Naturalist*, *143* (5), 907–931.

Kreft, J. M. (2010). *Effects of forming multicellular colonies or attaching to surfaces on feeding rates of the choanoflagellate Salpingoeca rosetta.* University of California, Berkeley.

Kuhn, G., Hijri, M., & Sanders, I. R. (2001). Evidence for the evolution of multiple genomes in arbuscular mycorrhizal fungi. *Nature*, *414*(6865), 745–748.

LaPaglia, C., & Hartzell, P. L. (1997). Stress-induced production of biofilm in thehyperthermophile *Archaeoglobus fulgidus. Applied and Environmental Microbiology*, *63*(8), 3158–3163. http://www.pubmedcentral.nih.gov/articlerender.fcgi?artid=1389226&tool =pmcentrez&rendertype=abstract

Laundon, D., Larson, B. T., McDonald, K., King, N., & Burkhardt, P. (2019). The architecture of cell differentiation in choanoflagellates and sponge choanocytes. *PLoS Biology*, *17*(4). https://doi.org/10.1371/journal.pbio.3000226

Leadbeater, B. S. C. (2015). *The choanoflagellates.* Cambridge University Press, Cambridge, UK.

Long, C., & Chilton, B. (2019). *Chimeric Fluidity: a case study of a male stem cell/bone marrow transplant patient. 30th International Symposium on Human Identification*, Palm Springs, CA. https://vb6ykw2twb15uf9341ls5n11-wpengine.netdna-ssl.com/wp-content/uploads/2019/07/6.-ChimericFluidityWCSOFSD.pdf

Lüring, M., Van Donk, E., Lürling, M., & Van Donk, E. (1997). Morphological changes in Scenedesmus induced by infochemicals released in situ from zooplankton grazers. *Limnology and Oceanography*, *42*(4), 783–788. https://doi.org/10.4319/lo.1997.42.4.0783

Lürling, M. (1999a). Grazer-induced coenobial formation in clonal cultures of Scenedesmus obliquus (Chlorococcales, Chlorophyceae). *Journal of Plankton Research*, *23*, 19–23. https://doi.org/10.1046/j.1529-8817.1999.3510019.x

Lürling, M. (1999b). Grazer-induced coenobial formation in clonal cultures of *Scenendesmus obliquus* (Chlorococcales, Chlorophyceae). *Journal of Phycology*, *35*(1), 19–23.

Lürling, M. (2001). Grazing-associated infochemicals induce colony formation in the green alga *Scenedesmus*. In *Protist* (Vol. 152, Issue 1, pp. 7–16). https://doi.org/10.1078/1434-4610-00038

Lürling, M. (2020). Grazing resistance in phytoplankton. *Hydrobiologia*, 1–13.

Lurling, M., & Beekman, W. (2006). Palmelloids formation in *Chlamydomonas reinhardtii*: defence against rotifer predators? *Annales de Limnologie-International Journal of Limnology, 42*(2), 65–72.

Lürling, M., Van Donk, E., Donk, E., & Van Donk, E. (1996). Zooplankton-induced unicell-colony transformation in *Scenedesmus acutus* and its effect on growth of herbivore Daphnia. *Oecologia, 108*(3), 432–437. https://doi.org/10.1007/BF00333718

Lyons, N. A., & Kolter, R. (2015). On the evolution of bacterial multicellularity. *Current Opinion in Microbiology, 24*, 21–28. https://doi.org/10.1016/j.mib.2014.12.007

Marleau, J., Dalpé, Y., St-Arnaud, M., & Hijri, M. (2011). Spore development and nuclear inheritance in arbuscular mycorrhizal fungi. *BMC Evolutionary Biology, 11*(1), 1–11.

Matt, G., & Umen, J. (2016). Volvox: a simple algal model for embryogenesis, morphogenesis and cellular differentiation. *Developmental Biology, 419*(1), 99–113.

Mayerhofer, L. E., Macario, A. J. L., Conway, E., & Macario, D. E. (1992). Lamina, a Novel Multicellular Form of Methanosarcina mazei S-6. *Journal of Bacteriology* 174(1), 309–314. http://jb.asm.org/

Maynard Smith, J., & Szathmary, E. (1995). *The major transitions in evolution.* W.H. Freeman Spektrum, Oxford.

Medawar, P. B. (1948). Immunity to homologous grafted skin. III. The fate of skin homographs transplanted to the brain, to subcutaneous tissue, and to the anterior chamber of the eye. *British Journal of Experimental Pathology, 29*(1), 58.

Mehdiabadi, N. J., Jack, C. N., Farnham, T. T., Platt, T. G., Kalla, S. E., Shaulsky, G., Queller, D. C., & Strassmann, J. E. (2006). Social evolution: kin preference in a social microbe. *Nature, 442*(7105), 881–882.

Meissner, M., Stark, K., Cresnar, B., Kirk, D. L., & Schmitt, R. (1999). *Volvox* germline-specific genes that are putative targets of RegA repression encode chloroplast proteins. *Current Genetics, 36*(6), 363–370. https://doi.org/10.1007/s002940050511

Metzger, M. J., Villalba, A., Carballal, M. J., Iglesias, D., Sherry, J., Reinisch, C., Muttray, A. F., Baldwin, S. A., & Goff, S. P. (2016). Widespread transmission of independent cancer lineages within multiple bivalve species. *Nature, 534*(7609), 705–709. https://doi.org/10.1038/nature18599

Michod, R E. (2003). *Cooperation and conflict mediation in the evolution of multicellularity. Genetic and Cultural Evolution of Cooperation. MIT Press, Cambridge, Massachusetts*, 291–308.

Michod, R. E. (2007). Evolution of individuality during the transition from unicellular to multicellular life. *Proceedings of the National Academy of Sciences of the United States of America, 104*, 8613–8618. https://doi.org/10.1073/pnas.0701489104

Michod, R. E., Roze, D., & Biolog, E. (1997). Transitions in individuality. *Proceedings. Biological Sciences/The Royal Society, 264*(1383), 853–857. https://doi.org/10.1098/rspb.1997.0119

Michod, R. E. (1999). Individuality, immortality, and sex. *Levels of Selection in Evolution, 66*, 53.

Michod, R. E., Nedelcu, A. M., & Roze, D. (2003). Cooperation and conflict in the evolution of individuality: IV. Conflict mediation and evolvability in *Volvox carteri*. *BioSystems, 69*(2–3), 95–114.

Michod, R. E., Viossat, Y., Solari, C. A., Hurand, M., & Nedelcu, A. M. (2006). Life-history evolution and the origin of multicellularity. *Journal of Theoretical Biology, 239*(2), 257–272.

Mikhailov, K. V., Konstantinova, A. V., Nikitin, M. A., Troshin, P. V., Rusin, L. Y., Lyubetsky, V. A., Panchin, Y. V., Mylnikov, A. P., Moroz, L. L., Kumar, S., & Aleoshin, V. V. (2009). The origin of Metazoa: a transition from temporal to spatial cell differentiation. *BioEssays, 31*(7), 758–768. https://doi.org/10.1002/bies.200800214

Muehlenbachs, A., Bhatnagar, J., Agudelo, C. A., Hidron, A., Eberhard, M. L., Mathison, B. A., Frace, M. A., Ito, A., Metcalfe, M. G., Rollin, D. C., Visvesvara, G. S., Pham, C. D., Jones, T. L., Greer, P. W., Hoyos, A. V., Olson, P. D., Diazgranados, L. R., & Zaki, S. R. (2015). Malignant transformation of *Hymenolepis nana* in a human host. *New England Journal of Medicine*, *373*(19), 1845–1852. https://doi.org/10.1056/NEJMoa1505892

Murchison, E. P. (2009). Clonally transmissible cancers in dogs and Tasmanian devils. *Oncogene*, *27*(S2), S19–S30. https://doi.org/10.1038/onc.2009.350

Murchison, E. P., Wedge, D. C., Alexandrov, L. B., Fu, B., Martincorena, I., Ning, Z., Tubio, J. M. C. C., Werner, E. I., Allen, J., De Nardi, A. B., Donelan, E. M., Marino, G., Fassati, A., Campbell, P. J., Yang, F., Burt, A., Weiss, R. A., & Stratton, M. R. (2014). Transmissible dog cancer genome reveals the origin and history of an ancient cell lineage. *Science*, *343*(6169), 437–440. https://doi.org/10.1126/science.1247167

Nabout, J. C., da Silva Rocha, B., Carneiro, F. M., & Sant'Anna, C. L. (2013). How many species of Cyanobacteria are there? Using a discovery curve to predict the species number. *Biodiversity and Conservation*, *22*(12), 2907–2918.

Nagy, L. G., Varga, T., Csernetics, Á., & Virágh, M. (2020). Fungi took a unique evolutionary route to multicellularity: seven key challenges for fungal multicellular life. *Fungal Biology Reviews*, *34*(4), 151–169. https://doi.org/10.1016/j.fbr.2020.07.002

Nichols, S. A., Dayel, M. J., & King, N. (2009). Genomic, phylogenetic, and cell biological insights into metazoan origins. In *Animal Evolution: Genomes, Fossils and Trees*, Oxford University Press, Oxford, 24–32.

Niklas, K. J. (2014). The evolutionary-developmental origins of multicellularity. *American Journal of Botany*, *101*(1), 6–25. https://doi.org/10.3732/ajb.1300314

Niklas, K. J., & Newman, S. A. (2013). The origins of multicellular organisms. *Evolution & Development*, *15*(1), 41–52.

O'Donoghue, K., Chan, J., de la Fuente, J., Kennea, N., Sandison, A., Anderson, J. R., Roberts, I. A. G., & Fisk, N. M. (2004). Microchimerism in female bone marrow and bone decades after fetal mesenchymal stem-cell trafficking in pregnancy. *The Lancet*, *364*(9429), 179–182.

Olsson, P. A., Jakobsen, I., & Wallander, H. (2002). Foraging and resource allocation strategies of mycorrhizal fungi in a patchy environment. In *Mycorrhizal ecology* (pp. 93–115). Springer, Berlin, Heidelberg.

Owen, R. D. (1945). Immunogenetic consequences of vascular anastomoses between bovine twins. *Science*, *102*(2651), 400–401.

Pandey, M. K., Rani, R., & Agrawal, S. (2005). An update in recurrent spontaneous abortion. *Archives of Gynecology and Obstetrics*, *272*(2), 95–108. https://doi.org/10.1007/s00404-004-0706-y

Pearse, V. B., & Voigt, O. (2007). Field biology of placozoans (*Trichoplax*): distribution, diversity, biotic interactions. *Integrative and Comparative Biology*, *47*(5), 677–692.

Ploug, H., Stolte, W., & Jørgensen, B. B. (1999). Diffusive boundary layers of the colony-forming plankton alga, *Phaeocystis* sp. – implications for nutrient uptake and cellular growth. *Limnology and Oceanography*, *44*(8), 1959–1967.

Queller, D. C., & Strassmann, J. E. (2009). Beyond society: the evolution of organismality. *Philosophical Transactions of the Royal Society B: Biological Sciences*, *364*(1533), 3143–3155.

Ratcliff, W. C., Denison, R. F., Borrello, M., & Travisano, M. (2012). Experimental evolution of multicellularity. *Proceedings of the National Academy of Sciences of the United States of America*, *109*(5), 1595–1600. https://doi.org/10.1073/pnas.1115323109

Reynolds, C. S. (1984). *The ecology of freshwater phytoplankton*. Cambridge University Press, Cambridge, UK.

Richards, T. A., Leonard, G., & Wideman, J. G. (2017). What defines the "kingdom" fungi? *The Fungal Kingdom*, 57–77.

Richter, D. J., & King, N. (2013). The genomic and cellular foundations of animal origins. *Annual Review of Genetics*, 47, 509–537. https://doi.org/10.1146/annurev-genet-111212-133456

Roper, M., Dayel, M. J., Pepper, R. E., & Koehl, M. A. R. (2013). Cooperatively generated stresslet flows supply fresh fluid to multicellular choanoflagellate colonies. *Physical Review Letters*, *110*(22), 1–5. https://doi.org/10.1103/PhysRevLett.110.228104

Rossetti, V., Schirrmeister, B. E., Bernasconi, M. V., & Bagheri, H. C. (2010). The evolutionary path to terminal differentiation and division of labor in cyanobacteria. *Journal of Theoretical Biology*, *262*(1), 23–34. https://doi.org/10.1016/j.jtbi.2009.09.009

Roze, D., & Michod, R. E. (2001). Mutation, multilevel selection, and the evolution of propagule size during the origin of multicellularity. *The American Naturalist*, *158*(6), 638–654.

Ruthmann, A., & Terwelp, U. (1979). Disaggregation and reaggregation of cells of the primitive metazoon *Trichoplax adhaerens*. *Differentiation*, *13*(3), 185–198.

Salvini-Plawen, L. V. (1978). On the origin and evolution of the lower Metazoa. *Journal of Zoological Systematics and Evolutionary Research*, *16*(1), 40–87.

Schierwater, B, & Eitel, M. (2019). *World Placozoa Database in the Catalogue of Life*. The Catalogue of Life Partnership.

Schierwater, B., & DeSalle, R. (2018). Placozoa. *Current Biology*, *28*(3), R97–R98.

Schirrmeister, B. E., Antonelli, A., & Bagheri, H. C. (2011). The origin of multicellularity in cyanobacteria. *BMC Evolutionary Biology*, *11*(1), 45.

Schirrmeister, B. E., de Vos, J. M., Antonelli, A., & Bagheri, H. C. (2013). Evolution of multicellularity coincided with increased diversification of cyanobacteria and the Great Oxidation Event. *Proceedings of the National Academy of Sciences of the United States of America*, *110*(5), 1791–1796. https://doi.org/10.1073/pnas.1209927110

Scott, T. W., Kiers, E. T., Cooper, G. A., dos Santos, M., & West, S. A. (2019). Evolutionary maintenance of genomic diversity within arbuscular mycorrhizal fungi. *Ecology and Evolution*, *9*(5), 2425–2435. https://doi.org/10.1002/ece3.4834

Shelton, D. E., Desnitskiy, A. G., & Michod, R. E. (2012). Distributions of reproductive and somatic cell numbers in diverse *Volvox* (Chlorophyta) species. *Evolutionary Ecology Research*, *14*, 707.

Shelton, D. E., & Michod, R. E. (2010). Philosophical foundations for the hierarchy of life. *Biology & Philosophy*, *25*(3), 391–403.

Signorovitch, A. Y., Dellaporta, S. L., & Buss, L. W. (2006). Caribbean placozoan phylogeography. *The Biological Bulletin*, *211*(2), 149–156.

Smith, G. M. (1944). A comparative study of the species of *Volvox*. *Transactions of the American Microscopical Society*, *63*(4), 265–310.

Smith, J., Queller, D. C., & Strassmann, J. E. (2014). Fruiting bodies of the social amoeba *Dictyostelium discoideum* increase spore transport by Drosophila. *BMC Evolutionary Biology*, *14*(1), 105.

Strassmann, J. E. (2010). *Dictyostelium*, the social amoeba. In *Encyclopedia of Animal Behavior* (Vol. 1, pp. 13–519). Edited by Breed, M. D., & Moore, J. Academic Press, Oxford.

Strassmann, J. E., & Queller, D. C. (2011). How social evolution theory impacts our understanding of development in the social amoeba *Dictyostelium*. *Development, Growth & Differentiation*, *53*(4), 597–607.

Strassmann, J. E., Zhu, Y., & Queller, D. C. (2000). Altruism and social cheating in the social amoeba *Dictyostelium discoideum*. *Nature*, *408*(6815), 965–967.

Tollrian, R., & Dodson, S. I. (1999). Inducible defences in cladocera: constraints, costs, and multipredator environments. *The Ecology and Evolution of Inducible Defenses (Ed TRHCD)*. Princeton University Press, Princeton, NJ.

Trainor, F. R. (1993). Cyclomorphosis in *Scenedesmus subspicatus* (Chlorococcales, Chlorophyta): stimulation of colony development at low temperature. *Phycologia*, *32*(6), 429–433.

Trainor, F. R. (1992). Cyclomorphosis in *Scenedesmus armatus* (Chlorophyta): an ordered sequence of ecomorph development 1. *Journal of Phycology*, *28*(4), 553–558.

Trainor, F. R. (1998). *Biological aspects of Scenedesmus (Chlorophyceae)-Phenotypic Plasticity* (Vol. 117). *Nova Hedwigia*, Beiheft, *117*, J. Cramer, Berlin, 367.

Tur-Torres, M. H., Garrido-Gimenez, C., & Alijotas-Reig, J. (2017). Genetics of recurrent miscarriage and fetal loss. *Best Practice and Research: Clinical Obstetrics and Gynaecology* (Vol. 42, pp. 11–25), *42*, 11–25 https://doi.org/10.1016/j.bpobgyn.2017.03.007

Turin, L., Invernizzi, P., Woodcock, M., Grati, F. R., Riva, F., Tribbioli, G., & Laible, G. (2007). Bovine fetal microchimerism in normal and embryo transfer pregnancies and its implications for biotechnology applications in cattle. *Biotechnology Journal: Healthcare Nutrition Technology*, *2*(4), 486–491.

Umen, J. G. (2020). *Volvox* and volvocine green algae. *EvoDevo*, *11*(1), 1–9.

Valentine, J. W., & Marshall, C. R. (2015). Fossil and transcriptomic perspectives on the origins and success of metazoan multicellularity. In *Evolutionary transitions to multicellular life* (pp. 31–46). Springer, Dordrecht.

Van Dijk, B. A., Boomsma, D. I., & De Man, A. J. M. (1996). Blood group chimerism in human multiple births is not rare. *American Journal of Medical Genetics*, *61*(3), 264–268.

Van Donk, E., Hessen, D. O., & Van Donk, E. (1993). Morphological changes in *Scenedesmus* induced by substances released from Daphnia. *Archiv Für Hydrobiologie*, *127*, 129–140.

Voigt, O., Collins, A. G., Pearse, V. B., Pearse, J. S., Ender, A., Hadrys, H., & Schierwater, B. (2004). Placozoa–no longer a phylum of one. *Current Biology*, *14*(22), R944–R945.

West, S. A., & Gardner, A. (2013). Adaptation and inclusive fitness. *Current Biology*, *23*(13), R577–R584. https://doi.org/10.1016/j.cub.2013.05.031

West, S. A., & Cooper, G. A. (2016). Division of labour in microorganisms: an evolutionary perspective. *Nature Reviews Microbiology*, *14*(11), 716.

West, S. A., Fisher, R. M., Gardner, A., & Kiers, E. T. (2015). Major evolutionary transitions in individuality. *Proceedings of the National Academy of Sciences of the United States of America*, *112*(33), 10112–10119. https://doi.org/10.1073/pnas.1421402112

West, S. A., Griffin, A. S., & Gardner, A. (2007a). Social semantics: altruism, cooperation, mutualism, strong reciprocity and group selection. *Journal of Evolutionary Biology*, *20*(2), 415–432. https://doi.org/10.1111/j.1420-9101.2006.01258.x

West, S. A., Griffin, A. S., & Gardner, A. (2007b). Evolutionary explanations for cooperation. *Current Biology*, *17*(16), R661–72. https://doi.org/10.1016/j.cub.2007.06.004

Whiteway, M., & Bachewich, C. (2007). Morphogenesis in *Candida albicans. Annu. Rev. Microbiol.*, *61*, 529–553.

Yan, Z., Lambert, N. C., Guthrie, K. A., Porter, A. J., Loubiere, L. S., Madeleine, M. M., Stevens, A. M., Hermes, H. M., & Nelson, J. L. (2005). Male microchimerism in women without sons: quantitative assessment and correlation with pregnancy history. *The American Journal of Medicine*, *118*(8), 899–906.

Youssoufian, H., & Pyeritz, R. E. (2002). Mechanisms and consequences of somatic mosaicism in humans. *Nature Reviews Genetics*, *3*(10), 748–758. https://doi.org/10.1038/nrg906

Zhang, J., Del Aguila, R., Schneider, C., & Schneider, B. L. (2005). The importance of being big. *Journal of Investigative Dermatology Symposium Proceedings*, *10*(2), 131–141.

11 Group Maintenance in Clonal Multicellularity

Controlling Intra-organismal Evolution

Aurora M. Nedelcu
Biology Department, University of New Brunswick,
Fredericton, NB, Canada

Alexander N. May
Research Casting International, Quinte West, ON, Canada

CONTENTS

11.1 INTRODUCTION

Multicellularity has evolved independently in many lineages from both the prokaryotic and eukaryotic domains of life (e.g., Grosberg and Strathmann 2007). Generally, multicellular phenotypes can evolve via two very distinct pathways: cell aggregation

DOI: 10.1201/9780429351907-14

(such as in myxobacteria and social amoebae) or failure to separate following cell division (e.g., in filamentous bacteria and most eukaryotic multicellular groups). The latter evolutionary strategy is known as clonal multicellularity, as the constituent cells are, by definition, clonal. Extant multicellular lineages exhibit very different levels of complexity – from simple multicellular forms without specialized cells to very complex organisms with hundreds of cell types and diverse developmental patterns and life histories. Remarkably, complex multicellular organisms with large bodies and many specialized cell types are only known among clonal multicellular lineages (e.g., land plants and animals).

Several theoretical frameworks (including kin selection/inclusive fitness, multi-level selection, cooperation/cheating, conflict/conflict mediation, self-limitation/limitation of exploitation from inside; e.g., Michod and Roze 2001; Libby and Rainey 2013; Bourke 2019; Aktipis 2020) and mechanistic views (mostly in the context of cancer suppression; e.g., Aktipis et al. 2015; Nedelcu and Caulin 2016; Nedelcu 2020) have been used to address how multicellular groups can be maintained and evolve into complex multicellular organisms. Most commonly, the increased evolutionary success of clonal multicellularity – in terms of prevalence and complexity, is thought to have been facilitated by the high relatedness of cells in clonal multicellular organisms (e.g., Fisher et al. 2013). Increased relatedness allows kin selection to operate and promotes cooperative and altruistic behaviors among cells (e.g., Queller 2000; Bourke 2019). However, as in all cooperative behaviors, cheaters – that is, cells that enjoy the benefits of cooperation without paying the costs, are still expected to occur even in clonal multicellular organisms (Buss 1987; Michod 1996; Queller 2000; Aktipis et al. 2015). Thus, to facilitate the evolutionary stability of multicellular lineages, the emergence and success of selfish/cheater cells have to be limited (Michod 1996).

Here, we are taking a *first-principles approach* to explore different aspects involved in the evolutionary stability of clonal multicellular lineages. Specifically, we are using the framework of evolutionary transitions in individuality (Michod 1998) and consider multicellular groups as units of evolution – used here to imply both that (i) they are levels of selection and (ii) adaptations occur at the group level. For that to be the case, multicellular groups need to possess heritable variation in fitness at *their level of organization* (Michod 2007). However, because multicellular individuals evolved from groups of previously independent units of evolution (i.e., single-celled entities) that still possess the necessary conditions to evolve (heritable variation in fitness), variation can still occur and selection can still act at the cell level. Thus, for selection to act at the group level and for multicellular groups to be maintained and become stable evolutionary units, within-group variation and selection have to be lower than among-group variation and selection (Michod 1997). In other words, intra-organismal evolution needs to be *limited/controlled*.

At the mechanistic level, controlling intra-organismal evolution requires both reducing the incidence of mutations (limiting genetic variation within the group) and lowering the negative effects of such mutations by decreasing their selective advantage (limiting cell-level selection). Many different processes have likely contributed to decreasing intra-organismal evolution and increasing group stability in clonal multicellular systems. Nevertheless, a series of pre-conditions, constraints, and life-history traits specific to each lineage can also affect the type of mechanisms involved and the outcome in terms of the evolutionary stability and evolvability of the extant multicellular lineages. Here, we explore the relative contribution of these factors, both during the early evolution of multicellularity as well as in lineages that achieved high levels of morphological and developmental complexity, *with a focus on animal, green algal and plant lineages*, which have been studied more extensively.

11.2   WITHIN-GROUP VARIATION: FACTORS AND MECHANISMS

By definition, cells in clonal multicellular organisms are genetically related, and thus within-group variation is expected to be low. However, different *factors* can affect variation within groups, and a series of *mechanisms* are thought to have evolved specifically to lower within-group variation.

11.2.1   Mode of Reproduction

Although the evolution of clonal multicellularity is predicated on the inability of daughter cells to separate following the division of a single cell, a distinction has to be made between the origin of clonal multicellular organisms and their subsequently evolved modes of reproduction and development (i.e., life cycles). This is because although all clonal multicellular lineages evolved from single-celled ancestors, during their life cycles, the offspring can develop from either a single cell (spore or zygote) or multiple cells (propagule) (Figure 11.1A and C). These two distinct strategies result in marked differences in the potential for within-group genetic variation in the offspring (Figure 11.1B and D) and the fate of mutations during the early evolution of multicellularity (Ratcliff et al. 2017).

Single-cell reproduction has been proposed to be an adaptation to ensure that the clonality of cells in a multicellular organism is restored at the start of each generation (Szathmáry and Maynard Smith 1995; Grosberg and Strathmann 1998; Kuzdzal-Fick et al. 2011). The passage through a *"single-cell bottleneck"* is considered critical for the evolutionary stability of multicellular organisms by ensuring high cell relatedness, which is thought to be very important both in the early evolution of multicellular groups and as a means to prevent genetic conflicts in each generation, especially in multicellular lineages that evolved large body sizes and/or long lifespans (discussed later). But going through a single-celled stage every generation can also contribute to the elimination of deleterious mutations from the population by segregating and exposing the cell variants to inter-organismal selection in the next generation (Grosberg and Strathmann 1998). Such variants can include mutations that negatively affect both the cell and the multicellular group (uniformly deleterious; Roze and Michod 2001) as well as mutations that increase the fitness of a cell lineage at a cost to the group (selfish; Queller 2000) (Figure 11.1B and D).

Nevertheless, single-celled stages have also been proposed to be necessary for complex development (Wolpert and Szathmáry 2002) or as means to maximize fecundity and population growth (Pichugin et al. 2017). The latter is supported by evidence from the experimental evolution of a multicellular life cycle with a single-cell bottleneck in the unicellular alga *Chlamydomonas reinhardtii*; the evolution of a unicellular bottleneck prior to any genetic conflicts in the multicellular group

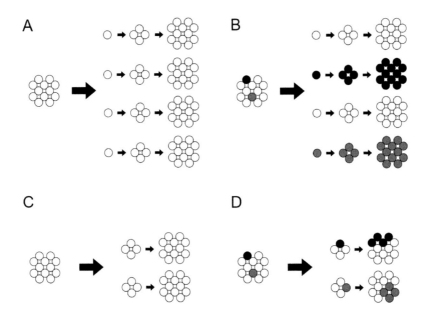

FIGURE 11.1 Simplified representation of two modes of asexual reproduction involving single-celled stages (A and B) and multicell propagules (C and D), highlighting the advantages of unicellular bottlenecks both in terms of fecundity (A versus C) as well as lowering intra-organismal variation and purging of uniformly deleterious (gray circles) and selfish (black circles) mutations (B versus D).

can imply that single-cell propagation is a preadaptation that was later co-opted for conflict suppression (Ratcliff et al. 2013). Furthermore, a series of life cycle models developed by Ratcliff et al. (2017) suggested that genetic bottlenecks and clonal development can also link the selection on heritable multicellular traits to that of the genes that affect them, and maximize the variance of group-level traits.

Although single-cell bottlenecks can be involved in many evolutionary processes (see Grosberg and Strathmann 1998 for a discussion), it is important to note that for lineages that reproduce sexually, the single-cell stage is a *de facto* phase unrelated to the evolution of multicellularity. That is, by definition, sexual reproduction requires passing through single-cell phases – the gametes and the zygote. Thus, the question of the role of the unicellular stage in the life cycle of obligately sexual multicellular lineages cannot be fully dissociated from the question of the role of sexual reproduction. Notably, many multicellular lineages are facultatively sexual, reproducing asexually for much of their life cycle (e.g., algae, sponges, cnidarians). Therefore, if single-cell bottlenecks are required to maintain the evolutionary stability of multicellular phenotypes in terms of reducing intra-organismal evolution, asexual reproduction should also employ a single-celled stage in facultatively sexual multicellular lineages.

However, in many clonal multicellular lineages, asexual reproduction can involve groups of cells (e.g., fragmentation in green, red or brown algae; stolons in land plants; budding in cnidarians) whose relatedness coefficient can vary depending on development and cell division patterns (discussed below). Accordingly, all else being equal, the potential for within-group variation in the developing offspring should be higher in these lineages. The potential for purging deleterious mutations is also expected to decrease, especially when propagules are large and/or the initiating cells in the propagule are distantly related (Kondrashov 1994; see Roze and Michod 2001 for a discussion). Nevertheless, if within-organism selection is stronger than selection among individuals, the mutation load can decrease as propagule size increases (Otto and Orive 1995). The fitness effect of mutations (uniformly deleterious vs selfish) can also influence the propagule size (which can affect the fitness of the offspring), with smaller propagules being favored when mutations are selfish, despite the cost in terms of smaller offspring (Roze and Michod 2001).

Many multicellular lineages (including among animals and plants) are known to successfully employ reproductive modes involving fragmentation or budding during their asexual phase or as their only means of reproduction over many generations. While for some lineages their eventual passage through a single cell stage during the sexual phase would reset the clonality within the multicellular group and "purge" them of deleterious mutations (Grosberg and Strathmann 1998), it is unclear if the unicell sexual stage is in fact an adaptation to decrease intra-organismal evolution or a by-product of sexual reproduction. In this context, it has also been suggested that unicellular bottlenecks are in fact "exaptations conferring immunity to future cell-cell conflicts rather than being adaptations *per se*" (Niklas and Newman 2020). Furthermore, in some instances (e.g., filamentous cyanobacteria and some green algae) both single and multicell asexual propagules can be produced, arguing that the two modes of reproduction are adaptations to selective pressures unrelated to preventing within-group variation (Singh and Montgomery 2011). Thus, it is possible that the frequent use of single-celled stages in the life cycle of most multicellular lineages is the result of multiple distinct selective forces. Moreover, even in lineages that always go through a single-celled stage, the potential for intra-organismal variation still exists, and additional mechanisms are required to control variation acquired during ontogeny. Overall, although a single-cell bottleneck is definitely an efficient way to decrease intra-organismal evolution, other factors and mechanisms are also equally important for the early evolutionary stability of clonal multicellularity (Libby et al. 2016; Queller and Strassmann 2009).

11.2.2 Developmental Modes

One of the most fundamental differences among multicellular phenotypes is related to the presence of specialized cells and in particular reproductive/germ and non-reproductive/somatic cells. The *absence of a germ-soma separation* in simple multicellular lineages (e.g., cyanobacteria, some green algae) implies that all intra-organismal variation will also be transmitted to the offspring. Whether that variation will affect the offspring depends on the reproductive mode. In lineages that always go through a single-cell bottleneck during the asexual phase (i.e., each cell in the group will produce a multicellular offspring; such as in multicellular volvocine green algae) each generation will start from a single cell founder. However, in lineages that reproduce by fragmentation (e.g., filamentous cyanobacteria and many green, red and brown algae) the fate of intra-organismal variation will be dependent on the cell division pattern and growth mode (apical, intercalary, lateral; see below).

Nevertheless, since multicellular organisms without a germ-soma separation have simple developmental and growth patterns, the cells in the propagule are expected to be closely related (Ratcliff et al. 2017).

In multicellular lineages with a defined soma and germline, the fate of the within-group variation will be affected by both the reproductive and developmental modes (Figure 11.2). The main difference in developmental modes is with respect to the timing of *germline segregation* (Figure 11.2). In the so-called "ancestral mode of development" (Buss 1987), the germline is segregated late in development. In groups with this mode of development (e.g., sponges, cnidarians, land plants), the somatic cell lineages are incapable of continuous division or re-differentiation and

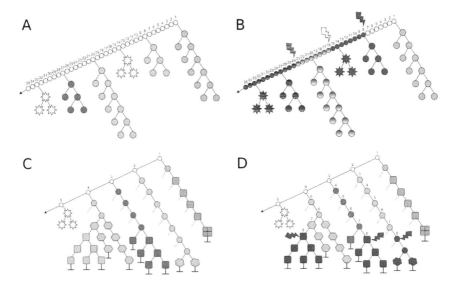

FIGURE 11.2 Simplified representation of the "ancestral" (A and B) and "derived" (C and D) modes of development (Buss 1987), highlighting the different impact that mutations have on intra-organismal variation in the two developmental modes (B versus D). (A) In the "ancestral" mode (such as in cnidarians and plants, for instance), a pluripotent lineage (white circles; numbers indicate cell divisions) gives rise to both differentiated cells (various coloured circles) and gametes (stars) throughout lifetime (and to offspring, during vegetative reproduction; arrow). (B) Mutations (lightning signs) that occur in this pluripotent lineage (for simplicity and to allow comparison of effects, mutation events are indicated every 8th cell division) will be inherited in all subsequent differentiated cell lineages (including gametes and asexual offspring) and can have cumulative effects (indicated by changes in colour). (C) In the "derived" mode (such as in animals and *V. carteri*, for example) a totipotent lineage (white circles; numbers indicate cell divisions) gives rise to several multipotent stem lineages that self-renew (various colours and shapes with dashed lines) and produce differentiated cells with limited replication potential (blunt arrows), and then terminally differentiates into gametes (stars) early in development. (D) Mutations (lightning signs) occurring with the same frequency as in panel B (every 8th cell division) will largely only affect the terminally differentiated cells and will be removed from the group as cells senesce and die.

thus they have to be replenished from one or a few pluripotent lineages that remain mitotically active throughout ontogeny, and can later differentiate into germ cells (Figure 11.2A). These multicellular lineages are also capable of vegetative reproduction via fragmentation or budding, which allows somatic variation to be transmitted to offspring. In contrast, in the "derived mode of development" (Buss 1987) – such as in most animals and some (but not all) multicellular volvocine algae (e.g., *Volvox carteri*), multipotent stem cells with various degrees of mitotic capacity (approaching immortality in some stem cell lineages) and/or potential for differentiation are produced from a totipotent lineage which then differentiates into germ cells early in the development (Figure 11.2C). The evolution of an early segregated germline is thought to mediate potential conflicts among cell lineages – including cheaters, in terms of access to the germline and representation in the next generation (Michod 1996; Michod et al. 2003). Nevertheless, since multicellular lineages with an early-segregated germline do also go through a bottleneck, such mutants will ultimately be removed from the population through among-group selection (as a multicellular group composed exclusively of cheaters will be less fit) (Figure 11.1B).

Important to controlling intra-organismal variation is also the *cell division potential* associated with the various developmental modes. For instance, in the ancestral mode of development, all cells in the multicellular organism (including the germ cells) are descendants from one or a few pluripotent cell lineages that divide continuously throughout ontogeny. Furthermore, in multicellular groups that reproduce through fragmentation, budding or stolons, the offspring inherits these long-lived proliferative lineages; thus, the potential for variation in the vegetatively produced offspring could be high (Figure 11.2B). On the other hand, in organisms with an early-segregated germline (i.e., the "derived mode of development"), somatic cells are descendants of a limited number of stem cell lineages that "delegate" proliferative tasks to a battery of progenitor/amplifying cells that replenish the terminally differentiated cells as needed (Figure 11.2C). This pattern of distribution of the cell proliferation potential is thought to represent an adaptation to limit the number of cell divisions (and thus potential for mutation) in the stem cells as a means to reduce the occurrence of oncogenic somatic mutations (DeGregori 2011).

Multicellular lineages also differ vastly in their embryonic development and ontogeny in terms of *cell division and multicellular growth patterns*. In multicellular organisms that can reproduce vegetatively and/or do not have an early-segregated germline, these aspects have the potential to influence the genetic composition of the multicellular offspring. For instance, plants employ complex cell division and growth patterns (even in the same individual), including apical and lateral cell division as well as intermediate/indeterminate growth (i.e., throughout life; e.g., roots, shoots) and determinate growth (i.e., stops when a final size is reached; e.g., leaves, flowers). Indeterminate growth (also characterizing some animal lineages; e.g., corals, many fishes, amphibians, snakes), by definition, will result in an increasing number of cell divisions and thus potential for high intra-organismal genetic variation. Nevertheless, the distribution of this intra-organismal variation is also under selection, depending on the tissue's contribution to the next generation and longevity.

For instance, in perennial (but not annual) plants, the rate of mutation accumulation (per unit time) in shoot apical meristems is lower than that in root apical tissues (Wang et al 2019). But even when growth is limited to a final size (such as in insects, mammals), cell divisions do occur during ontogeny to replace damaged as well as senescent cells, which can result in intra-organismal variation. A notable exception is among some invertebrates (e.g., *Caenorhabditis elegans;* Pearson and Sánchez Alvarado 2008) and volvocine algae (Kirk 1998), in which there are no post-embryonic cell divisions; in these lineages, the potential for intra-organismal variation is restricted to the embryonic stage.

Another difference in developmental modes that can influence the potential for intra-organismal genetic variation is the ability of cells to move within the group. The lack of cell mobility in plant lineages decreases the potential for variants to invade nearby tissues and/or migrate to distant locations, which might explain both the absence of malignant tumors and the high incidence of vegetative reproduction (including fragmentation, stolons/runners, bulbs) in many plant lineages. On the other hand, the ability of animal cells to move both during embryonic development as well as in adults reflects in the high incidence of malignancy and might contribute to the low incidence of vegetative reproduction involving fragmentation and budding in animal lineages.

11.2.3   BODY SIZE

The potential for intra-organismal genetic variation is also expected to increase with group/body size. That is because – all else being equal, more cells require more cell divisions, and more cells also equal more targets for mutation. This correlation is generally expressed in an expected increase in cancer incidence in multicellular lineages with large *body sizes.* However, this expected correlation has not been confirmed, a situation known as Peto's paradox (Peto et al. 1975). For instance, large organisms such as elephants and whales do not show the high cancer incidence expected based on their body size (Caulin and Maley 2011). The lack of correlation between body size and cancer rates is commonly attributed to the evolution of better/additional tumor suppression mechanisms in larger animals (Tollis et al. 2017). Nevertheless, cancer-unrelated life-history traits and pressures associated with the evolution of a large body size (e.g., low metabolic rates that could reflect in less oxidative damage; late maturation and low fecundity that would result in, or require, increased investment in somatic maintenance) could also have affected mutation rates or shaped the development in these lineages in a way that has resulted in lower than expected (based on size alone) cancer rates (Brown et al. 2015; Nedelcu and Caulin 2016; Møller et al. 2017; Nedelcu 2020).

Interestingly, large body sizes are achieved both in lineages with single-cell reproductive modes and early segregated germline (animals) as well as in lineages that do not segregate their germline early in development and do not necessarily go through a single-cell bottleneck in every generation (some plants). Notably, large long-lived plants also do not show the expected (based on size alone) increase in the per-generation mutation rate and are assumed to have evolved mechanisms to reduce mutation rates per unit growth (Orr et al. 2020).

11.2.4 LIFE SPAN

Increased *life span* is expected to increase intra-organismal variation as well since cell lineages go through a higher number of cell divisions during the lifetime of a long-lived individual. For instance, long-lived plants (such as conifers) have been shown to have among the highest per-generation mutation rates for any eukaryote, in spite of their remarkably low annual somatic base substitution rate (Hanlon et al. 2019; Hofmeister et al. 2020). Similarly, the number of somatic mutations in normal human liver was shown to increase with age, with up to 3.3 times more mutations per cell in aged humans than in young individuals (Brazhnik et al. 2020). This increase in the number of cell divisions and mutations with age should result in higher incidence of cancer in longer-lived multicellular lineages. The lack of correlation between lifespan and cancer incidence is another side of Peto's paradox, which is also commonly explained in terms of better cancer suppression mechanisms in long-lived species (Tollis et al. 2020). Nevertheless, as for body size, the lower than expected (based on lifespan alone) incidence of cancer can be an indirect by-product of life-history traits and adaptations unrelated to suppressing cancer (Nedelcu and Caulin 2016). Interestingly, an increased intra-organismal mutation load in long-lived trees is considered adaptive as it generates important genetic variation that enables selection both among offspring (as such mutations can be inherited since plants do not have a segregated germline) and among cell lineages within individual trees (e.g., a branch can acquire resistance to herbivory) (Padova et al. 2013; Hanlon et al. 2019).

11.2.5 SOMATIC MUTATION SUPPRESSION MECHANISMS

The potential for intra-organismal variation is dependent on the incidence of somatic mutations. Because of the impact such mutations have on the development of cancer, most studies on this topic are centered around animal systems. Although not all somatic mutations are associated with cancer (see below), it is generally believed that a series of mechanisms had to evolve specifically to prevent the initiation and progression of cancers in all multicellular lineages. These mechanisms are generally referred to as tumor suppression mechanisms. They include both *"caretakers"* (e.g., DNA damage sensing and DNA repair) and *"gatekeepers"* (premature senescence and apoptosis) (Kinzler and Vogelstein 1997; Hooper 2006). However, caretakers also function in indispensable cellular processes, and many, including p53 – the most frequently mutated tumor suppressor gene in human cancers, have actually evolved in single-celled lineages (Domazet-Lošo and Tautz 2010). On the other hand, the evolution of gatekeeper genes is thought to largely overlap with the emergence of metazoans and has been interpreted to reflect the need for both increased cooperation and cheating prevention (Domazet-Lošo and Tautz 2010). Yet, both premature senescence and apoptosis-like phenomena have been found in unicellular lineages, suggesting that they also predate the origin of multicellularity (Nedelcu et al. 2011; Nedelcu and Caulin 2016). Also, many tumor suppressor genes (including p53) do not seem to be involved in tumor suppression in invertebrates (Pearson and Sánchez Alvarado 2008). However, direct correlations between tumor suppressing mechanisms and the ability to decrease cancer potential have been reported. For instance, the lower than expected (based on size and lifespan) cancer incidence in elephants was correlated with the finding of extra copies of the tumor suppressor gene p53, which is thought to result in increased sensitivity to DNA damage-induced apoptosis (and thus elimination of potentially oncogenic cells) in elephants (Sulak et al. 2016). Nevertheless, additional p53 gene copies have not been found in the humpback whale, suggesting that if additional tumor suppressing mechanisms are indeed required to control cancer (and intra-organismal evolution) in large/long-lived multicellular bodies, they are lineage specific (Tollis et al. 2019). Lower per-year somatic mutations in long-lived angiosperms (such as poplar) – compared to annual plants, are also thought to be the result of mechanisms that can decrease the potential for somatic mutations (and thus intra-organismal variation). These include limiting the number of meristematic cell divisions and evolving ways to protect meristematic cells from DNA-damaging factors such as UV radiation (Hofmeister et al. 2020).

11.2.6 HOMEOSTASIS

In addition to DNA replication/repair and metabolically-induced mutations, environmental factors can also result in DNA damage and mutations (Figure 11.3). One of the proposed advantages of group living is homeostasis, which can provide protection from environmental stressors (Smukalla et al. 2008). Developmental modes that ensure stem cell lineages are protected from environmental challenges can also limit the potential of DNA damage-induced mutations and intra-organismal variation.

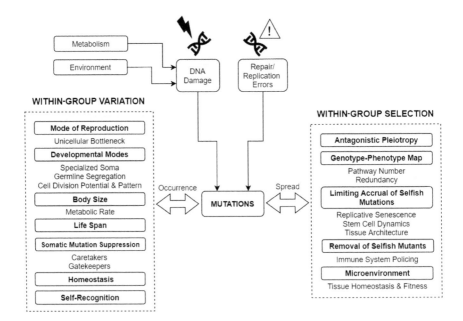

FIGURE 11.3 Simplified summary of the various factors and mechanisms that affect within-group variation and selection (see text for details and discussion).

These include, for instance, the location of stem cells in crypts (in animals) and of meristems in buds (in plants).

11.2.7 SELF-RECOGNITION

For multicellular groups that developed clonally but faced a continual threat of chimerism, the evolution of a self-recognition system would have had obvious benefits against germline-invading or fitness-reducing cells from other individuals (Fernández-Busquets et al. 2009). However, as in the case of the unicellular bottleneck, the timing of the evolution of self-recognition systems may determine whether they evolved as a specific anti-cheating mechanism or were later co-opted into that role. Sponges are good model-system for studying the role of adhesive/recognition mechanisms during the evolution of animal multicellularity (Fernández-Busquets et al. 2009; Vilanova et al. 2016).

The ability to discriminate against non-self can also be beneficial to recognize internal self-cell variants and thus act to suppress intra-organismal variation emerging during ontogeny. This aspect is extremely relevant in the context of the recognition of malignant cells by the immune system (see below). In animals, immune systems are thought to be efficient mechanisms to both eliminate intra-organismal genetic variants as well as provide a strong barrier to the inter-organismal transmission of malignant cells. The few special cases of transmissible cancer are thought to be facilitated by low non-self-recognition systems (Belov 2012).

11.3 WITHIN-GROUP SELECTION: LIMITING THE ADVANTAGE OF SELFISH MUTANTS

Mechanisms that reduce intra-organismal variation will affect the incidence and distribution of *all* types of mutations. But the ultimate fate of mutations will be determined by the effect (positive +; negative −; or neutral ~) they have on the fitness of both the cell (C) and the multicellular group (M). In multicellular organisms with a differentiated soma, somatic mutations can be (i) deleterious only at the cell level (C−/M~), (ii) deleterious or advantageous at both the cell and organism level (C−/M−, or C+/M+), (iii) altruistic (C−/M+), (iv) selfish (C+/M−), or (v) uniformly neutral (C~/M~). C−/M~ mutants, by definition, will likely be eliminated through negative selection at the cell level (Otto and Orive 1995; Otto and Hastings 1998). However, uniformly deleterious mutations can accumulate in some cell types and have been associated with human diseases, neurodegeneration and aging (see Gonzalez-Perez et al. 2019 and Brazhnik et al. 2020 for examples and references). Thus, mechanisms that reduce the occurrence and spread of C−/M− could be favored, if they affect the reproductive fitness of the multicellular individual. Nevertheless, in long-lived woody plants, such deleterious somatic mutations can accumulate, and this has been hypothesized to favor outcrossing by reducing the survival of inbred progeny (Bobiwash et al 2013).

The fate of cell-level advantageous mutants will be determined by their effect on organismal fitness. In plants, due to their developmental and reproduction modes, cell-level beneficial mutations occurring in the apical meristems can be selected for and transmitted to offspring, which might then affect inter-individual variation. Thus, in plants, intra-organismal variation can have implications for their ability to adapt to changing ecological conditions and, ultimately, for plant speciation (Hanlon et al. 2019; Schoen and Schultz 2019; Orr et al. 2020). However, in animals, mutants that gain selective advantages at the cell level are more likely to be costly at the organism level (Frank and Nowak 2004). Cancer is a reflection of such mutants, resulting in selection at the cell level overriding selection at organism level. If these selfish variants negatively affect the fitness of the multicellular group during the reproductive phase, they can affect the evolutionary stability of the lineage and have the potential to drive the groups to extinction directly (through decreasing their fitness) or indirectly through gaining access to the germline. Thus, several factors and mechanisms are known or have been proposed, to limit or reduce the selective advantage of selfish mutants (Figure 11.3), especially in the context of cancer.

11.3.1 Antagonistic Pleiotropy

One mechanism that can decrease the selective advantage of selfish mutants (and thus enforce cooperation) – especially during the early evolution of multicellularity, is antagonistic pleiotropy. Specifically, if cooperative genes are linked to advantageous individual-level traits, mutations in such genes will also negatively affect the fitness of the selfish mutants (Foster et al. 2004). A related evolutionary mechanism – coined as "type 1 ratcheting," has been proposed by Libby et al. (2016). This scenario envisions that reversion to unicellularity is hindered by the fact that the accumulation of mutations that increase cell-level fitness in a multicellular context are also costly in a single-celled context.

Pleiotropy has been shown to stabilize cooperation in aggregative multicellularity (Foster et al. 2004), but less is known about its role in the evolution of clonal multicellularity. However, at least in the volvocine alga *Volvox carteri*, selfish mutants that evade the developmental control of cell proliferation are also more sensitive to stress (Konig and Nedelcu 2020). Notably, the increased proliferation of cancer cells is also known to be linked to a lower ability (compared to normal somatic cells) to withstand nutrient stress due to their failure to trade-off cell proliferation for maintenance in stressful environments (Raffaghello et al. 2008; Lee et al. 2012). Similar trade-offs and pleiotropic effects are likely to have contributed to the evolutionary stability of early multicellular groups and have played roles in the evolution of other clonal multicellular lineages.

11.3.2 Genotype-phenotype Map Re-organization

During the transition to multicellularity, a new genotype-phenotype map (i.e., the relationship between genotype and phenotype; Alberch 1991) had to evolve to reflect both the loss of unicellular traits ("type 2 ratcheting"; Libby et al. 2016) and the emergence of new traits at the group level. How a new map is established can influence the stability of the group and the evolvability of the lineage (Nedelcu and Michod 2004). For instance, the differentiation of somatic cells in *V. carteri* is achieved through the induction of a single gene whose expression suppresses both the immortality and totipotency of these cells. Mutations in this gene alone result in somatic cells gaining full proliferative and reproductive abilities, with drastic negative effects for the group. Thus, having such a simple genetic architecture (that can be entirely lost via single mutations) can endanger the stability of the group and limit the evolutionary potential of the lineage.

The genotype-phenotype map is also very important for the stability of complex multicellular organisms. In the context of cancer, this aspect is reflected in the *number of genes/pathways* that need to be inactivated to induce malignancy, known as transformation stages. Interestingly, the number of stages differs between species. For instance, the transformation of fibroblasts requires that six signal pathways be affected in humans, compared to only two in mice (Rangarajan et al. 2004). Also, the development of retinoblastoma (an eye cancer that begins in the retina and mostly affects children) requires the inactivation of only one locus (Rb) in humans but two (Rb and p107) in mice (see discussion in Leroi et al. 2003). Furthermore, human cells require more mutations than mouse cells to create immortalized cultures; both the Rb and p53 pathways must be knocked out to immortalize human fibroblasts while mouse cells require only the p53 pathway to be inactivated (Hahn and Weinberg 2002).

Added *redundancy* in the form of extra tumor suppressor genes can also limit the selective advantage of selfish mutants. In this scenario, mutations in all copies would be required to result in malignancy. In support of this possibility, transgenic mice that contain an extra copy of p53 (including its regulatory elements) gain an increased resistance to cancer (García-Cao et al. 2002). Redundancy in tumor suppressor genes is also thought to be responsible for the lower than-expected cancer incidence in large animals (Nunney 1999; Leroi et al. 2003). For instance, there are at least 19 copies of p53 in the African elephant genome, and though 18 of these appear to be a result of retrotransposition events, some are expressed and are thought to contribute to improved cancer suppression in elephants (Abegglen et al. 2015).

11.3.3   LIMITING THE POTENTIAL FOR THE PROGRESSIVE ACCRUAL OF SELFISH MUTATIONS

In many multicellular lineages with an early segregated germline and determinate growth, most somatic cell lineages have limited proliferation potentials. In animals, this phenomenon is known as *replicative senescence* and is induced by telomere shortening (via repression of telomerase activity). In this way, the accumulation of mutations that could provide somatic lineages with selective advantages is limited (Campisi 2001). On the other hand, animals with indeterminate growth express telomerases in the tissues of adults (e.g., American lobster and rainbow trout; Klapper et al. 1998a, b). However, exceptions do exist. For instance, in mouse, somatic cells express telomerases, have very long telomeres and do not exhibit replicative senescence (senescence of these cells is thought to be a stress response; e.g., Seluanov et al. 2008). Furthermore, rodent species differ as far as telomerase activities in their somatic cells, and replicative senescence appears to correlate with body mass (Seluanov et al. 2007, 2008). Interestingly, the long-lived naked mole rat does express telomerase activities in its somatic cells, and despite its increased longevity relative to other rodents, spontaneous neoplasms have never been reported in this species (Buffenstein 2005). Notably, although it involves a different mechanism, replicative senescence is also known in yeast (Steinkraus et al. 2008) and the simple multicellular green alga, *Volvox carteri* (Shimizu et al. 2002). Similarly, although plants do express a form of replicative senescence known as mitotic senescence, this process does not involve shortening of telomeres and is involved in curtailing cell proliferation in germline-like apical meristems and during early stages of fruit development (Gan 2003).

In animals with an early segregated germline, many features of *stem cell dynamics* are also thought to be adaptations to reduce the selective advantage of potentially selfish mutants. These include: asymmetric divisions, preserving an immortal DNA strand in the self-renewed stem cell, limiting the number of stem cells, interposing a series of transiently amplifying cells between the stem cells and the terminally differentiated cells, and the imposition of differentiation on proliferating stem cell progeny (Potten et al. 2002; Frank and Nowak 2004; Caussinus and Gonzalez 2005; DeGregori 2011). Overall, stem cells are indeed known to experience reduced spontaneous mutation loads compared with differentiated cells, although the mechanisms are not fully understood (Brazhnik et al. 2020).

Additional ways that can reduce the impact of selfish mutants on the fitness of the multicellular organism include traits associated with *tissue organization and architecture* – such as the organization of epithelial tissues in crypts, microenvironmental signals and niche or stromal matrix contacts, and serial differentiation (Cairns 1975; Frank and Nowak 2004; Gatenby et al. 2010; DeGregori 2011). Differences in tissue architecture could influence the frequency of mutant cell lineages (and cancers) depending on the number of stem cells or the dynamics of the tissue itself (Leroi et al. 2003). For example, it has been suggested that under a model of serial differentiation it is possible to increase the number of cells and the amount of cell turnover per organism without increasing the number or proliferative activity of somatic cells, simply by increasing the number of non-stem stages (Pepper et al. 2007).

Tissue architecture and development also affect intra-organismal selection in plants. For instance, the large number of apical initials in conifers allows efficient selection among cells within the meristem, but the highly structured nature of their apical meristems might limit the potential for cells with higher fitness to remain within the meristem. Furthermore, conifers and angiosperms differ in both development and physical architecture, with most conifers having one dominant stem, little bifurcating branching, and a single layer of apical initials in a relatively simple meristem, all of which are thought to limit somatic selection (Hanlon et al. 2019). Generally, patterns of stem cell divisions – such as limiting the number of cell divisions between the meristem and the new branch, are thought to contribute to the low per-year somatic mutation rates and longevity in perennial plans (Burian et al. 2016).

11.3.4 REMOVAL OF POTENTIALLY SELFISH MUTANTS

Policing is a well-documented strategy to decrease the selective advantage of selfish mutants in all social groups. In many animal lineages, this role is performed by the *immune system* which, in addition to recognizing pathogens, can also identify and remove abnormal self-cells, including mutant cells that have the potential to become selfish (Dunn et al. 2004). The low incidence of cancer in Decapoda is credited in part to their immune system (Vogt 2008). Similarly, the immune system of verte-brates can recognize and eliminate primary developing tumors (Shankaran et al. 2001). However, following clonal escape and tumor formation, chronic activation of innate immune cells can also promote tumor growth (De Visser et al. 2006).

11.3.5 TISSUE MICROENVIRONMENT AND FITNESS

As in any system, the selective advantage of a cell variant is dependent on (or relative to) a specific environment. Recently, the role of *tissue microenvironment and fitness* in suppressing cancer (and thus, controlling intra-organismal evolution) has been acknowledged and is receiving a lot of attention. Healthy tissues are known to be able to provide a strong barrier to selection of mutant clones. For instance, *NOTCH1* mutant clones (often associated with cancer) have been found to increase with age in the human esophageal tissue and can coexist with normal clones (Martincorena et al. 2018). However, changes in the tissue microenvironment during aging as well as in response to environmental stimuli (e.g., smoking) and chronic inflammation can select for cancer clones better fit to those conditions and promote cancer develop-ment (Casás-Selves and DeGregori 2011; DeGregori 2018).

11.4   CONCLUSION

The capacity of multicellular groups to become stable evolutionary units is depen-dent on their ability to control intra-organismal evolution. For such groups to become units of evolution, within-group variation has to be lower than among-group varia-tion such that selection at the group level overrides selection at the cell level. That is, mechanisms to control both intra-organismal genetic variation and the selective advantage of within-group variants have to evolve (Figure 11.3). However, the spe-cific mechanisms and their relative contributions are dependent on the genetic and structural background on which multicellularity evolved.

For instance, in multicellular plant lineages, the potential for oxidative DNA damage due to their photosynthetic activities as well as the increased potential for UV-induced DNA damage associated with the sessile lifestyle of land plants are expected to increase mutation levels, and thus intra-organismal variation. Nevertheless, the presence of a cell wall inherited from their single-celled ances-tors (reflected in the strong connections between plant cells) decreases the selective advantage of mutant cells by limiting their ability to spread. These are in contrast to the high mobility of animal cells and the increased range of physiological and behavioral adaptations that animals evolved to cope with environmental stress. Also, in addition to differences inherited from their unicellular ancestors, plant and animal lineages also differ in their developmental and reproduction modes as well as life-history traits and strategies, which together are expected to have distinct effects on intra-organismal evolution.

A number of mechanisms have been proposed to have evolved to control intra-organismal evolution during the evolution of clonal multicellularity. However, although in most cases their current contribution to the evolutionary stability of a multicellular lineage is obvious, it is not always clear whether they evolved specifically to limit/control cell-level variation and selection. In other words, it is unclear whether other selective pressures or life-history traits shaped these mechanisms that in turn also allowed a better control of intra-organismal evolution.

The astonishing diversity that independently-evolved clonal multicellular lineages achieved in a relatively short evolutionary time reflects independently-evolved successful strategies to control intra-organismal evolution. A full understanding of the mechanisms underlying the success of clonal multicellularity in terms of evolutionary stability and increased complexity requires a comparative approach that must take into account both the evolutionary history of the lineages and the specific selective pressures and life-history traits that shaped the evolution of each multicellular lineage.

REFERENCES

Abegglen LM, Caulin AF, Chan A, et al. (2015). Potential mechanisms for cancer resistance in elephants and comparative cellular response to DNA Damage in Humans. JAMA – J Am Med Assoc 314:1850–1860.

Aktipis A (2020). The cheating cell: How evolution helps us understand and treat cancer. Princeton University Press, Princeton and Oxford.

Alberch P (1991). From genes to phenotype: dynamical systems and evolvability. Genetica 84:5–11.

Aktipis CA, Boddy AM, Jansen G, et al. (2015). Cancer across the tree of life: Cooperation and cheating in multicellularity. Philos Trans R Soc B Biol Sci 370:201402019.

Belov K (2012). Contagious cancer: Lessons from the devil and the dog. BioEssays 34:285–292.

Bobiwash K, Schultz ST, Schoen DJ. (2013). Somatic deleterious mutation rate in a woody plant: estimation from phenotypic data. Heredity 111:338–344.

Bourke AF (2019). Inclusive fitness and the major transitions in evolution. Curr. Opin. Insect Sci. 34:61–67.

Brazhnik K, Sun S, Alani O, et al. (2020). Single-cell analysis reveals different age-related somatic mutation profiles between stem and differentiated cells in human liver. Sci Adv 6:eaax2659.

Brown JS, Cunningham JJ, Gatenby R (2015). The multiple facets of Peto's paradox: A life-history model for the evolution of cancer suppression. Philos Trans R Soc B Biol Sci 370:20140221.

Buffenstein R (2005). The naked mole-rat: A new long-living model for human aging research. J. Gerontol Ser A Biol Sci Med Sci 60:1369–1377.

Burian A, Barbier de Reuille P, Kuhlemeier C. 2016. Patterns of stem cell divisions contribute to plant longevity. Curr Biol 26:1385–1394.

Buss LW (1987). The Evolution of Individuality. Princeton University Press. Princeton University Press, Princeton, New Jersey.

Cairns J (1975). Mutation selection and the natural history of cancer. Nature 255:197–200

Campisi J (2001). Cellular senescence as a tumor-suppressor mechanism. Trends Cell Biol. 11:S27–31.

Casás-Selves M, DeGregori J (2011). How cancer shapes evolution, and how evolution shapes cancer. Evolution 4:624–634.

Caulin AF, Maley CC (2011). Peto's Paradox: Evolution's prescription for cancer prevention. Trends Ecol Evol 26:175–182.

Caussinus E, Gonzalez C (2005). Induction of tumor growth by altered stem-cell asymmetric division in *Drosophila melanogaster*. Nat Genet 37:1125–1129.

De Visser KE, Eichten A, Coussens LM (2006). Paradoxical roles of the immune system during cancer development. Nat. Rev. Cancer 6:24–37.

DeGregori J (2011). Evolved tumor suppression: Why are we so good at not getting cancer? Cancer Res. 71:3739–3744.

DeGregori J (2018). Adaptive oncogenesis, A new understanding of how cancer evolve inside us. Harvard University Press, Cambridge, Massachusetts and London, England.

Domazet-Lošo T, Tautz D (2010). Phylostratigraphic tracking of cancer genes suggests a link to the emergence of multicellularity in metazoa. BMC Biol 8:66.

Dunn GP, Old LJ, Schreiber RD (2004). The immunobiology of cancer immunosurveillance and immunoediting. Immunity 21:137–148.

Fernández-Busquets X, Körnig A, Bucior I, et al. (2009). Self-recognition and Ca2+-dependent carbohydrate-carbohydrate Cell adhesion provide clues to the Cambrian explosion. Mol Biol Evol 26:2551–2561.

Fisher RM, Cornwallis CK, West SA (2013). Group formation, relatedness, and the evolution of multicellularity. Curr Biol 23:1120–1125.

Foster KR, Shaulsky G, Strassmann JE, et al. (2004). Pleiotropy as a mechanism to stabilize cooperation. Nature 431:693–696.

Frank SA, Nowak MA (2004). Problems of somatic mutation and cancer. BioEssays 26:291–299.

Gan S (2003). Mitotic and postmitotic senescence in plants. Sci Aging Knowledge Environ 2003:RE7.

García-Cao I, García-Cao M, Martín-Caballero J, et al. (2002). "Super p53" mice exhibit enhanced DNA damage response, are tumor resistant and age normally. EMBO J 21:6225–6235.

Gatenby RA, Gillies RJ, Brown JS (2010). The evolutionary dynamics of cancer prevention. Nat Rev Cancer 10:526–527.

Gonzalez-Perez A, Sabarinathan R, Lopez-Bigas N (2019). Local determinants of the mutational landscape of the human genome. Cell 177:101–114.

Grosberg RK, Strathmann RR (2007). The evolution of multicellularity: A Minor Major Transition? Annu Rev Ecol Evol Syst 38:621–654.

Grosberg RK, Strathmann RR (1998). One cell, two cell, red cell, blue cell: The persistence of a unicellular stage in multicellular life histories. Trends Ecol Evol 13:112–116.

Hahn WC, Weinberg RA (2002). Rules for making human tumor cells. N Engl J Med 347:1593–1603.

Hanlon VCT, Otto SP, Aitken SN (2019). Somatic mutations substantially increase the per-generation mutation rate in the conifer *Picea sitchensis*. Evol Lett 3:348–358.

Hofmeister BT, Denkena J, Colome-Tatche M, et al. (2020). A genome assembly and the somatic genetic and epigenetic mutation rate in a wild long-lived perennial *Populus trichocarpa*. Genome Biology 21:259.

Hooper ML (2006). Tumor suppressor genes. In: Encyclopedia of Life Sciences. e0006005, John Wiley & Sons, Ltd, Chichester, UK.

Kinzler KW, Vogelstein B (1997). Gatekeepers and caretakers. Nature 386:761–763.

Kirk DL (1998). Volvox: Molecular-Genetic origins of multicellularity. Cambridge University Press, Cambridge.

Klapper W, Heidorn K, Kühne K, et al. (1998a). Telomerase activity in "immortal" fish. FEBS Lett 434:409–412.

Klapper W, Kühne K, Singh KK, et al. (1998b). Longevity of lobsters is linked to ubiquitous telomerase expression. FEBS Lett 439:143–146.

Kondrashov AS (1994). Mutation load under vegetative reproduction and cytoplasmic inheritance. Genetics 137:311–318.

Konig SG, Nedelcu AM (2020). The genetic basis for the evolution of soma: Mechanistic evidence for the co-option of a stress-induced gene into a developmental master regulator. Proc R Soc Lond Ser B Biol Sci 287:20201414.

Kuzdzal-Fick JJ, Fox SA, Strassmann JE, Queller DC (2011). High relatedness is necessary and sufficient to maintain multicellularity in *Dictyostelium*. Science 334:1548–1551.

Lee C, Raffaghello L, Brandhorst S, et al. (2012). Fasting cycles retard growth of tumors and sensitize a range of cancer cell types to chemotherapy. Sci Transl Med 4:124ra27.

Leroi AM, Koufopanou V, Burt A (2003). Cancer selection. Nat Rev Cancer 3:226–231.

Libby E, Rainey PB (2013). A conceptual framework for the evolutionary origins of multicellularity. Phys Biol 10:035001.

Libby E, Conlin PL, Kerr B, Ratcliff WC (2016). Stabilizing multicellularity through ratcheting. Philos Trans R Soc B Biol Sci 371:20150444.

Martincorena I, Fowler JC, Wabik A, et al. (2018). Somatic mutant clones colonize the human esophagus with age. Science (80-) 362:911–917.

Michod RE (1996). Cooperation and conflict in the evolution of individuality. II. Conflict mediation. Proc R Soc London Ser B Biol Sci 263:813–822.

Michod RE (1998). Darwinian Dynamics. Princeton University Press, Princeton, New Jersey.

Michod RE (2007). Evolution of individuality during the transition from unicellular to multicellular life. Proc Natl Acad Sci USA 104:8613–8618.

Michod RE (1997). Cooperation and conflict in the evolution of individuality. I. Multilevel selection of the organism. Am Nat 149:607–645.

Michod RE, Nedelcu AM, Roze D (2003). Cooperation and conflict in the evolution of individuality: IV. Conflict mediation and evolvability in *Volvox carteri*. BioSystems 69:95–114.

Michod RE, Roze D (2001). Cooperation and conflict in the evolution of multicellularity. Heredity 86:1–7.

Møller AP, Erritzøe J, Soler JJ (2017). Life history, immunity, Peto's paradox and tumours in birds. J Evol Biol 30:960–967.

Nedelcu AM (2020). The evolution of multicellularity and cancer: Views and paradigms. Biochem Soc Trans 48:1505–1518.

Nedelcu AM, Caulin AF (2016). The evolution of cancer suppression mechanisms. In Maley CC, Greaves M (eds) Frontiers in Cancer Research. Springer, pp 217–246.

Nedelcu AM, Driscoll WW, Durand PM, et al. (2011). On the paradigm of altruistic suicide in the unicellular world. Evolution 65:3–20.

Nedelcu AM, Michod RE (2004). Evolvability, modularity, and individuality during the transition to multicellularity in Volvocalean green algae. In: Schlosser G, Wagner GP (eds) Modularity in Development and Evolution. Oxford University Press, Oxford, pp 466–489.

Nedelcu AM, Michod RE (2006). The evolutionary origin of an altruistic gene. Mol Biol Evol 23:1460–1464.

Niklas KJ, Newman SA (2020). The many roads to and from multicellularity. J Exp Botany 71:3247–3253.

Nunney L (1999). Lineage selection and the evolution of multistage carcinogenesis. Proc R Soc B Biol Sci 266:493–498.

Orr AJ, Padovan A, Kainer D, et al. (2020). A phylogenomic approach reveals a low somatic mutation rate in a long-lived plant. Proc R Soc B Biol Sci 287:20192364.

Otto SP, Hastings IM (1998). Mutation and selection within the individual. Genetica 102-103:507–524.

Otto SP, Orive ME (1995). Evolutionary consequences of mutation and selection within an individual. Genetics 141:1173–1187.

Padovan A, Lanfear R, Keszei A, et al. (2013). Differences in gene expression within a striking phenotypic mosaic *Eucalyptus* tree that varies in susceptibility to herbivory. BMC Plant Biol 13:29.

Pearson BJ, Sánchez Alvarado A (2008). Regeneration, stem cells, and the evolution of tumor suppression. In: Cold Spring Harbor Symposia on Quantitative Biology. Cold Spring Harb Symp Quant Biol, pp 565–572.

Pepper JW, Sprouffske K, Maley CC (2007). Animal cell differentiation patterns suppress somatic evolution. PLoS Comput Biol 3:2532–2545.

Peto H, Roe FJC, Lee PN, et al. (1975). Cancer and ageing in mice and men. Br J Cancer 32:411–426.

Pichugin Y, Peña J, Rainey PB, Traulsen A (2017). Fragmentation modes and the evolution of life cycles. PLoS Comput Biol 13:e1005860.

Potten CS, Owen G, Booth D (2002). Intestinal stem cells protect their genome by selective segregation of template DNA strands. J Cell Sci 115:2381–2388.

Queller DC (2000). Relatedness and the fraternal major transitions. Philos Trans R Soc London Ser B Biol Sci 355:1647–1655.

Queller DC, Strassmann JE (2009). Beyond society: The evolution of organismality. Philos Trans R Soc B Biol Sci 364:3143–3155.

Raffaghello L, Lee C, Safdie FM, et al. (2008). Starvation-dependent differential stress resistance protects normal but not cancer cells against high-dose chemotherapy. Proc Natl Acad Sci 105:8215–8220.

Rangarajan A, Hong SJ, Gifford A, Weinberg RA (2004). Species- and cell type-specific requirements for cellular transformation. Cancer Cell 6:171–183.

Ratcliff WC, Herron M, Conlin PL, Libby E (2017). Nascent life cycles and the emergence of higher-level individuality. Philos Trans R Soc B Biol Sci 372:20160420.

Ratcliff WC, Herron MD, Howell K, et al. (2013). Experimental evolution of an alternating uni- and multicellular life cycle in *Chlamydomonas reinhardtii*. Nat Commun 4:2742.

Roze D, Michod RE (2001). Mutation, multilevel selection, and the evolution of propagule size during the origin of multicellularity. Am Nat 158:638–654.

Schoen DJ, Schultz ST (2019). Somatic mutation and evolution in plants. Annu Rev Ecol Evol Syst 50:49–73.

Seluanov A, Chen Z, Hine C, et al. (2007). Telomerase activity coevolves with body mass not lifespan. Aging Cell 6:45–52.

Seluanov A, Hine C, Bozzella M, et al. (2008). Distinct tumor suppressor mechanisms evolve in rodent species that differ in size and lifespan. Aging Cell 7:813–823.

Shankaran V, Ikeda H, Bruce AT, et al. (2001). IFNγ, and lymphocytes prevent primary tumour development and shape tumour immunogenicity. Nature 410:1107–1111.

Shimizu T, Inoue T, Shiraishi H (2002). Cloning and characterization of novel extensin-like cDNAs that are expressed during late somatic cell phase in the green alga *Volvox carteri*. Gene 284:179–187.

Singh SP, Montgomery BL (2011). Determining cell shape: Adaptive regulation of cyanobacterial cellular differentiation and morphology. Trends Microbiol. 19:278–285.

Smukalla S, Caldara M, Pochet N, et al. (2008). *FLO1* is a variable green beard gene that drives biofilm-like cooperation in budding yeast. Cell 135:726–737.

Steinkraus KA, Kaeberlein M, Kennedy BK (2008). Replicative aging in yeast: The means to the end. Annu. Rev. Cell Dev. Biol. 24:29–54.

Sulak M, Fong L, Mika K, et al. (2016). TP53 copy number expansion is associated with the evolution of increased body size and an enhanced DNA damage response in elephants. eLife 5:e11994.

Szathmáry E, Maynard Smith J (1995). The major evolutionary transitions. Nature 374:227–232.

Tollis M, Boddy AM, Maley CC (2017). Peto's Paradox: How has evolution solved the problem of cancer prevention? BMC Biol. 15:60.

Tollis M, Robbins J, Webb AE, et al. (2019). Return to the sea, get huge, beat cancer: An analysis of cetacean genomes including an assembly for the humpback whale (Megaptera novaeangliae). Mol Biol Evol 36:1746–1763.

Tollis M, Schneider-Utaka A, Maley C (2020). The evolution of human cancer gene duplications across mammals. Mol Biol Evol 37:2875–2886.

Vilanova E, Santos GRC, Aquino RS, et al. (2016). Carbohydrate-carbohydrate interactions mediated by sulfate esters and calcium provide the cell adhesion required for the emergence of early metazoans. J Biol Chem 291:9425–9437.

Vogt G (2008). How to minimize formation and growth of tumours: Potential benefits of decapod crustaceans for cancer research. Int. J. Cancer 123:2727–2734.

Wang L, Ji Y, Hu Y, et al. (2019). The architecture of intra-organism mutation rate variation in plants. PLoS Biol 17:e3000191.

Wolpert L, Szathmáry E (2002). Evolution and the egg: Multicellularity. Nature 420:745.

12 Group Transformation
Life History Trade-offs, Division of Labor, and Evolutionary Transitions in Individuality

Guilhem Doulcier
Institut de Biologie de l'Ecole Normale
Supérieure (IBENS), Université Paris Sciences
et Lettres, CNRS, INSERM, Paris, France
Department of Evolutionary Theory, Max Planck
Institute for Evolutionary Biology, Plön, Germany

Katrin Hammerschmidt
Institute of Microbiology, Kiel University, Kiel, Germany

Pierrick Bourrat
Philosophy Department, Macquarie University,
Department of Philosophy & Charles Perkins Centre,
The University of Sydney, New South Wales, Australia

CONTENTS

DOI: 10.1201/9780429351907-15

12.1   INTRODUCTION

Division of labor is an old concept. One can find the basic idea in the Republic of Plato:

> Things are produced more plentifully and easily and of better quality when one man does one thing which is natural to him and does it in the right way, and leaves other things.

Ever since Plato, numerous theorists have proposed variations on this theme with different degrees of sophistication. In *The Wealth of Nations*, Adam Smith (1776) tells us that 10 men in a pin factory can produce approximately 48,000 pins in a single day, whereas he estimated that they would only produce less than 20 each individually or even none if 10 untrained men were performing all the 18 necessary steps to produce a pin on their own. This is because each man is *specialized* in one or two steps of the pin-producing process and, thus, performs the steps more efficiently without the need to switch between tasks than a man performing all the steps sequentially. Without this difference in efficiency and task-switching, there would be no advantage for a man to become a specialist because it only suffices that a single one of the 18 "types" of men is unavailable for no pins to be produced at all. However, in conditions where an individual can be confident in finding other men with each of the 17 other specializations (or with the ability to switch from one to another), it becomes advantageous to specialize in one of the steps for producing pins. This example illustrates the point that a division of labor entails "trade-offs." First, for dividing labor to pay off, an individual performing all the steps must be unable to produce an outcome with the same efficiency and at the same time as a specialist. Second, in situations where the men's interactions are limited or the number of men is too low, becoming a specialist must lead to a worse outcome than being a generalist. The idea of division of labor, like several other concepts in economics, has made its way to biological theory. Biological entities at all levels of organization exhibit division of labor, resulting in various degrees of specialization. However, in contrast to economic theory, division of labor is posited in evolutionary theory as an outcome of natural selection rather than rational decision.

One fundamental trade-off faced by all biological entities is the investment in maintenance (e.g., escaping predators, foraging, repairing damage) and reproduction (e.g., investing in gametes, finding a mate). It represents a particular kind of division—namely, *reproductive* division of labor. Multicellular organisms present intriguing examples of reproductive division of labor and a high degree of cellular differentiation. A widely accepted theory even suggests that germ-soma specialization has been key in the evolutionary transition from cellular groups to multicellular individuals (Buss, 1987; Simpson, 2012). Such transitions, where multiple preexisting entities form a new level of organization, are examples of *major evolutionary transitions* (Maynard Smith & Szathmáry, 1995) or *evolutionary transitions in individuality* (ETIs) (Buss, 1987; Michod, 2000). This chapter focuses on a theoretical model addressing the role of a trade-off between life history traits in selecting for a reproductive division of labor during the transition from unicellular to multicellular organisms (Michod, 2005; 2007; Michod et al., 2006; Michod & Herron, 2006), hereafter referred to as the life history trade-off model for the emergence of division of labor, or "life history model" (*LHM*) (Chapter 3).

The *LHM* has been inspired by the volvocine green algae, a taxonomic group where contemporary species range from unicellular over simple multicellular to fully differentiated (Kirk, 1998). These phototrophic eukaryotes use flagella to remain in the photic zone of freshwater environments where photosynthesis is possible. The best studied unicellular representative of this group is *Chlamydomonas reinhardtii*, which can be observed to possess its two flagella only for parts of its life cycle, during the growth phase. For cell division (i.e., reproduction), the flagella must be absorbed as cells face the functional constraint of simultaneous cell division and flagellation. This constraint necessitates a fundamental trade-off between swimming and cell division (Koufopanou, 1994).

The constraint that simultaneously bears upon viability (i.e., swimming) and reproduction has been "solved" in the closely related multicellular species *Volvox carteri*, where these incompatible functions are segregated into two different cell types. Its spherical colonies move around in the water column due to approximately 2,000 cells that look very similar to the cells of *C. reinhardtii* in that they each possess two flagella. Crucially, these cells never lose their flagella and cannot divide— they are the irreversibly differentiated soma. Reproduction is carried out by a few germ cells, called gonidia, which do not possess flagella but do possess the ability to divide. In contrast to the uncellular *C. reinhardtii*, for which these two functions are separated temporally during its life cycle, the multicellular *V. carteri* displays a spatial rather than temporal separation of somatic and reproductive functions with two cell types, which is characteristic of a reproductive division of labor.

This example of the origin of the division of labor in the volvocine green algae illustrated here is not unique. In fact, the need to accommodate two incompatible processes is also thought to drive the origin of the reproductive division of labor in other multicellular groups—for example, in metazoans, the incompatibility between cell division and flagellation (King, 2004) and, in cyanobacteria, the incompatibility between fixation of atmospheric N_2 and photosynthesis (Rossetti et al., 2010; Hammerschmidt et al., 2021).

The *LHM* relies on five key assumptions concerning the relationship between fitness and life history traits (i.e., viability and fecundity) of cells and collectives: (1) fitness is the product of viability and fecundity; (2) collective traits are linear functions (sum or average) of their cell counterparts; (3) there is a trade-off between a cell's viability and its fecundity; (4) cell traits are optimal in the sense that they display the traits that ensure the highest contribution to collective fitness; and (5) the viability–fecundity trade-off is convex for large collectives due to the initial cost of reproduction. Assumptions 1–2 are summarized in Figure 12.1, Assumptions 3–4 in Figure 12.3, and Assumption 5 in Figure 12.4. The definitions of the symbols used are presented in Table 12.1; the assumptions are summarized in Table 12.2.

In this chapter, we pursue two aims. First, we provide a step-by-step guide to these assumptions for the reader to build an intuitive understanding of the *LHM*. In doing so, we highlight some strengths and limits of the model and provide directions

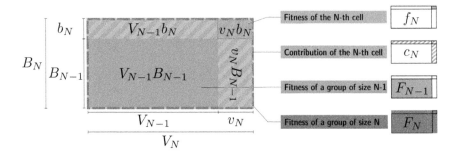

FIGURE 12.1 Geometric representation of collective fitness. As a consequence of Assumption 1 (i.e., fitness is the product of fecundity and viability) and Assumption 2 (i.e., collective traits are linear functions of their cell counterparts), it is possible to represent the fitness of a collective geometrically—as the area of a rectangle whose sides are its fecundity and viability—and to decompose it into the contribution of its constituent cells. Symbols are defined in Table 12.1.

to explore. Second, we present two interpretations of the *LHM* in the context of ETIs, with a particular focus on the metaphorical notion of "fitness transfer" and its limitations. Throughout, we illustrate our points with biological examples.

12.2   A STEP-BY-STEP GUIDE TO THE LIFE HISTORY MODEL OF DIVISION OF LABOR

12.2.1   FITNESS IS VIABILITY TIMES FECUNDITY

Assumption 1 of the *LHM* is that the value of two life history traits characterizes any entity (e.g., a cell or a collective): their viability (v for cells and V for collectives, which measures their propensity to survive) and their fecundity (b and B,

---

TABLE 12.1
Definitions of the Symbols

Description	Symbol	Formula
Effort of the i-th cell toward viability	e_i	(parameter)
Cell viability, cell fecundity (of the i-th cell)	v_i, b_i	$v_i := v(e_i); b_i := b(e_i)$ v and b are an increasing and decreasing function of e_i, respectively.
Cell fitness (of the i-th cell)	f_i	$f_i := v_i b_i$
Average cell fitness (in a collective of size N)	$\bar{f}$	$\bar{f} := N^{-1} \sum_{i=1}^{N} f_i$
Collective fitness (of a collective of size N)	F_N	$F_N := V_N B_N$
Viability, fecundity of a collective of size N	V_N, B_N	$V_N := \alpha \sum_{i=1}^{N} v_i; B_N := \alpha \sum_{i=1}^{N} b_i$
Contribution of the i-th cell to collective fitness	c_i	$c_i := b_i V_{i-1} + B_{i-1} v_i + b_i v_i$

TABLE 12.2

Summary of Modeling Assumptions 1–5

1	Fitness is viability times fecundity.
2	Collective traits are linear functions of their cell counterparts.
3	There is a trade-off between cell viability and fecundity.
4	Cell contribution to the collective is optimal.
5	There is an initial reproductive cost in large collectives.

which measures their propensity to reproduce). Fitness can be defined as the product of those two components ($f = vb$ and $F = VB$). The effect of fitness components[1] on the evolutionary success of organisms lies at the center of life history theory (Stearns, 1992), notably through the study of the constraints that link them together (see Stearns, 1989; Assumption 3). Other components of fitness exist in life history theory; however, the LHM focuses solely on viability and fecundity.

Taking the viability and fecundity product to compute fitness is a common assumption in the literature (Sober, 2001). There are at least two ways to justify this choice: phenomenologically and mechanistically. First, the product of two quantities *phenomenologically* characterizes the way these two components interact in the context of fitness. One can visualize fitness geometrically as the area of a rectangle whose sides' lengths are v and b (Figure 12.1). This representation helps to illustrate why 1) to be maximal, a multiplicative function requires a "strong balance" (Michod et al., 2006) between the two components; and 2) if one of the two sides is smaller, the marginal benefit (the surface gain) of increasing the other side is also relatively small. Additionally, if either side (fecundity or viability) has length zero, the area (fitness) is nil (we return to this point below).

Second, the product between two terms measuring fecundity and viability also arises naturally in various *mechanistic* models of population dynamics. As an example, consider a simple deterministic two-stage model with newborns and fertile adults that all share the same traits. Further, consider that whether proportion v of individuals reach the reproductive stage is given by their viability v ($0 < v < 1$), and that all adults leave a number of offspring equal to their fecundity ($b > 0$). It follows naturally that, on average, an individual will have vb offspring and the population size will grow geometrically with ratio vb in each generation (provided that generations do not overlap). This growth rate vb is also called the *Malthusian parameter* of the population and is commonly identified as a fitness measure (Fisher, 1930, p. 22). However, in a real biological situation, there is typically no fixed proportion of the population dying at each generation, and individuals leave a varying number of offspring. Despite this, the product fecundity–viability (or equivalent ratio of fecundity to mortality) is not just a feature of simple models. It also appears in more complex, stochastic models, where those fluctuations are taken into account (Kot, 2001; Haccou et al., 2007).

In the LHM, any relevant entity is characterized by fitness, which is broken down into its components. To study the two-level system of cells and collectives of interest for the evolution of multicellularity, one must describe how those two levels relate to each other. This is the purpose of Assumption 2.

12.2.2 Collective Viability and Fecundity Are a Linear Function of Cell Viability and Fecundity

Assumption 2 of the *LHM*, which is perhaps the most controversial, is that a collective's viability and fecundity are considered proportional to the sum or average of its component cells' viability and fecundity, respectively. In other words, the relationship between cell and collective fitness components is considered linear. Thus, a collective composed of N cells indexed $1, 2 \ldots N$ will have the viability $V_N = \alpha \sum_{i=1}^{N} v_i$ and the fecundity $B_N = \alpha \sum_{i=1}^{N} b_i$, where α is a coefficient of proportionality. If we assume $\alpha = 1$, the collective trait is the sum of the individual traits (Michod et al., 2006). If $\alpha = N^{-1}$, the collective trait is the average individual trait (Michod, 2006). The value of the coefficient is a matter of simplifying expressions and is irrelevant for most results unless comparing collectives of different sizes. For ease of presentation throughout the rest of the chapter, let $\alpha = 1$. The assumption of linearity is of great significance in the construction of the *LHM* because it qualifies the relationship of traits (and, thus, fitness components) between the lower level (cells) and the higher level (collectives). Therefore, it permits the unambiguous definition of the *contribution* of the N-th cell to collective fitness (c_N).

Assuming $F_N = F_{N-1} + c_N$, we have:

$$c_N := b_N V_{N-1} + B_{N-1} v_N + b_N v_N, \tag{12.1}$$

where F_{N-1}, V_{N-1}, and B_{N-1} respectively refer to the fitness, viability, and fecundity of a collective composed only of the cells 1 to $N-1$. We can see that c_N is the sum of three terms. The first term on the right-hand side is the effect of the focal cell's fecundity in the context of the remainder of the collective's viability ($b_N V_{N-1}$). The second term is the effect of the focal cell's viability in the context of the remainder of the collective's fecundity ($B_{N-1} v_N$). Finally, the last term is the N-th cell's fitness ($b_N v_N = f_N$). These three terms can be visualized as the sum of the three blue rectangles in Figure 12.1, and the contribution as the hatched orange area.

It follows from Assumption 2 that the only way a cell can affect collective viability and fecundity (and, ultimately, fitness) is by its own viability and fecundity. Thus, the indexing order and relative position of cells are irrelevant in this model. Further, since the cells' indexing is purely formal, any cell's contribution to the collective can be computed in the same fashion. However, a cell's contribution is not limited to its own fitness (the third term on the right-hand side of Equation 12.1). This is because it depends on the traits carried by the remainder of the collective (the first and second terms in Equation 12.1). Thus, the contribution of a cell can be higher if one of its components "compensates" for the weakness of the other component at the collective level (or, more accurately, the $N-1$ other cells). As previously, this can be visualized as in Figure 12.1. The same v quantity leads to a larger area with the orange dotted border (the cell contribution) if B is large and V is small, compared to a large V and a small B.

One consequence of Assumption 2 is that a cell with nil fitness ($vb = 0$) does not necessarily make a nil contribution toward collective fitness. For instance, consider a cell with zero viability and fecundity of one. In this situation, only the last two terms of Equation 12.1 are nil. This result might appear puzzling at first, particularly considering that cells with nil fitness might never exist or subsist in the population (provided further implicit but standard assumptions regarding population dynamics) (Godfrey-Smith 2011, Bourrat 2015b). However, it means that, even if a cell with nil fitness (or which tends toward zero) would quickly die, its contribution to collective fitness does not necessarily tend toward zero.

A cell's viability/fecundity contribution to collective fitness can be visualized by drawing isolines of fitness in the trait-space (see orange lines in Figure 12.3). An isoline of fitness is a curve in the space v, b that corresponds to a fixed value of collective fitness. An isoline can be thought of as the contour lines of a map. This allows us to visualize the potential contribution of any cell (i.e., any pair v, b) to an already existing collective. Note that the isolines are convex (see Box 12.1) and, provided that traits of the other cells are "balanced," form a "hill" with its crest following the first diagonal, when the two traits are balanced ($v = b$), and a valley close to the two axes when one of the traits is close to zero. The minimum contribution is the point $(0,0)$ where it is null and, thus, $F_N = F_{N-1}$.

Another way to visualize how cells with low fitness can "compensate" for one another and yield a high collective fitness is through what has been named the *group covariance effect* (Michod, 2006). Rewriting the terms of the definition of collective fitness (Table 12.1) shows the relationship between F_N and the average cell fitness ($\bar{f} := N^{-1} \sum_{i=1}^{N} f_i = N^{-1} \sum_{i=1}^{N} v_i b_i$):

$$F_N = N^2 \left[\bar{f} - cov\,(\boldsymbol{v},\boldsymbol{b}) \right] \tag{12.2}$$

Equation 12.2 shows that collective fitness is not simply proportional to the average of cell fitnesses $\bar{f}$, but that there is a corrective term due to the interplay of cells that can be identified as the sample covariance between the fecundity and viability of the N cells, which is defined as:

$$cov(\boldsymbol{v},\boldsymbol{b}) := \frac{1}{N} \sum_{i=1}^{N} (v_i - \bar{v})(b_i - \bar{b}) = \overline{vb} - \bar{v} \times \bar{b},$$

with $\bar{v} := N^{-1} \sum_{i=1}^{N} v_i$, $\bar{b} := N^{-1} \sum_{i=1}^{N} b_i$ and $\overline{vb} := \bar{f}$ and noting that $\bar{v} \times \bar{b} = N^{-2} F_N$.

Equation 12.2 shows that, when covariance is nil, such as when all cells are phenotypically indistinguishable (Figure 12.2a) or have independent trait values, collective fitness is directly proportional to the sum of its constituent cells' fitnesses. However, when cell fecundity and viability are not independent of one another, covariance is not nil—it is either positive or negative. If it is positive (Figure 12.2b), cells with a high v also have a high b, resulting in what we call "all-or-nothing cells." The opposite is true if it is negative (Figure 12.2c), resulting in specialized germ or soma-like cells.

Cell–cell interactions compensate for cell heterogeneities only when the covariance is negative. This can be seen by tallying the relative weight of "individual effects" of the cells on collective fitness (i.e., the direct sum of cell fitnesses, in color in Figure 12.2) and "interaction effects" due to the cross product between cell traits (the rest, in white in Figure 12.2). When this is done, it becomes apparent that

BOX 12.1   TRADE-OFF CONVEXITY

A trade-off is a relationship linking two quantities that cannot simultaneously be maximal; often, if one increases, the other must decrease. In this chapter, those two quantities are the life history traits of an individual (viability v and fecundity b).

   This relationship can be due to a variety of phenomena. A trade-off between size and nutrient intake might result from physical laws (e.g., diffusion), or the trade-off may arise from the resource allocation of an organism (with a given quantity of nutrient, only so many molecules might be synthesized, creating a natural trade-off between structural molecules, housekeeping, and reproductive machinery). Trade-offs may also be caused by the underlying genetic structure of the organism (e.g., a single regulator molecule acting on two pathways, making regulation of one and the other correlated, or the functional constraint of simultaneous cell division and flagellation in the case of *C. reinhardtii*), or through interaction with other species (the expression of a useful transporter might render the cell vulnerable to a certain type of virus). Consequently, trade-offs themselves might change during the evolutionary history of organisms.

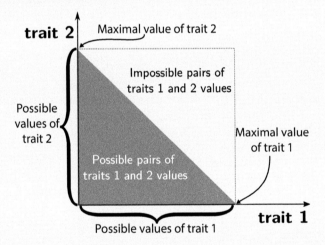

FIGURE 12.B1   Example of a trade-off between trait 1 and trait 2 with two degrees of freedom, represented by the purple surface.

   This box gives a short introduction to the simple, deterministic one-dimensional trade-offs used in the LHM for the division of labor in multicellularity. Additional resources can be found in life history theory textbooks, such as Flatt and Heyland (2012). Two-dimensional trade-offs are represented conveniently by placing the two measures on the axis of a plane and shading the area of pairs of values that are possible within the confine of the trade-off (Figure 12.B1). This may result in a surface (two degrees of freedom) or a curve (one degree of freedom) depending on the number of free dimensions the trade-off allows.

Such a trade-off provides a straightforward definition of a "specialist" organism (with a maximal or close to the maximal value in a trait and, accordingly, a lower value for the other trait) and a "generalist" organism (with an intermediate value in both traits).

A particularly useful graphical way of analyzing a trade-off is to consider its position with respect to any segment defined by any two couple of trait values. The trade-off curve (or the edge of the surface) might coincide (Figure 12.B2.a), go below (Figure 12.B2.b), above (Figure 12.B2.c), or cross (Figure 12.B2.d) these segments.

When the curve coincides with all segments, the trade-off is said to be linear. In this case, the relationship between the two traits is proportional, reducing the value of trait 1 by a quantity x, and increasing the value of trait 2 by ax (where a depends on the slope of the curve and may depend on the scaling of the trait values). When the curve is below all segments, the trade-off is said to be convex. A small reduction in trait 1 has a different effect on trait 2 if the trait is close to the maximum value (small effect when compared to the linear) or the minimum value (large effect). Similarly, if the trade-off is above all segments, it is said to be concave. In this case, a small reduction in an optimal trait has a large effect on the other trait, whereas a small increase in a low trait has a small effect on the other trait. If the curve is above some segments and is below or crosses others, it is neither convex nor concave but can be studied in part by focusing on the different regions.

Intuitively, if the trade-off is convex, being a specialist (i.e., being on either axis) is the only way to reach high trait values, while being a generalist is "costly" in the sense that it entails a large reduction in trait value. This is reversed if the trade-off is concave—generalists enjoy a less pronounced reduction of their trait values with respect to specialists (being a specialist can be considered "costly" in the sense that the marginal cost of increasing a high value trait is relatively high compared to the case of a convex trade-off).

However, note that the convexity (or even the shape) of a trade-off does not make a prediction about the outcome of the evolutionary process on its own. It simply delimits the set of possible organisms. To be able to predict the outcome of the evolutionary process from such a trade-off, one must make additional assumptions. For instance, one could assume, as we do in the LHM,

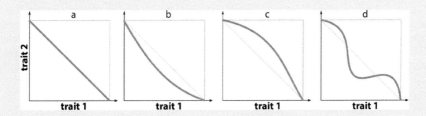

FIGURE 12.B2   a) Linear, b) convex, c) concave, d) composed trade-off.

that a fitness function F of the two traits exists and that the evolutionary dynamics reached an equilibrium state in which only the organisms with the highest fitness F are represented in the population. (In this case, one must determine the value within the set of possible individuals that gives the highest fitness). If density-dependent interactions are suspected to play a role, one possibility may be to define the invasion fitness of a rare mutant in a resident population for all pairs of points in the trade-off and look for evolutionary stable strategies, following the adaptive dynamics method (Geritz et al., 1998).

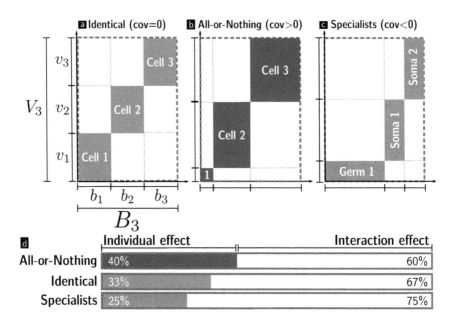

FIGURE 12.2 The covariance effect quantifies the extent to which collective fitness depends on intrinsic and interaction effects between cells. Three collectives composed of three cells are represented in the viability–fecundity space. The traits of each cell v,b are represented in color, and their fitness f is the area of a colored rectangle (green, red, orange). The three collectives have the same fitness $F_3 = V_3B_3$, represented by the area of the purple rectangle and the same average cell fitness $\bar{f}$. The collective fitness F_3 is the sum of the cell fitnesses (colored tiles, f) and of the interaction effects between cells (in white). **a)** a collective composed of three identical cells with equal traits (null covariance between v and b). **b)** a collective composed of "all-or-nothing" cells that would simultaneously have a high (low) viability and fecundity (positive covariance) (note that these cells are conceptual constructs; they are not biologically plausible). **c)** a collective composed of specialist cells with high fecundity and low viability (germ) or low fecundity and high viability (soma). **d)** the sum of cell fitnesses (colored area, labeled "individual effect") represents a larger fraction of the collective fitness (purple delimited area) when cells do not compensate for one another's weaknesses (all-or-nothing cells) as compared to when they do (specialists).

individual effects are relatively more important when the covariance is positive and relatively less important when the covariance is negative (Figure 12.2d). The converse is true for the interaction effects. Interaction effects are important because they explain how a collective can have high fitness, even if the fitness of its constituent cells is constrained to be low.

As stated earlier, Assumption 2 characterizes the relationship between cell and collective fitness in a more subtle way than simply taking the average. It also permits studying the combined effects of any set of cells (characterized by viability–fecundity pairs), as well as teasing apart a cell's direct contribution and its interactions with other cells in the collective by using the cell contributions (Equation 12.1) and the covariance effect (Equation 12.2). However, Assumption 2 is quite strong and, thus, comes at a steep price. In particular, it limits the range of phenomena that can be described satisfactorily by the model. As discussed below, biologically plausible scenarios of nonlinear and non-monotonic or, in general, higher-order interactions are impossible to describe within this framework due to this assumption. This limitation should be kept in mind by experimentalists and modelers alike.

A further limitation of Assumption 2 stems from the fact that it implies a monotonic relationship between cell and collective fitness. Thus, increasing the fecundity of one or all the cells of a collective of a given size is assumed to *always* increase the whole collective's fecundity by the same magnitude (up to the proportionality coefficient). In turn, this causes a net collective fitness increase, even though the return might be diminishing (when viability and fecundity are not well balanced). We can imagine that this assumption might not hold for all trait values. Increasing cell fecundity might increase collective fecundity by increasing the potential number of propagules the collective can produce. However, we might reasonably think that the fast proliferation of cells negatively interacts with the propagule-producing mechanisms when above a certain threshold.

Concerning the previous point, Assumption 2 also implies a kind of "beanbag" model of collectives, where the relative position and orders of cells cannot be captured. It might seem obvious for eukaryotes with sophisticated developmental dynamics and organ partitioning that a cell will have a different impact on the collective fate depending on its position and the nature of its neighboring cells. However, even relatively simple examples of multicellular organisms, such as heterocyst-forming filamentous cyanobacteria (Chapter 9), demonstrate why this is pervasive. In these species, the lack of combined nitrogen in the environment induces the formation of differentiated cells, heterocysts, which are devoted to the fixation of atmospheric N_2. Heterocysts exchange fixed nitrogen compounds for carbon products with the neighboring photosynthesizing (vegetative) cells of the filament. Crucially, heterocysts are not located at random spots in the filament; rather, they are spaced at regular intervals (Yoon & Golden, 1998). For example, in the model species *Anabaena* sp. PCC 7120, heterocysts are separated by 10–15 vegetative cells (Herrero et al., 2016). This ensures an adequate supply of fixed nitrogen compounds while maximizing the number of vegetative cells within a filament (Rossetti et al., 2010). Notably, while vegetative cells can divide and generate all other specialized cell types, heterocysts cannot divide and are terminally differentiated. Thus, we observe not only a metabolic division of labor but also a reproductive one, where heterocysts are comparable to the somatic and the vegetative cells to the germ cells in multicellular eukaryotic organisms (Rossetti et al., 2010). This structure cannot be described accurately in the original *LHM* (but see Yanni et al., 2020)

Assumption 2 implies that increasing any individual trait is bound to increase its collective counterpart. Assumption 3 prevents the simultaneous increase of both viability and fecundity.

12.2.3 TRADE-OFF BETWEEN CELL VIABILITY AND FECUNDITY

Assumption 3 of the *LHM* posits that a cell with a particular value for viability is necessarily constrained on its counterpart value for fecundity. Consequently, this reduces the number of free dimensions in the model—the two traits cannot vary independently.

This assumption covers the intuitive point that a cell cannot simultaneously be highly fecund and highly viable (i.e., an all-or-nothing cell) if it has a finite amount of energy to allow both of these (biological) functions. There are many ways to implement a trade-off in a model. The *LHM* does this using a relatively simple, deterministic, and one-dimensional method. Consider that, besides viability and fecundity, there is a third "hidden" trait for a cell, noted e, that quantifies the *effort* or *investment* toward one of the two traits. Then, by definition, the viability is an increasing function of the effort, $v_i = v(e_i)$, and the fecundity a decreasing function of the effort: $b_i = b(e_i)$. Here, v and b are (mathematical) functions that must be specified by the modeler. For instance, a simple linear trade-off can be defined as $v(e) = e$ and $b(e) = 1 - e$ for $e \in [0,1]$.

If the notion of effort is essential for understanding the logic of the trade-off, it can be abstracted graphically when representing the trade-off in the (v,b) plane introduced in Assumption 2. The trade-off can be represented as a curve (purple in Figure 12.3)

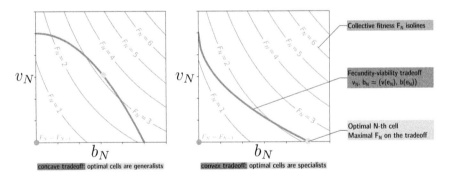

FIGURE 12.3 Isolines of fitness, trade-offs, and optimality of cell contributions. Representation in the plane (b_N, v_N) formed by the fecundity and viability of the N-th cell of a collective. As a consequence of Assumptions 1 and 2, collective fitness is a surface (represented by orange isolines) and, at the origin ($v_N = b_N = 0$), the collective fitness is minimal ($F_N = F_{N-1}$, green dot). As a consequence of Assumption 3, the values of v_N and b_N are constrained by a trade-off (purple line). As a consequence of Assumption 4, the model predicts that the traits favored by natural selection are those that yield the highest collective fitness (yellow disk) while satisfying the trade-off constraint. Concave (and linear) trade-offs (left) favor generalist cells (with balanced v,b), while convex trade-offs (right) favor specialist cells (with high v and low b, or vice versa).

constituting all the combinations of v_i, b_i given by all possible values of e_i. Different functional forms result in different trade-off shapes. Its shape and, in particular, its convexity are at the base of many strategies within the framework of life history theory (see Box 12.1 for a primer). We will return to this in discussing Assumption 5.

The notion of trade-offs in life history theory is an indubitably elegant way to incorporate an organism design's underlying constraints into a model. For instance, it can be used to account for the fact that the microtubule organizing center in the *Volvocaceae* cannot participate simultaneously in reproduction (through mitosis) and viability (through flagellar motility) (Koufopanou, 1994). While they are powerful theoretical tools, the existence of trade-offs is difficult to demonstrate, let alone quantify. One reason for this is that they can originate from many sources, such as physical (diffusion, buoyancy), genetic (metabolic pathways, regulations), or ecological (grazing, parasites) constraints. Moreover, trade-offs are not always set in stone. If physical constraints such as diffusion hardly change, mutation events can overturn other constraints—for instance, in the flagellate *Barbulanympha,* the microtubule organizing center can participate simultaneously in locomotion and reproduction (Buss, 1987). Note that, in the LHM, the shape of the trade-off changes with the size of the collective. This will be covered in more detail as part of Assumption 5.

Following Assumption 3, the set of all possible cells is reduced, as the trait of any new cell must be located on the trade-off curve. The model is not yet complete; natural selection acts on the organism in the context of these trade-offs, and its effect must be described. This is the purpose of Assumption 4.

12.2.4   CELL CONTRIBUTION TO THE COLLECTIVE IS OPTIMAL

So far, the role of natural selection has seldom been invoked in the LHM. We have only described the properties of cells and collectives and the diversity of traits they can exhibit, given some underlying constraints. Assumption 4 models the consequence of natural selection for this system—it assumes that all cells are optimal in terms of their contribution c to collective fitness. Formally, it means that the life history traits of any cell i within the collectives are such that the value of c_i is maximal:

$$e_i = argmax\, c_i$$

Note that optimality is an *assumption* rather than an outcome of the model.

Graphically, to find the values for a cell to contribute optimally to the collective, one must identify the intersection between the trade-off curve (purple in Figure 12.3) and the highest isoline of collective fitness (orange in Figure 12.3). This point is where a cell existing within the physiological constraints that link v and b has the highest fitness contribution. If the shape of the trade-off is sufficiently simple, there is a single optimal point and, thus, the model predicts the traits of any new cell based on Assumptions 1–4 plus the shape of the trade-off.

Of course, Assumption 4 could turn out to be incorrect if the cells are *not* optimal in their contribution to collective fitness. Cells might not be optimal for several reasons. For instance, the optimization of their traits might not occur independently from one another because they share the same underlying developmental program.

Alternatively, they might be "stuck" in another region of the trait space, in which case no viable mutation path would bring them to the optimal phenotype. Other reasons include that the trade-off's shape has recently changed due to changes in the environment or evolutionary forces (e.g., selection at another level or an evolutionary branching point) prevent the cells from reaching or remaining at the optimal phenotype.

Thus far, we have seen that the LHM assumes that the reproductive success of a collective depends on two fitness components (Assumption 1) that derive from their cell counterparts (Assumption 2), which are linked by underlying constraints (Assumption 3), and that natural selection is expected to favor optimal cells within this context (Assumption 4). The last piece of the puzzle is to qualify the shape of the trade-off—Assumption 5 does precisely this.

12.2.5 THERE IS AN INITIAL REPRODUCTIVE COST IN LARGE COLLECTIVES

Assumption 5 states that small collectives have a linear or concave trade-off—favoring generalist cells—while large collectives have a convex trade-off—favoring division of labor. The distinction between linear, concave, and convex trade-offs is presented in Box 12.1. The mechanism proposed to explain why large collectives have a convex trade-off is the initial cost of reproduction. This assumption is critical because it characterizes the underlying constraints that bear on cell traits, but also ties them to the collective, in particular to collective size.

To understand Assumption 5, consider a cell specialized in viability (i.e., with a low fecundity) (Figure 12.4a). The mechanism for the initial cost of reproduction hinges on the assumption that, if this cell was investing more in fecundity than it currently does, it would reduce its viability but would *not* increase its fecundity (Figure 12.4b) until a threshold is reached (Figure 12.4c), after which it would increase (Figure 12.4d) until the cell is fully specialized in fecundity (Figure 12.4e).

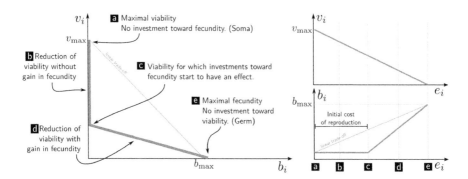

FIGURE 12.4 Initial reproductive cost as a model for a convex trade-off. Cell viability v_i and fecundity b_i are constrained, as represented by the purple curve corresponding to the possible combinations (v_i, b_i) under the trade-off modeled by the reproductive effort e_i. Starting from maximal investment toward viability (v_{max}, a), reduction in the viability effort has no effect on fecundity (b) until a threshold (c), where fecundity increases (d) up to the maximal fecundity allowed by the model (b_{max}, e). Contrast this with the simple linear trade-off (brown).

The relationship between group size and the shape of the trade-off between contribution to collective viability and fecundity is generally understood in terms of physical constraints. For instance, at the collective level, in the volvocine green algae, the enlargement of reproductive cells increases the downward gravitational force, increasing sinking; this is only overcome by the investment in more buoyant somatic cells (Solari et al., 2015). Thus, when colony size increases, a required initial investment toward buoyancy emerges that did not exist in unicellular organisms. This, in turn, explains how the trade-off, taken to be linear (or even concave) for single cells and small collectives, becomes convex when considering larger groups.

This is the last part of the *LHM*. As a consequence of Assumptions 1–5, large collectives favor the selection of specialist cells and, thus, division of labor.

12.3   DISCUSSION: FITNESS INTERPRETATIONS IN EVOLUTIONARY TRANSITIONS IN INDIVIDUALITY

The previous section presented a mechanism that promotes cell specialization between two life history traits (viability and fecundity) and, hence, a division of labor. This mechanism is based on the presence of a convex trade-off between the two traits due to the existence of an initial cost of reproduction in large collectives. This section places this model back in the broader context of ETIs by contrasting two interpretations of this phenomenon.

12.3.1   "Reorganization and Transfer of Fitness" Interpretation

A first interpretation of the *LHM* is based on the idea that the hallmark of an ETI is fitness reorganization/transfer/decoupling in the sense that cell specialization results in the lower-level cells "relinquish[ing] their autonomy in favor of the group" (Michod, 2005 p. 969; Michod et al., 2006, p. 258), resulting in a transfer of "fitness and individuality [from] the cell level to the group level" (ibid). We have tacitly assumed this interpretation throughout because it is the one with which the *LHM* was initially proposed when interpreting f_i as cell fitness and F_N as collective fitness.

This interpretation is rooted in the Multi-Level Selection 1–2 framework (Damuth & Heisler, 1988; Okasha, 2006), as cited in Michod (2005). In Multi-Level Selection 1 (MLS1) models, collective fitness is taken to be the average fitness of its members (or proportional to it), whereas in Multi-Level Selection 2 (MLS2), collective fitness cannot be defined in terms of particle survival and reproduction. Consequently, a new notion of fitness must be devised.

From this interpretation, the problem of explaining ETIs can be "reduced" to explaining the transition from an MLS1-like situation to an MLS2-like situation (Okasha, 2006, Chapter 8). Initially, this may appear to be an insurmountable hurdle because, in MLS1 (before the transition), cells are selected to have the highest cell fitness. In contrast, fully specialized cells in the model (after the ETI has occurred) have nil (or close to nil) fitness. The concept of *fitness transfer* (Michod, 2006) solves this problem by considering that, during an ETI, fitness between the two levels is reorganized—it is transferred from the lower level (the cells) to the higher level (the collective). The transfer is achieved through germ and soma specialization (which results from the trade-off's convexity). When cells become specialized, they relinquish their fitness to the benefit of the collective. Further, during the process, collective fitness transitions from a mere average of cell fitness (MLS1) to a quantity, which is no longer the cell fitness average in the collective (MLS2) due to the covariance effect.

Although this interpretation is appealing, there are some problems associated with it. The main problem is that it seems to imply that fitness is a material quantity that can be transferred from one entity to another, comparable to a liquid that can be poured from one container to another. While, at first glance, this analogy may seem helpful for obtaining an intuitive idea of the problem, it contradicts our modern understanding of fitness as a predictor of evolutionary success (Bourrat, 2015a, 2015b, Bourrat, 2021a, b; Doulcier et al., 2021). First, it implies that some cells with nil fitness (or close to it) are not dead, contradicting the principle of natural selection. Second, because the evolutionary fates of cells and collectives are tied (by virtue of being made of the same biological substrate), it is difficult to determine how one level could ever be favored at the other's expense (Black et al., 2020; Bourrat, 2021a, b). This point is known in the philosophical literature as the "causal-exclusion principle." If a phenomenon is explained or described exhaustively at the lower level, one cannot appeal to the higher level to explain this phenomenon further. Doing so is either a form of "double counting" or requires assuming that strongly emergent properties are created ex nihilo at the higher level. Assuming the existence of strongly emergent properties raises a new range of issues because they contradict materialism, the idea that all causes are physical in nature.

Finally, the fitness transfer interpretation implies that cells constituting a multicellular organism can have different fitness values. This conflicts with the fact that those cells are clones and should, thus, have the same (inclusive) fitness (Bourrat, 2015b). To clarify this point, while a queen and a worker bee have different reproductive outputs, they have the same (inclusive) fitness. Evolutionarily, it does not make sense to say that the queen is more successful than the worker. Similarly, it does not make sense to say that a liver cell in a multicellular organism is fitter than a brain cell.

Recent work has helped to solve these issues by proposing a new interpretation of fitness at different levels of organization (Shelton & Michod, 2014, 2020). Following this new interpretation, the term "cell fitness" does not represent the cell's fitness within the collective but rather the one it *would have* if it were without a collective (counterfactual fitness). When cells have the same fitness they would have in the absence of the collective, no transition has occurred. However, when cells have different fitness, a transition has occurred (or at least been initiated). This constitutes a reasonable argument toward deciding whether an ETI happened (i.e., the state of cells and collectives). Crucially, this says nothing about the mechanism of the transition. There is no actual "decoupling" or "transfer" of fitness, other than in the loose metaphorical sense that the purely theoretical counterfactual fitness aligns or not with the actual fitness.

Metaphors and analogies are incredibly useful in biology because they allow us to build intuition of complex mechanisms by drawing parallels with other systems. *Fitness transfer* or *reorganization* implies the physical transfer of a material quantity (with or without conservation). However, fitness is not transferred from one place (the cell) to another (the multicellular organism) in the way that heat, for instance, can be transferred. "Fitness transfer" might be used, but only in a loose metaphoric sense—that is, in the same way, teleological language in evolutionary biology is used in the context of a teleonomic explanation (Pittendrigh, 1958; Jacob, 1970). A further point worth mentioning is that metaphors may favor a specific interpretation that could obscure some aspects of the phenomena studied, such as how the sole focus on selection created the blind spots of the adaptationist program (Gould et al., 1979).

12.3.2 "PROPENSITY" INTERPRETATION

We favor an alternative interpretation of the *LHM*. This interpretation starts with the same mechanism—a convex trade-off between contribution to collective fecundity and collective viability will promote the emergence of individual specialist cells and, thus, a division of labor. It diverges from the previous one by its treatment of fitness, emphasizing how it emerges from cell *traits* rather than using it as a reified quantity of cells and collectives.

For the concept of fitness to qualify as a predictor in evolutionary biology, it cannot be reduced to an entity's actual success (i.e., its realized fitness). Instead, it must be tied to its *potential* success (or success in the long run). Without this point acknowledged, fitness is condemned to be tautological, as philosophers and biologists alike have long recognized (Manser, 1965; Popper, 1974; Smart, 1963; reviewed in Doulcier et al., 2021).[2] This conundrum has led to establishing several frameworks for the interpretation of fitness, one of which is the propensity interpretation of fitness (Brandon, 1978; Beatty, 1984; Pence & Ramsey, 2013). According to the propensity interpretation, fitness is a probabilistic property of entities summarizing their probability distribution of reproductive success (as defined by their demographic parameters: birth and death rates) in a given environment.

If we adopt this interpretation, the problems raised by the fitness transfer interpretation vanish. First, the problem of a collective's different (clonal) cells having different fitnesses disappears. Although their *realized* fitnesses (actual life history) might be different, their "true" fitnesses (potential life history) are equal because they relate to the potential success of the same genotype. Second, this interpretation does not appeal to fitness transfer or decoupling since cell and collective fitness are computed in expectation. Following the transfer of fitness interpretation, although this is not made explicit in the model, cell fitness and collective fitness are computed relative to different environments. In particular, collective-level demography (i.e., birth and death events of collectives) is typically ignored when computing cell fitness. Consequently, it becomes possible to define different values of fitness for the collective (F) and the cells (f). However, the fact that they are computed in different environments implies that they cannot legitimately be directly compared. When cell and collective fitnesses are computed in the same environment following the propensity interpretation—for instance, by factoring in collective events in the cell-level computation—they are necessarily equal (Bourrat, 2015a, 2015b; Bourrat et al., 2020). Cells and multicellular organisms are two levels of description of the same physical reality and cannot contradict one another, despite some claims to the contrary (e.g.,

Okasha, 2006).[3] Although conflicting processes might exist (e.g., segregation distortion locus, cancerous growth), fitness, properly computed to be comparable, must tally these conflicts and be coherent when referring to the same entity, regardless of the method of description.

Alternatively, one way to connect the interpretation we favor and the counterfactual fitness approach is to compare the fitness of free-living cells with cells within the collectives and observe *apparent decoupling* between these two environments: a decrease in free-living fitness and an increase in within-collective fitness (Bourrat, 2015a, 2015b, 2016; Bourrat et al., 2021; Bourrat, 2021a, 2021b). However, this apparent decoupling is rather a sign of *linkage* between the traits that contribute to free-living and within-collective fitness. The propensity interpretation of fitness can explain the same phenomena without invoking any "fitness transfer." If there is a "transfer," it is between the energetic investment of the cell toward different traits: from traits that provide no advantage to cells living in a collective (and potentially contributing to a free-living life cycle) toward traits that provide an advantage to the cells living in a collective (including vicarious advantages of cells with the same genotype).

Doing away with the reifying idea that fitness is something to be transferred and, more generally, treating the MLS1/MLS2 distinction as conventional—that is, two different ways to formalize the same idea—rather than as an evolutionary mechanism allows pursuing lines of inquiries that were more difficult to conceive within this framework. For instance, the focus on the relationship between cell and collective fitness leads naturally to the assumption that contributions to the collective fitness component are linear functions of their free-living counterparts (as was the case in Assumption 2). However, designing a mechanistic model naturally leads to relaxing this assumption. Traits of collectives are most certainly more complex than the arithmetic aggregation of individual quantities measured in the propagule or the fully developed collective. Rather, they are the result of internal developmental dynamics—that is, within-collective cellular ecological dynamics (Hammerschmidt et al., 2014; Rose et al., 2020). Selection of developmental mechanisms has long been recognized as a vital part of ETIs (Buss, 1987; Michod & Roze, 1997) and can be studied fully by models that describe the ecological dynamics within collectives explicitly (see Ikegami & Hashimoto, 2002; Williams & Lenton, 2007; Xie et al., 2019 for general cell communities; but see Doulcier et al., 2020 for an application to ETIs).

12.4 CONCLUSION

Division of labor is observed in complex organisms. The functions exhibited by multicellular organisms cannot be exhibited simultaneously by a single cell. Multicellularity solves this problem by allowing different subsets of cells to perform the different functions at once. The level of division of labor exhibited by a collective varies with the extent to which cells are specialized.

Natural selection favors specialist cells (hence, division of labor) if there is a convex trade-off between two equally important functions for cell fitness. The convexity of the trade-off is a consequence of two hypotheses: first, an energetic investment model in which a cell has limited energy to invest in two traits that contribute toward each function and, second, an initial investment cost whereby a small investment in a trait does not translate immediately to an improvement of the function. The *LHM* predicts that collectives constituted of cells investing less energy in traits that contribute toward free-living fecundity and viability but more in traits that contribute toward fecundity and viability of collectives will progressively outcompete other collectives and become widespread.

This phenomenon has been interpreted as a "transfer of fitness" in the sense that individual cells relinquish their autonomy (investing less in free-living traits) to participate in life history traits of collectives (investing more in contribution toward collective function). During this "reorganization of fitness," cell fitness has been proposed to decrease while collective fitness increases. The fact that cell fitness and collective fitness do not change in the same direction has been named "fitness decoupling." However, this interpretation can be misleading because it conflicts with the concept of fitness as used in evolutionary biology. To fully appreciate the relevance of the *LHM* to ETIs, two things must be stressed. First, trade-offs occur between traits, not between fitnesses at different levels of organization. Second, fitness can only be defined with respect to a given entity (cell or collective) in a given environment and cannot be incoherent between the whole and the part. Thus, fitness cannot literally be "transferred" from individuals to collectives, even if, in retrospect, the traits that are adaptive in a collective environment would be detrimental to a free-living organism.

Trade-offs between life history traits are valid mechanisms—independently of the interpretation in terms of fitness transfer or steady state propensity, or even any other kind of interpretation one might propose (e.g., inclusive fitness, game theory, altruism). The interpretation chosen only represents a useful narrative for placing ETIs in the broader context of the evolution of complexity and allowing us to pursue subsequent questions, such as regarding developmental programs. Nonetheless, invoking a fitness concept that is consistent with the broader use of this term represents the primary reason for preferring one interpretation of the *LHM* to the other.

ACKNOWLEDGMENTS

The authors are thankful to the Theory and Method in Biosciences group at the University of Sydney. GD's and PB's research was supported by a Macquarie University Research Fellowship and a Large Grant from the John Templeton Foundation (Grant ID 60811). KH is grateful for support from *The Hamburg Institute for Advanced Study* (HIAS) and the Joachim Hertz Foundation.

NOTES

1. Or life history traits—the two terms are often used indistinguishably (Flatt & Heyland, 2012).
2. The propensity interpretation of probability is contentious in philosophy (Hájek, 2012). It produces a number of problems, some of which are inherited by the propensity interpretation of fitness (Godfrey-Smith, 2009; Bourrat, 2017). In recent years, several alternative interpretations of probabilities that play the same role as propensities and solve the issues of the propensity account have been proposed (e.g., Rosenthal, 2010; Lyon, 2011; Strevens, 2011; Abrams, 2012). Addressing the differences between these various interpretations in the context of fitness is beyond the scope of the present work. For our

purpose, we use "propensity" loosely as an entity's dispositional property to produce offspring (or equivalent terms in the aforementioned interpretations) without committing to any particular probability interpretation.
3. This claim admits a few theoretical exceptions, which are not relevant to ETIs.

REFERENCES

Abrams, M. (2012). Mechanistic probability. *Synthese, 187*(2), 343–375. https://doi.org/10.1007/s11229-010-9830-3

Beatty, J. H. (1984). Chance and natural selection. *Philosophy of Science, 51*(2), 183–211.

Black, A. J., Bourrat, P., & Rainey, P. B. (2020). Ecological scaffolding and the evolution of individuality. *Nature Ecology & Evolution, 4*, 426–436. https://doi.org/10.1038/s41559-019-1086-9

Bourrat, P. (2015a). Levels of selection are artefacts of different fitness temporal measures. *Ratio, 28*(1), 40–50.

Bourrat, P. (2015b). Levels, time and fitness in evolutionary transitions in individuality. *Philosophy & Theory in Biology, 7*. http://doi.org/10.3998/ptb.6959004.0007.001

Bourrat, P. (2016). Generalizing contextual analysis. *Acta Biotheoretica, 64*(2), 197–217.

Bourrat, P. (2017). Explaining drift from a deterministic setting. *Biological Theory, 12*(1), 27–38.

Bourrat, P. (2021a). Transitions in individuality: A formal analysis. *Synthese, 198*, 3699–3731. https://doi.org/10.1007/s11229-019-02307-5.

Bourrat, P (2021b). *Facts, conventions, and the levels of selection.* Cambridge: Cambridge University Press. https://doi.org/10.1017/9781108885812.

Bourrat, P., Doulcier, G., Rose, C. J., Rainey, P. B. & Hammerschmidt, K. (2020). Beyond fitness decoupling: Tradeoff-breaking during evolutionary transitions in individuality. *bioRxiv* https://doi.org/10.1101/2021.09.01.458526

Brandon, R. N. (1978). Adaptation and evolutionary theory. *Studies In History and Philosophy of Science Part A, 9*(3), 181–206.

Buss, L. W. (1987). *The evolution of individuality.* Princeton, NJ: Princeton University Press.

Damuth, J., & Heisler, I. L. (1988). Alternative formulations of multilevel selection. *Biology and Philosophy, 3*(4), 407–430. https://doi.org/10.1007/BF00647962

Doulcier, G., Lambert, A., De Monte, S., & Rainey, P. B. (2020). Eco-evolutionary dynamics of nested Darwinian populations and the emergence of community-level heredity. *eLife, 9*, e53433. https://doi.org/10.7554/eLife.53433

Doulcier, G., Takacs, P., & Bourrat, P. (2021). Taming fitness: organism-environment interdependencies preclude long-term fitness forecasting. *BioEssays, 43*(1), 2000157. https://doi.org/10.1002/bies.202000157

Fisher, R. A. & Heyland, A. (1930). *The genetical theory of natural selection.* Oxford, UK: Oxford University Press.

Flatt, T., & Heyland, A. (Eds.). (2012). Mechanisms of life history evolution: *The genetics and physiology of life history traits and trade-offs.* Oxford, UK: Oxford University Press.

Geritz, S. A. H., Kisdi, E., Meszéna, G., & Metz, J. A. J. (1998). Evolutionarily singular strategies and the adaptive growth and branching of the evolutionary tree. *Evolutionary Ecology, 12*(1), 35–57. https://doi.org/10.1023/A:1006554906681.

Godfrey-Smith, P. (2009). *Darwinian populations and natural selection.* New York: Oxford University Press.

Godfrey-Smith, P. (2011). Darwinian populations and transitions in individuality. In B. Calcott & K. Sterelny (Eds.), *The major transitions in evolution revisited* (pp. 65–81). Cambridge, MA: MIT Press.

Gould, S. J., Lewontin, R. C., Maynard Smith, J., & Holliday, R. (1979). The spandrels of San Marco and the Panglossian paradigm: A critique of the adaptationist programme. *Proceedings of the Royal Society of London. Series B. Biological Sciences, 205*(1161), 581–598. https://doi.org/10.1098/rspb.1979.0086

Haccou, P., Jagers, P., Vatutin, V. A. (2007). *Branching processes: Variation, growth, and extinction of populations.* Cambridge UK: Cambridge University Press.

Hájek, A. (2012). Interpretations of probability. In E.N. Zalta (Ed.), *The Stanford encyclopedia of philosophy* (Winter 2012). http://plato.stanford.edu/archives/win2012/entries/probability-interpret/

Hammerschmidt, K., Rose, C. J., Kerr, B., & Rainey, P. B. (2014). Life cycles, fitness decoupling and the evolution of multicellularity. *Nature, 515*(7525), 75–79.

Hammerschmidt, K., Landan, G., Tria, F. D. K., Alcorta, J., & Dagan, T. (2021). The order of trait emergence in the evolution of cyanobacterial multicellularity. *Genome Biology and Evolution, 13*(2), evaa249. https://doi.org/10.1093/gbe/evaa249

Herrero, A., Stavans, J., & Flores, E. (2016). The multicellular nature of filamentous heterocyst-forming cyanobacteria. *FEMS Microbiology Reviews, 40*(6), 831–854. https://doi.org/10.1093/femsre/fuw029

Ikegami, T., & Hashimoto, K. (2002). Dynamical systems approach to higher-level heritability. *Journal of Biological Physics, 28*(4), 799–804. https://doi.org/10.1023/A:1021215511897

Jacob, F. (1970). *La logique du vivant: Une histoire de l'hérédité.* Paris: Gallimard.

King, N. (2004). The unicellular ancestry of animal development. *Developmental Cell, 7*(3), 313–325. https://doi.org/10.1016/j.devcel.2004.08.010

Kirk, D. L. (1998). *Volvox: A search for the molecular and genetic origins of multicellularity and cellular differentiation.* Cambridge, UK: Cambridge University Press

Kot, M. (2001). *Elements of mathematical ecology.* Cambridge, UK: Cambridge University Press.

Koufopanou, V. (1994). The evolution of soma in the volvocales. *The American Naturalist, 143*(5), 907–931. https://doi.org/10.1086/285639

Lyon, A. (2011). Deterministic probability: Neither chance nor credence. *Synthese, 182*, 413–432.

Manser, A. R. (1965). The concept of evolution. *Philosophy, 40*(151), 18–34.

Maynard Smith, J., & Szathmáry, E. (1995). *The major transitions in evolution.* New York: W.H. Freeman.

Michod, R. E. (2000). *Darwinian dynamics: Evolutionary transitions in fitness and individuality.* Princeton, NJ: Princeton University Press.

Michod, R. E. (2005). On the transfer of fitness from the cell to the multicellular organism. *Biology and Philosophy, 20*(5), 967–987.

Michod, R. E. (2006). The group covariance effect and fitness trade-offs during evolutionary transitions in individuality. *Proceedings of the National Academy of Sciences, 103*(24), 9113–9117. https://doi.org/10.1073/pnas.0601080103

Michod, R. E. (2007). Evolution of individuality during the transition from unicellular to multicellular life. *Proceedings of the National Academy of Sciences, 104*(suppl 1), 8613–8618. https://doi.org/10.1073/pnas.0701489104

Michod, R. E., & Herron, M. D. (2006). Cooperation and conflict during evolutionary transitions in individuality. *Journal of Evolutionary Biology, 19*(5), 1406–1409. https://doi.org/10.1111/j.1420-9101.2006.01142.x

Michod, R. E., & Roze, D. (1997). Transitions in individuality. *Proceedings of the Royal Society of London. Series B: Biological Sciences, 264*(1383), 853–857. https://doi.org/10.1098/rspb.1997.0119

Michod, R. E., Viossat, Y., Solari, C. A., Hurand, M., & Nedelcu, A. M. (2006). Life-history evolution and the origin of multicellularity. *Journal of Theoretical Biology, 239*(2), 257–272. https://doi.org/10.1016/j.jtbi.2005.08.043

Okasha, S. (2006). *Evolution and the levels of selection* (vol. 16). Oxford, UK: Oxford University Press.

Pence, C. H., & Ramsey, G. (2013). A new foundation for the propensity interpretation of fitness. *The British Journal for the Philosophy of Science, 64*(4), 851–881. https://doi.org/10.1093/bjps/axs037

Pittendrigh, C. S. (1958). Adaptation, natural selection, and behavior. In A. Roe & G. G. Simpson (Eds.), *Behavior and evolution* (pp. 390–416). New Haven, CT: Yale University Press.

Popper, K. R. (1974). Intellectual autobiography. *The Philosophy of Karl Popper, 92*. https://ci.nii.ac.jp/naid/10004481309/

Rose, C. J., Hammerschmidt, K., Pichugin, Y., & Rainey, P. B. (2020). Meta-population structure and the evolutionary transition to multicellularity. *Ecology Letters, 23*(9), 1380–1390. https://doi.org/10.1111/ele.13570

Rosenthal, J. (2010). The natural-range conception of probability. In G. Ernst & A. Hüttemann (Eds.), *Time, chance, and reduction: Philosophical aspects of statistical mechanics* (pp. 71–90). Cambridge, UK; New York: Cambridge University Press.

Rossetti, V., Schirrmeister, B. E., Bernasconi, M. V., & Bagheri, H. C. (2010). The evolutionary path to terminal differentiation and division of labor in cyanobacteria. *Journal of Theoretical Biology, 262*(1), 23–34. https://doi.org/10.1016/j.jtbi.2009.09.009

Shelton, D. E., & Michod, R. E. (2014). Group selection and group adaptation during a major evolutionary transition: Insights from the evolution of multicellularity in the volvocine algae. *Biological Theory, 9*(4), 452–469.

Shelton, D. E., & Michod, R. E. (2020). Group and individual selection during evolutionary transitions in individuality: Meanings and partitions. *Philosophical Transactions of the Royal Society B: Biological Sciences, 375*(1797), 20190364. https://doi.org/10.1098/rstb.2019.0364

Simpson, C. (2012). The evolutionary history of division of labour. *Proceedings of the Royal Society B: Biological Sciences, 279*(1726), 116–121. https://doi.org/10.1098/rspb.2011.0766

Smart, J. J. C. (1963). *Philosophy and scientific realism*. London, UK: Routledge & Kegan Paul Ltd.

Smith, A. (1776). *An inquiry into the nature and causes of the wealth of nations*. London: W. Strahan and T. Cadell.

Sober, E. (2001). The two faces of fitness. In R. S. Singh, Costas B. Krimbas, Diane B. Paul, & John Beatty (Eds.), *Thinking about evolution: Historical, philosophical, and political perspectives* (vol. 2, p. 309-320). Cambridge, UK: Cambridge University Press.

Solari, C. A., Kessler, J. O., & Michod, R. E. (2015). A hydrodynamics approach to the evolution of multicellularity: Flagellar motility and germ-soma differentiation in volvocalean green algae. *The American Naturalist*. https://doi.org/10.1086/501031.

Stearns, S. C. (1989). Trade-offs in life-history evolution. *Functional Ecology, 3*(3), 259–268. JSTOR. https://doi.org/10.2307/2389364.

Stearns, S. C. (1992). *The evolution of life histories*. New York: Oxford University Press.

Strevens, M. (2011). Probability out of determinism. In C. Beisbart & S. Hartman (Eds.), *Probabilities in physics* (pp. 339–364). New York: Oxford University Press.

Williams, H. T. P., & Lenton, T. M. (2007). Artificial selection of simulated microbial ecosystems. *Proceedings of the National Academy of Sciences, 104*(21), 8918–8923. https://doi.org/10.1073/pnas.0610038104

Xie, L., Yuan, A. E., & Shou, W. (2019). Simulations reveal challenges to artificial community selection and possible strategies for success. *PLOS Biology, 17*(6), e3000295. https://doi.org/10.1371/journal.pbio.3000295

Yanni, D., Jacobeen, S., Márquez-Zacarías, P., Weitz, J. S., Ratcliff, W. C., & Yunker, P. J. (2020). Topological constraints in early multicellularity favor reproductive division of labor. *eLife, 9*, e54348. https://doi.org/10.7554/eLife.54348

Yoon, H. S., & Golden, J. W. (1998). Heterocyst pattern formation controlled by a diffusible peptide. *Science, 282*(5390), 935–938. https://doi.org/10.1126/science.282.5390.935

Section 4

Life Cycles and Complex Multicellularity

13 The Single-Celled Ancestors of Animals
A History of Hypotheses

Thibaut Brunet
Howard Hughes Medical Institute and the
Department of Molecular and Cell Biology,
University of California, Berkeley, CA, USA
Department of Cell Biology and Infection,
Institut Pasteur, Paris, France

Nicole King
Howard Hughes Medical Institute and the
Department of Molecular and Cell Biology,
University of California, Berkeley, CA, USA

CONTENTS

DOI: 10.1201/9780429351907-17

251

This chapter has been made available under a CC-BY-NC-ND license.

13.1   ORIGIN OF THE QUESTION: THE CELL THEORY AND THE CONCEPT OF COMMON DESCENT

The question of the single-celled ancestor of animals only makes sense in the light of two concepts that are now central to biology and that emerged in parallel in the second half of the 19th century: (1) the cell theory, which posits that all living beings are composed of cells (some of many, some of only one) (Schleiden, 1839; Schwann, 1839); and (2) the theory of common descent, which posits that all living species – unicellular or multicellular – descended from a single common ancestor (Darwin and Wallace, 1858).

That all living beings are made of cells is the first fact many of us learned about biology and is so familiar that we sometimes take it for granted. But the cellular organization of all life forms was not initially obvious, and it took a full 250 years after the invention of the microscope for this idea to gain general acceptance. Two of the first people to observe microorganisms (van Leeuwenhoek, 1677; Müller, 1786) indiscriminately used the terms "infusorians" or "animalcules" to describe what we now think was a *mélange* of unicellular protists (e.g., ciliates, heliozoans, amoebae, and flagellates) and small multicellular animals (e.g., rotifers and flatworms). Multicellular organization was first described in 1665 by Robert Hooke based on his observations of dead plant tissue in the form of a bottle cork. Hooke was intrigued by the structures he was later to name "cells," but had no idea he had discovered a general phenomenon, and considered them a structural peculiarity of cork. An additional 170 years of research and many additional observations were needed before the official "birth date" of the cell theory, often attributed to Schwann (1838) and Schleiden (1839) (reviewed in (Morange, 2016). Once the cell theory was accepted, several early cell biologists (including Meyen (1839), Dujardin (1841), Barry (1843) and von Siebold (1845)) took the leap to posit that the simplest life forms might consist of only one cell (reviewed in Leadbeater and McCready, 2002).

The theory of evolution emerged in parallel with the cell theory. The first elaborate theory of evolution, proposed in 1809 by the French biologist Jean-Baptiste de Lamarck (1744–1829), assumed that life started with the spontaneous generation of "infusorians"– including both protists and small animals (Lamarck, 1809). Infusorians were then inferred to have gradually evolved into all other organisms through a progressive increase in size and complexity, with no individual step that would have clearly paralleled our modern concept of a transition to multicellularity. Lamarck's ideas attracted attention and criticism, but the concept of common descent did not become widely accepted until after the debate spurred by the theory of evolution through natural selection proposed by Charles Darwin (1809–1882) and Alfred Russel Wallace (1823–1913) (Darwin and Wallace, 1858; Darwin, 1859). Their theory was the first to propose a plausible mechanism for descent with modification and thus brought new credibility to the concept of evolution.

By the end of the 19th century, the scientific stage was set for considering the origin of animals: both evolution and cell theory had gained widespread acceptance, and three of the most abundant and charismatic groups of single-celled organisms

TABLE 13.1

Timeline of Hypotheses on the Single-celled Precursor of Animals

Nature of the Hypothesized Ancestor	Proposed as Early as	References
Amoeba	1876	(Haeckel, 1876, 1914), (Reutterer, 1969), (Hanson, 1977)
Ciliate	1882	(Kent, 1882), (Sedgwick, 1895), (Hadzi, 1953, 1963), (Steinböck, 1963), (Hanson, 1963, 1977)
Flagellate	1884	(Bütschli, 1884), (Metchnikoff, 1886), (Nielsen and Norrevang, 1985), (King, 2004), (Nielsen, 2008), (Cavalier-Smith, 2017)
Fucus-like syncytial brown alga	1924	(Franz, 1924)
Amoeboflagellate or complex ancestor	1949	(Zakhvatkin, 1949), (Sachwatkin, 1956), (Willmer, 1971), (Mikhailov et al., 2009), (Arendt et al., 2015), (Sebé-Pedrós, Degnan and Ruiz-Trillo, 2017), (Brunet et al., 2021)
Volvox-like alga	1953	(Hardy, 1953)
Prokaryote	1974	(Pflug, 1974)

Note: Hypotheses are organized in chronological order. The four most influential hypotheses (the amoeboid hypothesis, flagellate hypothesis, ciliate hypothesis, and amoeboflagellate hypothesis) are underlined in the table and discussed in specific sections in the text (Sections 13.2, 13.3/13.4, 13.3/13.4 and 13.7/13.8 respectively). Note that proponents of the ciliate hypothesis usually thought that animals were polyphyletic, with sponges having evolved from flagellates and all other animals from ciliates.

had been identified – flagellates, ciliates and amoebae (Table 13.1). Quickly, all three were considered potential ancestors of animals.

13.2　HAECKEL'S HYPOTHESIS: AMOEBAE AS ANCESTORS

The first researcher to attempt to reconstruct the unicellular progenitor of animals was the German biologist Ernst Haeckel (1834–1919), arguably one of Darwin's most high-profile supporters in continental Europe (Richards, 2008). While Haeckel's name is most often mentioned today in the context of his now-obsolete theory of recapitulation (according to which development directly recapitulated evolution [Gould, 1977]) or for his controversial drawings of vertebrate embryos (Pennisi, 1997; Richards, 2009), his contributions to biology were much broader, and one can get an idea of their scope by considering that he coined the words "ecology," "ontogeny," "phylogeny," and "gastrulation" among many others.

　　Haeckel had an exceptionally ambitious research program: organizing all of life's diversity into a phylogenetic framework and – if that was not enough – reconstituting the extinct ancestors that occupied the most important nodes of that tree. In his attempt to reconstitute the single-celled progenitor of animals, he inferred it was an amoeba based on two independent sources of evidence: (1) his theory of recapitulation; and (2) *Magosphaera planula*, a mysterious organism that he considered the "missing link" between protists and animals.

13.2.1 HAECKEL'S EMBRYOLOGICAL ARGUMENTS FOR AN AMOEBOID ANIMAL ANCESTOR

Haeckel's case for an amoeboid ancestor started with embryology (Haeckel, 1874, 1876, 1914). He noted that the egg cells of animals lack a flagellum but are often contractile. Moreover, he observed that in sponges, the unfertilized eggs are *bona fide*

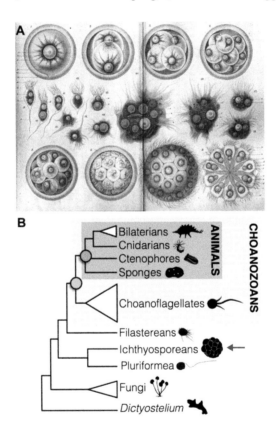

FIGURE 13.1 Mystery organism 1: *Magosphaera planula*, a facultatively multicellular amoeba described by Haeckel but never re-observed, which he thought resembled the ancestor of animals. (A) Haeckel's depiction of the life history of *M. planula* (Haeckel, 1870). The first and third row depict the cleavage of the large, spherical cell initially observed by Haeckel, resulting in a swimming sphere of multiciliated cells. The second row shows the cells produced by dissociation of that sphere, which are first multiciliated and then become amoeboid. (B) Modern phylogenetic tree showing interrelationships between animals and their closest relatives (modified from Nichols, Dayel and King [2009]). Ichthyosporeans (red arrow) belong to a lineage considered among the closest relatives of animals and form spherical masses of spores by cleavage at constant volume, which evokes *M. planula*.

**BOX 13.1   HAECKEL'S AMOEBOID
HYPOTHESIS IN HIS OWN WORDS**

*The existence of this single-celled Amoeba-like primary form of the
whole animal kingdom is proved by the extremely important fact that
the fructified egg of all animals, of the sponge and the worm up to that
of an ant and to that of man, is a simple cell. The ripe eggs of different animals frequently present very different shapes, accordingly as
they may be enclosed by variously formed coverings or burdened with
nutritive yolk. But the youthful egg-cells are still naked and without any
membrane, of the simplest construction, and at times they even creep
about in the body like an Amoeba – thus, for instance, in Sponges; they
were formerly, in this case, even considered to be parasitic Amoebae.
(The History of Creation, 1914, p. 149–150)*

crawling amoeboid cells (later confirmed by Franzen, 1988; Ereskovsky, 2010). After
fertilization, the sponge zygote divided to give rise to a ball of non-ciliated cells (the
morula) that only later acquired cilia and collectively formed an internal space (thus
becoming a blastula). According to Haeckel, future feeding cavities then formed
during gastrulation. Seen through a recapitulationist lens, these developmental facts
told a compelling evolutionary story: animals had evolved from free-living amoebae that had first formed balls of cells before acquiring ciliation, an internal cavity
and then, eventually, evolving a gut (Haeckel 1874, 1914; Figure 13.1; Box 13.1).
Embryology might have been enough to convince Haeckel of the amoeboid origin of
animals. But he thought he had another critical piece of evidence: a "missing link."[1]

13.2.2   *Magosphera planula*: Haeckel's "Missing Link"
between Amoebae and Animals

Haeckel's purported encounter with *M. planula* (Haeckel, 1870, reviewed in
Reynolds and Hülsmann 2008; see also Levit *et al.*, 2020) occurred in 1869 off the
coast of Bergen, Norway. In a seaweed sample, Haeckel observed tiny round capsules
(~70 µm large in diameter) that resembled egg cells, with a single central nucleus.
These egg-like structures then started dividing at constant cell volume – like the
cleavage of an early animal embryo – and gave rise to spheres of cells, each of which
then acquired a covering of motile cilia. These ciliated spheres started swimming
around, but did not develop further; instead, they fell to the bottom and dissociated
into individual amoeboid cells that crawled around. Haeckel did not observe the
further development of these amoebae but assumed they would eventually increase
in volume to give rise to another spherical cell, thus completing the cycle (Haeckel,
1870) (Figure 13.2A).

   In Haeckel's view, *Magosphaera* provided an important window into animal origins. Its amoeboid single-celled form matched his recapitulation-inspired view of
the animal ancestor. It had facultative multicellularity, which it reached by a cleavage

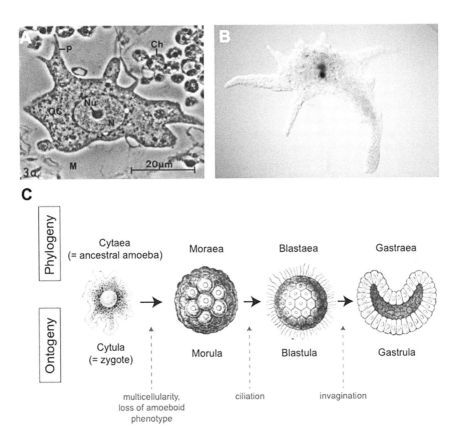

FIGURE 13.2 Haeckel's amoeboid hypothesis of animal origins. (A) the amoeboid egg cell of a sponge (from *Sycon raphanus* [Franzen, 1988]). (B) a free-living amoeba, *Chaos carolinense* (Creative commons license; https://commons.wikimedia.org/wiki/File:Chaos_carolinense.jpg). (C) Haeckel's model of animal origins, in which ontogeny (development) parallels phylogeny (evolution). In both courses, the starting point was depicted as an amoeboid cell, similar to the sponge oocyte and free-living amoebae. Proliferation of such amoeboid cells was inferred to first result in the formation of a non-ciliated sphere, the Moraea/morula. Ciliation in this scenario only arose after the evolution/development of multicellularity, at the Blastaea/blastula stage. The sketch of the cytula stage is a sponge egg cell (Haeckel, 1872) and sketches of the following stages are coral embryos (Haeckel, 1914).

process similar to animal embryos. It was a concrete, living embodiment of the Moraea stage of his evolutionary timeline.

Yet, *M. planula* is shrouded in mystery. Haeckel saw it only once and modern efforts to re-isolate it has failed (Reynolds and Hülsmann, 2008). It is an interesting exercise to take Haeckel's description at face value and wonder what he might have seen. The closest parallel to *Magosphaera* may be found among the ichthyosporeans, a lineage of unicellular opisthokonts closely related to animals and choanoflagellates (Figure 13.2B). Ichthyosporeans are free-living amoebae or flagellates that, like *Magosphaera*, can grow into large round cysts that divide at constant cell volume and finally dissociate back into single cells (Mendoza, Taylor and Ajello, 2002; Glockling, Marshall and Gleason, 2013; Suga and Ruiz-Trillo, 2013). They differ from *Magosphaera* in three main ways: (1) the spherical multicellular form of known ichthyosporeans is never swimming or multiciliated; (2) almost all known ichthyosporeans are commensal or parasitic, not free-living; (3) the large ichthyosporean cysts are multinucleated and divide by simultaneous global cellularization around pre-existing nuclei (Dudin et al., 2019), rather by serial binary division of a large, initially mononucleated cell (as in animal zygotes). However, and interestingly, an environmental metabarcoding study has suggested the existence of undescribed free-living ichthyosporeans (Del Campo and Ruiz-Trillo, 2013) and three apparently free-living species have since been isolated (belonging to the genera *Chromosphaera* [Grau-Bové et al., 2017] and *Sphaeroforma* [Hassett, López and Gradinger, 2015]).

If Haeckel's description was accurate, he could be the first person to have seen a free-living ichthyosporean, that would have (unlike other described ichthyosporeans) undergone serial binary cleavage from a mononucleated cyst. Alternatively, he might have misinterpreted – or exaggerated – what he saw.

For all of Haeckel's fame, his amoeboid hypothesis never seems to have gained followers. The reasons for this are unclear, but his hypothesis may have suffered from the rise of a worthy competitor: the flagellate hypothesis of animal origins.

13.3 METCHNIKOFF'S HYPOTHESIS: CHOANOFLAGELLATES AS ANCESTORS

Colonies of flagellates have been known since van Leeuwenhoek first observed *Volvox* (van Leeuwenhoek, 1677). The similarity of such colonies to the blastula stage of animal development (which impressed even Haeckel [1914]) seemed to suggest a plausible evolutionary path from flagellates to animals – an idea that emerged shortly after Haeckel's amoeboid hypothesis of animal origins.

A first piece of evidence was the striking similarity between choanoflagellates and the feeding cells of sponges, the choanocytes. Both have a near-identical appearance with a flagellum surrounded by a ring of microvilli, together forming a "collar complex" (Brunet and King, 2017) (Figure 13.3A, B). This resemblance was already evident to some of the earliest choanoflagellate observers, Henry James-Clark (1826–1873) and William Saville-Kent (1845–1908) (James-Clark, 1867; Kent, 1882). Both authors erroneously concluded that sponges were specialized choanoflagellates and not animals at all. In support of his hypothesis, Saville-Kent described facultative multicellular colonies in several choanoflagellates (Kent, 1882), suggesting a possible path to complex multicellularity. This idea was extended by Otto Bütschli (1848–1920), who suggested that sponges had evolved from choanoflagellates, while other animals had evolved from another (unidentified) flagellate group (Bütschli, 1884).

The Russian biologist Elie Metchnikoff[2] (1845–1916; better known for having later discovered macrophages), inspired by Haeckel's inference that sponges were *bona fide* animals, took seriously the similarity of sponges to both choanoflagellates and to other animals. On this basis, he suggested that all animals, including sponges, might have evolved from a choanoflagellate-like ancestor (Metchnikoff, 1886)[3]

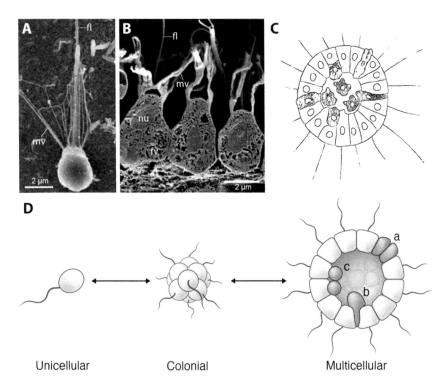

FIGURE 13.3 The flagellate hypothesis of animal origins. (A) the choanoflagellate *S. rosetta* (Dayel et al., 2011). (B) sponge choanocytes (Leys and Hill, 2012). Note the similarity of the apical collar complex between choanoflagellates and choanocytes, composed of a flagellum (fl) surrounded by microvilli (mv). nu: nucleus, fv: food vacuole. (C) Metchnikoff's postulated flagellate colony, with some cells internalizing to balance the flagellation/motility constraint (Metchnikoff, 1886). (D) a contemporary illustration of the flagellate model, redrawn after (King, 2004). Cells can either divide on the surface of the colony (a) or ingress inside the colony (b) and divide internally (c). (Drawing: Debbie Maizels)

(Figure 13.3C–D; Box 13.2). To explain the apparent absence of collar cells in animals other than sponges, Metchnikoff suggested that the microvillous collar had been lost in these lineages, and pointed out that it is retracted in some phases of the choanoflagellate life cycle (Leadbeater, 2015). (In the 20th century, it would be discovered that collar cells are in fact widespread in the animal kingdom and not restricted to sponges – see section 13.5 below).

The flagellate hypothesis was easy to combine with Haeckel's Blastaea theory: one just had to replace Haeckel's amoeboid ancestor with a flagellate. This made the resulting hypothesis more parsimonious, as it no longer required convergent evolution of flagella in protists and in animals. Perhaps because this synthesis appeared so intuitive, the concept of a flagellate ancestor has often been erroneously attributed to Haeckel himself in textbooks and in review papers of the 20th and 21st centuries, including by ourselves (Hyman, 1940; Willmer, 1990; Wainright et al., 1993; Brunet and King, 2017; Sogabe et al., 2019).

**BOX 13.2   METCHNIKOFF'S FLAGELLATE
HYPOTHESIS IN HIS OWN WORDS**

*The hypothesis which supposes that colonies of flagellate Infusoria were
transformed into primitive Metazoa explains very clearly the most important phenomena of metazoan development. On this view, the segmentation
of the egg, and especially the more primitive total segmentation, has been
derived from the division which the Flagellata undergoes in building up a
colony. In like manner the fact that the cells of so many blastospheres are
ciliated is probably due to inheritance from the Flagellata. This hypothesis (...) enables us, as Bütschli first pointed out, to comprehend the origin
of sexual multiplication. As a fact most embryologists, Ray Lankester and
Balfour among others, have adopted this (...) hypothesis, and after a pro-
longed trial it has become a basis for further speculations. Having progressed
this far, we should ask ourselves whether it is not possible, with the help of
our present knowledge, to determine more or less exactly the nature of those
Flagellate colonies from which the Metazoa are descended. Bütschli believes
the Metazoa have had a double origin: the Sponges he derives from colonies of the Choano-Flagellata, the rest of the Metazoa from colonies of true
Flagellata. Aside from the fact that there is very little ground for such a ven-
turesome assumption, we must remember that the two groups (of Flagellata)
are not sharply separated, and that the collar, which constitutes the main
point of difference, is in some cases entirely retracted.* ([Metchnikoff 1886]
translated in Wilson [1887]).

13.4   SAVILLE-KENT'S POLYPHYLETIC HYPOTHESIS OF ANIMAL ORIGINS: SPONGES FROM FLAGELLATES AND BILATERIANS FROM CILIATES

William Saville-Kent and Henry James-Clark – two of the first choanoflagellate
experts – agreed with Metchnikoff on the evolution of sponges from choanoflagellate-like ancestors. But they disagreed (collegially) with Metchnikoff and (much
more passionately) with Haeckel on the connection of sponges to animals (reviewed
in Leadbeater [2015]). This led Saville-Kent to conclude that animals had a dual
origin: sponges had evolved from choanoflagellates, while all other animals had
evolved from ciliates.

Haeckel initially thought of sponges as protists rather than animals (Haeckel,
1876) but changed his mind after he discovered that they went through a gastrula
stage (Haeckel, 1872) – an observation that was doubted for more than a century but
was confirmed in 2005 (Leys and Eerkes-Medrano, 2005). Saville-Kent, on the other
hand, strongly objected to the concept of sponges as animals, apparently because
he thought that it conflicted with their connection to choanoflagellates. From the
modern perspective, it seems clear that sponges can be related both to other animals
(through exclusive common ancestry) and to the sister group of animals, the choanoflagellates (as was evident to Metchnikoff). However, both Haeckel and Saville-Kent

seem to have strongly felt that sponges could either be related to one or the other, not both.

Saville-Kent's hostility toward Haeckel often got personal, and his comments on Haeckel's work contained a surprising density of personal attacks (e.g., Kent, 1878; reviewed in Leadbeater, 2015). Saville-Kent thought the strongest blow to Haeckel's views was the discovery of his own "missing link" (see Footnote 1) – a living species that he felt was the perfect intermediate between choanoflagellates and sponges. Out of sheer spite[4], Saville-Kent named that organism after his nemesis: *Proterospongia haeckelii*. *P. haeckelii* occupied a similar place in Saville-Kent's mind as *Magosphaera* did in Haeckel's: it was the keystone – and the concrete proof – of his hypothesis. It was also similar in another way: no one else ever managed to observe it, and to this day, we still don't know if it was real (Figure 13.4).

P. haeckelii was a flat colony of choanoflagellates with a unique feature: just like a sponge, it had spatially differentiated cells. Collar cells positioned on the outside of the colony (similar to sponge choanocytes) coexisted with amoeboid cells on the inside (similar to sponge archeocytes). All cells were embedded in a shared flat layer of extracellular matrix. The classification of these two cell types within the same

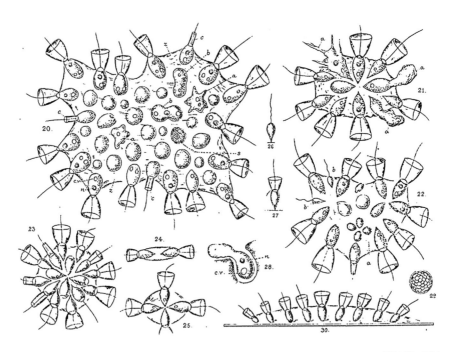

FIGURE 13.4 Mystery organism 2: *Proterospongia haeckelii* (from Kent [1882]). This purported colonial choanoflagellate was reported to contain both flagellated collar cells and amoeboid cells. All panels shown depict different developmental stages – arranged by Saville-Kent to minimize space on the page, and thus not in developmental order, which is as follows: 24, 25, 23, 21, 22 and 20. 30 is a side-view of a mature colony (same stage as 20). 28 is a close-up of a cell in the process of becoming amoeboid. 26/27 and 29 are respectively the thecate and spore forms of *P. haeckelii*.

species of choanoflagellate was supported by observed interconversions between both. Saville-Kent's drawing of a mature colony with differentiated cells (Figure 13.4, panel 20) found its way into textbooks and has been widely reproduced since (Buss, 1987; Brusca and Brusca, 2003), but he also produced many lesser-known illustrations covering the complete developmental trajectory of *P. haeckelii*, including a single collar cell serially dividing into 2, 4 and 8 cells, after which amoeboid cells started differentiating (Figure 13.4).

Saville-Kent was a thorough and careful microscopist, with his meticulous sketches anticipating structures that have been consistently detected and verified using modern techniques in microscopy. It is therefore unlikely that he would have simply misunderstood or misobserved an isolated specimen (such as a sponge larva). Instead, his description of the life history of *P. haeckelii* implies a detailed and extensive familiarity with multiple specimens, followed over an extended period of time. He was also a generally reliable observer, and his descriptions of other protists have been largely confirmed. Even though choanoflagellates have recently been shown to switch to an amoeboid form under confinement (Brunet et al., 2021), it is unlikely that Saville-Kent would have accidentally confined his samples: indeed, in the same book in which he described *P. haeckelii* (Kent, 1882), he reported the retraction of the choanoflagellate collar complex under confinement and its regeneration after confinement release.

If an honest mistake is ruled out, then *P. haeckelii* might have been real – and close to Saville-Kent's description. However, efforts to re-isolate *P. haeckelii* from the source location in Kew Gardens by one of us (T.B., together with Barry Leadbeater) have failed so far[5]. Given the personal rivalry between Saville-Kent and Haeckel, an alternative interpretation is that the description of *P. haeckelii* by Saville-Kent was either partly or entirely fabricated, possibly to get back at Haeckel. As with *Magosphaera*, the existence of *P. haeckelii* remains a mystery.

As significant as Saville-Kent thought *P. haeckelii* was, he only considered it relevant to the origin of sponges, but not of other animals. Instead, he proposed that (most) animals had evolved from ciliates – not just once, but many times, with different ciliates giving rise to different animal lineages (Kent, 1882). Saville-Kent was struck by the similarity in size, shape, and behavior between ciliates and small animals (both meiofaunal species – like rotifers or flatworms – and planktonic larvae; Figure 13.5A). His idea initially drew skepticism (Lankester, 1883) but had a few early supporters (Sedgwick, 1895). It would, however, make a spectacular comeback and then recede again in the 20th century.

13.5   20th CENTURY: THE RISE AND FALL OF THE CILIATE HYPOTHESIS

13.5.1   Similarities between Acoels and Ciliates and the Rise of the Ciliate Hypothesis

Saville-Kent was correct on one point: the similarities between ciliates and small animals of the interstitial fauna are striking (reviewed in Leander, 2008; Rundell and Leander, 2010). At first sight, one could easily mistake *Paramecium* for an acoel

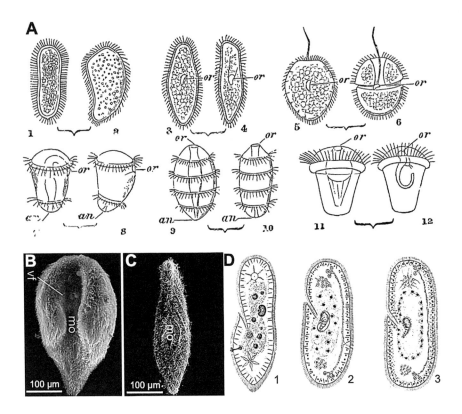

FIGURE 13.5 The ciliate hypothesis of animal origins. (A) Saville-Kent's depicted parallels between small animals (left) and ciliates (right). 1: Cnidarian planula larva; 2: *Opalina*; 3: flatworm; 4: *Paramecium*; 5: nemertean larva; 6: *Melodinium*; 7: annelid trochophore larva; 8: *Telotrochidium*; 9: echinoderm larva; 10: *Didinium*; 11: bryozoan larva; 12: *Vorticella*. from (Kent, 1882). (B) The acoel *Convolutriloba longifissura. mo*: mouth, *vf*: ventral folds. from (Hejnol and Martindale, 2008). (C) the ciliate *Paramecium sonneborni*, from (Aufderheide, Daggett and Nerad, 1983). *mo*: cytostome ("cellular mouth"). (D) Hadzi's cellularization hypothesis, from (Willmer, 1990) after (Hadzi, 1963). (1) a *Paramecium*-like ciliate hypothesized to be ancestral to animals, with multiple nuclei, pulsatile vacuoles, a cytostome ("cellular mouth") and periodic invaginations of the plasma membrane. (2) nuclei migrate to the periphery of the cell to be lodged underneath the plasma membrane and between the invaginations. The cytostome becomes more elaborate and acquires features of a pharynx. (3) an acoel-like hypothetical ancestor of animals, with different body parts (epidermis, gut, nephridia) having evolved by partial or complete cellularization from ciliate structures.

worm (Figure 13.5B, C). Both are elongated, bilaterally symmetrical, nearly half a millimeter long, and densely covered in motile cilia. Acoel worms are minute animals of extreme simplicity (long believed to be flatworms, but now known to belong to a separate bilaterian lineage (Ruiz-Trillo et al., 1999; Cannon et al., 2016; Marlétaz, 2019; Philippe et al., 2019). They lack excretory organs, an anus, and even a proper gut. Early histological studies emphasized that simplicity and many observers went so far as to erroneously conclude that acoels lacked separate cells (except perhaps in the epidermis) and instead represented a single large syncytium containing floating nuclei. Uncertainty around this point persisted from the 1880s to the 1960s, when electron microscopy finally demonstrated that acoels were actually almost entirely cellular (to the exception of their digestive cellular mass, which is genuinely syncytial; see Delage, 1886; Pedersen, 1964 for reviews).

In the meantime, however, the supposedly syncytial organization of acoels, together with their overall similarity to ciliates prompted a revival of Saville-Kent's ciliate hypothesis of bilaterian origins. The idea was proposed independently by Jovan Hadži (1884–1972) and Otto Steinböck (1893–1969), and further elaborated by Earl D. Hanson (1927–1993) (Hadzi, 1953, 1963; Hanson, 1963, 1977; Steinböck, 1963). These authors identified many purported homologies among ciliates and acoels: ciliary arrays of the former were homologized to the ciliated epidermis of the latter; the contractile infraciliary lattice of ciliates was inferred to represent an antecedent of acoel musculature; the digestive vacuoles were proposed to be equivalent to the acoel digestive mass and pulsatile vacuoles in ciliates were considered homologous to nephridia (excretory organs that are absent in acoels but found in flatworms). The fact that ciliates only have two nuclei (a micronucleus and a macronucleus) and do not display a multicellular or even syncytial organization was countered by pointing to *Opalina*, a protist then considered to be a ciliate which possessed many nuclei underneath its cell membrane (and which is now known to be a heterokont that only convergently resembles ciliates [Cavalier-Smith and Chao, 2006]). Like Saville-Kent, supporters of the ciliate hypothesis explained the similarity between choanoflagellates and choanocytes by hypothesizing that sponges were specialized choanoflagellates, and thus unrelated to other animals. Animals were thus assumed to have had at least two independent origins in the protistan world, and maybe even three (with cnidarians possibly descending from amoebae [Hanson, 1977]).

13.5.2 THE FALL OF THE CILIATE HYPOTHESIS

The hypothesis of the syncytial nature of acoels was finally disproved by electron microscopy in the mid-1960s (Pedersen, 1964), but the ciliate hypothesis of animal origins had by then taken a life of its own and survived the loss of its former central argument (Hanson, 1977). As late as the 1980s–1990s, the ciliate hypothesis and the polyphyletic origin of animals were still often presented as the likeliest hypotheses of animal origins in popular texts and textbooks. In his best-seller *Wonderful Life*, Stephen Jay Gould wrote: "*The vernacular term* animal *itself probably denotes a polyphyletic group, since sponges (almost surely), and probably corals and their allies as well, arose separately from unicellular ancestors – while all other animals of our ordinary definition belong to a third distinct group*" (Gould, 1989). Similar statements could be found in many contemporary zoology textbooks (Mitchell, Mulmor and Dolphin, 1988; Willmer, 1990; Miller and Harley, 1999), although a few were critical (Brusca and Brusca, 1990). Surprisingly, the ciliate hypothesis survived the first molecular phylogenies as well: early studies included only a few genes analyzed with simple, similarity-based algorithms and often failed to recover the monophyly of the animal kingdom, thus apparently lending credence to multiple independent origins of animals from several protist groups (Field et al., 1988;

Lake, 1990; Christen et al., 1991). It was only with larger datasets and better models of sequence evolution that a consistent picture of monophyletic animals closely related to choanoflagellates finally emerged, with ciliates relegated to a very distant branch (Wainright et al., 1993), making the ciliate hypothesis untenable. Unsurprisingly, the hypothesized homologies also eventually failed to withstand molecular scrutiny. For example, the infraciliary contractile lattice of *Paramecium* was found to be made of centrins, a family of contractile proteins unrelated to actin and myosin, the contractile proteins of animal musculature (Levy et al., 1996).

With the benefit of hindsight, many of the arguments underlying the ciliate hypothesis appear contrived. Yet, it convinced many – if not most – experts for nearly 30 years. We now know that its proponents were misled by an impressive suite of morphological convergences between metazoans, ciliates, and additional protists like *Opalina*. While the ciliate hypothesis has now been dismissed as inconsistent with the modern eukaryotic phylogeny, it serves as a reminder of how much complexity – in morphology, patterning, and behavior – can be achieved by a single cell (Marshall, 2020). The animal-like behaviors of ciliates, which fascinated scientists and philosophers at the turn of the 20th century (Schloegel and Schmidgen, 2002), are currently undergoing a renaissance as a research topic (Coyle et al., 2019; Dexter, Prabakaran and Gunawardena, 2019; Mathijssen et al., 2019; Wan and Jékely, 2020), as are the mechanisms of their patterning and morphogenesis (Marshall, 2020). Properly understood as an independent and unique evolutionary experiment in achieving levels of size and morphological complexity that rival those of small animals, ciliates remain as fascinating as ever.

13.6 20th CENTURY: THE COLLARED FLAGELLATE/ CHOANOBLASTAEA MODEL

Although it had to compete with the ciliate hypothesis for part of the 20th century, Metchnikoff's concept of a choanoflagellate-like ancestor for all animals – and not just for sponges – was continuously supported by some authors (Hyman, 1940; Rieger, 1976; Salvini-Plawen, 1978; Nielsen and Norrevang, 1985). These researchers were each convinced about the monophyly of animals based on shared features such as sperm and eggs, epithelia, and gastrulation. This implied that all animals had evolved from a single lineage of protist, of which choanoflagellates were considered the most plausible living representative as their similarity to choanocytes was so strong. This view received further support from the discovery of choanocyte-like collar cells by electron microscopy in diverse animal phyla other than sponges (Nerrevang and Wingstrand, 1970; Lyons, 1973; Rieger, 1976; Brunet and King, 2017). Claus Nielsen named this revised Blastaea model – starting from a collared ancestor – the "Choanoblastaea" (Nielsen, 2008) (Figure 13.6).

13.6.1 MOLECULAR PHYLOGENIES AND THE RISE OF THE CHOANOBLASTAEA MODEL

While early molecular studies initially contradicted the Choanoblastaea hypothesis and suggested animal polyphyly (see section 13.4 above), improved analyses with more

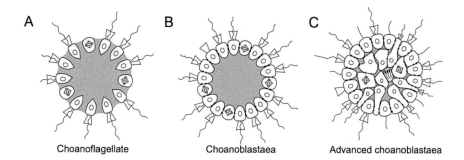

FIGURE 13.6 The choanoblastaea model of animal origins (Nielsen, 2008). (A) a modern choanoflagellate rosette colony proposed to resemble early stem-animals. Cells are arranged as a sphere surrounding a shared core of extracellular matrix (dark grey). (B) a hypothetical later stem-animal ("Choanoblastaea"), in which cells have become adjacent and have evolved intercellular junctions and now form a sealed epithelial sphere. (C) a later hypothetical stem-animal ("Advanced choanoblastaea") in which some cells have become amoeboid and populated the inner space of the colony (compare *P. haeckelii*, Figure 13.4). Note that cell division is now restricted to those inner cells.

data and better statistical models of sequence evolution ended up consistently supporting the monophyly of animals and their sister-group relationship to choanoflagellates (Wainright et al., 1993; King and Carroll, 2001; Lang et al., 2002; King, Hittinger and Carroll, 2003; Rokas et al., 2003; King et al., 2008; Ruiz-Trillo et al., 2008). Unlike hypothesized homologies between ciliates and animals, the inferred homology of the collar complex in animals and choanoflagellates survived molecular and biochemical analyses, which confirmed that the collar is composed of homologous cytoskeletal filaments in both choanoflagellates, sponges, and other animals (reviewed in Leadbeater, 2015; Brunet and King, 2017). The hypothesis of the homology of the collar complex – proposed on morphological grounds in the 19th century – thus appears to have been predictive (Colgren and Nichols, 2020) and is now accepted by many authors (but see Mah, Christensen-Dalsgaard and Leys, 2014; Sogabe et al., 2019 for exceptions and Brunet and King, 2017; Myers, 2019; Colgren and Nichols, 2020 for responses).

13.6.2 THE LIMITS OF THE CHOANOBLASTAEA MODEL

Despite its support from the data, the Choanoblastaea model leaves some questions unresolved. One is the similarity of crawling amoeboid cells, widespread in animals, to the amoeboid motility of diverse protists. While some authors explicitly ascribed that similarity to evolutionary convergence (Cavalier-Smith, 2017), few directly recognized or addressed the issue. While one solution could have been to revive Haeckel's amoeboid hypothesis, a strict interpretation of his hypothesis had clearly become incompatible with structural information that had emerged in the 20th century showing the homology of flagella in animals and diverse protists (reviewed in Margulis, 1981). Instead, one parsimonious way to account for all the data has been to reconstruct the progenitor of animals as a shape-shifter: sometimes flagellate, sometimes amoeba, and maybe more.

13.7   20th CENTURY: THE AMOEBOFLAGELLATE
MODEL AND THE SYNZOOSPORE MODEL

Complex life cycles in protists have been known since the 19th century. In 1898, the British medical doctor Ronald Ross (1857–1932) described the different life stages of the unicellular parasite that causes malaria, *Plasmodium falciparum* (reviewed in Cox, 2002). A year later, the Austrian biologist Franz Schardinger (1853–1920) discovered *Naegleria gruberi* (then named *Amoeba gruberi*), a free-living amoeba that had the unusual ability to transdifferentiate into a flagellate form (Schardinger, 1899; Fulton, 1977, 1993).

The transition between the amoeboid and the flagellate forms of *Naegleria* is reminiscent of the reversible transdifferentiation between the flagellated choanocytes and the amoeboid archeocytes of sponges (Figure 13.7) that was already known to Saville-Kent (Kent, 1882) and later confirmed by modern studies (Nakanishi, Sogabe and Degnan, 2014; Sogabe et al., 2019).

In spite of this parallel, shape-shifting protists such as *Naegleria* were apparently never considered relevant to animal origins before the mid-20th century,

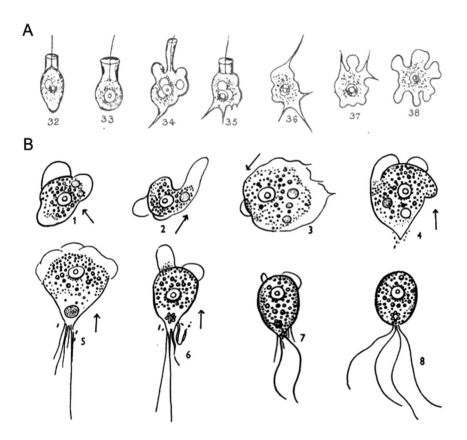

FIGURE 13.7   Interconversions between flagellate and amoeboid phenotypes in sponge cell transdifferentiation. (A) (Kent, 1882) and in *Naegleria* development (B) (Willmer, 1971).

when the Soviet biologist Alexey Zakhvatkin (1906–1950, alternatively spelled Sachwatkin) and the British biologist E. N. Willmer (1902–2001) independently hypothesized that the elaborate life cycles of animals might have their roots in the unicellular world. In his treatise *Comparative embryology of the low invertebrates* (Zakhvatkin, 1949 for the Russian original and Sachwatkin, 1956 for the German translation, which we consulted), Zakhvatkin explicitly compared the complex life cycles of protists and animals. By comparing animal cell differentiation with the reversible amoeboid/flagellate switches of *Naegleria gruberi*[6] and of *Polytomella citri* (a parasitic green alga[7]; Kater, 1925), Zakhvatkin suggested animals evolved from an amoeboflagellate. He also noted that cleavage at constant volume of the animal zygote (a process called "palintomy") had parallels in several protists, including dinoflagellates and green algae, in which it resulted in a mass of flagellated "zoospores" that eventually dissociated and underwent dispersal. Zakhvatkin suggested that the morula stage of animal development might have evolved from zoospores that failed to separate – a "synzoospore."

Because Zakhvatkin's work was only available in Russian and in German, it did not immediately reach the English-speaking world. It is thus independently of Zakhvatkin and based on his own studies of *Naegleria*, that Willmer came to remarkably similar conclusions and proposed an amoeboflagellate ancestry for animals in his 1971 book *Cytology and Evolution* (Willmer, 1971). While he did not believe that *Naegleria* was directly related to animals, he thought it gave an idea of what animal ancestors might have looked like.

Zakhvatkin's and Willmer's ideas seem to have gone mostly unnoticed in their time, and debates regarding animal origins remained dominated by the ciliate hypothesis and the flagellate hypothesis. It is only in the last decade – the 2010s – that the concept of a protist ancestor with a complex life history has undergone a revival.

13.8 21st CENTURY: HOW COMPLEX WAS THE METAZOAN PRECURSOR?

In 2009, Zahkvatkin's ideas were shared with a broader audience thanks to a review paper that presented his hypothesis in English and named it the "temporal-to-spatial transition" model of animal origins (Mikhailov et al., 2009). Nearly at the same time, molecular phylogenies revealed that the previously enigmatic filasterans and ichthyosporeans (Ruiz-Trillo et al., 2008) are the closest known living relatives of choanozoans (the clade formed by choanoflagellates and animals; Figure 13.2). Together, choanozoans, ichthyosporeans and filasterans form the clade Holozoa[8]. Interestingly, single-celled holozoans assume diverse cellular forms (including flagellates, amoebae, and cystic forms), and many of them have complex life histories with multiple phenotypes (as do choanoflagellates, which have sessile, swimming and colonial flagellate forms, and often spores as well; Leadbeater, 2015).

Several studies have investigated the cellular and molecular basis for the complex life histories of unicellular holozoans (Fairclough, Dayel and King, 2010; Dayel et al., 2011; Sebé-Pedrós et al., 2013; Suga and Ruiz-Trillo, 2013). Remarkably, many

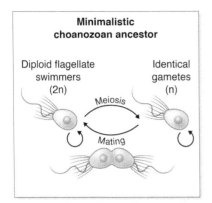

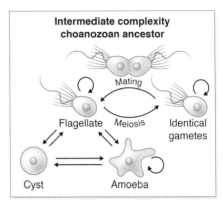

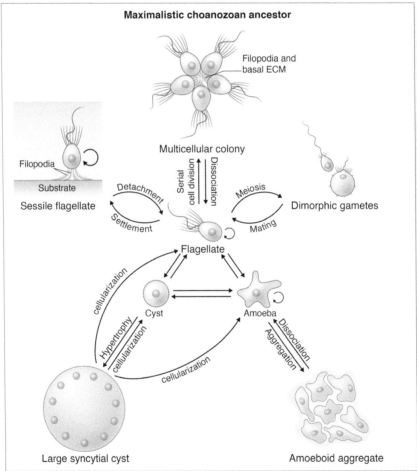

FIGURE 13.8 Current minimalistic, intermediate and maximalistic concepts of the last common ancestor of animals and choanoflagellates. The minimalistic ancestor (upper left) is reconstructed as a collared flagellate with a sexual cycle, but no multicellularity and no other

of these cell type transitions correlate with chromatin remodeling and pre- and post-transcriptional regulation (Fairclough et al., 2013; Sebé-Pedrós, Ballaré, et al., 2016; Sebé-Pedrós, Peña, et al., 2016; Dudin et al., 2019), suggesting the existence of shared mechanisms with animal cell differentiation. Adding to this picture of generally dynamic cell phenotypes, novel amoeboflagellate species were recently discovered among holozoans (Hehenberger et al., 2017; Tikhonenkov, Hehenberger, et al., 2020). Finally, choanoflagellates themselves turned out to be able to reversibly switch to an amoeboid phenotype in response to spatial confinement (Brunet et al., 2021), thus reviving Saville-Kent's concept of amoeboid phenotypes in choanoflagellates. Overall, these data converged to suggest that our ancestors along the holozoan stem-line – including the choanozoan ancestor – almost certainly had the ability to generate more cell phenotypes than just a collared flagellate, potentially paving the way to animal cell differentiation; and modern variants of the Choanoblastaea hypothesis have started to incorporate that idea (Arendt et al., 2015).

What did the choanozoan ancestor look like? Although we have made progress since Haeckel, Metchnikoff and Saville-Kent, many questions remain open. In Figure 13.8, we have illustrated two extreme options – a "minimalistic ancestor" (a simple collared flagellate without other phenotypes) and a "maximalistic ancestor" that combines several cell phenotypes frequently found in single-celled holozoans (most of which have an equivalent in animal biology) – along with an intermediate scenario that captures features we consider likely to have existed in the protistan ancestor of animals. The life cycle of this long extinct organism might have included "facultative features" such as amoeboid migration, encystment, clonal multicellularity (with or without palintomy) and aggregative multicellularity.

FIGURE 13.8 (*Continued*)

cell phenotype. The intermediate ancestor (upper right) is also assumed to have been obligately unicellular but can transdifferentiate into several forms, including a collared flagellate, an amoeba (similar to animal crawling cells such as macrophages), and a quiescent cyst (perhaps similar to animal quiescent stem cells or egg cells). The maximalistic ancestor (bottom) displays several forms of facultative multicellularity and combines several additional phenotypes known in single-celled relatives of animals, all of which have parallels among animal cell types and represent hypothetical evolutionary precursors of the latter. Spherical multicellular colonies of flagellates, similar to those of some choanoflagellates (Dayel et al., 2011), resemble the Morula stage of animal embryos. Sessile flagellated cells adhere to the substrate by a combination of filopodia and secreted extracellular matrix (ECM, green), as in modern choanoflagellates (Dayel et al., 2011) and in the filasteran *Capsaspora* (Parra-Acero et al., 2018, 2020). This might have prefigured the adhesion of animal epithelial cells to the basal lamina. Amoeboid cells are proposed to undergo aggregative multicellularity, similar to *Capsaspora* (Sebé-Pedrós et al., 2013) and to dissociated sponge cells (Dunham et al., 1983). Note that aggregation of flagellated cells has been observed in *Syssomonas* (Tikhonenkov, Hehenberger et al. 2020) and might also have been present in the last choanozoan common ancestor, though it is not depicted here. Finally, cysts are proposed to undergo hypertrophy by nuclear proliferation without cytokinesis, resulting in a syncytium that can cleave at constant volume to revert to a uninucleated state, as in modern ichthyosporeans (Suga and Ruiz-Trillo, 2013; Dudin et al., 2019) and chytrid fungi (Medina et al., 2020). This process could have been the evolutionary precursor to the cleavage of animal zygotes. (Drawing: Debbie Maizels)

Interestingly, comparative genomics has revealed that many genes thought to be animal-specific are present in their single-celled relatives – but often with a patchy and mosaic distribution, indicating rampant gene loss in most lineages (Suga et al., 2013; Richter et al., 2018). This suggests that the last choanozoan common ancestor possessed a mosaic of features that are not fully realized in any of its living relatives or descendants. We think this lends credibility to the possibility of a "maximalistic ancestor." Future work will help to refine the "checklist" of ancestral choanozoan features – which will not necessarily include all those we depicted in Figure 13.8, nor will necessarily be restricted to them.

13.9   CONCLUSION

The past has only left incomplete traces, and our understanding of it is inevitably simplified. There is, however, another force that often pushes us to simplification: the urge to summarize history as a linear narrative that leads to the present. In this review, we have strived to embrace the complexity of the past – both of our scientific predecessors, and of our evolutionary ancestors. We hope the winding history of our field is worth appreciating for itself and for the many small gems it contains, before trying to extract an – inevitably simplified – global message.

Nonetheless, a few general themes emerge. The diversity of historical hypotheses simultaneously reflects the complexity of the problem itself, the limited information available at the time, and the personal assumptions and preferences of their authors. On the one hand, morphological data were clearly confounded by multiple events of evolutionary convergence (such as between ciliates and animals), parallelism, and rampant loss. Solving the problem from morphology only was genuinely challenging (even after the advent of electron microscopy), and involved some degree of subjective judgment. On the other hand, many authors seemed to have made the task unnecessarily more difficult by assuming that the last single-celled ancestors of animals necessarily had an exact equivalent within living protists – while this ancestor likely had its own, unique combination of features that is not necessarily represented today. This point has become increasingly salient in the past few years, and we expect it to remain central to future research. Consistently, several species of single-celled holozoans with novel phenotypes have been newly described in the past few years (Hehenberger et al., 2017; Brunet et al., 2019; Tikhonenkov, Hehenberger, et al., 2020; Tikhonenkov, Mikhailov, et al., 2020), and metagenomic surveys have provided evidence for the existence of additional undiscovered holozoan lineages (Del Campo and Ruiz-Trillo, 2013; Arroyo et al., 2020). Further exploration of single-celled biodiversity thus holds the promise to enrich our reconstitution of animal ancestors – and eventually maybe even to clarify the mysteries of *Magosphaera planula* and *Proterospongia haeckelii*. Depending on their phylogenetic position, these species – if they exist and if the original descriptions were accurate – might provide stronger evidence for a pre-metazoan origin of palintomy and spatial cell differentiation, respectively.

Another point of interest is the way in which past controversies were resolved. Many debates could only be settled after the invention of new techniques; yet, technical innovations alone were rarely sufficient. The first molecular phylogenies, for example, were rather inaccurate. Consensus was only reached after commonly accepted standards of evidence were agreed upon, and once multiple independent, technically solid studies converged toward the same answer. At a time where a new wealth of molecular data (notably from single-cell techniques) is promising to bring an unprecedented quantity of evidence to bear on the study of the evolution of cell phenotypes, we hope that our historical summary can be read both as a cautionary tale and as a reason for optimism.

ACKNOWLEDGMENTS

We are especially indebted to the following historical reviews: (Jägersten, 1955; Salvini-Plawen, 1978; Willmer, 1990; Leadbeater and McCready, 2002; Nielsen, 2008, 2012; Leadbeater, 2015). We thank Debbie Maizels for help with the figures; Elliott Smith of the UC Berkeley library for access to scans of Haeckel books; Iñaki Ruiz-Trillo for feedback on Figure 13.8; Hiral Shah, Gautam Dey and Omaya Dudin for "PreLights" coverage, comments on the preprint version of this manuscript, and discussions on ichthyosporeans and palintomy; David Booth, the editors of this book, and two anonymous reviewers for comments on the text; and the entire King lab for stimulating discussions.

NOTES

1. The concept of the extant "missing link" that represents the ancestral condition is no longer considered valid, but it accurately captures Haeckel's views, which often envisioned some living groups as identical to the ancestors of other living groups.
2. Alternatively spelled Metschnikoff or Mechnikov.
3. See (Wilson, 1887) for an English translation.
4. Naming species out of spite is a venerable tradition that dates back to Linnaeus, who named an especially smelly weed, *Siegesbeckia*, after one of his detractors. For a list, see https://www.science-shenanigans.com/species-named-out-of-spite/
5. As a caveat, the reisolation of even a well-studied choanoflagellate species can be challenging. For example, the laboratory model species *Salpingoeca rosetta* (Dayel *et al.*, 2011) has been isolated only once and we have been unable to re-isolate it from its source location despite repeated attempts.
6. Which he referred to as *Vahlkampfia gruberi*.
7. *P. citri* is a secondarily non-photosynthetic, parasitic green alga, and thus belongs to a lineage whose sequenced representatives have lost regulators of cell crawling such as SCAR/WAVE (Fritz-Laylin, Lord and Mullins, 2017) and myosin II (Sebé-Pedrós *et al.*, 2014). While the genome of *P. citri* itself has not been sequenced, it is interesting to wonder how amoeboid mobility could function in this species if it also lacks those genes.
8. A few additional lineages (such as corallochytrids) have since been added (Figure 2B).

REFERENCES

Arendt, D. *et al.* (2015). 'Gastric pouches and the mucociliary sole: Setting the stage for nervous system evolution,' *Philosophical Transactions of the Royal Society B: Biological Sciences*, 370(1684), p. 20150286. doi: 10.1098/rstb.2015.0286.

Arroyo, A. S. *et al.* (2020). 'Gene similarity networks unveil a potential novel unicellular group closely related to animals from the tara oceans expedition,' *Genome Biology and Evolution*, 12(9), pp. 1664–1678. doi: 10.1093/gbe/evaa117.

Aufderheide, K. J., Daggett, P. M. M. and Nerad, T. A. (1983). '*Paramecium sonneborni* n. sp., a New Member of the *Paramecium aurelia* Species-Complex,' *The Journal of Protozoology*, 30(1), pp. 128–131. doi: 10.1111/j.1550-7408.1983.tb01046.x.

Brunet, T. *et al.* (2019). 'Light-regulated collective contractility in a multicellular choanoflagellate,' *Science*, 366(6463), pp. 326–334. doi: 10.1126/science.aay2346.

Brunet, T. *et al.* (2021). 'A flagellate-to-amoeboid switch in the closest living relatives of animals,' *eLife*, 2021(10), p. e61037. doi: 10.7554/elife.61037.

Brunet, T. and King, N. (2017). 'The origin of animal multicellularity and cell differentiation.,' *Developmental cell*. United States, 43(2), pp. 124–140. doi: 10.1016/j.devcel.2017.09.016.

Brusca, R. C. and Brusca, G. J. (1990). *Invertebrates*. 1st edn. Sunderland, Mass.: Sinauer Associates.

Brusca, R. C. and Brusca, G. J. (2003). *Invertebrates*. 2nd edn. Sunderland, MA: Sinauer Associates, Inc.

Buss, L. W. (1987). *The evolution of individuality*. Princeton, NJ: Princeton University Press.

Bütschli, O. (1884). 'Bemerkungen zur Gastraea-Theorie,' *Morph. Jahrb.*, 9, pp. 415–427.

Del Campo, J. and Ruiz-Trillo, I. (2013). 'Environmental survey meta-analysis reveals hidden diversity among unicellular opisthokonts,' *Molecular Biology and Evolution*, 30(4), pp. 802–805. doi: 10.1093/molbev/mst006.

Cannon, J. T. *et al.* (2016). 'Xenacoelomorpha is the sister group to Nephrozoa,' *Nature*. doi: 10.1038/nature16520.

Cavalier-Smith, T. (2017). 'Origin of animal multicellularity: Precursors, causes, consequences – the choanoflagellate/sponge transition, neurogenesis and the Cambrian explosion,' *Philosophical Transactions of the Royal Society B: Biological Sciences*, 372(1713), p. 20150476. doi: 10.1098/rstb.2015.0476.

Cavalier-Smith, T. and Chao, E. E. Y. (2006). 'Phylogeny and megasystematics of phagotrophic heterokonts (kingdom Chromista),' *Journal of Molecular Evolution*, 62(4), pp. 388–420. doi: 10.1007/s00239-004-0353-8.

Christen, R. *et al.* (1991). 'An analysis of the origin of metazoans, using comparisons of partial sequences of the 28S RNA, reveals an early emergence of triploblasts,' *EMBO Journal*, 10(3), pp. 499–503. doi: 10.1002/j.1460-2075.1991.tb07975.x.

Colgren, J. and Nichols, S. A. (2020). 'The significance of sponges for comparative studies of developmental evolution,' *Wiley Interdisciplinary Reviews: Developmental Biology*, 9(2), p. e359. doi: 10.1002/wdev.359.

Cox, F. E. G. (2002). 'History of human parasitology,' *Clinical Microbiology Reviews*, 15(4), pp. 595–612. doi: 10.1128/CMR.15.4.595-612.2002.

Coyle, S. M. *et al.* (2019). 'Coupled active systems encode an emergent hunting behavior in the unicellular predator *Lacrymaria olor*,' *Current Biology*, 29(22), pp. 3838–3850. doi: 10.1016/j.cub.2019.09.034.

Darwin, C. (1859). *The Origin of Species by means of Natural Selection; or the Preservation of Favoured Races in the Struggle for Life*. London: Murray.

Darwin, C. and Wallace, A. (1858). 'On the tendency of species to form varieties; and on the perpetuation of varieties and species by natural means of selection.,' *Journal of the Proceedings of the Linnean Society of London. Zoology*, 3(9), pp. 46–62. Doi: 10.1111/j.1096-3642.1858.tb02500.x.

Dayel, M. J. *et al.* (2011). 'Cell differentiation and morphogenesis in the colony-forming choanoflagellate *Salpingoeca rosetta*,' *Developmental Biology*, 357(1), pp. 73–82. Doi: 10.1016/j.ydbio.2011.06.003.

Delage, Y. (1886). 'Etudes histologiques sur les planaires rhabdocoeles acoeles.,' *Archives de Zoologie Expérimentale et Générale*, II(4), pp. 109–160.

Dexter, J. P., Prabakaran, S. and Gunawardena, J. (2019). 'A complex hierarchy of avoidance behaviors in a single-cell eukaryote,' *Current Biology*, 29(24), pp. 4323–4329. Doi: 10.1016/j.cub.2019.10.059.

Dudin, O. *et al.* (2019). 'A unicellular relative of animals generates a layer of polarized cells by actomyosin-dependent cellularization,' *eLife*, 2019(8), p. e49801. Doi: 10.7554/eLife.49801.

Dunham, P. *et al.* (1983). 'Stimulus-response coupling in sponge cell aggregation: Evidence for calcium as an intracellular messenger,' *Proceedings of the National Academy of Sciences of the United States of America*, 80(15), pp. 4756–4760. Doi: 10.1073/pnas.80.15.4756.

Ereskovsky, A. V. (2010). *The comparative embryology of sponges*. Hedeilberg, London, New York: Springer. doi: 10.1007/978-90-481-8575-7.

Fairclough, S. R. *et al.* (2013). 'Premetazoan genome evolution and the regulation of cell differentiation in the choanoflagellate *Salpingoeca rosetta*,' *Genome Biology*, 14(2), pp. 1–15. doi: 10.1186/gb-2013-14-2-r15.

Fairclough, S. R., Dayel, M. J. and King, N. (2010). 'Multicellular development in a choanoflagellate,' *Current Biology*, 20(20), pp. R875–R876. doi: 10.1016/j.cub.2010.09.014.

Field, K. G. *et al.* (1988). 'Molecular phylogeny of the animal kingdom,' *Science*, 239(4841), pp. 748–753. doi: 10.1126/science.3277277.

Franz, V. (1924). *Geschichte der Organismen*. Jena: Fischer.

Franzen, W. (1988). 'Oogenesis and larval development of *Scypha ciliata* (Porifera, Calcarea),' *Zoomorphology*, 107(6), pp. 349–357. doi: 10.1007/BF00312218.

Fritz-Laylin, L. K., Lord, S. J. and Mullins, R. D. (2017). 'WASP and SCAR are evolutionarily conserved in actin-filled pseudopod-based motility,' *Journal of Cell Biology*, 216(6), pp. 1673–1688. doi: 10.1083/jcb.201701074.

Fulton, C. (1977). 'Cell differentiation in *Naegleria gruberi*,' *Annual Review of Microbiology*, 31, pp. 597–627. doi: 10.1146/annurev.mi.31.100177.003121.

Fulton, C. (1993). '*Naegleria*: A research partner for cell and developmental biology,' *Journal of Eukaryotic Microbiology*. doi: 10.1111/j.1550-7408.1993.tb04945.x.

Glockling, S. L., Marshall, W. L. and Gleason, F. H. (2013). 'Phylogenetic interpretations and ecological potentials of the Mesomycetozoea (Ichthyosporea),' *Fungal Ecology*. doi: 10.1016/j.funeco.2013.03.005.

Gould, S. J. (1977). *Ontogeny and Phylogeny*. Cambridge, MA: Belknap Press.

Gould, S. J. (1989). *Wonderful Life: The Burgess Shale and the Nature of History*. New York: W. W. Norton & Co.

Grau-Bové, X. *et al.* (2017). 'Dynamics of genomic innovation in the unicellular ancestry of animals,' *eLife*. doi: 10.7554/eLife.26036.

Hadzi, J. (1953). 'An attempt to reconstruct the system of animal classification,' *Systematic Zoology*, 2(4), pp. 145–154. doi: 10.2307/2411558.

Hadzi, J. (1963). *The evolution of the metazoa*. Oxford: Pergamon Press. doi: 10.5962/bhl.title.6821.

Haeckel, E. (1870). *Biologische Studien. Erstes Heft: Studien über Moneren und andere Protisten*. Leipzig, Engelmann.

Haeckel, E. (1872). *Die Kalkschwämme. Eine Monographie*. Berlin: G. Reimer.

Haeckel, E. (1874). *The evolution of man: a popular exposition of the principal points of human ontogeny and phylogeny*. Watts & Co, London. doi: 10.5962/bhl.title.61275.

Haeckel, E. (1876). *The History of Creation, or the Development of the Earth and its Inhabitants by the Action of Natural Causes, volume 2*. 1st edn. London: Henry S. King & Co.

Haeckel, E. (1914). *The History of Creation, or the Development of the Earth and its Inhabitants by the Action of Natural Causes, volume 2*. 6th edn. New York: D. Appleton & Co.

Hanson, E. D. (1963). 'Homologies and the ciliate origin of the Eumetazoa,' in Dougherty, E. C. et al. (eds) *The Lower Metazoa*. Cambridge: Cambridge University Press.

Hanson, E. D. (1977). *The Origin and Early Evolution of Animals*. London: Pitman.

Hardy, A. C. (1953). 'On the origin of the metazoa,' *Journal of Microscopical Science*, 94(4), pp. 441–443.

Hassett, B. T., López, J. A. and Gradinger, R. (2015). 'Two new species of marine saprotrophic sphaeroformids in the mesomycetozoea isolated from the sub-arctic Bering sea,' *Protist*. doi: 10.1016/j.protis.2015.04.004.

Hehenberger, E. *et al.* (2017). 'Novel predators reshape holozoan phylogeny and reveal the presence of a two-component signaling system in the ancestor of animals,' *Current Biology*, 27(13), pp. 2043–2050. doi: 10.1016/j.cub.2017.06.006.

Hejnol, A. and Martindale, M. Q. (2008). 'Acoel development indicates the independent evolution of the bilaterian mouth and anus,' *Nature*, 456(7220), pp. 382–386. doi: 10.1038/nature07309.

Hyman, L. H. (1940). *The Invertebrates: Protozoa through Ctenophora*. New York, NY: McGraw-Hill Book Company.

Jägersten, G. (1955). 'On the early phylogeny of the Metazoa: the Bilaterogastraea theory,' *Zool. Bidrag*, 30, pp. 321–254.

James-Clark, H. (1867). 'On the Spongiae Ciliatae as Infusoria Flagellata: or observations on the structure, animality and relationship of *Leucosolenia botryoides* Bowerbank,' *Memoirs of the Boston Society of Natural History*, 1, pp. 305–340.

Kater, J. M. (1925). 'Morphology and life history of *Polytomella citri* sp. nov,' *The Biological Bulletin*, 49(3), pp. 213–236. doi: 10.2307/1536462.

Kent, W. S. (1878). 'Observations upon Prof. Ernst Haeckel's "Physemaria" and on the affinity of the sponges,' *Annals and Magazine of Natural History*, 5(1), pp. 1–17.

Kent, W. S. (1882). *A Manual of the Infusoria*. London: D. Bogue.

King, N. (2004). 'The unicellular ancestry of animal development,' *Developmental Cell*, 7(3), pp. 313–325. doi: 10.1016/j.devcel.2004.08.010.

King, N. *et al.* (2008). 'The genome of the choanoflagellate *Monosiga brevicollis* and the origin of metazoans,' *Nature*, 451(7180), pp. 783–788. doi: 10.1038/nature06617.

King, N. and Carroll, S. B. (2001). 'A receptor tyrosine kinase from choanoflagellates: Molecular insights into early animal evolution,' *Proceedings of the National Academy of Sciences of the United States of America*, 98(26), pp. 15032–15037. doi: 10.1073/pnas.261477698.

King, N., Hittinger, C. T. and Carroll, S. B. (2003). 'Evolution of key cell signaling and adhesion protein families predates animal origins,' *Science*, 301(5631), pp. 361–363. doi: 10.1126/science.1083853.

Lake, J. A. (1990). 'Origin of the metazoa,' *Proceedings of the National Academy of Sciences of the United States of America*, 87(2), pp. 763–766. doi: 10.1073/pnas.87.2.763.

Lamarck, J.-B. (1809). *Philosophie zoologique, ou Exposition des considérations relatives à l'histoire naturelle des animaux*. Paris: Musée d'Histoire Naturelle.

Lang, B. F. *et al.* (2002). 'The closest unicellular relatives of animals,' *Current Biology*, 12(20), pp. 1773–1778. doi: 10.1016/S0960-9822(02)01187-9.

Lankester, E. R. (1883). 'A manual of the infusoria; including a description of all known flagellate, ciliate, and tentaculiferous protozoa (book review),' *Nature*, 27, pp. 601–603. doi: 10.1038/027601a0.

Leadbeater, B. S. C. (2015). *The Choanoflagellates*. Cambridge: Cambridge University Press. doi: 10.1017/cbo9781139051125.

Leadbeater, B. S. C. and McCready, S. M. M. (2002). 'The flagellates: Historical perspectives,' in Leadbeater, B. S. C. and Green, J. C. (eds) *The Flagellates. Unity, Diversity, and Evolution*. Milton Park: Taylor & Francis, p. 414.

Leander, B. S. (2008). 'A hierarchical view of convergent evolution in microbial eukaryotes,' *Journal of Eukaryotic Microbiology*, 55(2), pp. 59–68. doi: 10.1111/j.1550-7408.2008.00308.x.

van Leeuwenhoek, A. (1677). 'Observations, communicated to the publisher by Mr. Antony van Leeuwenhoeck, in a dutch letter of the 9th Octob. 1676. here English'd: concerning little animals by him observed in rain-well-sea- and snow water; as also in water wherein pepper had lain infus,' *Philosophical Transactions of the Royal Society of London*, 12(133), pp. 821–831. doi: 10.1098/rstl.1677.0003.

Levit, G. S. *et al.* (2020). 'The biogenetic law and the gastraea theory: From Ernst Haeckel's discoveries to contemporary views,' *preprints.* doi: 10.20944/PREPRINTS202006.0215. V1.

Levy, Y. Y. *et al.* (1996). 'Centrin is a conserved protein that forms diverse associations with centrioles and MTOCs in *Naegleria* and other organisms,' *Cell Motility and the Cytoskeleton,* 33(4), pp. 298–323. doi: 10.1002/(SICI)1097-0169(1996)33: 4<298::AID-CM6>3.0.CO;2-5.

Leys, S. P. and Eerkes-Medrano, D. (2005). 'Gastrulation in calcareous sponges: In search of Haeckel's Gastraea,' *Integrative and Comparative Biology,* 45(2), pp. 342–351. doi: 10.1093/icb/45.2.342.

Leys, S. P. and Hill, A. (2012). 'The physiology and molecular biology of sponge tissues,' in *Advances in Marine Biology.* doi: 10.1016/B978-0-12-394283-8.00001-1.

Lyons, K. M. (1973). 'Collar cells in planula and adult tentacle ectoderm of the solitary coral *Balanophyllia regia* (Anthozoa eupsammiidae),' *Zeitschrift für Zellforschung und Mikroskopische Anatomie,* 145(1), pp. 57–74. doi: 10.1007/BF00307189.

Mah, J. L., Christensen-Dalsgaard, K. K. and Leys, S. P. (2014). 'Choanoflagellate and choanocyte collar-flagellar systems and the assumption of homology,' *Evolution and Development,* 16(1), pp. 25–37. doi: 10.1111/ede.12060.

Margulis, L. (1981). *Symbiosis in Cell Evolution.* New York: W.H. Freeman & Co Ltd.

Marlétaz, F. (2019). 'Zoology: Worming into the origin of bilaterians,' *Current Biology,* 29(12), pp. R577–R579. doi: 10.1016/j.cub.2019.05.006.

Marshall, W. F. (2020). 'Pattern formation and complexity in single cells,' *Current Biology,* 30(10), pp. R544–R552. doi: 10.1016/j.cub.2020.04.011.

Mathijssen, A. J. T. M. *et al.* (2019). 'Collective intercellular communication through ultrafast hydrodynamic trigger waves,' *Nature,* 571(7766), pp. 560–564. doi: 10.1038/s41586-019-1387-9.

Medina, E. M. *et al.* (2020). 'Genetic transformation of *Spizellomyces punctatus,* a resource for studying chytrid biology and evolutionary cell biology,' *eLife,* 11(9), p. e52741. doi: 10.7554/eLife.52741.

Mendoza, L., Taylor, J. W. and Ajello, L. (2002). 'The class Mesomycetozoea: A heterogeneous group of microorganisms at the animal-fungal boundary,' *Annual Review of Microbiology.* doi: 10.1146/annurev.micro.56.012302.160950.

Metchnikoff, E. (1886). *Embryologische Studien an Medusen : Ein Beitrag zur Genealogie der Primitiv-organe.* Vienna: A. Hölder. doi: 10.5962/bhl.title.5982.

Mikhailov, K. V. *et al.* (2009). 'The origin of Metazoa: A transition from temporal to spatial cell differentiation,' *BioEssays,* 31(7), pp. 758–768. doi: 10.1002/bies.200800214.

Miller, S. A. and Harley, J. P. (1999). *Zoology.* 4th edn. Boston: McGraw-Hill Book Company.

Mitchell, L. A., Mulmor, J. A. and Dolphin, W. D. (1988). *Zoology.* Menlo Park, California: The Benjamin/Cummings Publishing Company.

Morange, M. (2016). *Une histoire de la biologie.* Paris: Editions du Seuil.

Müller, O. F. (1786). *Animalcula infusoria fluviatilia et marina Quae Detexit, Systematice Descripsit et Ad Vivum Delineari Curavit.* Hauniae: Mölleri.

Myers, P. Z. (2019). *Actually, I fail to see a single thing in this paper that would require any textbook rewriting at all, Pharyngula.* Available at: https://freethoughtblogs.com/pharyngula/2019/06/13/actually-i-fail-to-see-a-single-thing-in-this-paper-that-would-require-any-textbook-rewriting-at-all/.

Nakanishi, N., Sogabe, S. and Degnan, B. M. (2014). 'Evolutionary origin of gastrulation: Insights from sponge development,' *BMC Biology,* 12, p. 26. doi: 10.1186/1741-7007-12-26.

Nerrevang, A. and Wingstrand, K. G. (1970). 'On the occurrence and structure of choanocyte-like cells in some echinoderms,' *Acta Zoologica,* 51(3), pp. 249–270. doi: 10.1111/j.1463-6395.1970.tb00436.x.

Nichols, S. A., Dayel, M. J. and King, N. (2009). 'Genomic, phylogenetic, and cell biological insights into metazoan origins,' in *Animal Evolution: Genomes, Fossils, and Trees*. doi: 10.1093/acprof:oso/9780199549429.003.0003.

Nielsen, C. (2008). 'Six major steps in animal evolution: Are we derived sponge larvae?,' *Evolution and Development*, 10(2), pp. 241–257. doi: 10.1111/j.1525-142X .2008.00231.x.

Nielsen, C. (2012). *Animal Evolution. Interrelationships of the living phyla*. 3rd edn. Oxford University Press.

Nielsen, C. and Norrevang, A. (1985). 'The trochaea theory: an example of life cycle phylogeny,' in Conway Morris, S. et al. (eds) *The Origins and Relationships of Lower Invertebrates*. Oxford: Clarendon Press, pp. 28–41.

Parra-Acero, H. *et al.* (2018). 'Transfection of *Capsaspora owczarzaki*, a close unicellular relative of animals,' *Development*, 145(10), p. dev162107. doi: 10.1242/dev.162107.

Parra-Acero, H. *et al.* (2020). 'Integrin-mediated adhesion in the unicellular holozoan *Capsaspora owczarzaki*,' *Current Biology*, 30(21), pp. 4270–4275. doi: 10.1101/ 2020.02.27.967653.

Pedersen, K. J. (1964). 'The cellular organization of *Convoluta convoluta*, an acoel turbellarian: A cytological, histochemical and fine structural study,' *Zeitschrift für Zellforschung und Mikroskopische Anatomie*, 20(64), pp. 655–687. doi: 10.1007/BF01258542.

Pennisi, E. (1997). 'Haeckel's embryos: Fraud rediscovered,' *Science*, 277(5331), p. 1435. doi: 10.1126/science.277.5331.1435a.

Pflug, H. D. (1974). 'Vor- und Fruh-geschichte der Metazoen,' *N. Jahrb. Geol. Palaeont. Abh.*, 145, pp. 328–374.

Philippe, H. *et al.* (2019). 'Mitigating anticipated effects of systematic errors supports sister-group relationship between Xenacoelomorpha and Ambulacraria,' *Current Biology*. doi: 10.1016/j.cub.2019.04.009.

Reutterer, A. (1969). 'Zum problem der metazoenabstammung,' *Z. Zool. Syst. Evol.*, 7, pp. 30–53.

Reynolds, A. and Hülsmann, N. (2008). 'Ernst Haeckel's discovery of *Magosphaera planula*: A vestige of metazoan origins?,' *History and Philosophy of the Life Sciences*, 30(3–4), pp. 339–386.

Richards, R. J. (2008). *The Tragic Sense of Life: Ernst Haeckel and the Struggle over Evolutionary Thought*. University of Chicago Press.

Richards, R. J. (2009). 'Haeckel's embryos: Fraud not proven,' *Biology & Philosophy*, (24), pp. 147–154.

Richter, D. J. *et al.* (2018). 'Gene family innovation, conservation and loss on the animal stem lineage,' *eLife*, 7, p. e34226. doi: 10.7554/eLife.34226.

Rieger, R. M. (1976). 'Monociliated epidermal cells in Gastrotricha: Significance for concepts of early metazoan evolution,' *Journal of Zoological Systematics and Evolutionary Research*, 14(3), pp. 198–226. doi: 10.1111/j.1439-0469.1976.tb00937.x.

Rokas, A. *et al.* (2003). 'Conflicting phylogenetic signals at the base of the metazoan tree,' *Evolution and Development*, 5(4), pp. 346–359. doi: 10.1046/j.1525-142X.2003 .03042.x.

Ruiz-Trillo, I. *et al.* (1999). 'Acoel flatworms: Earliest extant bilaterian metazoans, not members of platyhelminthes,' *Science*. doi: 10.1126/science.283.5409.1919.

Ruiz-Trillo, I. *et al.* (2008). 'A phylogenomic investigation into the origin of Metazoa,' *Molecular Biology and Evolution*, 25(4), pp. 664–672. doi: 10.1093/molbev/ msn006.

Rundell, R. J. and Leander, B. S. (2010). 'Masters of miniaturization: Convergent evolution among interstitial eukaryotes,' *BioEssays*, 32(5), pp. 430–437. doi: 10.1002/ bies.200900116.

Sachwatkin, A. A. (1956). *Vergleichende Embryologie der niederen Wirbellosen: Ursprung und Gestaltungswege der individuellen Entwicklung der Vielzeller.* Berlin: VEB Deutscher Verlag der Wissenschaften.

Salvini-Plawen, L. V. (1978). 'On the origin and evolution of the lower Metazoa,' *Journal of Zoological Systematics and Evolutionary Research*, 16(1), pp. 40–88. doi: 10.1111/j.1439-0469.1978.tb00919.x.

Schardinger, F. (1899). 'Entwicklungskreis einer Amoeba lobosa (Gymnamoeba): *Amoeba Gruberi.* Sitzb Kaiserl,' *Akad. Wiss. Wien Abt.*, 1, pp. 713–734.

Schleiden, M. (1839). 'Beiträge zur Phytogenesis,' *Archiv für Anatomie, Physiologie und wissenschaftliche Medicin*, 1838, pp. 137–176.

Schloegel, J. J. and Schmidgen, H. (2002). 'General physiology, experimental psychology, and evolutionism. Unicellular organisms as objects of psychophysiological research, 1877-1918.,' *Isis*, 93, pp. 614–645. doi: 10.1086/375954.

Schwann, T. (1839). *Mikroskopische Untersuchungen über die Uebereinstimmung in der Struktur und dem Wachsthum der Thiere und Pflanzen.* Berlin: Sander.

Sebé-Pedrós, A. *et al.* (2013). 'Regulated aggregative multicellularity in a close unicellular relative of metazoa,' *eLife*, 24(2), p. e01287. doi: 10.7554/eLife.01287.

Sebé-Pedrós, A. *et al.* (2014). 'Evolution and classification of myosins, a paneukaryotic whole-genome approach,' *Genome Biology and Evolution*, 6(2), pp. 290–305. doi: 10.1093/gbe/evu013.

Sebé-Pedrós, A., Peña, M. I., *et al.* (2016). 'High-throughput proteomics reveals the unicellular roots of animal phosphosignaling and cell differentiation,' *Developmental Cell*, 39(2), pp. 186–197. doi: 10.1016/j.devcel.2016.09.019.

Sebé-Pedrós, A., Ballaré, C., *et al.* (2016). 'The dynamic regulatory genome of *Capsaspora* and the origin of animal multicellularity,' *Cell*, 165(5), pp. 1224–1237. doi: 10.1016/j.cell.2016.03.034.

Sebé-Pedrós, A., Degnan, B. M. and Ruiz-Trillo, I. (2017). 'The origin of Metazoa: A unicellular perspective,' *Nature Reviews Genetics*, 18(8), pp. 498–512. doi: 10.1038/nrg.2017.21.

Sedgwick, A. (1895). 'Memoirs: Further remarks on the cell-theory, with a reply to Mr. Bourne,' *Journal of Cell Science*, 38, pp. 331–337.

Sogabe, S. *et al.* (2019). 'Pluripotency and the origin of animal multicellularity,' *Nature*, 570(7662), pp. 519–522. doi: 10.1038/s41586-019-1290-4.

Steinböck, O. (1963). 'Origin and affinities of the lower Metazoa: the "acoeloid" ancestry of the Eumetazoa,' in Dougherty, E. C. et al. (eds) *The Lower Metazoa*. University of California Press, Berkeley, pp. 45–54.

Suga, H. *et al.* (2013). 'The *Capsaspora* genome reveals a complex unicellular prehistory of animals,' *Nature Communications*, 4, p. 2325. doi: 10.1038/ncomms3325.

Suga, H. and Ruiz-Trillo, I. (2013). 'Development of ichthyosporeans sheds light on the origin of metazoan multicellularity,' *Developmental Biology*, 377(1), pp. 284–292. doi: 10.1016/j.ydbio.2013.01.009.

Tikhonenkov, D. V., Hehenberger, E., *et al.* (2020). 'Insights into the origin of metazoan multicellularity from predatory unicellular relatives of animals,' *BMC Biology*, 18(1), p. 39. doi: 10.1186/s12915-020-0762-1.

Tikhonenkov, D. V., Mikhailov, K. V., *et al.* (2020). 'New lineage of microbial predators adds complexity to reconstructing the evolutionary origin of animals,' *Current Biology*. doi: 10.1016/j.cub.2020.08.061.

Wainright, P. O. *et al.* (1993). 'Monophyletic origins of the metazoa: An evolutionary link with fungi,' *Science*, 260(5106), pp. 340–342. doi: 10.1126/science.8469985.

Wan, K. and Jékely, G. (2020). 'Origins of eukaryotic excitability,' *arXiv*. Available at: https://arxiv.org/abs/2007.13388v1.

Willmer, E. N. (1971). *Cytology and Evolution*. 2nd edn. Cambridge, MA: Academic Press.

Willmer, P. (1990). *Invertebrate Relationships, Patterns in animal evolution.* Cambridge: Cambridge University Press. doi: 10.1017/cbo9780511623547.

Wilson, H. V. (1887). 'Metschnikoff on germ-layers,' *The American Naturalist*, 21(4), pp. 334–350. doi: 10.1086/274457.

Zakhvatkin, A. (1949). *The comparative embryology of the low invertebrates. Sources and method of the origin of metazoan development.* Moscow: Soviet Science.

14 Convergent Evolution of Complex Multicellularity in Fungi

László G. Nagy
Synthetic and Systems Biology Unit, Biological
Research Centre, Szeged, Hungary
Department of Plant Anatomy, Institute of Biology,
Eötvös Loránd University, Budapest, Hungary

CONTENTS

14.1 INTRODUCTION – WHAT IS COMPLEX MULTICELLULARITY?

The evolution of multicellularity has been one of the most significant transitions in the history of life. Multicellularity comes in many forms, ranging from simple aggregation of cells to the largest macroscopic organisms that dominate the visible world. Some of these are reminiscent of primordial states of multicellularity that hardly go beyond unicellularity in terms of function, whereas others feature sophisticated, highly integrated bodies and structures, structural and functional differentiation, or complex behaviors. Whereas simple cell aggregations, colonies, filaments or other simple solutions to multicellularity have emerged many times during evolution (Grosberg and Strathmann 2007), the highest levels of complexity

DOI: 10.1201/9780429351907-18

evolved on only a few occasions. To facilitate discussion, therefore, the continuum of complexity levels is often classified into simple and complex multicellularity. Knoll (2011) provided a detailed but simple, operational distinction between simple and complex multicellularity. Complex multicellular organisms grow three-dimensional structures or bodies, in which adhesion and a sophisticated division of labor between cells takes place and the shape and size of the organism is determined by a genetically encoded developmental program. A key trait of complex multicellularity is that not all cells are in direct contact with the environment, necessitating cellular or intercellular mechanisms for transporting oxygen and nutrients to inner cells of the organism. Complex organisms have evolved circulatory and respiratory structures to circumvent this obstacle, a trait that is not seen in simple cell aggregations or filaments.

This chapter focuses on convergent events in multicellular evolution in fungi. Fungi represent one of the most diverse multicellular lineages, with approximately 130,000 described and potentially several times more undescribed species (Willis 2018). The vast majority of fungi are multicellular through most of their life cycle, whereas a minority of species, in particular, early-diverging fungi, are unicellular, similar to related opisthokont protists. Multicellular fungi form tubular, elongate filaments, called hyphae, which grow apically and branch to form an intricate filamentous thallus. Yeasts, although mostly considered as unicellular, secondarily evolved reduced complexity and are capable of switching to diverse multicellular behaviors (biofilms, hyphae) in response to diverse environmental cues or at given time points in their life cycle. Transitions in levels of complexity show a great deal of convergence in fungi, which is not commonly seen in other frequently studied organisms. Here, we discuss evidence for potential convergent transitions in cellularity level, with a particular focus on complex multicellularity.

14.2 MULTICELLULARITY IN FUNGI

14.2.1 THE DRIVING FORCE FOR THE EVOLUTION OF MULTICELLULARITY IN FUNGI

Multicellularity in fungi differs from that of other lineages in many respects, which is probably a result of the combination of fungi's unique growth mode via hyphae and the unique selection pressures that drove their evolution. As sessile heterotrophic organisms, fungi forage for nutrients by apically growing hyphae, which extend at their apex in response to various chemical cues. It has been speculated that the rigid cell wall, which evolved in early, unicellular fungal ancestors, constrained the trajectory of multicellular evolution (Kiss et al. 2019; Nagy et al. 2020). The cell wall evolved in early fungal ancestors may have offered protection to the cells, but rendered vegetative, feeding cells of early fungi sessile (note that motile cell types in such taxa are restricted to sexual reproduction), which perhaps allowed for efficient feeding in aquatic habitats, where most unicellular fungi live. However, the transition to terrestrial habitats, where nutrient sources are patchy, probably required a strategy that allowed for efficient foraging across larger distances. Under such circumstances and with a rigid cell wall, an apically growing, filamentous thallus might have been the most, or only, optimal solution for fungi, which may explain the evolution of hyphae. Thus, the cell wall and patchy terrestrial nutrient sources together may have led to the evolution of hyphae in fungi (Kiss et al. 2019; Heaton et al. 2020). Of note, very similar tubular hyphae evolved convergently in the Oomycota (Stramenopila), a group that shares a heterotrophic, sessile nature, the decomposition of solid food sources and uptake of nutrients by osmotrophy as well as a cell wall with fungi (Diéguez-Uribeondo et al. 2004; Money et al. 2004).

Compared to the hyphae of extant, early-diverging fungi (e.g., Mucoromycota), early fungal hyphae were likely compartmentalized syncytia, in which the flow of nuclei and organelles were little regulated (Spatafora et al. 2016). Compartmentalization is achieved by cross-walls, called septa, that range from incomplete in early-diverging fungi to highly structured closures between neighboring cells (Jedd 2011; Riquelme et al. 2018). Septation regulates the flow of cytoplasm, organelles and nuclei. Hyphal growth and septation also have a bearing on how intra-organismal conflict is handled in fungi. Conflict naturally arises in the evolution of multicellularity, due to unicells having to align individual behaviors in the interest of a higher, organism-level fitness level (Michod and Roze 2001; Ratcliff et al. 2012; Rainey and Monte 2014). In contrast to most multicellular lineages, that evolved via colonial intermediates (e.g., animals; Sebé-Pedrós et al. 2017, Chapter 13), in fungi multicellular hyphae evolved by what Niklas referred to as a 'direct' route, from siphonous tubular to multicellular (Niklas 2014), which required that fungi handled conflict also in a different way (discussed in Nagy et al. 2020 and Kiss et al. 2019).

14.2.2   SIMPLE VERSUS COMPLEX MULTICELLULARITY IN FUNGI

In terms of complexity, hyphal growth resembles other lineages in which the predominant multicellular forms are filamentous (e.g., algae, cyanobacteria) or colonial (e.g., choanoflagellates). This is often referred to as simple multicellularity, which is defined as the level of organization in which all cells are in contact with the environment and cell-to-cell differentiation is limited (Knoll 2011; Nagy et al. 2018; Kiss et al. 2019). Fungi also evolved complex multicellular structures, however, these evolved significantly later than hyphal multicellularity (Kiss et al. 2019). Complex multicellularity in fungi is most frequently discussed in the context of sexual fruiting bodies, structures that aid the protection, development, and dispersal of sexual spores (Figure 14.1). Fruiting bodies are three-dimensional structures whose development follows a genetically encoded program and results in species-specific shapes and sizes. Fruiting bodies come in several shapes and sizes, from simple, crust-like morphologies to the most complex mushroom shapes known in fungi (Figure 14.1d,f). The evolution of fruiting body morphologies follows some well-known trends. For example, Basidiomycota fruiting bodies evolved from crust-like forms towards the typical 'toadstool' morphology, i.e., those with cap, stipe and gills (Varga et al. 2019). In the Ascomycota, fruiting bodies evolved from open types, which bear the spore-producing surface on the upper side open to the environment (called apothecium-type), towards closed forms in which spore-producing cells (asci) develop internally and shoot spores into the air through various pores/channels (called perithecium-type; Liu and Hall 2004).

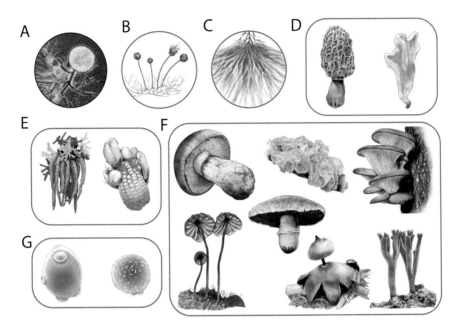

FIGURE 14.1 Example representatives of complexity levels in fungi. Fungi produce a plethora of different structures that are plesiomorphically unicellular (a: *Spizellomyces punctatus*), simple multicellular (b: hyphal colony with conidiophores; c: hyphal network), complex multicellular (d: a morel fungus, *Morchella* and *Neolecta irregularis*; e: fruiting bodies [*Gymnosporangium* sp.] and galls [*Ustilago maydis*] of complex multicellular rust and smut fungi, respectively; f: representatives of mushroom-forming fungi, Agaricomycotina left to right, top to bottom: *Cortinarius sp, Tremella mesenterica, Pleurotus ostreatus, Agaricus bisporus, Mycena sp, Geastrum sp, Clavulinopsis helvola*) and secondarily unicellular yeasts (g: *Saccharomyces cerevisiae* and *Cryptococcus neoformans*).

It should be noted, however, that a large number of other complex multicellular structures also evolved in fungi, these include sclerotia, ectomycorrhizae or asexual fruiting bodies, to name a few (Nagy et al. 2018). It is important to note that growth remains hyphal even in fruiting bodies, but cell shape is modified, in some cases to an extent that resembles isodiametric (polyhedral) cell morphologies (Lord and Read 2011).

14.2.3   BLURRED LINES BETWEEN SIMPLE AND COMPLEX MULTICELLULARITY IN FUNGI

Whereas the distinction between simple and complex multicellularity is relatively straightforward in most lineages (Knoll 2011), the diversity of fungal morphologies often blurs lines between these categories. Fungi evolved a range of forms that lie intermediate between simple and complex multicellularity and which has also led to confusion about the use of these terms in the mycological literature (see Nagy et al 2020 for more discussion). For example, asexual spores are produced on specialized structures, called conidiophores (Figure 14.1b), that grow out of vegetative mycelia and are made up of individual cells that are in contact with the environment but follow a genetically determined developmental program and grow determinately. Thus, while conidiophores are not real three-dimensional structures, they do show several attributes of complex multicellularity as defined by Knoll (2011).

Similar, but less well-known cases can be encountered in sexual reproductive structures also. In both the Asco- and Basidiomycota, early-diverging species produce sporogenous cells called asci and basidia, respectively, on the bare surface of the substrate, without an enclosing fleshy fruiting body. Asci and basidia are sexual sporangia and resemble conidiophores, in that their shapes and sizes are species-specific and genetically encoded and, in the simplest cases, they grow naked on the substrate. It is conceivable that such 'naked' asci and basidia represent the ancestral condition in the Asco- and Basidiomycota, respectively (Hibbett 2004; Varga et al. 2019), although reductive evolution has not been completely ruled out in many of the cases (Wynns 2015). Nevertheless, a gradient in complexity levels can be found in early-diverging Asco- and Basidiomycota, with a gradual thickening of hyphal layers that eventually enclose asci and basidia into a thick, protective tissue that supports spore production. These evolutionary processes might explain some of the pervasiveness of convergence fungal fruiting bodies (see below).

14.2.4 REDUCTION OF MULTICELLULARITY IN FUNGI

Along with the emergence of multicellularity, fungi display repeated reductions in complexity level (Figure 14.2). The best-known examples are yeasts, which are secondarily simplified organisms that spend much of their life cycle in a unicellular form. This is a remarkable case of reduced complexity, given that most lineages seem to evolve towards increased complexity (O'Malley et al. 2016) or that most research attention is directed towards that. Losses of multicellularity are rare across the tree of life (but seen in cancerous cells; Chen et al. 2015), especially in lineages, like fungi, which transitioned to stable multicellular organization hundreds of millions of years ago (losses were less surprising in lineages that show transient or primordial forms of multicellularity). The loss of multicellularity also may seem surprising given that it is considered to offer selective advantages to the organisms in diverse environments, for example, through increased cell size which helps to avoid predation (Rokas 2008). How can we then explain the unicellularity of yeasts?

The term yeasts refers to a polyphyletic assemblage of fungi that are observed primarily as unicells (Nagy et al. 2017) and that evolved convergently from more complex, hyphal ancestors. The best-known yeasts are those found in the Saccharomycotina subphylum, such as baker's yeast *Saccharomyces cerevisiae* or *Candida* spp, which are causative agents of often life-threatening mycoses. Another widely known yeast is the fission yeast *Schizosaccharomyces pombe* (Taphrinomycotina subphylum), which is best known for its role as a model organism. However, there are several other yeast-like lineages in fungi as well (Nagy et al. 2017). The term yeast-like fungus is used in this review for species that spend the majority of their life cycle as walled unicells that feed by osmotrophy and divide by

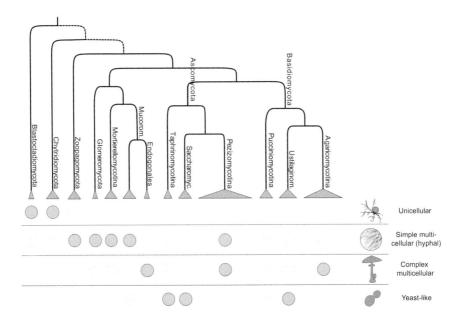

FIGURE 14.2 The diversity and patchiness of complexity levels in fungi. A schematic representation of the fungal tree is shown with main complex multicellular and yeast-like clades highlighted. The phylogenetic distribution of unicellular, simple and complex multicellular as well as yeast-like fungi is shown on the bottom panel.

budding or fission. The Basidiomycota comprises three yeast-like lineages, the class Tremellomycetes, which includes the opportunistic pathogen *Cryptococcus neoformans*, the Ustilaginomycotina subphylum, which includes plant pathogenic smuts (e.g., *Ustilago maydis*) and the Pucciniomycotina, which comprises several groups rust species and red yeasts (e.g., *Rhodotorula*). Fungi further include several dimorphic fungi, which, by definition are able to switch between unicellular and multicellular morphologies (Boyce and Andrianopoulos 2015).

To understand how surprising the reduction of multicellularity in yeasts is, it is important to reiterate that for fungi, the selective advantage of multicellularity might have been different, perhaps to move from liquid niches to terrestrial habitats. In view of this, if we examine the niche that yeasts fill, we find that they share properties with aquatic niches of plesiomorphically unicellular fungi. Yeast-like fungi are often found in liquid niches, such as, for example, nectars, insect guts, or various external or internal surfaces of mammals, including humans. Simple sugars are readily available in these niches, which makes active foraging for spatially heterogeneous nutrients dispensable. It is, thus, not surprising that yeasts produce unicellular morphologies that divide mitotically either by budding or fission. A widely held view assumes that, collaterally, the ability to form multicellular hyphae has also been lost in yeast species. In support of this hypothesis, yeasts experienced rampant gene loss during their evolution (Nagy et al. 2014) and recent analyses of yeast genomes suggested a complex ancestor of the budding yeast lineage (Saccharomycotina; Shen et al. 2018). To test this hypothesis, we tested whether the massive gene loss event that five independent yeast-like lineages experienced in their evolution affected hypha morphogenesis genes more, less or to a similar extent than it did metabolic and other gene functions (Kiss et al. 2019). Surprisingly, genes related to hyphal growth were significantly depleted among lost genes, suggesting that such genes are preferentially retained by yeasts despite having lost ~3–5,000 genes since their last common ancestors. This argues against the presumed losses, or reductions in multicellularity and, rather, reinforces the view that multicellularity, once acquired, is maintained through evolution.

It is important to note that yeasts display a number of multicellular traits, such as communication between cells (e.g., via quorum sensing), synchronization of cellular behaviors or cell differentiation. Most yeast species are capable of producing (reduced) hyphae and they can switch between unicellular and hyphal growth in response to environmental stimuli. For example, opportunistic pathogen *Candida albicans*, which is a commensal resident of the mucosal and genital surfaces in healthy human individuals, can form true hyphae and invade tissues upon changing environmental or host conditions. This, again, underscores the finding that yeasts, despite their most commonly seen morphology, retained multicellular growth form in their evolution.

14.2.5 COMPLEX MULTICELLULAR FUNGI ARE PHYLOGENETICALLY SCATTERED

Complex multicellularity represents the highest level of morphological organization that evolved across the tree of life. It refers to organisms or structures in which cells are tightly packed, show extensive morphological and functional differentiation and are organized into higher-level functional units, such as tissues or organs (Knoll 2011). In contrast to simple multicellularity, which evolved in >25 lineages (Grosberg and Strathmann 2007), complex multicellularity is limited to five major groups: metazoans, green plants, brown and red algae as well as fungi. Species in these groups dominate the visible world and have evolved to fill diverse niches on Earth. Whereas research on multicellular behaviors of metazoans and green plants has deep roots and goes back to over a century, our knowledge on complex multicellularity and how it evolved in fungi is more limited. For example, while cell type diversity is well-understood in several animal and plant species, and current research aims to refine cell type definitions and classifications (e.g., by single-cell transcriptomics), our estimates of cell-type diversity in complex multicellular fungi are based on the counting of morphologically distinct cellular morphologies. Even using such simple approaches, up to 28–30 morphologically distinct cell types were recognized in sexual fruiting bodies of fungi (Bistis et al. 2003; Lord and Read 2011; Kües and Navarro-González 2015), but this number is expected to rise significantly as single-cell approaches get deployed for complex multicellularity fungi.

Complex multicellularity is rare across the tree of life; this accords well with the hypothesis that lineages need to overcome a significant number of genetic obstacles to evolve it (Knoll 2011). Complex multicellular fungi are not monophyletic, rather, they display a patchy phylogenetic distribution. The best-known complex multicellularity fungi are found in the Agarico- and Pezizomycetes, which belong to the Basidio- and Ascomycota respectively and are separated by at least 600 million years from each other (Kohler et al. 2015). These include the best-known mush-room-forming fungi, such as truffles and morels in the Pezizomycotina and aga-rics, puffballs, bracket fungi, among others, in the Agaricomycotina (Figure 14.1f). However, a number of other complex multicellular lineages are also found in fungi, some of which comprise a single genus with less than a handful of species (*Neolecta*, Taphrinomycotina or *Modicella*, Mortierellomycotina, see Figure 14.1d), whereas others are highly diverse. Estimates for the species diversity and age of these lineages range from 2 to over 32,000 species and from very recent origins to clades that are >500 million years old (Nagy et al. 2018). Altogether, we identified 8–11 clades of complex multicellular clades across fungi, although their number may change slightly in the future as some phylogenetic uncertainties in the fungal tree of life get resolved. These clades are separated from each other by simple multicellular and facultatively unicellular, yeast-like lineages. This scattered nature of complex multicellular fungal lineages suggests an intricate evolutionary history either with multiple origins or with a single origin and several losses.

14.2.6 FUNGI OFFER AN UNPARALLELED MODEL SYSTEM FOR STUDYING COMPLEX MULTICELLULARITY

Given the disparate occurrence of complex multicellularity in fungi, we were interested in what genetic mechanisms might underlie its evolution. To this end, we used developmental transcriptomes – gene expression data that can reveal multicellularity-related genes in fungi. Fungi offer a great model system to study transitions in multicellularity because they undergo a transition from simple to complex multicellularity as part of their life cycle. The vegetative mycelium, which is an underground fungal network of filaments that is adapted to explore the substrate and assimilate nutrients, is simple multicellular, whereas sexual fruiting bodies, which enclose developing spores into a resistant, reproductive organ, are complex multicellular (Figure 14.1a–f). Because the transition from simple to complex multicellularity happens within the same, extant fungal species, it allows us to assay what genes become activated upon the onset of the complex multicellular stage of the life cycle, offering a window into the genetic bases of complex multicellularity.

Fruiting body formation can be induced under laboratory conditions by changes in environmental variables (nutrient availability, light). Accompanying changes in gene expression can be assayed in real-time for example, by RNA-Seq, providing an unparalleled model system not available for studying complex multicellularity in other lineages. This is not available in animal or plant model systems that evolved complex multicellularity once and exist in that state throughout their life cycle (except for a unicellular bottleneck in gametes) and therefore, research on the evolution of their multicellularity has to rely on looking into the past, using comparative genomic techniques (Sebé-Pedrós et al. 2017). The existence of facultative multicellular developmental stages in fungi is surprisingly similar to that seen in aggregative multicellular amoebozoans, such as the slime mold *Dictyostelium*, which produces very simple fruiting bodies and has been subject to multicellularity research for decades (Eichinger et al. 2005; Du et al. 2015; Glöckner et al. 2016). Notably, both fungi and slime molds respond to starvation and use conserved cAMP signal transduction pathways during this process. However, this probably reflects conserved genetic circuits underlying sexual reproduction – which is induced when nutrients become limiting – rather than homologies among fungal and slime mold fruiting bodies, given the clearly independent origins of the latter.

Researchers have taken advantage of the ability of fungi to produce fruiting bodies and used high-throughput -omics techniques in both the Asco- and Basidiomycota to unravel the genes that get activated or downregulated upon the start of complex multicellular development. For this, RNA-Seq offers a suitable approach (Nowrousian 2018), which, combined with the ever increasing number of sequenced fungal genomes (Grigoriev et al. 2014) provides a window into how complex multicellular development takes place in fungi. RNA-Seq data on sexual fruiting bodies started to accumulate from the two largest complex multicellular clades. In the Ascomycota, some classic model systems of sexual fruiting body development, for which RNA-Seq data have been published, include *Neurospora crassa* (Lehr et al. 2014; Wang et al. 2014; Trail et al. 2017), *Aspergillus nidulans* (Han 2009), *Sordaria macrospora* (Nowrousian et al. 2010; Teichert et al. 2020) along with more recently emerging models such as *Pyronema confluens* (Traeger et al. 2013), *Fusarium* spp (Kim et al. 2019) or *Botrytis cinerea* (Rodenburg et al. 2018). In the Agaricomycotina, the most researched model species are *Coprinopsis cinerea* (Kues 2000; Stajich et al. 2010; Krizsan et al. 2019) and *Schizophyllum commune* (Ohm et al. 2010), with fruiting body transcriptomes available for both (Ohm et al. 2011; Muraguchi et al. 2015; Almási et al. 2019; Krizsan et al. 2019). Fruiting body development of a number of other Agaricomycotina species has also been examined, including that of *Lentinula edodes* (Sakamoto et al. 2017), *Flammulina velutipes* (Park et al. 2014), among others, to a large extent motivated by interest in informing rational strain development for commercially produced mushroom species.

14.3   THE GENETIC BASES OF TRANSITIONS TO COMPLEX MULTICELLULARITY

14.3.1   TRANSITIONS TO COMPLEX MULTICELLULARITY MAY NOT BE SO HARD FOR FUNGI

It has been hypothesized that complex multicellularity is rare across the tree of life because organisms have to overcome a significant number of genetic/physiological obstacles to evolve it (Knoll 2011). This is in line with complex multicellularity being limited to animals, plants, green and brown algae, as well as fungi. However, it appears that in fungi, a large number of independent complex multicellular clades exist, raising questions on convergent evolution and on to how complicated it might be for fungi to evolve complex multicellularity.

A particularly relevant group of organisms for examining this question is the genus *Neolecta*, which contains species that form yellow, tongue-like fruiting bodies in temperate forests. Phylogenetically, this genus is nested in the Taphrinomycotina, which primarily contains simplified, yeast-like or filamentous fungi, such as the fission yeast *Schizosaccharomyces pombe*. The genome of *Neolecta irregularis* has recently been published (Nguyen et al. 2017). This has revealed that its genome contains ~5,500 genes, which is an unusually low gene count for a complex multicellular fungus (Nagy 2017). Most filamentous and complex multicellular fungal genomes encode on average 12,000–15,000 protein coding genes, whereas yeasts and yeast-like fungi have small, highly streamlined genomes, containing 5,000 to 7,000 protein coding genes. Further evidence for the similarity between *Neolecta* and yeast genomes is that yeast genomes and that of *Neolecta* evolved predominantly by gene loss and are depleted in introns (Nguyen et al. 2017), whereas genomes of complex multicellular fungi did not experience massive gene loss events, are rich in intron/exon boundaries and are known to be alternatively spliced (Marshall et al. 2013; Gehrmann et al. 2016).

The finding that *Neolecta* is able to produce complex multicellular fruiting bodies suggested, first, that the gene count difference between yeast-like and complex multicellular fungi may not be relevant to differences in complexity level, (ii) that complex multicellularity can exist in organisms with highly streamlined, small genomes, and (iii) that, in terms of protein coding capacity, it does not require a lot more to build complex multicellular fruiting bodies than to build hyphae or yeast cells. A broader look across complex multicellular fungi revealed that there are several complex multicellular clades with reduced genomes, particularly in the Basidiomycota (Nagy 2017). Several species in the reduced-genome Puccinio- and Ustilaginomycotina, as well as the Tremellomycetes produce complex multicellular fruiting bodies. For example, the golden jelly fungus (*Tremella mesenterica*), which forms gelatinous, yellow fruiting bodies on dead branches of trees, possesses ca. 8,000 genes and, like *Neolecta*, is a close relative of morphologically secondarily simplified yeast-like fungi, such as *Cryptococcus neoformans*.

The streamlined genomes and close phylogenetic affinity of certain complex multicellular fungi to yeast-like species, combined with the phylogenetic patchiness of complex multicellularity in fungi suggests that complex multicellularity might be easy to evolve for fungi. Given the multiple origins model of fungal complex multicellularity, these fruiting body-forming fungi must have evolved in the middle of otherwise simplified fungi that have already undergone severe genome contractions (Nagy et al. 2014). It also follows from these that the genome contractions that yeasts have experienced must have left much of the genetic toolkit required for complex multicellularity largely intact, otherwise, these complex multicellular fungi could not have evolved from within yeast-like clades. It may seem imperative to revisit, in light of these observations, the single origin model, which could more logically explain how complex multicellular fungi can be nested in secondarily simplified clades. In that scenario, complex multicellularity would be the ancestral condition, which was retained by some species (Figure 14.3), whereas others have lost complex multicellularity (and even hyphae) and reversed to a facultatively unicellular lifestyle. However, species richness data and topologies of phylogenetic trees speak against the single origin model in both the Tremellomycetes and the Taphrinomycotina, where *Tremella* and *Neolecta* belong, respectively. Both clades are composed primarily of simple organisms, with an estimated diversity of 400 and 200 species in the Tremellomycetes and the Taphrinomycotina, of which only <100 and 3 are complex multicellular, respectively. Thus, considering trait gain/loss at higher resolution, it becomes clear that a complex multicellular ancestry in these

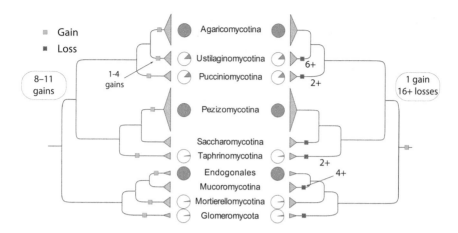

FIGURE 14.3 Contrasting scenarios of the multiple and single origin model describing the evolution of complex multicellularity in fungi. Acquisitions and losses of complex multicellularity under two contrasting models are shown by green and blue squares, respectively. The multiple (convergent) origins model (left) requires 8–11 steps to explain the phylogenetic distribution of complex multicellular fungi, whereas the single origin model (right) implies a single origin and at least 16 losses. Pie charts mark clades containing complex multicellular species; solid section denotes the approximate proportion of complex multicellular species.

clades would entail an excessively high number of loss events, which makes the single origin model even less likely.

14.3.2 REPEATED CO-OPTION OF GENES IN COMPLEX MULTICELLULAR FUNGI

As explained above, developmental transcriptomes offer a window into the genetic bases of the formation of complex multicellular fruiting bodies in fungi. These can be used to address questions about the origins of fruiting body formation as well. We used transcriptomes from five Asco- and four Basidiomycota species to test plausible models of the evolution of complex multicellularity in fungi (Merényi et al. 2020). More specifically, in the Ascomycota, we focused on the Pezizomycotina, whereas in the Basidiomycota we sampled species from the Agaricomycotina. These two sub-phyla represent the largest complex multicellular clades in fungi and include several species that can complete their life cycle under laboratory conditions.

Asco- and Basidiomycota fruiting bodies show no evidence for homology, which could be because of independent origins, or because, after a single origin, they have diverged so much that traces of homology are not detectable anymore. In line with the lack of discernible homology, ancestral character state reconstructions unequivocally supported the independent origins model, with fruiting bodies having originated independently in the most recent common ancestor of the Agaricomycotina and in that of the Pezizomycotina. However, phylogenetic methods have little power for distinguishing between potential models of trait evolution. While phylogenetic methods build on homology hypotheses at the phenotypic level, looking for homologies at the level of the genetic toolkit behind complex multicellularity might provide

a more resolved view on how complex multicellularity evolved and help the evaluation of alternative models of trait evolution.

We analyzed the extent to which developmentally regulated genes were shared
among Asco- and Basidiomycota fruiting bodies. After reanalyzing published developmental transcriptome data, we identified developmentally regulated genes, i.e.,
genes that show considerable (in this case ≥4 fold) expression dynamics throughout
fruiting body development. Given the convergent nature of fruiting bodies in these
clades, we expected little overlap among developmentally regulated genes. Yet, we
identified 1,026 gene families that were developmentally regulated in ≥75% of the
species in either the Asco-, the Basidiomycota or both. These break down to 314,
439, and 273 families that were developmentally regulated in ≥7 of 9 species overall,
≥3 of 4 Basidiomycota, and ≥4 of 5 Ascomycota species, respectively (Figure 14.4).

To understand the evolution of complex multicellularity at a higher resolution, we
reconstructed the evolution of developmental gene families along the phylogeny (see
Figure 14.4a). Of the 1,026 conserved developmental families, 297 and 169 families
that are taxonomically restricted to the Agarico- and Pezizomycotina, respectively.
On the other hand, 560 families predate the origin of complex multicellularity, indicating that they were likely co-opted for multicellularity-related functions (Figure
14.4b). These ancient families were significantly more often developmentally regulated in both clades than expected by chance (314 out of 560 ancient families,
$P < 10^{-4}$, permutation-test), indicating that parallel co-option of these families has
been rampant. On the contrary, the frequency of clade-specific co-option was low,
with only about 7.5% and 12% in the Agarico- and Pezizomycotina (42 and 67 families). The combination of limited clade-specific co-option with rampant parallel co-
option suggests that genes with suitable biochemical properties were predominantly
recruited into the genetic toolkit underlying fungal complex multicellularity. It is
also consistent with the hypothesis that genes suitable for any given trait are rare and
thus they mostly end up being recruited for convergent traits (Christin et al. 2010).

The observed evolutionary patterns of developmental gene families can be
explained by two hypotheses. First, the simplest model of convergent evolution can
explain gene families with clade-specific taxonomic distribution or clade-specific
co-option. Under this scenario, complex multicellularity emerged independently
in the Agarico- and Pezizomycotina and different genes evolved *de novo*, or were
co-opted into the genetic toolkit. Similarly, shared developmentally regulated gene
families could have been parallelly co-opted for complex multicellularity in the
Agarico- and Pezizomycotina. However, this assumes a large number of parallel
co-option events. As a more parsimonious explanation, these families could also
encode functions that the Dikarya ancestor possessed and that were integrated into
complex multicellular fruiting bodies of the Agarico- and Pezizomycotina as single
units or developmental modules. This scenario does not assume independent co-
option events for each of those genes and is thus more parsimonious, from a purely
phylogenetic perspective, than a simple model that assumes convergence at all levels.
Processes related to sexual spore formation represent probable candidates for such
functions: they are conserved in sporulating tissues across fungi, and fruiting bodies
evolved to provide a protected environment to sporulating tissues. As a consequence,
genes associated with spore formation show expression peaks within fruiting bodies

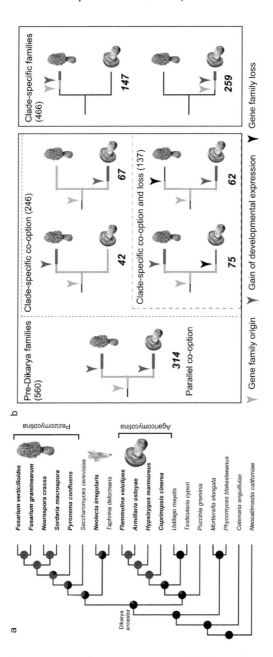

FIGURE 14.4 Phylogenetic relationships of complex multicellularity (a) and the fate of conserved developmentally regulated gene families (b). (a) two independent clades of complex multicellular species are marked, and typical fruiting body morphologies are shown. Pie charts at nodes indicate the ancestral states inferred by Maximum Likelihood (proportional likelihoods of complex multicellular [red] and non-complex multicellular [black]). Complex multicellular species are shown by boldface font. (b) developmentally regulated gene families grouped by evolutionary conservation and history. Adapted from Merenyi et al. (2020).

and will show up among the developmentally regulated genes we identified. There are other plesiomorphic functions that may have been present in the last common ancestor of the Agarico- and Pezizomycotina, such as adhesion or defense-related gene families that may have proven useful for complex multicellularity, explaining their parallel co-option.

14.3.3 Plausible Genetic Models for Phylogenetically Patchy Traits

Complex multicellularity in fungi is a typical phylogenetically patchy trait: it occurs in disparate clades across the tree, yet certain components of its genetic bases show homologies across independent occurrences. Such traits are not uncommon in biology. For example, the formation of nodules that harbor nitrogen-fixing bacteria occurs in several clades of leguminous plants and its phylogenetic distribution suggests convergent origins. Yet, nodule development shares a surprisingly high fraction of its genetic components across clades (van Velzen et al. 2018). Griesmann et al addressed the origins of root nodule symbiosis with nitrogen-fixing bacteria using a broad sampling of plant genomes (Griesmann et al. 2018) and found that, although the NODULE INCEPTION (NIN) gene, which is a key regulator of nodule development, was shared across a wide phylogenetic spectrum of plants and was lost repeatedly in multiple non-nodulating leguminous plants, its conservation can't explain the convergent origins of nodule formation in plants.

Traits with patchy phylogenetic distributions pose a challenge for evolutionary biologists to explain. They often display homologies in terms of genetic composition, in conflict with the classic model of convergent evolution, which implies that traits arose independently via different genetic changes to non-homologous genetic precursors (Figure 14.5). On the other hand, their patchiness across the phylogenetic tree speaks against homology because explaining their contemporary distribution on the tree would require an excessive number of gene losses to be assumed (cf. Figure 14.3).

Intermediate models that incorporate elements of both homology and independent origins have been proposed recently (Figure 14.5/b–c). The latent homology model posits that precursor traits underlie several phylogenetically patchy or convergent characters (Nagy et al. 2014; Nagy 2018). Precursors are traits that have been in place in the common ancestors of species that show the trait and were convergently modified, resulting in the repeated emergence of the patchy trait. Such conserved precursor traits can underlie the convergent emergence of new traits if they can easily be co-opted for new functions by simple genetic changes. Such changes could include but are not limited to, tweaks to the regulation of the precursor, which results in its expression in a new context. For example, a few mutations affecting the regulation of a master regulator that orchestrates the expression of an entire developmental module should be sufficient to cause the expression of this module in a new context. Precursor traits could be developmental modules, gene regulatory circuits or other modular gene assemblages, among others, that are mutationally accessible for being deployed for a new function. Thus, latent homologies can reduce the mutational target size for evolution, by providing ready access for evolution to new cellular outputs via a few mutations.

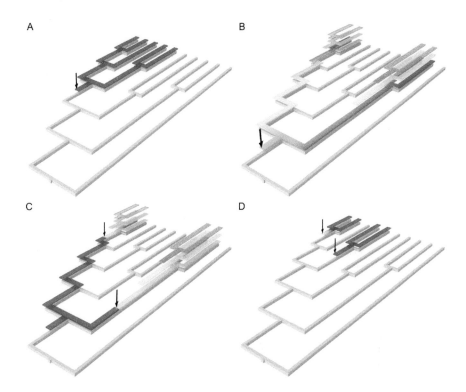

FIGURE 14.5 Models of trait evolution. The simplest models include a single origin followed by divergence (a). Traits share deep homology if divergence blurs similarity and common origins completely (b). Phylogenetically patchy traits can evolve convergently, building on non-homologous genetic bases (d) or via latent homologies (c) where similar function arises convergently by independent co-option of the homologous genes or genetic modules for the same, new function. Black arrows mark the origin of similar traits.

This increases the likelihood of convergence and makes convergent scenarios more parsimonious than if we had to assume convergent or parallel mutations to each of the member genes in that module. Examples where a genetic module is only a few mutations away from being expressed in a new context are abundant in the literature, and there is a documented history of such traits in developmental biology (Carroll 2008); however, how frequently such cases underlie convergent traits has not been explored yet. This way, latent homologies can increase the likelihood of convergent evolution without assuming highly unparsimonious evolutionary scenarios.

The latent homology model was first proposed to explain the convergent evolution of yeasts and yeast-like fungi (Nagy et al. 2014). As explained above, yeast-like fungi are secondarily and facultatively unicellular organisms that spend much of their life cycle in a unicellular form. They emerged independently in fungi several times and include some of the most important human pathogens. They evolved presumably as adaptations to liquid niches, such as nectars, insect intestinal tracts, or other habitats rich in simple sugars. Based on the phylogenetically widespread occurrence of yeast-like fungi, we suggested that some part of the genetic toolkit for yeast-like growth, the putative precursor trait, is ancient and conserved in fungi. In support of this hypothesis, we reconstructed convergent changes in transcription factor families along stem branches leading to yeast-like clades (Nagy et al 2014). Although the identity of the precursor trait remains unknown, the latent homology model can explain how the yeast morphology may have evolved repeatedly at various phylogenetic depths and clades and why the morphology of yeast-like fungi shows surprisingly high similarity across clades.

In light of these considerations, it is imperative to examine if precursor traits might underlie the convergent evolution of complex multicellularity as well in fungi. The presence of such precursor traits in non-complex multicellular ancestors would have predisposed lineages for evolving complex multicellularity by providing stepping stones for evolution, leading to a higher likelihood of phenotypic convergence and explaining why complex multicellularity is so common across fungi. To understand if any functions and which might have served as precursors to complex multicellularity, we looked at the genetic makeup of the last common ancestor of the Agarico- and Pezizomycotina. Although this ancestral species most likely did not have fruiting bodies, we reasoned that its ancestral genome composition could shed light on whether predisposition by precursor traits provides a plausible hypothesis for the evolution of complex multicellularity in fungi. We reconstructed 989 genes in the 314 shared developmental families (Figure 14.4) of the Dikarya ancestor, which showed an enrichment of genes related to several multicellular and developmental functions of fungi (see Merényi et al. 2020), reminiscent of general functions required for fungal development. These include gene regulatory circuits related to sexual reproduction, mating partner recognition, light, nutrient and starvation sensing, fungal cell wall synthesis and modification, cell-to-cell signaling, and morphogenesis. It also includes genetic circuits related to adhesion and cell-differentiation-related genes that are used by non-complex multicellular fungi for pathogenicity or asexual reproduction, although the hypothesis that these were precursors to adhesion and cell differentiation in complex multicellular fruiting bodies remains challenging to test experimentally.

Taken together, the patchy phylogenetic distribution of complex multicellularity makes the fungal kingdom an oddball in the history of the evolution of multicellularity. This patchiness manifests in morphologically diverse fruiting bodies and other structures and led to the formulation of a range of genetic and evolutionary hypotheses. Addressing these from a mechanistic standpoint will require further research and the integration of suitable model systems with genetic, evo-devo and -omics techniques, all of which, fortunately, are at the disposal of fungal evolutionary biologists and just need to be combined in effective ways to understand one (or many?) of the most spectacular transitions in evolution.

ACKNOWLEDGMENTS

Sándor Kocsube is thanked for his help preparing figures. The author acknowledged support from the European Research Council (Grant # 758161) and the Hungarian Academy of Sciences' Momentum Program (contract No. LP2019-13/2019).

REFERENCES

Almási É., Sahu N., Krizsán K., Bálint B., Kovács G.M., Kiss B., Cseklye J., Drula E., Henrissat B., Nagy I., Chovatia M., Adam C., LaButti K., Lipzen A., Riley R., Grigoriev I. V., Nagy L.G. 2019. Comparative genomics reveals unique wood-decay strategies and fruiting body development in the Schizophyllaceae. New Phytol. 224:902–915.

Bistis G.N., Perkins D.D., Read N.D. 2003. Different cell types in *Neurospora crassa*. Fungal Genet. Rep. 50: 17–19.

Boyce K.J., Andrianopoulos A. 2015. Fungal dimorphism: the switch from hyphae to yeast is a specialized morphogenetic adaptation allowing colonization of a host. FEMS Microbiol. Rev. 39:797–811.

Carroll S.B. 2008. Evo-devo and an expanding evolutionary synthesis: a genetic theory of morphological evolution. Cell. 134:25–36.

Chen H., Lin F., Xing K., He X. 2015. The reverse evolution from multicellularity to unicellularity during carcinogenesis. Nat. Commun 6:6367.

Christin P.A., Weinreich D.M., Besnard G. 2010. Causes and evolutionary significance of genetic convergence. Trends Genet. 26:400–405.

Diéguez-Uribeondo J., Gierz G., Bartnicki-García S. 2004. Image analysis of hyphal morphogenesis in Saprolegniaceae (Oomycetes). Fungal Genet. Biol. 41:293–307.

Du Q., Kawabe Y., Schilde C., Chen Z.H., Schaap P. 2015. The evolution of aggregative multicellularity and cell-cell communication in the Dictyostelia. J. Mol. Biol. 427:3722–3733.

Eichinger I., Pachebat J.A., Glöckner G., Rajandream M.A., Sucgang R., Berriman M., Song J., Olsen R., Szafranski K., Xu Q., Tunggal B., Kummerfeld S., Madera M., Konfortov B.A., Rivero F., Bankier A.T., Lehmann R., Hamlin N., Davies R., Gaudet P., Fey P., Pilcher K., Chen G., Saunders D., Sodergren E., Davis P., Kerhornou A., Nie X., Hall N., Anjard C., Hemphill L., Bason N., Farbrother P., Desany B., Just E., Morio T., Rost R., Churcher C., Cooper J., Haydock S., Van Driessche N., Cronin A., Goodhead I., Muzny D., Mourier T., Pain A., Lu M., Harper D., Lindsay R., Hauser H., James K., Quiles M., Madan Babu M., Saito T., Buchrieser C., Wardroper A., Felder M., Thangavelu M., Johnson D., Knights A., Loulseged H., Mungall K., Oliver K., Price C., Quail M.A., Urushihara H., Hernandez J., Rabbinowitsch E., Steffen D., Sanders M., Ma J., Kohara Y., Sharp S., Simmonds M., Spiegler S., Tivey A., Sugano S., White B., Walker D., Woodward J., Winckler T., Tanaka Y., Shaulsky G., Schleicher M., Weinstock G., Rosenthal A., Cox E.C., Chisholm R.L., Gibbs R., Loomis W.F., Platzer M., Kay R.R., Williams J., Dear P.H., Noegel A.A., Barrell B., Kuspa A. 2005. The genome of the social amoeba *Dictyostelium discoideum*. Nature. 435:43–57.

Gehrmann T., Pelkmans J.F., Lugones L.G., Wösten H.A.B., Abeel T., Reinders M.J.T. 2016. *Schizophyllum commune* has an extensive and functional alternative splicing repertoire. Sci. Rep. 6.:33640.

Glöckner G., Lawal H.M., Felder M., Singh R., Singer G., Weijer C.J., Schaap P. 2016. The multicellularity genes of dictyostelid social amoebas. Nat. Commun. 7:12085.

Griesmann M., Chang Y., Liu X., Song Y., Haberer G., Crook M.B., Billault-Penneteau B., Lauressergues D., Keller J., Imanishi L., Roswanjaya Y.P., Kohlen W., Pujic P., Battenberg K., Alloisio N., Liang Y., Hilhorst H., Salgado M.G., Hocher V., Gherbi H., Svistoonoff S., Doyle J.J., He S., Xu Y., Xu S., Qu J., Gao Q., Fang X., Fu Y., Normand P., Berry A.M., Wall L.G., Ané J.-M., Pawlowski K., Xu X., Yang H., Spannagl M., Mayer K.F.X., Wong G.K.-S., Parniske M., Delaux P.-M., Cheng S. 2018. Phylogenomics reveals multiple losses of nitrogen-fixing root nodule symbiosis. Science (80). 361:eaat1743.

Grigoriev I. V, Nikitin R., Haridas S., Kuo A., Ohm R., Otillar R., Riley R., Salamov A., Zhao X., Korzeniewski F., Smirnova T., Nordberg H., Dubchak I., Shabalov I. 2014. MycoCosm portal: gearing up for 1000 fungal genomes. Nucleic Acids Res. 42:D699–704.

Grosberg R.K., Strathmann R.R. 2007. The evolution of multicellularity: a minor major transition? Annu. Rev. Ecol. Evol. Syst. 38:621–654.

Han K.-H. 2009. Molecular Genetics of *Emericella nidulans* sexual development. Mycobiology. 37:171–182.

Heaton L.L.M., Jones N.S., Fricker M.D. 2020. A mechanistic explanation of the transition to simple multicellularity in fungi. Nat. Commun. 11:1–9.

Hibbett D.S. 2004. Trends in morphological evolution in homobasidiomycetes inferred using maximum likelihood: a comparison of binary and multistate approaches. Syst. Biol. 53:889–903.

Jedd G. 2011. Fungal evo-devo: organelles and multicellular complexity. Trends Cell Biol. 21:12–9.

JW S., MC A., IV G., F M., JE S., M B. 2017. The fungal tree of life: from molecular systematics to genome-scale phylogenies. The Fungal Kingdom. American Society of Microbiology. p. 3–34.

Kim W., Cavinder B., Proctor R.H., O'Donnell K., Townsend J.P., Trail F. 2019. Comparative genomics and transcriptomics during sexual development gives insight into the life history of the cosmopolitan fungus *Fusarium neocosmosporiellum*. Front. Microbiol. 10:1247.

Kiss E., Hegedüs B., Virágh M., Varga T., Merényi Z., Kószó T., Bálint B., Prasanna A.N., Krizsán K., Kocsubé S., Riquelme M., Takeshita N., Nagy L.G. 2019. Comparative genomics reveals the origin of fungal hyphae and multicellularity. Nat. Commun. 10:4080.

Knoll A.H. 2011. The Multiple Origins of Complex Multicellularity. Annu. Rev. Earth Planet. Sci. 39:217–239.

Kohler A., Kuo A., Nagy L.G., Morin E., Barry K.W., Buscot F., Canbäck B., Choi C., Cichocki N., Clum A., Colpaert J., Copeland A., Costa M.D., Doré J., Floudas D., Gay G., Girlanda M., Henrissat B., Herrmann S., Hess J., Högberg N., Johansson T., Khouja H.R., Labutti K., Lahrmann U., Levasseur A., Lindquist E.A., Lipzen A., Marmeisse R., Martino E., Murat C., Ngan C.Y., Nehls U., Plett J.M., Pringle A., Ohm R.A., Perotto S., Peter M., Riley R., Rineau F., Ruytinx J., Salamov A., Shah F., Sun H., Tarkka M., Tritt A., Veneault-Fourrey C., Zuccaro A., Tunlid A., Grigoriev I. V, Hibbett D.S., Martin F. 2015. Convergent losses of decay mechanisms and rapid turnover of symbiosis genes in mycorrhizal mutualists. Nat. Genet. 47:410–415.

Krizsan K., Almasi E., Merenyi Z., Sahu N., Viragh M., Koszo T., Mondo S., Kiss B., Balint B., Kues U., Barry K., Cseklye J., Hegedus B., Henrissat B., Johnson J., Lipzen A., Ohm R.A., Nagy I., Pangilinan J., Yan J., Xiong Y., Grigoriev I. V., Hibbett D.S., Nagy L.G. 2019. Transcriptomic atlas of mushroom development highlights an independent origin of complex multicellularity. Proc. Natl. Acad. Sci. USA 116:7409–7418.

Kues U. 2000. Life history and developmental processes in the basidiomycete *Coprinus cinereus*. Microbiol Mol Biol Rev. 64:316–353.

Kües U., Navarro-González M. 2015. How do Agaricomycetes shape their fruiting bodies? 1. Morphological aspects of development. Fungal Biol. Rev. 29:63–97.

Lehr N.A., Wang Z., Li N., Hewitt D.A., López-Giráldez F., Trail F., Townsend J.P. 2014. Gene expression differences among three *Neurospora* species reveal genes required for sexual reproduction in *Neurospora* crassa. PLoS One. 9:e110398.

Liu Y.J., Hall B.D. 2004. Body plan evolution of ascomycetes, as inferred from an RNA polymerase II phylogeny. Proc. Natl. Acad. Sci. USA 101:4507–12.

Lord K.M., Read N.D. 2011. Perithecium morphogenesis in *Sordaria macrospora*. Fungal Genet. Biol. 48:388–399.

Marshall A.N., Montealegre M.C., Jiménez-López C., Lorenz M.C., van Hoof A. 2013. Alternative splicing and subfunctionalization generates functional diversity in fungal proteomes. PLoS Genet. 9:e1003376.

Merényi Z., Prasanna A.N., Wang Z., Kovács K., Hegedüs B., Bálint B., Papp B., Townsend J.P., Nagy L.G. 2020. Unmatched level of molecular convergence among deeply divergent complex multicellular fungi. Mol. Biol. Evol. 37:2228–2240.

Michod R.E., Roze D. 2001. Cooperation and conflict in the evolution of multicellularity. Heredity (Edinb). 86:1–7.

Money N.P., Davis C.M., Ravishankar J.P. 2004. Biomechanical evidence for convergent evolution of the invasive growth process among fungi and oomycete water molds. Fungal Genet. Biol. 41:872–876.

Muraguchi H., Umezawa K., Niikura M., Yoshida M., Kozaki T., Ishii K., Sakai K., Shimizu M., Nakahori K., Sakamoto Y., Choi C., Ngan C.Y., Lindquist E., Lipzen A., Tritt A., Haridas S., Barry K., Grigoriev I. V, Pukkila P.J. 2015. Strand-specific RNA-seq analyses of fruiting body development in *Coprinopsis cinerea*. PLoS One. 10.

Nagy L.G. 2017. Evolution: complex multicellular life with 5,500 genes. Curr. Biol. 27:R609–R612.

Nagy L.G. 2018. Many roads to convergence. Science 361:125–126.

Nagy L.G., Kovács G.M., Krizsán K. 2018. Complex multicellularity in fungi: evolutionary convergence, single origin, or both? Biol. Rev. Camb. Philos. Soc. 93:1778–1794.

Nagy L.G., Ohm R.A., Kovács G.M., Floudas D., Riley R., Gácser A., Sipiczki M., Davis J.M., Doty S.L., de Hoog G.S., Lang B.F., Spatafora J.W., Martin F.M., Grigoriev I. V, Hibbett D.S. 2014. Latent homology and convergent regulatory evolution underlies the repeated emergence of yeasts. Nat. Commun. 5:4471.

Nagy L.G., Tóth R., Kiss E., Slot J., Gácser A., Kovács G.M. 2017. Six key traits of fungi: their evolutionary origins and genetic bases. The Fungal Kingdom. American Society of Microbiology. p. 35–56.

Nagy L.G., Varga T., Csernetics Á., Virágh M. 2020. Fungi took a unique evolutionary route to multicellularity: seven key challenges for fungal multicellular life. Fungal Biol. Rev. 34:151–169.

Nguyen T.A., Cissé O.H., Yun Wong J., Zheng P., Hewitt D., Nowrousian M., Stajich J.E., Jedd G. 2017. Innovation and constraint leading to complex multicellularity in the Ascomycota. 8:14444.

Niklas K.J. 2014. The evolutionary-developmental origins of multicellularity. Am. J. Bot. 101:6–25.

Nowrousian M. 2018. Genomics and transcriptomics to study fruiting body development: An update. Fungal Biol. Rev. 32:231–235.

Nowrousian M., Stajich J.E., Chu M., Engh I., Espagne E., Halliday K., Kamerewerd J., Kempken F., Knab B., Kuo H.C., Osiewacz H.D., Pöggeler S., Read N.D., Seiler S., Smith K.M., Zickler D., Kück U., Freitag M. 2010. De novo assembly of a 40 Mb eukaryotic genome from short sequence reads: *Sordaria macrospora*, a model organism for fungal morphogenesis. PLoS Genet. 6:e1000891.

O'Malley M.A., Wideman J.G., Ruiz-Trillo I. 2016. Losing complexity: the role of simplification in macroevolution. Trends Ecol. Evol. 31:608–621.

Ohm R.A., de Jong J.F., de Bekker C., Wösten H.A.B., Lugones L.G. 2011. Transcription factor genes of *Schizophyllum commune* involved in regulation of mushroom formation. Mol. Microbiol. 81:1433–1445.

Ohm R.A., De Jong J.F., Lugones L.G., Aerts A., Kothe E., Stajich J.E., De Vries R.P., Record E., Levasseur A., Baker S.E., Bartholomew K.A., Coutinho P.M., Erdmann S., Fowler T.J., Gathman A.C., Lombard V., Henrissat B., Knabe N., Kües U., Lilly W.W., Lindquist E., Lucas S., Magnuson J.K., Piumi F., Raudaskoski M., Salamov A., Schmutz J., Schwarze F.W.M.R., Vankuyk P.A., Horton J.S., Grigoriev I. V, Wösten H.A.B. 2010. Genome sequence of the model mushroom *Schizophyllum commune*. Nat. Biotechnol. 28:957–963.

Park Y.-J., Baek J.H., Lee S., Kim C., Rhee H., Kim H., Seo J.-S., Park H.-R., Yoon D.-E., Nam J.-Y., Kim H.-I., Kim J.-G., Yoon H., Kang H.-W., Cho J.-Y., Song E.-S., Sung G.-H., Yoo Y.-B., Lee C.-S., Lee B.-M., Kong W.-S. 2014. Whole genome and global gene expression analyses of the model mushroom *Flammulina velutipes* reveal a high capacity for lignocellulose degradation. PLoS One. 9:e93560.

Rainey P.B., Monte S. De. 2014. Resolving conflicts during the evolutionary transition to multicellular life. Annu. Rev. Ecol. Evol. Syst. 45:599–620.

Ratcliff W.C., Denison R.F., Borrello M., Travisano M. 2012. Experimental evolution of multicellularity. Proc. Natl. Acad. Sci. USA 109:1595–1600.

Riquelme M., Aguirre J., Bartnicki-García S., Braus G.H., Feldbrügge M., Fleig U., Hansberg W., Herrera-Estrella A., Kämper J., Kück U., Mouriño-Pérez R.R., Takeshita N., Fischer R. 2018. Fungal morphogenesis, from the polarized growth of hyphae to complex reproduction and infection structures. Microbiol. Mol. Biol. Rev. 82:e00068–17.

Rodenburg S.Y.A., Terhem R.B., Veloso J., Stassen J.H.M., Van Kan J.A.L. 2018. Functional analysis of mating type genes and transcriptome analysis during fruiting body development of *Botrytis cinerea*. MBio. 9:01939–17.

Rokas A. 2008. The origins of multicellularity and the early history of the genetic toolkit for animal development. Annu. Rev. Genet. 42:235–251.

Sakamoto Y., Nakade K., Sato S., Yoshida K., Miyazaki K., Natsume S., Konno N. 2017. *Lentinula edodes* genome survey and postharvest transcriptome analysis. Appl. Environ. Microbiol. 83:e02990–16.

Sebé-Pedrós A., Degnan B.M., Ruiz-Trillo I. 2017. The origin of Metazoa: a unicellular perspective. Nat. Rev. Genet. 18:498–512.

Shen X.-X., Opulente D.A., Kominek J., Zhou X., Steenwyk J.L., Buh K. V., Haase M.A.B., Wisecaver J.H., Wang M., Doering D.T., Boudouris J.T., Schneider R.M., Langdon Q.K., Ohkuma M., Endoh R., Takashima M., Manabe R., Čadež N., Libkind D., Rosa C.A., DeVirgilio J., Hulfachor A.B., Groenewald M., Kurtzman C.P., Hittinger C.T., Rokas A. 2018. Tempo and mode of genome evolution in the budding yeast subphylum. Cell. 175:1533–1545.e20.

Spatafora J.W., Chang Y., Benny G.L., Lazarus K., Smith M.E., Berbee M.L., Bonito G., Corradi N., Grigoriev I., Gryganskyi A., James T.Y., O'Donnell K., Roberson R.W., Taylor T.N., Uehling J., Vilgalys R., White M.M., Stajich J.E. 2016. A phylum-level phylogenetic classification of zygomycete fungi based on genome-scale data. Mycologia. 108:1028–1046.

Stajich J.E., Wilke S.K., Ahren D., Au C.H., Birren B.W., Borodovsky M., Burns C., Canback B., Casselton L.A., Cheng C.K., Deng J., Dietrich F.S., Fargo D.C., Farman M.L., Gathman A.C., Goldberg J., Guigo R., Hoegger P.J., Hooker J.B., Huggins A., James T.Y., Kamada T., Kilaru S., Kodira C., Kues U., Kupfer D., Kwan H.S., Lomsadze A., Li W., Lilly W.W., Ma L.-J., Mackey A.J., Manning G., Martin F., Muraguchi H., Natvig D.O., Palmerini H., Ramesh M.A., Rehmeyer C.J., Roe B.A., Shenoy N., Stanke M., Ter-Hovhannisyan V., Tunlid A., Velagapudi R., Vision T.J., Zeng Q., Zolan M.E., Pukkila P.J. 2010. Insights into evolution of multicellular fungi from the assembled chromosomes of the mushroom *Coprinopsis cinerea* (*Coprinus cinereus*). Proc. Natl. Acad. Sci. 107:11889–11894.

Teichert I., Pöggeler S., Nowrousian M. 2020. *Sordaria macrospora*: 25 years as a model organism for studying the molecular mechanisms of fruiting body development. Appl. Microbiol. Biotechnol. 104:3691–3704.

Traeger S., Altegoer F., Freitag M., Gabaldon T., Kempken F., Kumar A., Marcet-Houben M., Pöggeler S., Stajich J.E., Nowrousian M. 2013. The genome and development-dependent transcriptomes of *Pyronema confluens*: a window into fungal evolution. PLoS Genet. 9:e1003820.

Trail F., Wang Z., Stefanko K., Cubba C., Townsend J.P. 2017. The ancestral levels of transcription and the evolution of sexual phenotypes in filamentous fungi. PLOS Genet. 13:e1006867.

Varga T., Krizsán K., Földi C., Dima B., Sánchez-García M., Sánchez-Ramírez S., Szöllősi G.J., Szarkándi J.G., Papp V., Albert L., Andreopoulos W., Angelini C., Antonín V., Barry K.W., Bougher N.L., Buchanan P., Buyck B., Bense V., Catcheside P., Chovatia M., Cooper J., Dämon W., Desjardin D., Finy P., Geml J., Haridas S., Hughes K., Justo A., Karasiński D., Kautmanova I., Kiss B., Kocsubé S., Kotiranta H., LaButti K.M., Lechner B.E., Liimatainen K., Lipzen A., Lukács Z., Mihaltcheva S., Morgado L.N., Niskanen T., Noordeloos M.E., Ohm R.A., Ortiz-Santana B., Ovrebo C., Rácz N., Riley R., Savchenko A., Shiryaev A., Soop K., Spirin V., Szebenyi C., Tomšovský M., Tulloss R.E., Uehling J., Grigoriev I. V., Vágvölgyi C., Papp T., Martin F.M., Miettinen O., Hibbett D.S., Nagy L.G. 2019. Megaphylogeny resolves global patterns of mushroom evolution. Nat. Ecol. Evol. 3:668–678.

van Velzen R., Holmer R., Bu F., Rutten L., van Zeijl A., Liu W., Santuari L., Cao Q., Sharma T., Shen D., Roswanjaya Y., Wardhani T.A.K., Kalhor M.S., Jansen J., van den Hoogen J., Güngör B., Hartog M., Hontelez J., Verver J., Yang W.-C., Schijlen E., Repin R., Schilthuizen M., Schranz M.E., Heidstra R., Miyata K., Fedorova E., Kohlen W., Bisseling T., Smit S., Geurts R. 2018. Comparative genomics of the nonlegume *Parasponia* reveals insights into evolution of nitrogen-fixing rhizobium symbioses. Proc. Natl. Acad. Sci. U. S. A. 115:E4700–E4709.

Wang Z., Lopez-Giraldez F., Lehr N., Farré M., Common R., Trail F., Townsend J.P. 2014. Global gene expression and focused knockout analysis reveals genes associated with fungal fruiting body development in *Neurospora crassa*. Eukaryot. Cell. 13:154–169.

Willis K.J. 2018. State of the world's fungi 2018. Report. State of the world's fungi 2018. Report.

Wynns A.A. 2015. Convergent evolution of highly reduced fruiting bodies in Pezizomycotina suggests key adaptations to the bee habitat. BMC Evol. Biol. 15:145.

15 Genetic and Developmental Mechanisms of Cellular Differentiation in Algae

Susana M. Coelho
Department of Algal Development and Evolution,
Max Planck Institute for Developmental Biology,
Tübingen, Germany

J. Mark Cock
Laboratory of Integrative Biology of Marine Models, Station
Biologique de Roscoff, Sorbonne Université, Roscoff, France

CONTENTS

15.1 INTRODUCTION

The term 'algae' is generally used to group together all photosynthetic eukaryotes with the exception of land plants. The algae are therefore a highly polyphyletic group that includes organisms from many of the major eukaryotic lineages (Figure 15.1). The broad phylogenetic distribution of these organisms is explained by the evolutionary history of their photosynthetic organelles (plastids). With only one known exception (Marin et al., 2005), these organelles are all thought to be derived from a cyanobacterium captured by an ancestor of the Archaeplastida (the lineage that includes green and red algae; Figure 15.1; Keeling, 2013). Following this primary endosymbiosis, plastids were subsequently acquired by organisms in other eukaryotic supergroups

DOI: 10.1201/9780429351907-19

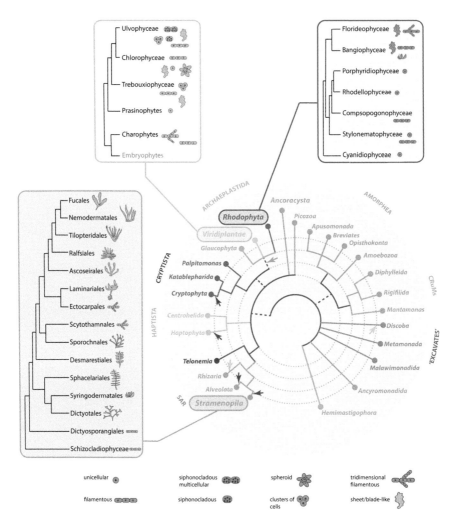

FIGURE 15.1 Schematic view of the eukaryotic tree of life and multicellular and morphological complexity in the brown, red and green algal lineages. The eukaryotic tree (adapted from Burki et al., 2020) is based on a consensus of recent phylogenomic studies. Broken lines indicate uncertainty about the monophyly of certain groups. Colored arrows indicate plastid endosymbiosis events (Bodył, 2018; Keeling, 2013). The blue arrow indicates the primary plastid endosymbiosis event that occurred in a common ancestor of glaucophytes, red (Rhodophyta) and green (Viridiplantae) algae. Red or green arrows indicate plastids acquired by endosymbiosis from algae of the red or green lineages, respectively. Note that these acquisitions could have been direct via a secondary endosymbiosis or indirect, involving two or more transfers of plastids between hosts (tertiary or quaternary endosymbioses). Note also that the presence of a colored arrow does not necessarily indicate that all the members of the lineage corresponding to that branch of the tree acquired a plastid. The simplified trees for the three major algal lineages are based on Leliaert et al. (2012) and De Clerck et al. (2012) for the green algae, Qiu et al. (2016) for the red algae and Silberfeld et al. (2014), Kawai et al. (2015) and Bringloe et al. (2020) for the brown algae. Only the brown algae, together with their sister clade, the Schizocladiophyceae, are shown for the Stramenopila.

over the course of evolution via secondary endosymbiosis events involving capture of red or green algae and integration of their plastid systems into the machinery of the host cell (Keeling, 2013). It was these events, which presumably involved unicellular ancestors in each lineage, that led to the very broad distribution of algae across the eukaryotic supergroups. For many of the lineages, the extant species are still unicellular (glaucophytes, dinoflagellates, chlorarachnids, euglenids, cryptophytes, haptophytes). In other lineages, different levels of multicellularity have evolved, ranging from the formation of simple colonial chains or clusters of cells (e.g., some diatom species) to the emergence of large, complex multicellular organisms with differentiated organs and multiple cell types (brown, red and green macroalgae; Figure 15.1).

This chapter will focus on the three macroalgal lineages, which are the algal groups that exhibit the highest level of multicellularity and are the only algal groups that are considered to have made the transition from simple multicellularity to complex multicellular body plans (along with the animals, land plants and fungi; Cock et al., 2010; Cock and Collén, 2015; Knoll, 2011). The chapter will describe what is known about developmental events in each lineage and will discuss the developmental features of each group that are currently under investigation in relation to the evolution of multicellularity.

15.2   GREEN ALGAE

The green algae and the land plants compose the green lineage (Viridiplantae), one of the major groups of oxygenic photosynthetic eukaryotes. Current hypotheses propose early divergence of two discrete clades from an ancestral green flagellate organism. One clade comprises the chlorophytes. The other clade, the streptophytes, includes the charophyte green algae from which the land plants evolved (Leliaert et al., 2012). Multicellular taxa have repeatedly evolved in the green algae, as well as macroscopically complex unicellular forms that show many of the features that are emblematic of multicellularity. We will focus here on two multicellular groups of chlorophytes, the volvocines and ulvophytes, whilst charophyte algae will be treated in another Chapter 16.

The chlorophytes diverged from the streptophytes approximately 900-1000 MYA (Bhattacharya et al., 2009; Hedges et al., 2004). Ancestral green algal species were probably small unicellular marine biflagellates (Leliaert et al., 2012) and this form is still prevalent among modern aquatic algae, such as the well-studied unicellular, biflagellate alga *Chlamydomonas reinhardtii*. Of the four classes of chlorophyte, three (Chlorophyceae, Trebouxiophyceae, and Ulvophyceae) include members that are multicellular during at least part of their life cycle. Transitions to multicellularity or macroscopic forms occurred several times independently (Lewis and McCourt, 2004; Mattox and Stewart, 1984; Chapter 9). These transitions gave rise to an exceptional range of morphological patterns, from relatively large algae (up to 1 m) with complex tissues and organs to several species that form small motile multicellular colonies (Van den Hoek, Mann and Jahns, 1995; Lewis and McCourt, 2004) (Figures 15.1 and 15.2).

In multicellular green algae, cells are connected by adhesives to maintain a coherent, physically attached body. Cells may also communicate through plasmodesmata,

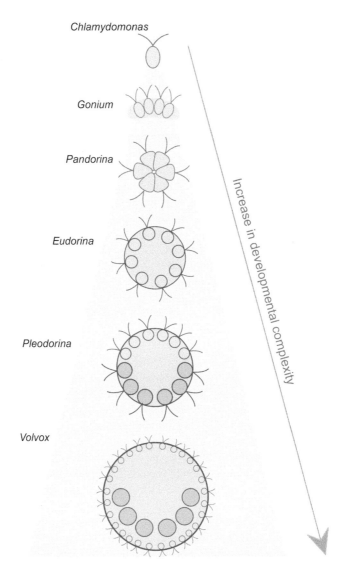

Chlamydomonas

Gonium

Pandorina

Eudorina

Pleodorina

Volvox

Increase in developmental complexity

FIGURE 15.2 Schema indicating different degrees of developmental complexity in volvocine algae, from *C. reinhardtii* to *V. carteri*. Different shades of green represent different cell types.

which ensure a supra-cellular organization at the level of the whole organism (Raven, 1997). Some multicellular green algae, however, are merely unconnected groups of cells embedded in an extracellular matrix, and the form of the alga can be extremely variable, from simple unbranched filamentous forms to organisms with, for example, highly branched thalli or leaf-like blades. Note that complex and macroscopic unicellular organization is found in giant unicellular algae (e.g. *Acetabularia*) that possess a single nucleus (siphonous), but also in coenocytic algae where large single cells contain several nuclei within a common cytoplasm (reviewed in Leliaert et al., 2012).

Although these organisms are not 'multicellular', they possess remarkable structures with specialized sub-domains. For example, *Caulerpa* has different 'tissues' that resemble roots, stems and leaves (Jacobs, 1970). Some green algae may also be both multicellular and coenocytic (siphonocladous) because they are composed of multiple cells, each of which contains multiple nuclei (e.g., *Cladophora*).

In the next sections, we will discuss both established and emerging model green algae and describe how these models are being used to explore the mechanisms underlying cell differentiation and the emergence of complex multicellular developmental patterns.

15.2.1 VOLVOCINE ALGAE (CHLOROPHYCEAE)

The volvocine algae, in the order Volvocales, are a monophyletic group of chlorophyceae algae that together form a fascinating study set for the emergence of multicellularity and evolution of developmental complexity. They include a series of species exhibiting increasing levels of cell-type specialization and developmental complexity, from unicellular forms such as *Chlamydomonas*, through multicellular groups with undifferentiated cells (e.g., *Gonium* and *Eudorina*) to relatively complex multicellular individuals that may have one (e.g., *Pleodorina*) or two (*Volvox*) specialized cell types.

The Volvocales include experimentally tractable model-systems for which a range of molecular tools are available (Merchant et al., 2007; Prochnik et al., 2010; Umen and Olson 2012; Umen, 2020). The genomes of three species that represent a broad range of developmental complexity —*Tetrabaena socialis*, *Gonium pectorale*, and *Volvox carteri*, together with the genome of a unicellular close relative, *C. reinhardtii*, have been sequenced and compared (Featherston et al., 2018; Hanschen et al., 2016; Merchant et al., 2007; Prochnik et al., 2010). Analysis of these genomes has revealed that, although the four species exhibit contrasting levels of morphological complexity, their genomes are relatively similar with the notable exception of the expansion of gene families involved in extracellular matrix (ECM) formation and cell-cycle regulation (Featherston et al., 2018; Hanschen et al., 2016; Prochnik et al., 2010). These comparative genomic analyses, together with classical genetic studies, have shed light on the mechanisms underlying the stepwise progression of increasing cell number, colony size and degree of cell-type specialization. Specifically for *V. carteri*, genetic screens have been a powerful tool in revealing the molecular pathways involved in developmental and morphological traits (e.g., Kirk, 1998, 2001).

The evolution of multicellularity in the Volvocales is thought to have involved a series of 12 specific innovations (Kirk, 2005; Umen, 2014). These innovations include, for example, genetic modulation of the number of cells per individual and modifications to the cell cycle program (Kirk, 1998, 2005; Umen and Olson, 2012). *C. reinhardtii* cells can divide as soon as they double in size, whereas cells in colonial species must grow to a specific size between reproductive cycles so that entire new colonies can be formed with the appropriate number of cells. In *C. reinhardtii*, the retinoblastoma (Rb) tumor suppressor pathway has been shown to control cell size and the number of cell divisions a cell undergoes before the daughter cells separate (Fang et al., 2006; Umen and Goodenough, 2001). It is thought that the Rb pathway may have been modified in colonial species to increase the minimum size threshold at which division can occur.

Another important feature of the evolution of multicellularity in this group is the transformation of the cell wall into an ECM and ECM expansion and complexification. For example, the walls of individual *Gonium* cells have an additional outer layer, compared to those of *C. reinhardtii*, that maintains colony cohesion (Nozaki, 1990). The more complex, spheroidal species *Pandorina*, *Eudorina*, *Pleodorina*, and *Volvox* also have expanded ECMs and additional modifications of this structure in these species include the formation of a colony boundary layer.

One of the best-studied innovations during the transition to multicellularity in the volvocines is the differentiation of germ and somatic cell lineages that is observed in the genera *Pleodorina* and *Volvox*. In *V. carteri*, germ-soma differentiation takes place after completion of embryonic cell division. Germ versus soma cell fate is established solely based on cell size and not by inheritance of cytoplasmic cell fate factors (Kirk et al., 1993). The nature of the cell size signal is unknown but it may be linked to the Rb pathway, which has a role in cell size control in *C. reinhardtii* (Fang et al., 2006; see above). Somatic cell fate determination involves differential expression of a master regulatory gene (somatic regenerator, *RegA*), which is thought to encode a transcriptional repressor of genes for growth and reproduction. Mutation of this gene results in somatic cells regaining reproductive capability (Kirk et al., 1999; Stark et al., 2001). *RegA* expression is restricted to somatic cells and its mRNA is first detectable shortly after embryogenesis is completed (Kirk et al., 1999). RegA belongs to a family of VARL (Volvocine algae RegA-like) proteins with 14 members in *V. carteri* and 12 in *C. reinhardtii* (Duncan et al., 2007). The closest homolog of *RegA* in *C. reinhardtii* is *RLS1*, and its expression is induced under nutrient and light limitation and during stationary phase (Nedelcu, 2009; Nedelcu and Michod, 2006). Based on these observations, it has been hypothesized that the evolution of somatic cells in *V. carteri* involved the co-option of an ancestral environmentally-induced *RLS1*-like gene by switching its regulation from a temporal/environmental to a spatial/developmental context.

Recent comparative transcriptomic approaches indicated that *V. carteri* orthologs of genes that are diurnally controlled in *C. reinhardtii* exhibit a strongly partitioned pattern of expression, with expression of dark-phase genes overrepresented in somatic cells and light-phase genes overrepresented in the gonidial cells of the germ cell lineage (Matt and Umen, 2018). This observation further supports the idea that cell-type programs in *V. carteri* arose by co-option of temporally or environmentally controlled genes from a unicellular ancestor that came under cell-type control in the *V. carteri* lineage.

The idea that large-scale changes in gene expression underlie the transition to a multicellular life cycle is further supported by experimental evolution approaches. Herron and colleagues used experimentally evolved *C. reinhardtii* to explore the genetics underlying a *de novo* origin of multicellularity. They identified changes in gene structure and expression that distinguish an evolved 'multicellular' clone from its unicellular ancestor. Among these, changes to genes involved in cell cycle and reproductive processes were overrepresented, supporting previous indications from genetic screens (see above; Herron et al., 2018). Isolation of new regulatory loci using genetic screens and further investigation of *V. carteri* and *C. reinhardtii* mutants will help to clarify the origins of cell-type specification in this lineage and to further deepen our understanding of how multicellularity emerged.

15.2.2 Ulvophyceae

The Ulvophyceae, or green seaweeds, represent an independent emergence of a macroscopic plant-like vegetative body. Members of the class Ulvophyceae display a fascinating morphological and cytological diversity, including multicellular organisms such as the sea lettuce *Ulva* but also species with macroscopic siphonous coenocytes (Figure 15.1).

Despite the interest of this group of green algae with regard to the transition to multicellular development and the evolution of the cytomorphological diversity, genomic resources for the Ulvophyceae have remained relatively scarce, except for a transcriptomic study of *U. linza* (Zhang et al., 2012), a description of the mating-type locus of *U. partita* (Yamazaki et al., 2017) and a study of the distribution of transcripts in the thallus of the siphonous green seaweed *Caulerpa* (Ranjan et al., 2015).

This situation is changing rapidly with the emergence of *Ulva* sp. as an exciting novel model organism for studies of green algal growth, development and morphogenesis as well as mutualistic interactions (Wichard et al., 2015). Several interesting aspects of *Ulva* biology, including cell cycle, cytology, life-cycle transition, induction of spore and gamete formation, and bacterial-controlled morphogenesis, have been studied in detail (reviewed in Wichard et al., 2015). The *Ulva* thallus has a relatively simple multicellular organization, with small uninucleate cells and a limited number of cell types. In natural environments, *Ulva* may exhibit two morphological patterns: either flattened blades that are two-cells thick or thalli that develop as tubes that are one-cell thick (Figure 15.3). These morphologies require the presence of symbiotic bacteria, without which only slow-growing, undifferentiated 'callus-like colonies' develop (Spoerner et al., 2012; Stratmann et al., 1996). The growth, cell differentiation and morphogenesis of *Ulva* depend on interactions with specific bacteria and the chemical mediators these bacteria produce, in particular the compound thalusin

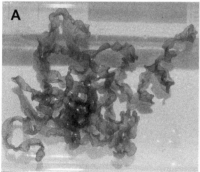 

FIGURE 15.3 Alternative morphological patterns in *Ulva* spp. Thalli either develop as tubes (A) or as flattened blades (B).

(Egan et al., 2013; Goecke et al., 2010; Wichard et al., 2015). The possibility of triggering morphogenesis by adding an engineered microbiome (Wichard et al., 2015) or a compound produced by these microbes is a major asset that may allow insights to be gained into molecular events during cellular differentiation. Therefore, *Ulva* is being increasingly used as a model organism for investigating morphogenesis (De Clerck et al., 2018; Wichard et al., 2015).

The first whole-genome sequence of an *Ulva* species, *U. mutabilis*, has recently opened new opportunities for the study of the emergence of multicellularity in the green lineage (De Clerck et al., 2018). Remarkably, analysis of the *U. mutabilis* genome has detected loss of genes that encode components of the retinoblastoma (RB)/E2F pathway and associated D-type cyclins. Comparative genomic studies of volvocine algae have revealed that the co-option of the RB cell-cycle pathway is key to the emergence of multicellularity in this lineage (see above). It appears therefore that the paths towards multicellularity in *U. mutabilis* and in volvocines progressed through different routes. Parallels can however still be drawn between ECM gene family expansions in the volvocine algae and the diversity of protein domains associated with the ECM in *U. mutabilis* relative to closely related unicellular taxa, especially given the proposed role of expanded volvocine ECM gene families in environmental signaling (De Clerck et al., 2018).

15.3   RED ALGAE

The red algae (Rhodophyta) are an ecologically important group of organisms, dominating many coastal environments, and are also of economic interest as a source of food and industrial colloids. They are a sister group to the green lineage (Viridiplantae) and acquired their plastids via the same primary endosymbiosis (Figure 15.1; Keeling, 2013). From an evolutionary point of view, red algae are also important because they donated plastids to several other eukaryotic supergroups via secondary endosymbioses (Figure 15.1; Keeling, 2013). Fossil evidence indicates that the red algae were the first eukaryotic lineage to evolve complex multicellularity, as early as 1600 Mya (Bengtson et al., 2017; Butterfield, 2000). Extant red algae exhibit a broad range of complexity, including species that are unicellular, colonial, grow as simple filaments or that form large, foliose thalli (Cock and Collén, 2015; Waaland, 1990). The latter can be either pseudoparenchymatous (made up of amalgamated filaments) or parenchymatous (true tissues formed by three-dimensional cell divisions). The larger multicellular red algae can possess several types of organs, including stems, holdfasts, bladders and bladelike fronds and can reach lengths of up to three meters. However, even the largest red algae exhibit significantly less morphological and cellular complexity than the most complex representatives of the brown algae.

In the 1970s and 1980s, genetic approaches were applied to the red algae with the objective, in part, of studying multicellular development. Red algal morphological mutants were isolated and analyzed genetically, but these studies did not go as far as identifying the underlying genetic loci (van der Meer et al., 1990). Mutant analyses did however provide some insights into developmental processes at the cellular level. For example, in several members of the Bangiales, meiosis coincides with the first

TABLE 15.1
Red Algal Genome Assemblies

Species	Multicellular or Unicellular	Genome Size (Mbp)	Repeat Sequences	Reference
Chondrus crispus	Multicellular	104.8	59%	(Collén et al., 2013)
Calliarthron tuberculosum	Multicellular	(51.1)*	nd	(Chan et al., 2011)
Gracilariopsis chorda	Multicellular	92.1	61%	(Lee et al., 2018)
Gracilariopsis lemaneiformis	Multicellular	81.2	55%	(Zhou et al., 2013)
Gracilaria changii	Multicellular	(35.8)*	nd	(Ho et al., 2018)
Porphyra umbilicalis	Multicellular	87.7	43%	(Brawley et al., 2017)
Pyropia yezoensis	Multicellular	(43.0)*	nd	(Nakamura et al., 2013)
Pyropia yezoensis	Multicellular	108	48%	(Wang et al., 2020)
Pyropia haitanensis	Multicellular	53.3	24%	(Cao et al., 2020)
Cyanidioschyzon merolae	Unicellular	16.5	20%	(Matsuzaki et al., 2004)
Galdieria sulphuraria	Unicellular	13.7	16%	(Schönknecht et al., 2013)
Porphyridium purpureum	Unicellular	19.7	4%	(Bhattacharya et al., 2013)

* Draft genomes and therefore probably incomplete assemblies. nd, not determined.

cell divisions of the gametophyte resulting in chimeric individuals descended from all four meiotic products. Using color mutants, this phenomenon was exploited to study cell lineage patterns in the developing gametophyte (Mitman and Meer, 1994; Niwa, 2010).

Interest in red algal multicellular development has been revived recently as genome data has become available for this lineage. Complete genome sequences, with varying qualities of assembly, have currently been reported for 11 red algal species, including eight multicellular taxa (Table 15.1). Comparison of these assemblies indicates that multicellular species tend to have larger genomes than unicellular species (50–100 Mbp compared with 16–20 Mbp; Table 15.1), a trend that has been observed in other multicellular lineages. It has been proposed that expansion of the genomes of multicellular organisms occurs due to a weakening of purifying selection as a result of reduced effective population sizes (Lynch and Conery, 2003). Genome expansion in multicellular red algae appears to have occurred principally as a result of the proliferation of transposable elements (Table 15.1) rather than by polyploidization (Collén et al., 2013; Lee et al., 2018). Collén *et al.* (2013) have proposed a general scenario for genome evolution in the red algae in which there was a marked reduction in genome size during the early evolution of the lineage, followed by genome expansion associated with the transition to multicellularity. The genomes of multicellular red algae have been analyzed for features that may be related to the transition to multicellularity, such as gene family expansions and the composition of signaling gene families including transcription factors (Brawley et al., 2017; Collén et al., 2013). Interestingly, it has been suggested that the constricted repertoire of cytoskeleton genes in red algae may have prevented the lineage from evolving highly complex multicellular body plans because of limitations on cell size and complexity (Brawley et al., 2017).

Genomic data, together with transcriptomics, is also being used to identify potential key developmental genes. For example, a recent study showed that a three amino acid loop extension homeodomain transcription factor (TALE HD TF) was upregulated in the conchosporangium of *Pyropia yezoensis*, suggesting a possible role in the regulation of life-cycle-related developmental processes (Mikami et al., 2019).

From an evolutionary point of view, it would clearly be of interest to have a deeper understanding of multicellular development in the red algae. Genome sequencing has provided the necessary gene catalogs on which to build such analyses but genetic tools are also required to directly investigate gene function. Another key factor is likely to be the selection of a model species and focusing of effort across the red algal community on that species. *Pyropia yezoensis* (then *Porphyra yezoensis*) was proposed as a model several years ago (Waaland et al., 2004) and this species remains a strong candidate, but other species should also be considered.

15.4   BROWN ALGAE

The brown algae (Phaeophyceae) are part of the stramenopile (or heterokont) supergroup (Figure 15.1), which also includes diatoms and oomycetes. Brown seaweeds have therefore had a very different evolutionary history to the green and red algae, having diverged from the Archaeplastida lineage (i.e., Glaucophyta, Viridiplantae and Rhodophyta) at the time of the eukaryotic crown radiation. The photosynthetic stramenopiles (ochrophytes) acquired their plastid via a secondary endosymbiosis involving a red alga (Figure 15.1; Keeling, 2013).

All brown algae are multicellular but they exhibit a broad range of complexities, from simple filamentous species to large complex organisms with distinct, parenchymatous tissues and organs (Charrier et al., 2012). Kelps of the order Laminariales, for example, rival land plants in their complexity, with well-defined organs such as the holdfast, which anchors the alga to its substratum, the stipe, which serves a similar function to a land plant stem, and the frond, which represents the main photosynthetic surface, equivalent to land plant leaves. Each of these structures is composed of several cell types, including epidermal structures, cortical tissues and other specialized cells. Of particular interest are the trumpet hyphal cells, which represent a primitive vascular system. *Macrocystis pyrifera* (giant kelp) is often cited as an example of the level of multicellular complexity that has been attained by the brown algae. This kelp can grow up to lengths of 50 meters and therefore represents one of the largest organisms on the planet.

Stramenopiles other than the brown algae are all unicellular or filamentous organisms, including the stramenopile taxa closest to the brown algae, the filamentous alga *Schizocladia*. Moreover, the thalli of the most basal group within the brown algae, the Discosporangiales, also consist of uniseriate filaments (chains of single cells) whereas the most complex brown algae tend to belong to recently evolved taxa such as the Laminariales or the Fucales. These observations suggest a gradual acquisition of multicellular complexity during the evolution of the brown algae. However, while this conclusion is probably correct overall, a closer look at the phylogeny indicates a slightly more complex story because, for example, some recently evolved groups such as the Ectocarpales exhibit quite simple filamentous morphologies and some basal groups, such as the Dictyotales, have more complex features. These observations suggest that there has either been independent, parallel evolution of complex features in different brown algal groups or that orders such as the Ectocarpales have experienced some loss of complexity over evolutionary time.

The following sections will discuss how model organisms have been used and are being used to investigate developmental processes in the brown algae.

15.4.1 *Fucus* and *Dictyota*

Fucoids are a group of brown algae that has attained a high level of multicellular complexity in terms of number of cell types and developmental complexity of tissues and organs (Figure 15.1). They are also the only brown algal group that has evolved diploid life cycles, where the gametophyte stage is inexistent, their life cycle resembling that of an animal (Heesch et al., 2019) (Figure 15.4).

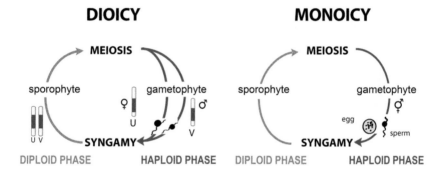

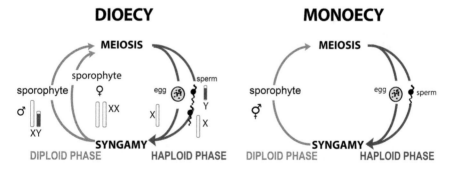

FIGURE 15.4 Schematic representation of the different types of sexual systems in the brown algae. The diversity of sexual systems include haploid or diploid sex determination, each associated with either co-sexuality (monoicy, monoecy) or separate sexes (dioicy, dioecy). Sex chromosomes are represented in the scheme as grey bars with the non-recombining region highlighted in dark grey. Note that some species have a large sexual dimorphism in terms of gamete size (oogamy, e.g., *Fucus* spp.) or a subtle difference in terms of male versus female gamete size (near-isogamy, e.g., *Ectocarpus*).

Brown algal zygotes and embryos have served as a system to explore early developmental events, such as polarity and asymmetric cell division because fertilization is external and embryos develop free from maternal tissue. This feature allows access to the cellular events underlying polarization, the first cell division and cell fate determination. Accordingly, *Fucus* zygotes and embryos have been employed to study polarity and asymmetry during early embryo development (reviewed in Brownlee et al., 2001). The initial asymmetric cell division produces two cells – a rhizoid and a thallus cell – with distinct morphologies and fates. The rhizoid cell generates the holdfast which will attach the alga to substrates, whereas the thallus cell will give rise to the stipe and the rest of the algal body (fronds, air bladders, and reproductive structures in *Fucus*). Fucoid eggs possess no intrinsic polarity and no cell wall. After fertilization, zygotes synthesize a cell wall within minutes and become polarized in response to external vectors, most frequently unilateral light. Photopolarization and germination are followed by the asymmetric division of the zygote, with subsequent divisions occurring in a highly ordered, spatial, and temporal pattern (Brownlee and Bouget, 1998). The cell wall was identified as a source of position-dependent information required for polarization and fate determination in the zygote and 2-celled embryo (Berger et al., 1994). Regeneration is regulated in a position-dependent manner and is strongly influenced by intercellular communication, likely involving transport or diffusion of inhibitory signals which appear to be essential for regulation of cell fate decisions (Bouget et al., 1998). Apoplastic diffusible gradients of unknown nature were proposed to be involved in pattern formation in the multicellular embryo (Bouget et al., 1998).

While a large amount of work has been done in the cell biology mechanisms underlying asymmetrical cell division, we still know very little about the molecular mechanisms that drive cell fate determination in fucoid algae, mostly because there are very few molecular tools developed for this group of organisms. Injection of double-stranded RNA has been shown to provide a potential means to knockdown gene expression by RNA interference but no stable transformation system has been described yet. Classical genetic approaches are not feasible due to the slow growth rate and long life cycle.

A more recent model for cell fate determination and evolution of multicellularity is the brown alga *Dictyota dichotoma* (Bogaert et al., 2020; Coelho and Cock, 2020). In contrast to the situation for *Fucus*, the life cycle of *D. dichotoma* can be completed under laboratory conditions, opening the possibility of applying classical genetic approaches (Bogaert et al., 2016). As in other brown algae, the establishment of the apical-basal multicellular pattern is achieved through an initial asymmetric cell division (Bogaert et al., 2017; De Smet and Beeckman, 2011; Peters et al., 2008) and recent studies have indicated a potential role for auxin in patterning multicellular development in *D. dichotoma* (Bogaert et al., 2019).

15.4.2 *Ectocarpus*

Both *Fucus* and *D. dichotoma* have provided important insights into the cellular events during early embryogenesis but neither model is currently adapted to the analysis of the molecular events underlying developmental processes. This situation may change for *D. dichotoma* in the future as the life cycle of this species can be completed in the laboratory (see above), but this is not the case for *Fucus*. Currently, the most powerful model system for exploring developmental processes in the brown algae at the molecular level is the filamentous alga *Ectocarpus* (Coelho et al., 2020; Coelho and Cock, 2020). As a filamentous alga, this species does not exhibit the same level of developmental complexity as the kelps, for example, but the Ectocarpales are a sister order to the Laminariales (kelps) and the genetic systems that control developmental processes are expected to be conserved between the two taxa.

Ectocarpus was proposed as a general model system several years ago (Peters et al., 2004) based on its potential for genetic and genomic analyses. *Ectocarpus* cultures can be easily maintained in the laboratory and earlier work, principally by Dieter Müller and colleagues at Konstanz University in Germany, had shown that genetic approaches could be successfully applied to this organism (e.g., Müller, 1964; Müller and Eichenberger, 1997). Since 2004, a broad range of tools has been developed for *Ectocarpus* (Coelho et al., 2020; Coelho and Cock, 2020) including notably a high-quality genome assembly (Cock et al., 2010; Cormier et al., 2017). These tools are currently being employed to investigate the genetic basis of developmental processes with the objective of comparing developmental programs in brown algae with those of other eukaryotic lineages that have evolved complex multicellularity.

Analysis of the *Ectocarpus* genome sequence, which was the first macroalgal genome to be described (Cock et al., 2010), allowed the identification of a number of genomic features that may be related to the evolution of multicellularity in the brown algal lineage. These included structural features, such as the presence of a considerable amount of repeated sequence and large numbers of introns, but also the identification of genes that may be important for the regulation of developmental processes, such as transcription factors, ion channels and microRNAs (Cock et al., 2010; Cock and Collén, 2015; Tarver et al., 2015). Interestingly, genome analysis revealed that the brown algae independently evolved membrane-localized receptor kinases, a feature they share with two other complex multicellular lineages: animals and land plants (Cock et al., 2010). An additional four complete brown algal genome sequences have been reported since the publication of the *Ectocarpus* genome, allowing these analyses to be extended to other brown algal species (Table 15.2).

TABLE 15.2
Brown Algal Genome Assemblies

Species	Multicellular or Unicellular	Genome Size (Mbp)	Repeat Sequences	Reference
Ectocarpus sp.	Multicellular	214	22.7%	(Cormier et al., 2017)
Cladosiphon okamuranus	Multicellular	140	4.1%	(Nishitsuji et al., 2016)
Nemacystus decipiens	Multicellular	154	8.8%	(Nishitsuji et al., 2019)
Saccharina japonica	Multicellular	537	40%	(Ye et al., 2015)
Undaria pinnatifida	Multicellular	511	34.2%	(Shan et al., 2020)

Whilst comparative genomics can provide important information about genes potentially linked to the transition to multicellularity, this approach has the weakness that it relies on predicting gene function based on sequence homology and, therefore, that it only allows the detection of features that are shared with the reference organisms that are used for the comparisons. A major recent advance in the brown algal field has been the development of forward genetic approaches, including positional cloning and cloning-by-sequencing, for *Ectocarpus* (Godfroy et al., 2017; Macaisne et al., 2017), which allow the identification of genes that play important roles in developmental processes based solely on mutant phenotypes.

The first brown algal developmental gene to be identified by a genetic approach was *IMMEDIATE UPRIGHT* (*IMM*; Macaisne et al., 2017). This gene is particularly interesting because it is a member of a large gene family that is completely absent from animal and land plant genomes and therefore corresponds to the type of gene that would not be identified by comparative genomic approaches. Mutation of *IMM* leads to a change in the morphology of the basal system of the sporophyte generation. *Ectocarpus* has a haploid diploid life cycle that involves alternation between a diploid sporophyte generation and haploid, male and female (dioicous) gametophytes (Figure 15.4). Both generations are multicellular, filamentous organisms and are of similar size but with distinct morphologies. In *imm* mutants, the normally extensive basal system of the sporophyte is replaced by a structure that resembles the gametophyte basal system, i.e., a small rhizoid-like structure. The IMM protein does not contain any known protein domains but includes a motif that resembles an atypical zinc-finger, suggesting a possible role in signaling. Mutations in a second gene, *DISTAG* (*DIS*), also affect the development of the basal system, but in this mutant, both the sporophyte and the gametophyte generation are affected and both completely lose their basal systems (Godfroy *et al.*, 2017). *DIS* encodes TBCCd1, a protein that has been implicated in cytoskeleton function and cellular architecture in diverse eukaryotic species (Andre et al., 2013; Feldman and Marshall, 2009; Goncalves et al., 2010). Interestingly, the existence of the sporophyte and gametophyte generations means that *Ectocarpus* has had to evolve developmental programs to construct two different multicellular body plans. Based on the phenotypes of the *imm* and *dis* mutants, this process appears to have involved both sharing of developmental components by the two generations (such as the *DIS* gene) and evolution of developmental components specific to each generation (such as the *IMM* gene).

Land plants also have haploid-diploid life cycles, and the deployment of the sporophyte generation has been shown to be under the control of a pair of TALE HD TFs (Horst et al., 2016; Sakakibara et al., 2013), orthologues of which also control the deployment of the diploid phase of the life cycle in unicellular green algae (Lee et al., 2008). Remarkably, deployment of the sporophyte developmental program in *Ectocarpus* has also been shown to be under the control of a pair of TALE HD TFs (called OUROBOROS and SAMSARA; Arun et al., 2019). Based on these observations, together with studies implicating HD TFs in life cycle regulation in other eukaryotic lineages (Hedgethorne et al., 2017; Nasmyth and Shore, 1987), we have suggested that these proteins are all derived from an ancient life cycle regulatory system that has been independently exapted to act as a developmental regulator in at least two complex multicellular lineages, the land plants and the brown algae.

This is a compelling example of a phenomenon referred to as latent homology (Merényi et al., 2020; Nagy et al., 2014) in which an evolutionary process, in this case, the emergence of complex multicellularity, is constrained by the shared, ancestral genetic tool kit, leading to convergent evolution of similar regulatory systems.

Although only a limited number of brown algal developmental genes have been characterized so far, these analyses have already provided several interesting insights into the evolution of developmental processes and multicellularity, including the identification of evolutionary novelties, convergent evolution constrained by an ancestral tool kit and differences between the developmental programs of body plans deployed at different generations of the life cycle. Ongoing efforts to use forward genetic approaches, coupled with complementary methodologies such as transcriptomics and network analysis, to identify additional key regulatory genes are expected to significantly expand our knowledge of brown algal regulatory gene networks in the coming years. Additional approaches of interest include recent studies aimed at characterizing chromatin modifications in brown algae (Bourdareau et al., 2020; Fan et al., 2020; Gueno et al., 2020), as epigenetic processes presumably play important roles in establishing and maintaining the differentiated cell states that underlie multicellular development.

15.5 CONCLUSION

The algae played a prominent role in the emergence of multicellularity over the course of evolution. Three of the five transitions to complex multicellularity involved algal lineages (the red, green and brown algae) and multiple additional algal taxa underwent transitions from unicellularity to simple forms of multicellularity. They also include the first eukaryotic lineage to undergo the transition to complex multicellularity, the red algae. As a consequence, algae potentially represent highly interesting model systems to understand many aspects of the evolution of multicellularity, including the two key transitions from a unicellular state to multicellularity and from simple to complex multicellularity. However, with the exception of long-standing model systems within the Volvocales (Matt and Umen, 2016), algae have remained a relatively underexploited resource, mainly due to a dearth of genomic information and a lack of effective genetic tools to explore gene function. This situation is changing as an increasing number of whole-genome sequences are being made available for a broad range of algal species and with the emergence of model species associated with powerful tools for forward and reverse genetic approaches (Brodie et al., 2017; Coelho and Cock, 2020). This process is expected to accelerate in the coming years with the completion of large-scale genome sequencing projects and adaptation of new tools such as CRISPR-Cas9 to algal systems (Badis et al., 2021; Nymark et al., 2016; Shin et al., 2016). With these advances, we expect algae to continue to make important contributions to multicellularity research.

ACKNOWLEDGMENTS

We thank Luna Van der Loos and Xiaojie Liu for *Ulva* spp. photos.

S.M.C. is supported by the ERC (grant 638240) and MC by the ANR (ANR-19-CE20-0028-01). This work was supported by the Centre National de la Recherche Scientifique and by Sorbonne University.

REFERENCES

Andre J, Harrison S, Towers K, Qi X, Vaughan S, McKean PG, Ginger ML. 2013. The tubulin cofactor C family member TBCCD1 orchestrates cytoskeletal filament formation. *J Cell Sci* **126**:5350–5356. doi:10.1242/jcs.136515

Arun A, Coelho SM, Peters AF, Bourdareau S, Pérès L, Scornet D, Strittmatter M, Lipinska AP, Yao H, Godfroy O, Montecinos GJ, Avia K, Macaisne N, Troadec C, Bendahmane A, Cock JM. 2019. Convergent recruitment of TALE homeodomain life cycle regulators to direct sporophyte development in land plants and brown algae. *eLife* **8**:e43101. doi:10.7554/eLife.43101

Badis Y, Scornet D, Harada M, Caillard C, Raphalen M, Gachon CMM, Coelho SM, Motomura T, Nagasato C, Cock, JM. 2021. Targeted CRISPR-Cas9-based gene knockouts in the model brown alga *Ectocarpus*. *New Phytol* **231**:2077–2091. doi:10.1111/nph.17525

Bengtson S, Sallstedt T, Belivanova V, Whitehouse M. 2017. Three-dimensional preservation of cellular and subcellular structures suggests 1.6 billion-year-old crown-group red algae. *PLoS Biol* **15**:e2000735. doi:10.1371/journal.pbio.2000735

Berger F, Taylor A, Brownlee C. 1994. Cell fate determination by the cell wall in early *Fucus* development. *Science* **263**:1421–1423.

Bhattacharya D, Price DC, Chan CX, Qiu H, Rose N, Ball S, Weber APM, Arias MC, Henrissat B, Coutinho PM, Krishnan A, Zäuner S, Morath S, Hilliou F, Egizi A, Perrineau M-M, Yoon HS. 2013. Genome of the red alga *Porphyridium purpureum*. *Nat Commun* **4**:1941. doi:10.1038/ncomms2931

Bhattacharya D, Yoon H, Hedges S, Hackett J. 2009. Eukaryotes In: Hedges S, Kumar S, editors. *The Timetree of Life*. New York: Oxford University Press. pp. 116–120.

Bodyl A. 2018. Did some red alga-derived plastids evolve via kleptoplastidy? A hypothesis. *Biol Rev* **93**:201–222. doi:10.1111/brv.12340

Bogaert, K., Beeckman, T., De Clerck, O. 2016. Abiotic regulation of growth and fertility in the sporophyte of *Dictyota dichotoma* (Hudson) J.V. Lamouroux (Dictyotales, Phaeophyceae). *J Appl Phycol* **28**: 2915–2924 (2016). https://doi.org/10.1007/s10811-016-0801-z

Bogaert KA, Beeckman T, De Clerck O. 2017. Two-step cell polarization in algal zygotes. *Nat Plants* **3**:16221. doi:10.1038/nplants.2016.221

Bogaert KA, Blommaert L, Ljung K, Beeckman T, De Clerck O. 2019. Auxin function in the brown alga *Dictyota dichotoma*. *Plant Physiol* **179**:280–299. doi:10.1104/pp.18.01041

Bogaert KA, Delva S, De Clerck O. 2020. Concise review of the genus *Dictyota* J.V. Lamouroux. *J Appl Phycol* **32**:1521–1543. doi:10.1007/s10811-020-02121-4

Bouget FY, Berger F, Brownlee C. 1998. Position dependent control of cell fate in the *Fucus* embryo: role of intercellular communication. *Development* **125**:1999–2008.

Bourdareau S, Tirichine L, Lombard B, Loew D, Scornet D, Wu Y, Coelho SM, Cock JM. 2021. Histone modifications during the life cycle of the brown alga *Ectocarpus*. *Genome Biol* **22**:1-27. doi:10.1186/s13059-020-02216-8

Brawley SH, Blouin NA, Ficko-Blean E, Wheeler GL, Lohr M, Goodson HV, Jenkins JW, Blaby-Haas CE, Helliwell KE, Chan CX, Marriage TN, Bhattacharya D, Klein AS, Badis Y, Brodie J, Cao Y, Collén J, Dittami SM, Gachon CMM, Green BR, Karpowicz SJ, Kim JW, Kudahl UJ, Lin S, Michel G, Mittag M, Olson BJSC, Pangilinan JL, Peng Y, Qiu H, Shu S, Singer JT, Smith AG, Sprecher BN, Wagner V, Wang W, Wang Z-Y, Yan J, Yarish C, Zäuner-Riek S, Zhuang Y, Zou Y, Lindquist EA, Grimwood J, Barry KW, Rokhsar DS, Schmutz J, Stiller JW, Grossman AR, Prochnik SE. 2017. Insights into the red algae and eukaryotic evolution from the genome of *Porphyra umbilicalis* (Bangiophyceae, Rhodophyta). *Proc Natl Acad Sci* **114**:E6361–E6370. doi:10.1073/pnas.1703088114

Bringloe TT, Starko S, Wade RM, Vieira C, Kawai H, Clerck OD, Cock JM, Coelho SM, Destombe C, Valero M, Neiva J, Pearson GA, Faugeron S, Serrão EA, Verbruggen H. 2020. Phylogeny and evolution of the brown algae. *Crit Rev Plant Sci* **39**:281–321. doi: 10.1080/07352689.2020.1787679

Brodie J, Chan CX, De Clerck O, Cock JM, Coelho SM, Gachon C, Grossman AR, Mock T, Raven JA, Smith AG, Yoon HS, Bhattacharya D. 2017. The algal revolution. *Trends Plant Sci* **22**:726–738. doi:10.1016/j.tplants.2017.05.005

Brownlee C, Bouget FY. 1998. Polarity determination in *Fucus*: from zygote to multicellular embryo. *Semin Cell Dev Biol* **9**:179–85. doi:10.1006/scdb.1997.0212

Brownlee C, Bouget FY, Corellou F. 2001. Choosing sides: establishment of polarity in zygotes of fucoid algae. *Semin Cell Dev Biol* **12**:345–51. doi:10.1006/scdb.2001.0262

Burki F, Roger AJ, Brown MW, Simpson AGB. 2020. The new tree of eukaryotes. *Trends Ecol Evol* **35**:43–55. doi:10.1016/j.tree.2019.08.008

Butterfield NJ. 2000. *Bangiomorpha pubescens* n. gen., n. sp.: Implications for the evolution of sex, multicellularity, and the Mesoproterozoic/Neoproterozoic radiation of eukaryotes. *Paleobiol* **26**:386–404.

Cao M, Xu K, Yu X, Bi G, Liu Y, Kong F, Sun P, Tang X, Du G, Ge Y, Wang D, Mao Y. 2020. A chromosome-level genome assembly of *Pyropia haitanensis* (Bangiales, Rhodophyta). *Mol Ecol Resour* **20**:216–227. doi:10.1111/1755-0998.13102

Chan CX, Yang EC, Banerjee T, Yoon HS, Martone PT, Estevez JM, Bhattacharya D. 2011. Red and green algal monophyly and extensive gene sharing found in a rich repertoire of red algal genes. *Curr Biol* **21**:328–33. doi:10.1016/j.cub.2011.01.037

Charrier B, Le Bail A, de Reviers B. 2012. Plant Proteus: brown algal morphological plasticity and underlying developmental mechanisms. *Trends Plant Sci* **17**:468–77. doi:10.1016/j.tplants.2012.03.003

Cock JM, Collén J. 2015. Independent emergence of complex multicellularity in the brown and red algae In: Ruiz-Trillo I, Nedelcu AM, editors. *Evolutionary Transitions to Multicellular Life, Advances in Marine Genomics*. Springer Verlag. pp. 335–361.

Cock JM, Sterck L, Rouzé P, Scornet D, Allen AE, Amoutzias G, Anthouard V, Artiguenave F, Aury J, Badger J, Beszteri B, Billiau K, Bonnet E, Bothwell J, Bowler C, Boyen C, Brownlee C, Carrano C, Charrier B, Cho G, Coelho S, Collén J, Corre E, Da Silva C, Delage L, Delaroque N, Dittami S, Doulbeau S, Elias M, Farnham G, Gachon C, Gschloessl B, Heesch S, Jabbari K, Jubin C, Kawai H, Kimura K, Kloareg B, Küpper F, Lang D, Le Bail A, Leblanc C, Lerouge P, Lohr M, Lopez P, Martens C, Maumus F, Michel G, Miranda-Saavedra D, Morales J, Moreau H, Motomura T, Nagasato C, Napoli C, Nelson D, Nyvall-Collén P, Peters A, Pommier C, Potin P, Poulain J, Quesneville H, Read B, Rensing S, Ritter A, Rousvoal S, Samanta M, Samson G, Schroeder D, Ségurens B, Strittmatter M, Tonon T, Tregear J, Valentin K, von Dassow P, Yamagishi T, Van de Peer Y, Wincker P. 2010. The *Ectocarpus* genome and the independent evolution of multicellularity in brown algae. *Nature* **465**:617–621. doi:10.1038/nature09016

Coelho S, Cock J. 2020. Brown algal model organisms. *Ann Rev Genet* **54**:71–92.

Coelho SM, Peters AF, Müller D, Cock JM. 2020. *Ectocarpus:* an evo-devo model for the brown algae. *EvoDevo* **11**:19. doi:10.1186/s13227-020-00164-9.

Collén J, Porcel B, Carré W, Ball SG, Chaparro C, Tonon T, Barbeyron T, Michel G, Noel B, Valentin K, Elias M, Artiguenave F, Arun A, Aury JM, Barbosa-Neto JF, Bothwell JH, Bouget FY, Brillet L, Cabello-Hurtado F, Capella-Gutiérrez S, Charrier B, Cladière L, Cock JM, Coelho SM, Colleoni C, Czjzek M, Da Silva C, Delage L, Denoeud F, Deschamps P, Dittami SM, Gabaldón T, Gachon CM, Groisillier A, Hervé C, Jabbari K, Katinka M, Kloareg B, Kowalczyk N, Labadie K, Leblanc C, Lopez PJ, McLachlan DH, Meslet-Cladiere L, Moustafa A, Nehr Z, Nyvall Collén P, Panaud O, Partensky F, Poulain J, Rensing SA, Rousvoal S, Samson G, Symeonidi A, Weissenbach J,

Zambounis A, Wincker P, Boyen C. 2013. Genome structure and metabolic features in the red seaweed *Chondrus crispus* shed light on evolution of the Archaeplastida. *Proc Natl Acad Sci USA* **110**:5247–52. doi:10.1073/pnas.1221259110.

Cormier A, Avia K, Sterck L, Derrien T, Wucher V, Andres G, Monsoor M, Godfroy O, Lipinska A, Perrineau M-M, Van De Peer Y, Hitte C, Corre E, Coelho SM, Cock JM. 2017. Re-annotation, improved large-scale assembly and establishment of a catalogue of noncoding loci for the genome of the model brown alga *Ectocarpus*. *New Phytol* **214**:219–232. doi:10.1111/nph.14321

De Clerck O, Bogaert KA, Leliaert F. 2012. Chapter two – Diversity and evolution of algae: Primary endosymbiosis In: Piganeau G, editor. *Advances in Botanical Research, Genomic Insights into the Biology of Algae*. Academic Press. pp. 55–86. doi:10.1016/B978-0-12-391499-6.00002-5

De Clerck O, Kao SM, Bogaert KA, Blomme J, Foflonker F, Kwantes M, Vancaester E, Vanderstraeten L, Aydogdu E, Boesger J, Califano G, Charrier B, Clewes R, Del Cortona A, D'Hondt S, Fernandez-Pozo N, Gachon CM, Hanikenne M, Lattermann L, Leliaert F, Liu X, Maggs CA, Popper ZA, Raven JA, Van Bel M, Wilhelmsson PKI, Bhattacharya D, Coates JC, Rensing SA, Van Der Straeten D, Vardi A, Sterck L, Vandepoele K, Van de Peer Y, Wichard T, Bothwell JH. 2018. Insights into the evolution of multicellularity from the sea lettuce genome. *Curr Biol CB* **28**:2921–2933.e5. doi:10.1016/j.cub.2018.08.015

De Smet I, Beeckman T. 2011. Asymmetric cell division in land plants and algae: the driving force for differentiation. *Nat Rev Mol Cell Biol* **12**:177–88. doi:10.1038/nrm3064

Duncan L, Nishii I, Harryman A, Buckley S, Howard A, Friedman NR, Miller SM. 2007. The *VARL* gene family and the evolutionary origins of the master cell-type regulatory gene, *regA*, in *Volvox carteri*. *J Mol Evol* **65**:1–11. doi:10.1007/s00239-006-0225-5

Egan S, Harder T, Burke C, Steinberg P, Kjelleberg S, Thomas T. 2013. The seaweed holobiont: understanding seaweed-bacteria interactions. *FEMS Microbiol Rev* **37**:462–476. doi:10.1111/1574-6976.12011

Fan X, Han W, Teng L, Jiang P, Zhang X, Xu D, Li C, Pellegrini M, Wu C, Wang Y, Kaczurowski MJS, Lin X, Tirichine L, Mock T, Ye N. 2020. Single-base methylome profiling of the giant kelp *Saccharina japonica* reveals significant differences in DNA methylation to microalgae and plants. *New Phytol* **225**:234–249. doi:10.1111/nph.16125

Fang SC, de los Reyes C, Umen JG. 2006. Cell size checkpoint control by the retinoblastoma tumor suppressor pathway. *PLoS Genet* **2**:e167. doi:10.1371/journal.pgen.0020167

Featherston J, Arakaki Y, Hanschen ER, Ferris PJ, Michod RE, Olson BJSC, Nozaki H, Durand PM. 2018. The 4-celled *Tetrabaena socialis* nuclear genome reveals the essential components for genetic control of cell number at the origin of multicellularity in the volvocine lineage. *Mol Biol Evol* **35**:855–870. doi:10.1093/molbev/msx332

Feldman JL, Marshall WF. 2009. ASQ2 encodes a TBCC-like protein required for mother-daughter centriole linkage and mitotic spindle orientation. *Curr Biol* **19**:1238–1243. doi:10.1016/j.cub.2009.05.071

Godfroy O, Uji T, Nagasato C, Lipinska AP, Scornet D, Peters AF, Avia K, Colin S, Mignerot L, Motomura T, Cock JM, Coelho SM. 2017. DISTAG/TBCCd1 is required for basal cell fate determination in *Ectocarpus*. *Plant Cell* **29**:3102–3122. doi:10.1105/tpc.17.00440

Goecke F, Labes A, Wiese J, Imhoff JF. 2010. Chemical interactions between marine macroalgae and bacteria. *Mar Ecol Prog Ser* **409**:267–299. doi:10.3354/meps08607

Goncalves J, Nolasco S, Nascimento R, Lopez Fanarraga M, Zabala JC, Soares H. 2010. TBCCD1, a new centrosomal protein, is required for centrosome and Golgi apparatus positioning. *EMBO Rep* **11**:194–200. doi:10.1038/embor.2010.5

Gueno J, Bourdareau S, Cossard G, Godfroy O, Lipinska A, Tirichine L, Cock JM, Coelho SM. 2020. Chromatin dynamics associated with sexual differentiation in a UV sex determination system. *bioRxiv* 2020.10.29.359190. doi:10.1101/2020.10.29.359190

Hanschen ER, Marriage TN, Ferris PJ, Hamaji T, Toyoda A, Fujiyama A, Neme R, Noguchi H, Minakuchi Y, Suzuki M, Kawai-Toyooka H, Smith DR, Sparks H, Anderson J, Bakarić R, Luria V, Karger A, Kirschner MW, Durand PM, Michod RE, Nozaki H, Olson BJSC. 2016. The *Gonium pectorale* genome demonstrates co-option of cell cycle regulation during the evolution of multicellularity. *Nat Commun* **7**:11370. doi:10.1038/ncomms11370

Hedges SB, Blair JE, Venturi ML, Shoe JL. 2004. A molecular timescale of eukaryote evolution and the rise of complex multicellular life. *BMC Evol Biol* **4**:2. doi:10.1186/1471-2148-4-2

Hedgethorne K, Eustermann S, Yang J-C, Ogden TEH, Neuhaus D, Bloomfield G. 2017. Homeodomain-like DNA binding proteins control the haploid-to-diploid transition in *Dictyostelium*. *Sci Adv* **3**:e1602937. doi:10.1126/sciadv.1602937

Heesch S, Serrano-Serrano M, Luthringer R, Peters AF, Destombe C, Cock JM, Valero M, Roze D, Salamin N, Coelho S. 2019. Evolution of life cycles and reproductive traits: insights from the brown algae. *bioRxiv* 530477. doi:10.1101/530477

Herron MD, Ratcliff WC, Boswell J, Rosenzweig F. 2018. Genetics of a *de novo* origin of undifferentiated multicellularity. *R Soc Open Sci* **5**:180912. doi:10.1098/rsos.180912

Ho C-L, Lee W-K, Lim E-L. 2018. Unraveling the nuclear and chloroplast genomes of an agar producing red macroalga, *Gracilaria changii* (Rhodophyta, Gracilariales). *Genomics* **110**:124–133. doi:10.1016/j.ygeno.2017.09.003

Horst NA, Katz A, Pereman I, Decker EL, Ohad N, Reski R. 2016. A single homeobox gene triggers phase transition, embryogenesis and asexual reproduction. *Nat Plants* **2**:15209. doi:10.1038/nplants.2015.209

Jacobs WP. 1970. Development and regeneration of the algal giant coenocyte *Caulerpa*. *NY Acad Sci Ann* **175**:732–748. doi:10.1111/j.1749-6632.1970.tb45188.x

Kawai H, Hanyuda T, Draisma SGA, Wilce RT, Andersen RA. 2015. Molecular phylogeny of two unusual brown algae, *Phaeostrophion irregulare* and *Platysiphon glacialis*, proposal of the Stschapoviales ord. nov. and Platysiphonaceae fam. nov., and a re-examination of divergence times for brown algal orders. *J Phycol* **51**:918–928. doi:10.1111/jpy.12332

Keeling PJ. 2013. The number, speed, and impact of plastid endosymbioses in eukaryotic evolution. *Annu Rev Plant Biol* **64**:583–607. doi:10.1146/annurev-arplant-050312-120144

Kirk D. 1998. *Volvox: Molecular-Genetic Origins of Multicellularity and Cellular Differentiation*. Cambridge, UK: Cambridge University Press.

Kirk DL. 2005. A twelve-step program for evolving multicellularity and a division of labor. *BioEssays News Rev Mol Cell Dev Biol* **27**:299–310. doi:10.1002/bies.20197

Kirk DL. 2001. Germ-soma differentiation in *Volvox*. *Dev Biol* **238**:213–223. doi:10.1006/dbio.2001.0402

Kirk MM, Ransick A, McRae SE, Kirk DL. 1993. The relationship between cell size and cell fate in *Volvox carteri*. *J Cell Biol* **123**:191–208. doi:10.1083/jcb.123.1.191

Kirk MM, Stark K, Miller SM, Müller W, Taillon BE, Gruber H, Schmitt R, Kirk DL. 1999. *regA*, a *Volvox* gene that plays a central role in germ-soma differentiation, encodes a novel regulatory protein. *Dev Camb Engl* **126**:639–647.

Knoll AH. 2011. The multiple origins of complex multicellularity. *Annu Rev Earth Planet Sci* **39**:217–239.

Lee J, Yang EC, Graf L, Yang JH, Qiu H, Zelzion U, Chan CX, Stephens TG, Weber APM, Boo GH, Boo SM, Kim KM, Shin Y, Jung M, Lee SJ, Yim HS, Lee JH, Bhattacharya D, Yoon HS. 2018. Analysis of the draft genome of the red seaweed *Gracilariopsis chorda* provides insights into genome size evolution in Rhodophyta. *Mol Biol Evol* **35**:1869–1886. doi:10.1093/molbev/msy081

Lee JH, Lin H, Joo S, Goodenough U. 2008. Early sexual origins of homeoprotein heterodimerization and evolution of the plant KNOX/BELL family. *Cell* **133**:829–840. doi:10.1016/j.cell.2008.04.028

Leliaert F, Smith DR, Moreau H, Herron MD, Verbruggen H, Delwiche CF, De Clerck O. 2012. Phylogeny and molecular evolution of the green algae. *Crit Rev Plant Sci* **31**:1–46. doi:10.1080/07352689.2011.615705

Lewis LA, McCourt RM. 2004. Green algae and the origin of land plants. *Am J Bot* **91**:1535–1556. doi:10.3732/ajb.91.10.1535

Lynch M, Conery JS. 2003. The origins of genome complexity. *Science* **302**:1401–4. doi:10.1126/science.1089370

Macaisne N, Liu F, Scornet D, Peters AF, Lipinska A, Perrineau M-M, Henry A, Strittmatter M, Coelho SM, Cock JM. 2017. The *Ectocarpus IMMEDIATE UPRIGHT* gene encodes a member of a novel family of cysteine-rich proteins with an unusual distribution across the eukaryotes. *Development* **144**:409–418. doi:10.1242/dev.141523

Marin B, Nowack EC, Melkonian M. 2005. A plastid in the making: evidence for a second primary endosymbiosis. *Protist* **156**:425–32. doi:10.1016/j.protis.2005.09.001

Matsuzaki M, Misumi O, Shin-I T, Maruyama S, Takahara M, Miyagishima SY, Mori T, Nishida K, Yagisawa F, Yoshida Y, Nishimura Y, Nakao S, Kobayashi T, Momoyama Y, Higashiyama T, Minoda A, Sano M, Nomoto H, Oishi K, Hayashi H, Ohta F, Nishizaka S, Haga S, Miura S, Morishita T, Kabeya Y, Terasawa K, Suzuki Y, Ishii Y, Asakawa S, Takano H, Ohta N, Kuroiwa H, Tanaka K, Shimizu N, Sugano S, Sato N, Nozaki H, Ogasawara N, Kohara Y, Kuroiwa T. 2004. Genome sequence of the ultrasmall unicellular red alga *Cyanidioschyzon merolae* 10D. *Nature* **428**:653–7. doi:10.1038/nature02398

Matt G, Umen J. 2016. *Volvox*: A simple algal model for embryogenesis, morphogenesis and cellular differentiation. *Dev Biol.* **419**:99-113 doi:10.1016/j.ydbio.2016.07.014

Matt GY, Umen JG. 2018. Cell-type transcriptomes of the multicellular green alga *Volvox carteri* yield insights into the evolutionary origins of germ and somatic differentiation programs. *G3* **8**:531–550. doi:10.1534/g3.117.300253

Mattox K, Stewart K. 1984. Classification of the green algae: a concept based on comparative cytology In: Irvine D, John D, editors. *The Systematics of Green Algae*. London: Academic Press. pp. 29–72.

Merchant, S. S., Prochnik, S. E., Vallon, O., Harris, E. H., Karpowicz, S. J., Witman, G. B., Terry, A., Salamov, A., Fritz-Laylin, L. K., Maréchal-Drouard, L., Marshall, W. F., Qu, L. H., Nelson, D. R., Sanderfoot, A. A., Spalding, M. H., Kapitonov, V. V., Ren, Q., Ferris, P., Lindquist, E., Shapiro, H., … Grossman, A. R. 2007. The *Chlamydomonas* genome reveals the evolution of key animal and plant functions. *Science* **318**:245–250. https://doi.org/10.1126/science.1143609

Merényi Z, Prasanna AN, Wang Z, Kovács K, Hegedüs B, Bálint B, Papp B, Townsend JP, Nagy LG. 2020. Unmatched level of molecular convergence among deeply divergent complex multicellular fungi. *Mol Biol Evol* **37**:2228–2240. doi:10.1093/molbev/msaa077

Mikami K, Li C, Irie R, Hama Y. 2019. A unique life cycle transition in the red seaweed *Pyropia yezoensis* depends on apospory. *Commun Biol* **2**:299. doi:10.1038/s42003-019-0549-5

Mitman GG, Meer JP. 1994. Meiosis, blade development, and sex determination in *Porphyra purpurea* (Rhodophyta). *J Phycol* **30**:147–159. doi:10.1111/j.0022-3646.1994.00147.x

Müller DG. 1964. Life-cycle of *Ectocarpus siliculosus* from Naples, Italy. *Nature* **26**:1402.

Müller DG, Eichenberger W. 1997. Mendelian genetics in brown algae: inheritance of a lipid defect mutation and sex alleles in *Ectocarpus siliculosus* (Ectocarpales, Phaeophyceae). *Phycologia* **36**:79–81.

Nagy LG, Ohm RA, Kovács GM, Floudas D, Riley R, Gácser A, Sipiczki M, Davis JM, Doty SL, de Hoog GS, Lang BF, Spatafora JW, Martin FM, Grigoriev IV, Hibbett DS. 2014. Latent homology and convergent regulatory evolution underlies the repeated emergence of yeasts. *Nat Commun* **5**:4471. doi:10.1038/ncomms5471

Nakamura Y, Sasaki N, Kobayashi M, Ojima N, Yasuike M, Shigenobu Y, Satomi M, Fukuma Y, Shiwaku K, Tsujimoto A, Kobayashi T, Nakayama I, Ito F, Nakajima K, Sano M, Wada T, Kuhara S, Inouye K, Gojobori T, Ikeo K. 2013. The first symbiont-free genome sequence of marine red alga, susabi-nori (*Pyropia yezoensis*). *PLoS One* **8**:e57122. doi:10.1371/journal.pone.0057122

Nasmyth K, Shore D. 1987. Transcriptional regulation in the yeast life cycle. *Science* **237**:1162–1170.

Nedelcu AM. 2009. Environmentally induced responses co-opted for reproductive altruism. *Biol Lett* **5**:805–808. doi:10.1098/rsbl.2009.0334

Nedelcu AM, Michod RE. 2006. The evolutionary origin of an altruistic gene. *Mol Biol Evol* **23**:1460–1464. doi:10.1093/molbev/msl016

Nishitsuji K, Arimoto A, Higa Y, Mekaru M, Kawamitsu M, Satoh N, Shoguchi E. 2019. Draft genome of the brown alga, *Nemacystus decipiens*, Onna-1 strain: Fusion of genes involved in the sulfated fucan biosynthesis pathway. *Sci Rep* **9**:4607. doi:10.1038/s41598-019-40955-2

Nishitsuji K, Arimoto A, Iwai K, Sudo Y, Hisata K, Fujie M, Arakaki N, Kushiro T, Konishi T, Shinzato C, Satoh N, Shoguchi E. 2016. A draft genome of the brown alga, *Cladosiphon okamuranus*, S-strain: a platform for future studies of "mozuku" biology. *DNA Res* **23**:561–570. doi:10.1093/dnares/dsw039

Niwa K. 2010. Genetic analysis of artificial green and red mutants of *Porphyra yezoensis* Ueda (Bangiales, Rhodophyta). *Aquaculture* **308**:6–12. doi:10.1016/j.aquaculture.2010.08.007

Nozaki H. 1990. Ultrastructure of the extracellular matrix of *Gonium* (Volvocales, Chlorophyta). *Phycologia* **29**:1–8. doi:10.2216/i0031-8884-29-1-1.1

Nymark M, Sharma AK, Sparstad T, Bones AM, Winge P. 2016. A CRISPR/Cas9 system adapted for gene editing in marine algae. *Sci Rep* **6**:24951. doi:10.1038/srep24951

Peters AF, Marie D, Scornet D, Kloareg B, Cock JM. 2004. Proposal of *Ectocarpus siliculosus* (Ectocarpales, Phaeophyceae) as a model organism for brown algal genetics and genomics. *J Phycol* **40**:1079–1088.

Peters AF, Scornet D, Ratin M, Charrier B, Monnier A, Merrien Y, Corre E, Coelho SM, Cock JM. 2008. Life-cycle-generation-specific developmental processes are modified in the *immediate upright* mutant of the brown alga *Ectocarpus siliculosus*. *Development* **135**:1503–1512. doi:10.1242/dev.016303

Prochnik SE, Umen J, Nedelcu AM, Hallmann A, Miller SM, Nishii I, Ferris P, Kuo A, Mitros T, Fritz-Laylin LK, Hellsten U, Chapman J, Simakov O, Rensing SA, Terry A, Pangilinan J, Kapitonov V, Jurka J, Salamov A, Shapiro H, Schmutz J, Grimwood J, Lindquist E, Lucas S, Grigoriev IV, Schmitt R, Kirk D, Rokhsar DS. 2010. Genomic analysis of organismal complexity in the multicellular green alga *Volvox carteri*. *Science* **329**:223–226. doi:10.1126/science.1188800

Qiu H, Yoon HS, Bhattacharya D. 2016. Red algal phylogenomics provides a robust framework for inferring evolution of key metabolic pathways. *PLoS Curr* **8**. doi:10.1371/currents.tol.7b037376e6d84a1be34af756a4d90846

Ranjan A, Townsley BT, Ichihashi Y, Sinha NR, Chitwood DH. 2015. An intracellular transcriptomic atlas of the giant coenocyte *Caulerpa taxifolia*. *PLoS Genet* **11**:e1004900. doi:10.1371/journal.pgen.1004900

Raven JA. 1997. Miniview: Multiple origins of plasmodesmata. *Eur J Phycol* **32**:95–101. doi:10.1080/09670269710001737009

Sakakibara K, Ando S, Yip HK, Tamada Y, Hiwatashi Y, Murata T, Deguchi H, Hasebe M, Bowman JL. 2013. KNOX2 genes regulate the haploid-to-diploid morphological transition in land plants. *Science* **339**:1067–1070. doi:10.1126/science.1230082

Schönknecht G, Chen WH, Ternes CM, Barbier GG, Shrestha RP, Stanke M, Bräutigam A, Baker BJ, Banfield JF, Garavito RM, Carr K, Wilkerson C, Rensing SA, Gagneul D, Dickenson NE, Oesterhelt C, Lercher MJ, Weber AP. 2013. Gene transfer from bacteria and archaea facilitated evolution of an extremophilic eukaryote. *Science* **339**:1207–10. doi:10.1126/science.1231707

Shan T, Yuan J, Su L, Li J, Leng X, Zhang Y, Gao H, Pang S. 2020. First genome of the brown alga *Undaria pinnatifida*: Chromosome-level assembly using PacBio and Hi-C technologies. *Front Genet* **11**:140. doi:10.3389/fgene.2020.00140

Shin SE, Lim JM, Koh HG, Kim EK, Kang NK, Jeon S, Kwon S, Shin WS, Lee B, Hwangbo K, Kim J, Ye SH, Yun JY, Seo H, Oh HM, Kim KJ, Kim JS, Jeong WJ, Chang YK, Jeong BR. 2016. CRISPR/Cas9-induced knockout and knock-in mutations in *Chlamydomonas reinhardtii*. *Sci Rep* **6**:27810. doi:10.1038/srep27810

Silberfeld T, Rousseau F, de Reviers B. 2014. An updated classification of brown algae (Ochrophyta, Phaeophyceae). *Cryptogam Algol* **35**:117–156.

Spoerner M, Wichard T, Bachhuber T, Stratmann J, Oertel W. 2012. Growth and thallus morphogenesis of *Ulva mutabilis* (Chlorophyta) depends on a combination of two bacterial species excreting regulatory factors. *J Phycol* **48**:1433–1447. doi:10.1111/j.1529-8817.2012.01231.x

Stark K, Kirk DL, Schmitt R. 2001. Two enhancers and one silencer located in the introns of *regA* control somatic cell differentiation in *Volvox carteri*. *Genes Dev* **15**:1449–1460. doi:10.1101/gad.195101

Stratmann J, Paputsoglu G, Oertel W. 1996. Differentiation of *Ulva Mutabilis* (chlorophyta) gametangia and gamete release are controlled by extracellular inhibitors. *J Phycol* **32**:1009–1021. doi:10.1111/j.0022-3646.1996.01009.x

Tarver JE, Cormier A, Pinzón N, Taylor RS, Carré W, Strittmatter M, Seitz H, Coelho SM, Cock JM. 2015. MicroRNAs and the evolution of complex multicellularity: identification of a large, diverse complement of microRNAs in the brown alga *Ectocarpus*. *Nucl Acids Res* **43**:6384–6398.

Umen, J.G. 2020. Volvox and volvocine green algae. *EvoDevo* **11**. https://doi.org/10.1186/s13227-020-00158-7

Umen JG. 2014. Green algae and the origins of multicellularity in the plant kingdom. *Cold Spring Harb Perspect Biol* **6**:a016170. doi:10.1101/cshperspect.a016170

Umen JG, Goodenough UW. 2001. Control of cell division by a retinoblastoma protein homolog in *Chlamydomonas*. *Genes Dev* **15**:1652–1661. doi:10.1101/gad.892101

Umen JG, Olson BJSC. 2012. Chapter six – genomics of volvocine algae In: Piganeau G, editor. *Advances in Botanical Research, Genomic Insights into the Biology of Algae*. Academic Press. pp. 185–243. doi:10.1016/B978-0-12-391499-6.00006-2

Van Den Hoek C, Mann DG, Jahns HM. Algae. An introduction to phycology. *Eur J Phycol* **32**:203–205. doi:10.1017/S096702629621100X

Van der Meer JP, Cole K, Sheath R. 1990. *Genetics Biology of the Red Algae*. Cambridge University Press. pp. 103–121.

Waaland JR, Stiller JW, Cheney DP. 2004. Macroalgal candidates for genomics. *J Phycol* **40**:26–33. doi:10.1111/j.0022-3646.2003.03-148.x

Waaland SD. 1990. Development In: Cole K, Sheath R, editors. *Biology of the Red Algae*. Cambridge University Press. pp. 259–273.

Wang D, Yu X, Xu K, Bi G, Cao M, Zelzion E, Fu C, Sun P, Liu Y, Kong F, Du G, Tang X, Yang R, Wang J, Tang L, Wang L, Zhao Y, Ge Y, Zhuang Y, Mo Z, Chen Y, Gao T, Guan X, Chen R, Qu W, Sun B, Bhattacharya D, Mao Y. 2020. *Pyropia yezoensis* genome reveals diverse mechanisms of carbon acquisition in the intertidal environment. *Nat Commun* **11**:4028. doi:10.1038/s41467-020-17689-1

Wichard T, Charrier B, Mineur F, Bothwell JH, Clerck OD, Coates JC. 2015. The green seaweed *Ulva*: a model system to study morphogenesis. *Front Plant Sci* **6**:72. doi:10.3389/fpls.2015.00072

Yamazaki T, Ichihara K, Suzuki R, Oshima K, Miyamura S, Kuwano K, Toyoda A, Suzuki Y, Sugano S, Hattori M, Kawano S. 2017. Genomic structure and evolution of the mating type locus in the green seaweed *Ulva partita*. *Sci Rep* **7**:11679. doi:10.1038/s41598-017-11677-0

Ye N, Zhang X, Miao M, Fan X, Zheng Y, Xu D, Wang J, Zhou L, Wang D, Gao Y, Wang Y, Shi W, Ji P, Li D, Guan Z, Shao C, Zhuang Z, Gao Z, Qi J, Zhao F. 2015. *Saccharina* genomes provide novel insight into kelp biology. *Nat Commun* **6**:6986. doi:10.1038/ncomms7986

Zhang X, Ye N, Liang C, Mou S, Fan X, Xu J, Xu D, Zhuang Z. 2012. De novo sequencing and analysis of the *Ulva linza* transcriptome to discover putative mechanisms associated with its successful colonization of coastal ecosystems. *BMC Genomics* **13**:565. doi:10.1186/1471-2164-13-565

Zhou W, Hu Y, Sui Z, Fu F, Wang J, Chang L, Guo W, Li B. 2013. Genome survey sequencing and genetic background characterization of *Gracilariopsis lemaneiformis* (Rhodophyta) based on next-generation sequencing. *PloS One* **8**:e69909. doi:10.1371/journal.pone.0069909

16 The Evolution of Complex Multicellularity in Streptophytes

Liam N. Briginshaw
School of Biological Sciences, Monash University,
Melbourne, VIC, Australia

John L. Bowman
School of Biological Sciences, Monash University,
Melbourne, VIC, Australia

CONTENTS

16.1 THE EARLY DIVERGING STREPTOPHYTES – UNICELLULAR AND SIMPLE MULTICELLULAR ALGAE

Embryophytes (land plants) are nested within a grade of charophyte algae, and together these taxa comprise the Streptophytes (Figure 16.1). Charophyte algae inhabit freshwater environments, and it is their gametophyte generations, which range from unicellular to complex multicellular, that are predominantly observed. Their diploid generation is always unicellular, and is often a resting, or dispersal, stage of the life cycle. A lineage uniting *Mesostigma* and *Chlorokybus* is positioned as sister to all other Streptophytes (Lemieux *et al.* 2007; Puttick *et al.* 2018; Timme *et al.* 2012; Wang *et al.* 2020; Wickett *et al.* 2014). *Mesostigma* is an aquatic

DOI: 10.1201/9780429351907-20

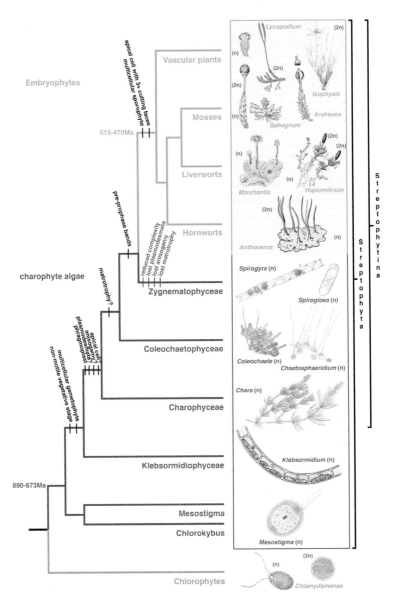

FIGURE 16.1 Phylogenetic relationships of the streptophytes. Streptophytes (green and dark blue branches) form a monophyletic group diverging as a sister group to Chlorophytes (light blue branch). Within Streptophytes, the Embryophytes (green branches) resolve as a monophyletic group nested within a phylogenetic grade of charophytes (dark blue branches). Putative character acquisitions (in black) and losses (in red) indicated at nodes are inferred by parsimony. The ploidy of life cycles stages for representative taxa are indicated in brackets (n, red; 2n, purple), with the unicellular diploid sporophytes of charophyte taxa not shown. Phylogenetic relationships follow Puttick *et al.* (2018) and node age estimates are based on Morris *et al.* (2018). (Images: Allen 1888; Collins 1909; Coupin 1911; De Bary 1858; Gottsche 1843; Hooker 1837; Klebahn 1914; Klebs 1896; Lauterborn 1899; Lendner 1911; Strasburger et al. 1911).

freshwater alga with a unicellular body plan, and *Chlorokybus* occupies alpine terrestrial habitats and consists of loose packets of cells (Geitler 1942; Lauterborn 1894; Marin and Melkonian 1999). Many features of *Mesostigma*, namely its unicellular motile vegetative phase and freshwater habitat are characters likely shared with the ancestral streptophyte (Leliaert *et al.* 2012; Umen 2014).

Along with land plants, extant members of all four other lineages of charophyte algae possess multicellular gametophytes, with flagella lost in the vegetative stages but maintained – with the exception of Zygnematophyceae taxa – in the zoospores (i.e. motile unicellular spores) (Stewart and Mattox 1975). Klebsormidiophyceae are usually placed as sister to a clade including a grade of three groups of charophyte algae (Charophyceae, Coleochaetophyceae and Zygnematatophyceae) and Embryophytes. Collectively this group has been named the Streptophytina (Lewis and McCourt 2004). Thus, simple multicellularity likely evolved in the common ancestor of the Klebsormidiophyceae and the Streptophytina (Figure 16.1).

Extant Klebsormidiophyceae taxa are characterized by unbranched filamentous, such as *Entransia* and *Klebsormidium* species, or sarcinoidal forms (i.e. three-dimensional packets of cells) as in some species of *Interfilum*, growing in shallow freshwater or on damp surfaces (Mikhailyuk *et al.* 2008, 2018; Rindi *et al.* 2008). Given that *Entransia* and *Hormidiella*, two unbranched filamentous genera, diverge from basal nodes within the Klebsormidiophyceae (Mikhailyuk *et al.* 2008, 2018), it is probable that the ancestor had an unbranched filamentous morphology, with sarcinoidal forms derived. Extant Klebsormidiophycean algae have a simple rather than complex multicellular form (following Knoll [2011]), with diffuse cell divisions, and with filaments generally lacking differentiated or specialized cells, apart from differentiation between reproductive and vegetative cell types (Cook and Graham 2017; Sluiman *et al.* 1989). Additionally, Klebsormidophyceae taxa lack plasmodesmata (Lokhorst 1996; Mikhailyuk *et al.* 2014; Stewart *et al.* 1973) – cytoplasmic connections between cells – suggesting that intercellular communication is less critical for development in this lineage. Given the taxonomic position of Klebsormidiophyceae, the earliest multicellular streptophytes – dating to around 800–600MYA (Morris *et al.* 2018) – were possibly comprised of simple unbranched filaments in the gametophyte generation, with this growth form retained by most extant Klebsormidiophyceae taxa (Leliaert *et al.* 2012; Umen 2014). The unbranched filamentous form may have been selectively advantageous, allowing for more efficient nutrient foraging via growth along vectors and additionally reducing the adverse effects of grazing predation by conferring greater size (Niklas 2000; Umen 2014).

16.2 THE STREPTOPHYTINA AND ANISOGAMY – A CLUE INTO THE ORIGINS OF MULTICELLULARITY

Recent phylogenetic analyses with broader taxon coverage place Zygnematophyceae as sister to Embryophytes, with Coleochaetophyceae sister to Zygnematophyceae + Embryophytes, and Charophyceae more distantly related (Figure 16.1) (Finet *et al.* 2010; Wickett *et al.* 2014). Extant Charophyceae, such as the stoneworts *Nitella* and *Chara*, have large gametophyte bodies composed of united branched filaments,

occupying deep and generally permanent freshwater habitats. The diversity of reproductive morphologies observed in the fossil record suggests that extant Charophyceae
only represent a small fraction of previous family diversity (Feist *et al.* 2005). Extant
Coleochaetophyceae are comprised of branched filamentous and discoidal forms
(Delwiche *et al.* 2002; Hall and Delwiche 2007; Thompson 1969), with the ancestral
form predicted to be a branched filament. Within Zygnematophyceae, there are both
unicellular and filamentous taxa, with filamentous forms represented by the basally
diverging genera *Spirogyra*, *Mougeotia* and *Mesotaenium*, and desmids, unicellular
algae composed of two rigid halves that share a nucleus, derived (Hall *et al.* 2008).
However, the recent placement of the unicellular genus *Spirogloea* as sister to all
other members raises the possibility that the ancestral Zygnematophyceae was unicellular (Cheng *et al.* 2019).

Along with land plants, multicellular gametophytes are present in extant members
of the four most recently diverging lineages of the charophyte algae grade (Figure
16.1), suggesting that multicellularity may have evolved just once in the ancestor of
Klebsormidiophyceae and Streptophytina. However, the potential for a unicellular
ancestral state of the Zygnematophyceae allows for the possibility of a unicellular
algal ancestor of land plants (Stebbins and Hill 1980).

Of the Streptophytina, Embryophytes, Charophyceae and Coleochaetophyceae
are anisogamous and oogamous, suggesting that these traits may be plesiomorphic
(i.e. ancestral), with both characters subsequently lost in the Zygnematophyceae,
which sexually reproduce via conjugation of cells between or within filaments
(reviewed in Mori *et al.* 2015). The transition from uni- or simple multicellularity to complex multicellularity in eukaryotic lineages has been proposed to drive
disruptive selection resulting in dimorphic gametes (i.e. anisogamy) (Bulmer and
Parker 2002; Hanschen *et al.* 2018; Parker *et al.* 1972). Most (but not all) unicellular
organisms are isogamous whereas multicellular organisms can more readily evolve
anisogamy (Bulmer and Parker 2002; Hanschen *et al.* 2018). If multicellularity
had evolved multiple times independently during the evolution of Streptophytina
lineages, then it might be expected that anisogamy would be a homoplasious (independently acquired) trait regulated by distinct genetic factors. However, the DUO
POLLEN1 (DUO1) transcription factors, which regulate sperm differentiation in
Embryophytes, are thought to have a homologous role in sperm differentiation of
the Charophyceae, since *Chara braunii DUO1* orthologs can rescue Mp*duo1* knockouts in the liverwort *Marchantia polymorpha* (Higo *et al.* 2018). *DUO1* orthologs
have been lost from Zygnematophyceae, which is consistent with these charophyte
algae having lost anisogamy, and instead sexual reproduce via conjugation of distinct mating-type cells (Higo *et al.* 2018; Hisanaga *et al.* 2019). Thus, given that
anisogamy is likely a homologous trait among the Streptophytina, multicellularity
probably evolved only once in the common ancestor of the Streptophytina, and by
parsimony in the common ancestor of the Streptophytina + Klebsormidiophyceae.
As discussed below, multicellularity in the Streptophytina – with the exception of
the Zygnematophyceae – is complex, with differentiation of specialized vegetative cell types in the gametophyte (as well as the sporophyte in Embryophyta).
Thus, complex multicellularity likely evolved in the gametophyte generation in the
ancestral Streptophytina.

16.3   THE STREPTOPHYTINA – INCREASING MORPHOLOGICAL COMPLEXITY: PHRAGMOPLASTS, BRANCHING, APICAL CELLS, SPECIALIZED TISSUES AND PLASMODESMATA

In *Mesostigma, Chlorokybus* and Klebsormidiophyceae, cytokinesis occurs via centripetal cleavage without the formation of a cell plate (Lokhorst *et al.* 1988; Manton and Ettl 1965; Pickett-Heaps 1975; reviewed in Buschmann and Zachgo 2016). In contrast, the Streptophytina possess efficient microtubule arrays called phragmoplasts, which assemble the cell plate and more effectively allow for shifts in the plane of cell division (Pickett-Heaps 1975; Pickett-Heaps *et al.* 1999). In Charophyceae, Coleochaetophyceae and some Zygnematophyceae (i.e. *Mougeotia*), the action of phragmoplasts facilitates the formation of branched filaments by rotations in the planes of division of filament cells (reviewed in Buschmann 2020). Additionally, preprophase bands of microtubules, which effectively mark the site of the next plane of cell division, originated in the ancestor of Zygnematophyceae and Embryophyta (Buschmann and Zachgo 2016). The acquisition of pre-prophase bands in the ancestor of Zygnematophyceae and Embryophyta would again suggest that their common ancestor was not unicellular, but rather multicellular, with cell divisions occurring in multiple dimensions.

Unlike the Klebsormidiophyceae, where growth is diffuse, Streptophytina have evolved localized growth from apical cells (Graham *et al.* 2000; Leliaert *et al.* 2012). However, as discussed later, it remains uncertain as to whether or not the apical cells of charophytes and Embryophytes are homologous. The ongoing development of a range of charophyte and Embryophyte model organisms (Domozych *et al.* 2016) provides the opportunity for determining whether a conserved genetic program regulates the apical cells of charophytes and Embryophytes. The apical cells of the Charophyceae, Zygnematophyceae and some Coleochaetophyceae divide anticlinally, and do not directly undergo rotations in plane division; instead, the subapical cells are the sites of lateral bulging and/or periclinal divisions giving rise to uniseriate branched filaments (Buschmann 2020; Delwiche *et al.* 2002; Graham *et al.* 2000). An exception is the majority of *Coleochaete* species, where the apical cell itself cuts from two faces to give rise to apically branching filaments or discoidal body plans (Delwiche *et al.* 2002; Graham *et al.* 2000). However, this character may be derived within Coleochaetophyceae rather than ancestral, given that early diverging lineages tend to display subapical branching (i.e., *Chaetosphaeridium* spp. and *Coleochaete irregularis*) (Delwiche *et al.* 2002; Thompson 1969). Embryophyte apical cells are unique among the Streptophytina in cutting from three or more faces and thus directly coordinating the establishment of three-dimensional tissues and body plan (Bowman *et al.* 2019; Campbell 1918; reviewed in Moody 2020).

Coinciding with the emergence of bodies controlled by apical cells in the Streptophytina is an increase in the number of specialized cell and tissue types (Figure 16.2). For example, the early diverging Charophyceae genera are branched filaments attached to the substrate by multicellular rhizoids, with filaments consisting of central axes of multinucleate internodal cells separating whorls of branchlets, which radiate from clusters of uninucleate nodal cells (Beilby and Casanova 2014; Wood and Imahori 1965). Additionally, members of the Coleochaetophyceae possess

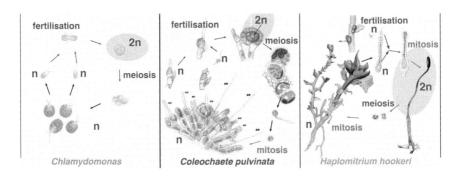

FIGURE 16.2 Increased morphological complexity in the life cycles of the streptophytes. Depicted are the life cycles of the unicellular chlorophyte *Chlamydomonas* sp., the charophyte *Coleochaete pulvinata* and the liverwort *Haplomitrium hookeri*. Vegetative gametophyte stages = n (red), gametes = n (pink) and sporophyte stages = 2n (purple). In *Chlamydomonas,* both stages of the life cycle are unicellular, with gametes fusing to produce a zygote that immediately undergoes meiosis to form spores that develop into the vegetative cells of the next generation. By parsimony, the ancestral streptophyte life cycle likely resembled the life cycle of *Chlamydomonas*. Members of the Streptophytina (Charophyceae, Coleochaetophyceae and Embryophyta) have evolved increased morphological complexity, with multiple specialized cell types and growth from apical cells. The charophyte alga *C. pulvinata* possesses a multicellular gametophyte, with growth controlled by localized apical cells, and produces specialized cell types such as seta (*se*) and gametangia (antheridia, *an* and oogonia, *oo*), with the latter producing dimorphic gametes. After fertilization, the enlarged zygotes are protected by specialized enveloping cells, with post-meiotic cell divisions producing 8–32 meiospores per fertilization event (Graham 1984). These meiospores disperse, and after settling on a substrate undergo mitoses (green) to generate a multicellular haploid plant. The liverwort *H. hookeri,* like all extant land plants, displays complex multicellularity in both generations, undergoing mitoses following both meiosis and fertilization. The gametophyte develops into a three-dimensional axial shoot, with gametangia producing sperm and eggs. After fertilization, an unbranched, axial sporophyte develops which is attached and nutritionally dependent on the gametophyte generation. (Images: Goroschankin 1891; Gottsche 1843; Pringsheim 1860.)

specialized seta cells, along with maternal cells that envelope the zygote following fertilization (Figure 16.2) (Marchant 1977; Pringsheim 1860). Likewise, some Zygnematophyceae taxa (*Mougeotia* and *Spirogyra* spp.) have rhizoids and/or rhizoid-like tip cells (Buschmann 2020; Inoue *et al.* 2002; Nagata 1973; Pascher 1906).

The evolution of localized apical cells and cell specialization suggests the requirement of increased cell-to-cell communication during development. All Streptophytina lineages except the Zygnematophyceae contain taxa with plasmodesmata – i.e. membrane-lined channels through cell walls (Cook *et al.* 1997; Marchant and Pickett-Heaps 1973; Pickett-Heaps 1967a, 1967b). Some authors suggest that charophyte algae and Embryophyte plasmodesmata are homologous (Graham *et al.* 2000; Nicolas *et al.* 2017; Raven 2005), while others propose that they evolved in parallel (Brunkard and Zambryski 2017). Plasmodesmata in bryophytes and vascular plants have central strands of compressed endoplasmic reticulum (i.e., desmotubules), a character not observed in studies of *Coleochaete* species (Cook *et al.* 1997;

Cook and Graham 1999; Stewart *et al.* 1973). Some studies have indicated that plasmodesmata in *Chara* cells occasionally have desmotubules or desmotubule-like central structures (Brecknock *et al.* 2011; Cook *et al.* 1997), while others have proposed that they are absent (Franceschi *et al.* 1994). The presence of plasmodesmata in the ancestral Streptophytina (or otherwise evolving in parallel shortly afterwards) would allow for metabolite and resource sharing, and additionally facilitate efficient transport of signaling molecules along spatial gradients, promoting cell specialization (Hernández-Hernández *et al.* 2012; Lucas and Lee 2004; Niklas and Newman 2013).

16.4   THE ADVENT OF A TERRESTRIAL FLORA

There exists evidence of a widespread land flora of Embryophyte affinity by the mid-Ordovician (470 MYA) in the form of meiotic cryptospores – fossilized spores with walls containing sporopollenin, but distinct from monolete and trilete spores and pollen grains – that in some respects resemble those produced by extant bryophytes (Edwards *et al.* 2014; Wellman and Gray 2000). Despite the abundant cryptospore assemblages, there is little fossil evidence of the plants that produced the cryptospores; however, it is likely they were produced inside complex multicellular sporophytes although it is possible that Embryophytic spores preceded the evolution of multicellular sporophytes (Brown and Lemmon 2011; Strother and Taylor 2018). Regardless, the common ancestor of all extant land plants possessed complex multicellular bodies in both haploid gametophyte and diploid sporophyte generations. Thus, in the 100 million or so years after their divergence from the most closely related lineage of extant charophycean algae, the lineage leading to land plants experienced two key innovations with respect to body plan. First, the evolution of a meristem containing an apical cell with three or more cutting faces enabled the development of complex multicellularity with three-dimensional tissues. A three-dimensional body plan with a differentiated epidermis surrounding parenchymatous tissues facilitated adaptation to the terrestrial environment where water is limiting. Second, complex multicellularity with three-dimensional tissues evolved in both gametophyte and sporophyte generations. It has been proposed that the adaptive significance of the evolution of a multicellular sporophyte is the potential generation of thousands of haploid spores from a single fertilization event, presumed to be limiting in the terrestrial environment, rather than just four haploid progeny per fertilization event in the ancestral alga (Bower 1908). These two key evolutionary innovations, along with the establishment of arbuscular mycorrhizal interactions with fungi, have been credited for the rapid radiation of early land plants (Field *et al.* 2015; Selosse and Le Tacon 1998).

That land plants evolved complex multicellularity in both gametophyte and sporophyte generations raises two key questions. First, what were the molecular genetic innovations that facilitated the evolution of complex multicellularity? Second, did the evolution of complex multicellularity evolve in one generation initially, with genetic programs subsequently co-opted by the second generation, or did complex multicellularity evolve *de novo* independently in each generation? While these two questions may be inextricably linked, we will discuss them in series below.

16.5   HOW DID COMPLEX TISSUE FORMATION EVOLVE?

As land plants evolved from an ancestral charophyte alga that possessed a life cycle similar to that of extant relatives, with a multicellular gametophyte generation and a single-celled (zygote) sporophyte generation, it has been proposed that the gametophyte generation is the 'older' multicellular generation (Bower 1908). While apical growth and apical cells evolved within the charophyte algal grade, given the sister relationship between the Zygnematophyceae and Embryophytes (Figure 16.1), it is not clear that they are homologous with apical cells in land plant gametophytes. In contrast, parsimony indicates gametophyte shoot apical meristems, and their apical cells, are homologous throughout Embryophytes (Bowman *et al.* 2019). As alluded to above, a key innovation was the evolution of apical cells with three or more cutting faces producing three-dimensional tissues (Graham *et al.* 2000). The apical cells and their immediate derivatives that also actively undergo division in multiple planes act as a pool of stem cells, a meristem, from which all the other tissues and organs of the plant body are ultimately derived. Development from apical meristems allows directional growth whose vector is influenced by light (phototropism) and gravity (gravitropism). Directional growth is another adaptation to terrestrial habitats, where substantial spatial differences in environmental stimuli exist, in contrast to the relative uniformity of aqueous environments.

Following spore germination, the gametophyte generations of bryophytes and ferns undergo a constitutive, or inducible depending upon light conditions, protonemal growth period, where one-dimensional algae-like filaments are formed, before transitioning to two-dimensional growth consisting either of branched protonemata or prothallus body plans (Campbell 1918; Goebel 1905). The duration of two-dimensional growth may be extended (mosses) or brief (most other lineages) following which three-dimensional growth is established. During this time, either in the two-dimensional prothallus or at the onset of three-dimensional growth, an apical cell with three or more cutting faces is specified. The precise anatomies of gametophyte shoot meristem apical cells vary among the different major lineages of land plants, but all involve shifts in division plane between successive cell divisions. The continual shifts in apical cell division planes and subsequent pattern formation imply substantial cell communication networks act between apical cells and other cells within the meristem. One form of communication is via plasmodesmata, which allow small cytoplasmic molecules to move between cells. In both fern and liverwort gametophytes, plasmolysis causes severance of plasmodesmatal connections (as well as other disruptions) resulting in cell dedifferentiation and subsequent reestablishment of new apical cells (e.g., Nagai 1919; Tilney *et al.* 1990). Likewise, if an apex including its apical cell is removed by decapitation, a new apical cell is specified nearby, indicating that signals emanating from the apical cell prevent other nearby cells from becoming apical cells (e.g., Nishihama *et al.* 2015; Vöchting 1885).

16.6   EVOLUTION OF NEW SIGNALLING PATHWAYS FACILITATED EVOLUTION OF THE GAMETOPHYTE SHOOT MERISTEM

A direct approach to uncover the genetic underpinnings of apical cell formation and function is to screen for mutants in which this process is disrupted. Such an approach has been undertaken in the moss *Physcomitrium* (*Physcomitrella*) *patens*, uncovering a number of genes whose activity is required for proper gametophyte apical cell function (Moody *et al.* 2018). As the functions of most of the genes identified have not been investigated in other taxa, and since this approach has been recently reviewed (Moody 2020; Véron *et al.* 2020), here we focus on a second approach – an investigation of candidate molecules based on the extensive knowledge of angiosperm sporophyte apical meristems.

Arguably the most important land plant hormone is auxin, with the auxin-mediated transcriptional responses as first characterized in angiosperms being a land plant innovation (Bowman *et al.* 2017; Mutte *et al.* 2018). In the liverwort *Marchantia polymorpha*, loss of the primary auxin biosynthetic pathway, that mediated by the Tryptophan Aminotransferase of Arabidopsis (TAA) and YUCCA enzymes, results in a complete loss of developmental patterning in the gametophyte, with the mutant plants reduced to globular masses of undifferentiated cells (Eklund *et al.* 2015). Both TAA1 and YUCCA2, the only gametophyte-expressed members of their respective gene families, are expressed in the apical cell and its immediate derivatives (Eklund *et al.* 2015). Likewise, in *P. patens*, gametophyte apical stem cells synthesize auxin, but the apical cell and its immediate derivatives themselves are insensitive to auxin, with auxin signaling present only more distally in cells that enter differentiation pathways (Landberg *et al.* 2020). These observations are consistent with the hypothesis that the evolution of auxin signaling was instrumental in the evolution of focal growth from meristems as opposed to more diffuse growth observed in many charophyte lineages (Flores-Sandoval *et al.* 2018). While local auxin synthesis and subsequent downstream signaling are critical for gametophyte shoot meristem formation and function, PIN-FORMED1 (PIN)-mediated polar transport of auxin has not yet been shown to be required for the formation or maintenance of the gametophyte shoot meristem, despite its importance in sporophyte shoot organization (Viaene *et al.* 2014). This implies that localized auxin synthesis and limited, perhaps diffusion-based, auxin movement along with a spatially localized response is sufficient for gametophyte shoot meristem function. Auxin-mediated signaling was integrated into a pre-existing transcriptional network and acts as a facilitator of other transcriptional networks rather than specifier of cellular fates in land plants (Bennett and Leyser 2014; Flores-Sandoval *et al.* 2015; Stewart and Nemhauser 2010). Hence, it is critical to identify the pre-existing networks with which it was integrated and those with which it co-evolved in order to understand the origin of the meristem.

A second class of signaling pathway crucial for meristem function also evolved in the ancestral land plant, with receptor kinase-peptide ligand signaling pathways greatly expanding in the ancestral land plant (Bowman *et al.* 2017). One class, composed of CLAVATA3/embryo surrounding region-related (CLE) peptides and associated receptors, was originally identified as negative regulators of stem cell activity in angiosperm sporophyte meristems (Somssich *et al.* 2016). Using a reverse genetic approach in *P. patens*, CLE peptide signaling was demonstrated to be critical for proper cell division plane reorientation that occurs during the transition from two- to three-dimensional growth (Whitewoods *et al.* 2018). It was reported that *P. patens* CLE loss-of-function mutants exhibit over-proliferation phenotypes, and application of exogenous CLE peptide to *P. patens* gametophyte shoots stunted their growth (Whitewoods *et al.* 2018). A similar reverse genetic approach undertaken in *M. polymorpha* also revealed a role for CLE signaling in gametophyte shoot meristem function. In *M. polymorpha*, application of exogenous MpCLE2 peptide causes an accumulation of stem cells in the shoot meristem, while Mp*cle2* loss-of-function alleles exhibit a reduction in meristem size (Hirakawa *et al.* 2020). These data clearly demonstrate that MpCLE2 signaling in the *M. polymorpha* meristem acts to stimulate stem cell proliferation (or restrict differentiation), an effect opposite to that observed in angiosperm sporophyte shoot meristems (Hirakawa *et al.* 2020). Furthermore, a key downstream target of CLE signaling in sporophyte meristems, a WUSCHEL-related homeobox (WOX) transcription factor, is not a downstream target of CLE signaling in bryophyte gametophyte shoot meristems (Hirakawa *et al.* 2020; Sakakibara *et al.* 2014). However, another class of transcription factor critical for sporophyte shoot meristem function, AINTEGUMENTA/PLETHORA/BABY BOOM (APB), does play a critical role in the establishment of *P. patens* three-dimensional growth, acting in the presumptive apical cell, perhaps in response to auxin signaling (Aoyama *et al.* 2012).

16.7 HOW ARE THE SHOOT APICAL MERISTEMS IN THE TWO GENERATIONS RELATED?

The question of whether, and if so, how, the evolution of multicellularity in the haploid gametophyte and diploid sporophyte are related can be rephrased as a question of whether the two generations are homologous or antithetic in origin (Bowman *et al.* 2016; Haig 2008). A homologous origin implies the ancestral land plant possessed isomorphic generations with the two generations subsequently morphologically diverging, while an antithetic origin implies the two generations were heteromorphic at their inception and had distinct evolutionary origins. Given that Embryophytes are nested within a grade of charophyte algae, and the most closely related extant algae possess a multicellular haploid generation and a single-celled diploid generation, this might then suggest that the origins of the two Embryophyte generations differ and that the origin of the alternation of generations is antithetic. This was the view of Bower when he considered the gametophyte generation to be older, with the sporophyte generation evolving by an intercalation of mitoses between gamete fusion and meiosis (Bower 1908). However, since current phylogenetic analyses suggest monophyletic bryophytes rather than a bryophyte grade, a homologous origin cannot be conclusively excluded. And while extant Embryophytes have distinctly heteromorphic alternations of generations, some Devonian Embryophytes possessed life cycles in which the two generations were more similar to one another than in any extant land plant (Kenrick 2018). While phylogenetic and fossil evidence are equivocal, developmental genetic analyses of shoot meristems in the two generations indicate that although they may share some common modules (e.g., auxin, APB, CLE signaling), some act differently in the two generations and other components crucial to sporophyte shoot meristems are not required for gametophyte shoot meristems (Bowman *et al.* 2019). Thus, although the shoot meristems in the two generations share some genetic components, it appears unlikely that there was a comprehensive co-option of a developmental regulatory network from one generation to the other as would be predicted if the generations had a homologous origin. If this is the case, two questions remain to be answered – how did complex multicellularity evolve in the gametophyte (assuming this is the older multicellular generation) and how much of the pre-existing gametophyte genetic machinery was co-opted during the evolution of complex multicellularity in the sporophyte generation.

16.8   CONCLUSION

Multicellularity arguably achieved its highest complexity in the Embryophyta and Metazoa lineages. Animal multicellularity can be traced back to a single unicellular lineage and is limited to the diploid phase of the life cycle. In contrast, in land plants complex multicellularity evolved in both the haploid and diploid generations. The haploid multicellularity of land plants has likely antecedents in related charophyte algal lineages, whereas the evolution of a multicellular diploid generation occurred in the ancestral land plant, likely as an adaptive response to ephemeral conditions for aquatic fertilization in terrestrial habitats. Major questions remain about the relationship between the two complex multicellular generations, whether genetics programs of the presumed older multicellular haploid generation were co-opted to regulate aspects of the multicellular diploid generation, or whether the evolution of complex multicellularity in the two generations utilized largely independent genetic programs.

ACKNOWLEDGMENTS

The authors wish to acknowledge funding from The Australian Research Council (DP170100049 and DP200100225 to JLB.) We apologize to those researchers whose work we were unable to cite owing to lack of space. For those whose work is cited, we assume full responsibility for any errors in interpretation or presentation.

REFERENCES

Allen TF (1888). '*The Characeae of America.*' (John C. Rankin, Jr: New York).

Aoyama T, Hiwatashi Y, Shigyo M, Kofuji R, Kubo M, Ito M, Hasebe M (2012). AP2-type transcription factors determine stem cell identity in the moss *Physcomitrella patens.* *Development* **139**, 3120–3129.

de Bary A (1858). '*Untersuchungen über die Familie der Conjugaten (Zygnemeen und Desmidieen).*' (A. Förstnersche Buchhandlung (Arthur Felix): Leipzig).

Beilby MJ, Casanova MT (2014). The characean plant. '*The Physiology of Characean Cells*'. (Eds MJ Beilby, MT Casanova) pp. 1–42. (Springer: Heidelberg) doi:10.1007/978-3-642-40288-3_1.

Bennett T, Leyser O (2014). The auxin question: a philosophical overview. '*Auxin and Its Role in Plant Development*'. (Eds E Zažímalová, J Petrášek, E Benková) pp. 3–19. (Springer Vienna: Vienna) doi:10.1007/978-3-7091-1526-8_1.

Bower F (1908). '*The origin of a land flora: a theory based upon the facts of alternation.*' (Macmillan: London).

Bowman JL, Briginshaw LN, Florent SN (2019). Chapter two – Evolution and co-option of developmental regulatory networks in early land plants. '*Current Topics in Developmental Biology*'. (Ed U Grossniklaus) pp. 35–53. (Academic Press) doi:10.1016/bs.ctdb.2018.10.001.

Bowman JL, Kohchi T, Yamato KT, Jenkins J, Shu S, Ishizaki K, Yamaoka S, Nishihama R, Nakamura Y, Berger F, Adam C, Aki SS, Althoff F, Araki T, Arteaga-Vazquez MA, Balasubrmanian S, Barry K, Bauer D, Boehm CR, Briginshaw L, Caballero-Perez J, Catarino B, Chen F, Chiyoda S, Chovatia M, Davies KM, Delmans M, Demura T, Dierschke T, Dolan L, Dorantes-Acosta AE, Eklund DM, Florent SN, Flores-Sandoval E, Fujiyama A, Fukuzawa H, Galik B, Grimanelli D, Grimwood J, Grossniklaus U, Hamada T, Haseloff J, Hetherington AJ, Higo A, Hirakawa Y, Hundley HN, Ikeda Y, Inoue K, Inoue S, Ishida S, Jia Q, Kakita M, Kanazawa T, Kawai Y, Kawashima T, Kennedy M, Kinose K, Kinoshita T, Kohara Y, Koide E, Komatsu K, Kopischke S, Kubo M, Kyozuka J, Lagercrantz U, Lin S-S, Lindquist E, Lipzen AM, Lu C-W, De Luna E, Martienssen RA, Minamino N, Mizutani M, Mizutani M, Mochizuki N, Monte I, Mosher R, Nagasaki H, Nakagami H, Naramoto S, Nishitani K, Ohtani M, Okamoto T, Okumura M, Phillips J, Pollak B, Reinders A, Rövekamp M, Sano R, Sawa S, Schmid MW, Shirakawa M, Solano R, Spunde A, Suetsugu N, Sugano S, Sugiyama A, Sun R, Suzuki Y, Takenaka M, Takezawa D, Tomogane H, Tsuzuki M, Ueda T, Umeda M, Ward JM, Watanabe Y, Yazaki K, Yokoyama R, Yoshitake Y, Yotsui I, Zachgo S, Schmutz J (2017). Insights into land plant evolution garnered from the *Marchantia polymorpha* genome. *Cell* **171**, 287–304.e15. doi:10.1016/j.cell.2017.09.030.

Bowman JL, Sakakibara K, Furumizu C, Dierschke T (2016). Evolution in the cycles of life. *Annual Review of Genetics* **50**, 133–154. doi:10.1146/annurev-genet-120215-035227.

Brecknock S, Dibbayawan TP, Vesk M, Vesk PA, Faulkner C, Barton DA, Overall RL (2011). High resolution scanning electron microscopy of plasmodesmata. *Planta* **234**, 749–758. doi:10.1007/s00425-011-1440-x.

Brown RC, Lemmon BE (2011). Spores before sporophytes: hypothesizing the origin of sporogenesis at the algal–plant transition. *New Phytologist* **190**, 875–881.

Brunkard JO, Zambryski PC (2017). Plasmodesmata enable multicellularity: new insights into their evolution, biogenesis, and functions in development and immunity. *Growth and development* **35**, 76–83. doi:10.1016/j.pbi.2016.11.007.

Bulmer MG, Parker GA (2002). The evolution of anisogamy: a game-theoretic approach. *Proceedings of the Royal Society of London Series B: Biological Sciences* **269**, 2381–2388.

Buschmann H (2020). Into another dimension: how streptophyte algae gained morphological complexity. *Journal of Experimental Botany.* **71**, 3279-3386 doi:10.1093/jxb/eraa181.

Buschmann H, Zachgo S (2016). The evolution of cell division: from streptophyte algae to land plants. *Trends in Plant Science* **21**, 872–883. doi:10.1016/j.tplants.2016.07.004.

Campbell DH (1918). '*Mosses and Ferns.*' (Macmillan Company: New York).

Cheng S, Xian W, Fu Y, Marin B, Keller J, Wu T, Sun W, Li X, Xu Y, Zhang Y, Wittek S, Reder T, Günther G, Gontcharov A, Wang S, Li L, Liu X, Wang J, Yang H, Xu X, Delaux P-M, Melkonian B, Wong GK-S, Melkonian M (2019). Genomes of subaerial zygnematophyceae provide insights into land plant evolution. *Cell* **179**, 1057–1067.e14. doi:10.1016/j.cell.2019.10.019.

Collins FS (1909). '*The green Algae of North America.*' (Tufts College: Massachusetts).

Cook ME, Graham LE (1999). Evolution of plasmodesmata. '*Plasmodesmata: Structure, Function, Role in Cell Communication*'. (Eds AJE van Bel, WJP Van Kesteren) pp. 101–117. (Springer Berlin Heidelberg: Berlin, Heidelberg) doi:10.1007/978-3-642-60035-7_7.

Cook ME, Graham LE (2017). Chlorokybophyceae, Klebsormidiophyceae, Coleochaetophyceae. *Handbook of the protists Cham: Springer International Publishing* 1–20.

Cook ME, Graham LE, Botha CEJ, Lavin CA (1997). Comparative ultrastructure of plasmodesmata of *Chara* and selected bryophytes: toward an elucidation of the evolutionary origin of plant plasmodesmata. *American Journal of Botany* **84**, 1169–1178. doi:10.2307/2446040.

Coupin H (1911). *'Album Général des Cryptogames.'* (E. Orlhac: Paris).

Delwiche CF, Karol KG, Cimino MT, Sytsma KJ (2002). Phylogeny of the genus *Coleochaete* (Coleochaetales, Charophyta) and related taxa inferred by analysis of the chloroplast gene *rbcl1*. *Journal of Phycology* **38**, 394–403. doi:10.1046/j.1529-8817.2002.01174.x.

Domozych D, Popper Z, Sorensen I (2016). Charophytes: evolutionary giants and emerging model organisms. *Frontiers in Plant Science* **7**, 1470. doi:10.3389/fpls.2016.01470.

Edwards D, Morris JL, Richardson JB, Kenrick P (2014). Cryptospores and cryptophytes reveal hidden diversity in early land floras. *New Phytologist* **202**, 50–78.

Eklund DM, Ishizaki K, Flores-Sandoval E, Kikuchi S, Takebayashi Y, Tsukamoto S, Hirakawa Y, Nonomura M, Kato H, Kouno M, Bhalerao RP, Lagercrantz U, Kasahara H, Kohchi T, Bowman JL (2015). Auxin produced by the indole-3-pyruvic acid pathway regulates development and gemmae dormancy in the liverwort *Marchantia polymorpha*. *The Plant Cell* **27**, 1650–1669. doi:10.1105/tpc.15.00065.

Feist M, Liu J, Tafforeau P (2005). New insights into Paleozoic charophyte morphology and phylogeny. *American Journal of Botany* **92**, 1152–1160. doi:10.3732/ajb.92.7.1152.

Field KJ, Pressel S, Duckett JG, Rimington WR, Bidartondo MI (2015). Symbiotic options for the conquest of land. *Trends in Ecology & Evolution* **30**, 477–486.

Finet C, Timme RE, Delwiche CF, Marlétaz F (2010). Multigene phylogeny of the green lineage reveals the origin and diversification of land plants. *Current Biology* **20**, 2217–2222. doi:10.1016/j.cub.2010.11.035.

Flores-Sandoval E, Eklund DM, Bowman JL (2015). A simple auxin transcriptional response system regulates multiple morphogenetic processes in the liverwort *Marchantia polymorpha*. *PLOS Genetics* **11**, e1005207. doi:10.1371/journal.pgen.1005207.

Flores-Sandoval E, Eklund DM, Hong SF, Alvarez JP, Fisher TJ, Lampugnani ER, Golz JF, Vázquez-Lobo A, Dierschke T, Lin SS, Bowman JL (2018). Class C ARFs evolved before the origin of land plants and antagonize differentiation and developmental transitions in *Marchantia polymorpha*. *New Phytologist* **218**, 1612–1630. doi:10.1111/nph.15090.

Franceschi VR, Ding B, Lucas WJ (1994). Mechanism of plasmodesmata formation in characean algae in relation to evolution of intercellular communication in higher plants. *Planta* **192**, 347–358. doi:10.1007/BF00198570.

Geitler L (1942). Morphologie, entwicklungsgeschichte und systematik neuer bemerkenswerter atmophytischer algen aus wien. *Flora oder Allgemeine Botanische Zeitung* **136**, 1–29. doi:10.1016/S0367-1615(17)31209-0.

Goebel K (1905). *'Organography of plants, especially of the Archegoniata and Spermaphyta.'* (Clarendon Press: Oxford).

Goroschankin JN (1891). Beiträge zur kenntniss dermorphologie und systematik der Chlamydomonaden. *Bull Soc Imp Nat Moscou* **5**, 101–142. doi:10.1016/S0367-1615(17)31611-7.

Gottsche K (1843). Anatomisch-physiologische untersuchungen über haplomitrium hookeri nve, mit vergleichung anderer lebermoose. *'Verhandlungen der Kaiserlichen Leopoldinisch-Carolinischen Akademie Der Naturforscher'*. (Breslau und Bonn: Bonn).

Graham LE (1984). *Coleochaete* and the origin of land plants. *American Journal of Botany* **71**, 603–608.

Graham LE, Cook ME, Busse JS (2000). The origin of plants: body plan changes contributing to a major evolutionary radiation. *Proceedings of the National Academy of Sciences* **97**, 4535–4540. doi:10.1073/pnas.97.9.4535.

Haig D (2008). Homologous versus antithetic alternation of generations and the origin of sporophytes. *The Botanical Review* **74**, 395–418.

Hall JD, Delwiche CF (2007). In the shadows of giants: systematics of the charophyte green algae. *'Unravelling the algae: the past, present and future of algal systematics'*. (Eds J Brodie, J Lewis) pp. 155–169. (CRC Press: Boca Raton, FL).

Hall JD, Karol KG, McCourt RM, Delwiche CF (2008). Phylogeny of the conjugating green algae based on chloroplast and mitochondrial nucleotide sequence data. *Journal of Phycology* **44**, 467–477. doi:10.1111/j.1529-8817.2008.00485.x.

Hanschen ER, Herron MD, Weins JJ, Nozaki H, Michod RE (2018). Multicellularity drives the evolution of sexual traits. *The American Naturalist* **192**, E93–105.

Hernández-Hernández V, Niklas KJ, Newman SA, Benítez M (2012). Dynamical patterning modules in plant development and evolution. *International Journal of Developmental Biology* **56**, 661–674.

Higo A, Kawashima T, Borg M, Zhao M, López-Vidriero I, Sakayama H, Montgomery SA, Sekimoto H, Hackenberg D, Shimamura M (2018). Transcription factor DUO1 generated by neo-functionalization is associated with evolution of sperm differentiation in plants. *Nature communications* **9**, 1–13.

Hirakawa Y, Fujimoto T, Ishida S, Uchida N, Sawa S, Kiyosue T, Ishizaki K, Nishihama R, Kohchi T, Bowman JL (2020). Induction of multichotomous branching by CLAVATA peptide in *Marchantia polymorpha*. *Current Biology* **30**, 3833–3840. e4.

Hisanaga T, Yamaoka S, Kawashima T, Higo A, Nakajima K, Araki T, Kohchi T, Berger F (2019). Building new insights in plant gametogenesis from an evolutionary perspective. *Nature Plants* **5**, 663–669.

Hooker WJ (1837). '*Icones plantarum or figures, with brief descriptive characters and remarks, of new or rare plants, selected from the author's herbarium.*' (Longman, Rees, Orme, Brown, Green, & Longman: London).

Inoue N, Yamada S, Nagata Y, Shimmen T (2002). Rhizoid differentiation in *Spirogyra*: position sensing by terminal cells. *Plant and Cell Physiology* **43**, 479–483. doi:10.1093/pcp/pcf056.

Kenrick P (2018). Changing expressions: a hypothesis for the origin of the vascular plant life cycle. *Philosophical Transactions of the Royal Society B: Biological Sciences* **373**, 20170149.

Klebahn H (1914). '*Die Algen, Moose und Farnpflanzen.*' (G.J. Göschen'sche: Berlin).

Klebs G (1896). '*Die Bedingungen der Fortpflanzung bei einigen Algen und Pilzen.*' (Gustav Fischer: Jena).

Knoll AH (2011). The multiple origins of complex multicellularity. *Annual Review of Earth and Planetary Sciences* **39**, 217–239.

Landberg K, Šimura J, Ljung K, Sundberg E, Thelander M (2020). Studies of moss reproductive development indicate that auxin biosynthesis in apical stem cells may constitute an ancestral function for focal growth control. *New Phytologist* **229**, 649–1177.

Lauterborn R (1894). Über die winterfauna einiger gewässer der oberrheinebene. *Biologisches Zentralblatt* **14**, 390–398.

Lauterborn R (1899). Protozoen-studien. lv. theil. flagellaten aus dem gebiete des oberrheins. *Zeitschrift für wissenschaftliche Zoologie* **65**, Tafel XVII–XVIII.

Leliaert F, Smith DR, Moreau H, Herron MD, Verbruggen H, Delwiche CF, De Clerck O (2012). Phylogeny and molecular evolution of the green algae. *Critical Reviews in Plant Sciences* **31**, 1–46. doi:10.1080/07352689.2011.615705.

Lemieux C, Otis C, Turmel M (2007). A clade uniting the green algae *Mesostigma viride* and *Chlorokybus atmophyticus* represents the deepest branch of the Streptophyta in chloroplast genome-based phylogenies. *BMC Biology* **5**, 1–17. doi:10.1186/1741-7007-5-2.

Lendner A (1911). Les Mucorinées de la Suisse. '*Beiträge zur Kryptogamenflora der Schweiz*'. (K.J. Wyss: Bern).

Lewis LA, McCourt RM (2004). Green algae and the origin of land plants. *American Journal of Botany* **91**, 1535–1556.

Lokhorst GM (1996). Comparative taxonomic studies on the genus *Klebsormidium* (Charophyceae) in Europe. *Cryptogamic Studies* **5**, 1–132.

Lokhorst GM, Sluiman HJ, Star W (1988). The ultrastructure of mitosis and cytokinesis in the sarcinoid *Chlorokybus atmophyticus* (Chlorophyta, Charophyceae) revealed by rapid freeze fixation and freeze substitution. *Journal of Phycology* **24**, 237–248.

Lucas WJ, Lee J-Y (2004). Plasmodesmata as a supracellular control network in plants. *Nature Reviews Molecular Cell Biology* **5**, 712–726. doi:10.1038/nrm1470.

Manton I, Ettl H (1965). Observations on the fine structure of *Mesostigma viride* Lauterborn. *Journal of the Linnean Society of London, Botany* **59**, 175–184.

Marchant HJ (1977). Ultrastructure, development and cytoplasmic rotation of seta-bearing cells of *Coleochaete scutata* (Chlorophyceae). *Journal of Phycology* **13**, 28–36.

Marchant HJ, Pickett-Heaps JD (1973). Mitosis and cytokinesis in Coleochaete scutata. *Journal of Phycology* **9**, 461–471. doi:10.1111/j.1529-8817.1973.tb04122.x.

Marin B, Melkonian M (1999). Mesostigmatophyceae, a new class of streptophyte green algae revealed by SSU rRNA sequence comparisons. *Protist* **150**, 399–417. doi:10.1016/S1434-4610(99)70041-6.

Mikhailyuk T, Holzinger A, Massalski A, Karsten U (2014). Morphology and ultrastructure of *Interfilum* and *Klebsormidium* (Klebsormidiales, Streptophyta) with special reference to cell division and thallus formation. *European Journal of Phycology* **49**, 395–412. doi:10.1080/09670262.2014.949308.

Mikhailyuk T, Lukešová A, Glaser K, Holzinger A, Obwegeser S, Nyporko S, Friedl T, Karsten U (2018). New taxa of streptophyte algae (Streptophyta) from terrestrial habitats revealed using an integrative approach. *Protist* **169**, 406–431. doi:10.1016/j.protis.2018.03.002.

Mikhailyuk TI, Sluiman HJ, Massalski A, Mudimu O, Demchenko EM, Kondratyuk SYa, Friedl T (2008). New streptophyte green algae from terrestrial habitats and an assessment of the genus *Interfilum* (Klebsormidiophyceae, Streptophyta). *Journal of Phycology* **44**, 1586–1603. doi:10.1111/j.1529-8817.2008.00606.x.

Moody LA (2020). Three-dimensional growth: a developmental innovation that facilitated plant terrestrialization. *Journal of Plant Research* **133**, 283–290. doi:10.1007/s10265-020-01173-4.

Moody LA, Kelly S, Rabbinowitsch E, Langdale JA (2018). Genetic regulation of the 2D to 3D growth transition in the moss *Physcomitrella patens*. *Current Biology* **28**, 473–478. e5.

Mori T, Kawai-Toyooka H, Igawa T, Nozaki H (2015). Gamete dialogs in green lineages. *Molecular Plant* **8**, 1442–1454. doi:10.1016/j.molp.2015.06.008.

Morris JL, Puttick MN, Clark JW, Edwards D, Kenrick P, Pressel S, Wellman CH, Yang Z, Schneider H, Donoghue PCJ (2018). The timescale of early land plant evolution. *Proceedings of the National Academy of Sciences* **115**, E2274. doi:10.1073/pnas.1719588115.

Mutte SK, Kato H, Rothfels C, Melkonian M, Wong GK-S, Weijers D (2018). Origin and evolution of the nuclear auxin response system. *eLife* **7**, e33399.

Nagai I (1919). Induced adventitious growth in the gemmae of *Marchantia*. *Shokubutsugaku Zasshi* **33**, 99–109.

Nagata Y (1973). Rhizoid differentiation in *Spirogyra* I. Basic features of rhizoid formation. *Plant and Cell Physiology* **14**, 531–541. doi:10.1093/oxfordjournals.pcp.a074889.

Nicolas WJ, Grison MS, Bayer EM (2017). Shaping intercellular channels of plasmodesmata: the structure-to-function missing link. *Journal of Experimental Botany* **69**, 91–103. doi:10.1093/jxb/erx225.

Niklas KJ (2000). The evolution of plant body plans – a biomechanical perspective. *Annals of Botany* **85**, 411–438. doi:10.1006/anbo.1999.1100.

Niklas KJ, Newman SA (2013). The origins of multicellular organisms. *Evolution & Development* **15**, 41–52. doi:10.1111/ede.12013.

Nishihama R, Ishizaki K, Hosaka M, Matsuda Y, Kubota A, Kohchi T (2015). Phytochrome-mediated regulation of cell division and growth during regeneration and sporeling development in the liverwort *Marchantia polymorpha*. *Journal of Plant Research* **128**, 407–421. doi:10.1007/s10265-015-0724-9.

Parker GA, Baker RR, Smith VGF (1972). The origin and evolution of gamete dimorphism and the male-female phenomenon. J Theor Biol. *J Theor Biol* **36**, 553.

Pascher AA (1906). Über auffallende rhizoid- und zweigbildungen bei einer mougeotia-art. *Flora oder Allgemeine Botanische Zeitung* **97**, 107–115. doi:10.1016/S0367-1615(17)32750-7.

Pickett-Heaps J (1967a). Ultrastructure and differentiation in *Chara* sp. I. Vegetative cells. *Australian Journal of Biological Sciences* **20**, 539–552.

Pickett-Heaps J (1967b). Ultrastructure and differentiation in *Chara* sp. II. Mitosis. *Australian Journal of Biological Sciences* **20**, 883–894.

Pickett-Heaps JD (1975). '*Green algae: structure, reproduction and evolution in selected genera.*' (Mass., Sinauer Associates).

Pickett-Heaps JD, Gunning BES, Brown RC, Lemmon BE, Cleary AL (1999). The cytoplast concept in dividing plant cells: cytoplasmic domains and the evolution of spatially organized cell division. *American Journal of Botany* **86**, 153–172. doi:10.2307/2656933.

Pringsheim N (1860). Beiträge zue morphologie und systematik der algen iii die coleochaeteen. *Jahrbücher für Wissenschaftliche Botanik* **2**, 1–38.

Puttick MN, Morris JL, Williams TA, Cox CJ, Edwards D, Kenrick P, Pressel S, Wellman CH, Schneider H, Pisani D, Donoghue PCJ (2018). The interrelationships of land plants and the nature of the ancestral embryophyte. *Current Biology* **28**, 733–745.e2. doi:10.1016/j.cub.2018.01.063.

Raven JA (2005). Evolution of plasmodesmata. *Annual Plant Reviews* **18**, 33–52.

Rindi F, Guiry MD, López-Bautista JM (2008). Distribution, morphology, and phylogeny of *Klebsormidium* (*Klebsormidiales, Charophyceae*) in urban environments in Europe. *Journal of Phycology* **44**, 1529–1540.

Sakakibara K, Reisewitz P, Aoyama T, Friedrich T, Ando S, Sato Y, Tamada Y, Nishiyama T, Hiwatashi Y, Kurata T (2014). WOX13-like genes are required for reprogramming of leaf and protoplast cells into stem cells in the moss *Physcomitrella patens*. *Development* **141**, 1660–1670.

Selosse MA, Le Tacon F (1998). The land flora: a phototroph-fungus partnership? *Trends in Ecology & Evolution* **13**, 15–20.

Sluiman HJ, Kouwets FAC, Blommers PCJ (1989). Classification and definition of cytokinetic patterns in green algae: sporulation versus (vegetative) cell division. *Archiv für Protistenkunde* **137**, 277–290.

Somssich M, Je BI, Simon R, Jackson D (2016). CLAVATA-WUSCHEL signaling in the shoot meristem. *Development* **143**, 3238–3248.

Stebbins GL, Hill GJC (1980). Did multicellular plants invade the land? *The American Naturalist* **115**, 342–353. doi:10.1086/283565.

Stewart JL, Nemhauser JL (2010). Do trees grow on money? Auxin as the currency of the cellular economy. *Cold Spring Harbor Perspectives in Biology* **2**, a001420.

Stewart KD, Mattox KR (1975). Comparative cytology, evolution and classification of the green algae with some consideration of the origin of other organisms with chlorophylls a and b. *The Botanical Review* **41**, 104–135.

Stewart KD, Mattox KR, Floyd GL (1973). Mitosis, cytokinesis, the distribution of plasmodesmata, and other cytological characteristics in the Ulotrichales, Ulvales, and Chaetophorales: phylogenetic and taxonomic considerations. *Journal of Phycology* **9**, 128–141.

Strasburger E, Noll F, Schenck H, Schimper AFW (1911). '*Lehrbuch der Botanik Für Hochschulen.*' (Gustav Fischer: Jena).

Strother PK, Taylor WA (2018). The evolutionary origin of the plant spore in relation to the antithetic origin of the plant sporophyte. '*Transformative Paleobotany*' (Eds M Krings, CJ Harper, NR Cúneo, GW Rothwell). pp. 3–20. (Elsevier: Amsterdam).

Thompson RH (1969). Sexual reproduction in *Chaetosphaeridium globosum* (Nordst.) Klebahn (Chlorophyceae) and description of a species new to science. *Journal of Phycology* **5**, 285–290. doi:10.1111/j.1529-8817.1969.tb02616.x.

Tilney LG, Cooke TJ, Connelly PS, Tilney MS (1990). The distribution of plasmodesmata and its relationship to morphogenesis in fern gametophytes. *Development* **110**, 1209–1221.

Timme RE, Bachvaroff TR, Delwiche CF (2012). Broad phylogenomic sampling and the sister lineage of land plants. *PloS one* **7**, e29696–e29696. doi:10.1371/journal.pone.0029696.

Umen JG (2014). Green algae and the origins of multicellularity in the plant kingdom. *Cold Spring Harbor Perspectives in Biology* **6**, a016170–a016170. doi:10.1101/cshperspect. a016170.

Véron E, Vernoux T, Coudert Y (2020). Phyllotaxis from a single apical cell. *Trends in Plant Science* **26**, 124–131.

Viaene T, Landberg K, Thelander M, Medvecka E, Pederson E, Feraru E, Cooper ED, Karimi M, Delwiche CF, Ljung K, Geisler M, Sundberg E, Friml J (2014). Directional auxin transport mechanisms in early diverging land plants. *Current Biology* **24**, 2786–2791.

Vöchting H (1885). Über die regeneration der Marchantieen. *Jahrbücher für wissenschaftliche Botanik* **16**, 367–414.

Wang S, Li L, Li H, Sahu SK, Wang H, Xu Y, Xian W, Song B, Liang H, Cheng S, Chang Y, Song Y, Çebi Z, Wittek S, Reder T, Peterson M, Yang H, Wang J, Melkonian B, Van de Peer Y, Xu X, Wong GK-S, Melkonian M, Liu H, Liu X (2020). Genomes of early-diverging streptophyte algae shed light on plant terrestrialization. *Nature Plants* **6**, 95–106. doi:10.1038/s41477-019-0560-3.

Wellman CH, Gray J (2000). The microfossil record of early land plants. *Philosophical Transactions of the Royal Society of London Series B: Biological Sciences* **355**, 717–732.

Whitewoods CD, Cammarata J, Venza ZN, Sang S, Crook AD, Aoyama T, Wang XY, Waller M, Kamisugi Y, Cuming AC (2018). *CLAVATA* was a genetic novelty for the morphological innovation of 3D growth in land plants. *Current Biology* **28**, 2365–2376. e5.

Wickett NJ, Mirarab S, Nguyen N, Warnow T, Carpenter E, Matasci N, Ayyampalayam S, Barker MS, Burleigh JG, Gitzendanner MA, Ruhfel BR, Wafula E, Der JP, Graham SW, Mathews S, Melkonian M, Soltis DE, Soltis PS, Miles NW, Rothfels CJ, Pokorny L, Shaw AJ, DeGironimo L, Stevenson DW, Surek B, Villarreal JC, Roure B, Philippe H, dePamphilis CW, Chen T, Deyholos MK, Baucom RS, Kutchan TM, Augustin MM, Wang J, Zhang Y, Tian Z, Yan Z, Wu X, Sun X, Ka-Shu Wong G, Leebens-Mack J (2014). Phylotranscriptomic analysis of the origin and early diversification of land plants. *Proceedings of the National Academy of Science* **111**, E4859–E4868. doi:10.1073/pnas.1323926111.

Wood RD, Imahori K (1965). 'A Revision of the Characeae: Monograph of the Characeae.' (J. Cramer: Weinheim).

17 Multi-Species Multicellular Life Cycles

Rebecka Andersson
Department of Mathematics and Mathematical Statistics and
Integrated Science Lab
Umeå University, Umeå, Sweden

Hanna Isaksson
Department of Mathematics and Mathematical Statistics and
Integrated Science Lab
Umeå University, Umeå, Sweden

Eric Libby
Department of Mathematics and Mathematical Statistics and
Integrated Science Lab
Umeå University, Umeå, Sweden and
Santa Fe Institute, Santa Fe, NM, USA

CONTENTS

17.1 GRAFTING, CHIMERAS, AND FRANKENSTEIN CREATIONS

In order to boost productivity, commercial agriculture embraces something unnatural (Lee et al., 2010). Like some creation of Dr. Frankenstein or a creature from Greek mythology, different plant species are fused together in a process called grafting. For example, the fruiting part of a watermelon may be fused to the root system of a pumpkin. The resulting chimeric plant produces watermelon fruit but is more resistant to disease and stress than a typical watermelon (Davis et al., 2008). Grafting allows farmers to harness beneficial aspects from different species but in a way that does not naturally occur. Watermelons and pumpkins do not normally detach parts in order to fuse with each other, even though grafting shows there could be potential gains from

DOI: 10.1201/9780429351907-21

doing so. So why do such multi-species chimeras need human intervention to come into existence—why are they more associated with mythology than biology?

At first glance, there seem to be significant potential benefits to forming multi-species chimeras. The extensive number and diversity of mutualisms within the biological world demonstrate that cooperation among different species can produce synergies either through division of labor or novel combinations of functionalities (Bronstein et al., 2006; Callaway, 1995; Gestel et al., 2015; Hay et al., 2004; Janzen, 1985; Knowlton and Rohwer, 2003). Moreover, combining different genomes into a single, fitter entity is also widespread within biology; it is the basis of hybrid vigor (Birchler et al., 2003) and a well-recognized benefit of sexual reproduction (Colegrave, 2002; Michod et al., 2008). Yet, unlike hybridization and sex, multi-species chimeras keep their different genomes distinct, which means they likely face fewer issues of genetic incompatibilities. Since multi-species chimeras are not limited by traditional species boundaries, they have many possible combinations and opportunities available to uncover potential synergistic benefits.

Of course, even if there are significant possible benefits of forming multi-species chimeras, there may be barriers that prevent their formation or persistence. If we return to the example of the grafted watermelon-pumpkin combination, one obvious barrier to them forming naturally is reproduction. The pumpkin roots have no way of producing offspring so there is no direct way for the watermelon-pumpkin chimera to produce watermelon-pumpkin offspring. However, unlike mules and other sterile hybrids that also cannot reproduce, the grafted watermelon-pumpkin entity does not fuse genetic material, so its sterility is not necessarily due to a failure of genetics but instead a failure of configuration. It seems plausible that a different configuration would allow successful reproduction of the whole entity. In this chapter, we consider possible configurations of multi-species chimeras that can reproduce and potentially evolve. Identifying such configurations may reveal the structure of existing multi-species chimeras or help explain their relative rarity.

17.2   MULTI-SPECIES "ORGANISMS"

Before we identify possible configurations of multi-species chimeras, it is useful to have some benchmark for what kind of entity we are after. We use the term "entity" because it is not clear if or when a group of multiple species might be an "organism" or "individual". Certainly, there are many groups of species that are simply populations living in a shared environment, e.g., an arbitrary group of animals in the Serengeti. Yet, there are other examples of groups formed by different species that function like organisms or individuals (Bourrat and Griffiths, 2018; Godfrey-Smith, 2009; Queller and Strassmann, 2009). Lichens, for instance, are formed by a partnership between fungi and photosynthetic species (algae and/or cyanobacteria) and differ from an arbitrary group of species in fundamental ways: they exhibit complex traits that are not expressed by either constituent species on its own, they interact with their environment and other species in novel ways, and they can reproduce (at least vegetatively) and gain adaptations (Nash, 2008). If lichens are organisms and arbitrary groups of animals in the Serengeti are not, then it raises the question: where is the boundary between a community and a multi-species organism?

One difficulty with this question lies in how to disentangle a species or organism from others in its environment (Bourrat and Griffiths, 2018). For example, a human is often considered to be an example of a single-species organism, but it is intimately associated with a large community of microbes that play a significant functional role. Disruptions of these communities can impair fitness or even development of their hosts (Gilbert et al., 2012). The associations have led some to describe such host-microbiome associations as new types of biological entities, called holobionts, similar in some ways to a lichen (Bordenstein and Theis, 2015; Gilbert et al., 2012; Rosenberg and Zilber-Rosenberg, 2016). Others view holobiont associations more as ecological communities of organisms, e.g., a human and many environmentally-associated microbial species (Douglas and Werren, 2016; Moran and Sloan, 2015; Queller and Strassmann, 2016; Skillings, 2016).

A simple approach to delineating between populations and multi-species organisms is to adopt a restrictive organism definition such that only entities formed by a single species (or genome) are organisms and everything else is a community. A problem with this approach is the classification of eukaryotes. Eukaryotes have intracellular organelles such as the mitochondria and chloroplasts that evolved from endosymbioses between different species. These organelles retain their own distinct DNA, which means eukaryotes violate the restrictive, single-species definition of an organism and would be considered communities rather than organisms.

Regardless of whether eukaryotes are communities or organisms, there is a broader, underlying issue that groups of species seem different in terms of their functional integration and evolvability (Godfrey-Smith, 2009). For example, contrast a unicellular eukaryote such as a yeast or algal cell with an arbitrary community of soil bacteria. Besides the unicellular eukaryote's higher level of functional integration, it can give rise to offspring that resemble the parent. Furthermore, cell walls make offspring distinct from their parents, which helps selection fix mutations that increase the fitness of the unicellular eukaryote. In contrast, the ability of a community of soil bacteria to reproduce depends on what it means for the community as a whole to reproduce—is it at least one cell of each species or do the ratios somehow matter? Moreover, without physical boundaries between parent and offspring soil communities, it can be difficult for selection to act and fix beneficial mutations.

The lack of a clear distinction between a community and a multi-species organism and the recognition that multi-species populations vary in terms of functionality and evolvability suggest a different approach. Strassmann and Queller (Queller and Strassmann, 2009) propose viewing the term "organism" as a spectrum, such that some entities are more organismal than others (Godfrey Smith also adopts a spectrum approach in (Godfrey-Smith, 2009) to classify "Darwinian populations"). The minimum requirement for any organismal entity is that it has the capacity to reproduce and gain adaptations. So, the watermelon-pumpkin grafting is not organismal because it cannot reproduce, but a lichen is organismal. After satisfying the minimal criteria, the extent to which an entity is organismal, its "organismality", depends on other factors. Strassmann and Queller consider cooperation and conflict between the entity's constituent parts such that more organismal entities have high cooperation and low conflict. The key idea for our purposes is that there is a spectrum of organismality, and multi-species entities may evolve within this spectrum to become more/less organismal.

A consequence of the spectrum approach to organismality is that it suggests an evolutionary process by which a multi-species community may become a multi-species "organism". If an initial configuration of multiple species can somehow form entities capable of reproduction and evolution, then it may become more organismal by gaining mutations that reduce conflict or enhance cooperation—though it is not a certainty that it will evolve into something more organismal. Based on studies of the evolution of multicellularity within single species, it is likely that the evolutionary trajectory of the multi-species entity will depend heavily on its initial configuration (Ratcliff et al., 2017). In the next sections, we consider the various starting configurations of organismal multi-species entities with an aim of understanding their evolutionary potential for organismality.

17.3   MULTI-SPECIES GROUP CONFIGURATIONS

We now consider how an organismal multi-species entity first arises within a community. Since a multi-species entity must keep its different genomes distinct, it is likely to resemble some form of multicellularity—assuming that cells from different species do not fuse. Thus, we draw upon studies of the evolutionary origins of multicellularity to understand how a population of cells (or species) may evolve into something organismal. In particular, we focus on two basic requirements of any multicellular entity's life cycle (Black et al., 2020; Libby and Rainey, 2013; Ratcliff et al., 2017; Van Gestel and Tarnita, 2017): 1. a group structure and 2. a mode of group reproduction. We consider these aspects separately in order to produce a combinatorial framework that exhaustively describes initial configurations of an organismal multi-species entity.

Although the concept of a group is useful in studies of multicellularity as a way to distinguish nascent multicellularity from its ancestral unicellular population, the nature of what constitutes a group varies greatly between studies. For example, in studies that use experimental evolution techniques to evolve multicellularity, groups often form because mutations cause cells to stay physically attached following reproduction (Herron et al., 2019; Ratcliff et al., 2012, 2013). In contrast, social evolution studies often consider groups that form through cells aggregating and temporarily binding (Kessin, 2001; Muñoz-Dorado et al., 2016; Strassmann et al., 2000; Velicer et al., 2000). Groups can also form without any direct binding or attachment, rather by enclosing cells within a membrane or boundary (Black et al., 2020; Doulcier et al., 2020). If groups can form through temporary interactions or by enclosure within a boundary, it can be difficult to specify precise requirements that distinguish groups from populations. For our purposes, we consider a broad interpretation of groups and only require some direct interaction between constituent cells. We also note that since an organismal group needs some mechanism for reproduction and evolution, the group structure should make this possible if not straightforward.

In terms of how groups form, there may be many possible factors that determine whether different species can form a group together. For example, lifestyle or chemical repertoire might play a role such that organisms that produce extracellular glues to attach to surfaces might be more likely to form multi-species groups, by adhering to other species (Niklas and Newman, 2013; Rokas, 2008). In this chapter,

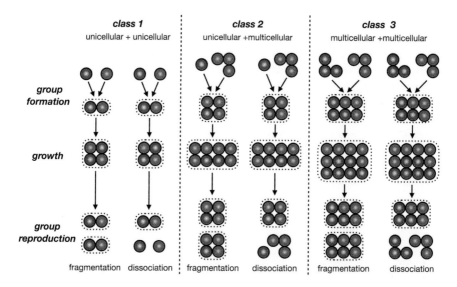

FIGURE 17.1 Multi-species multicellular life cycles. We outline a set of six life cycles that describe how an organismal multi-species group may form and generate group offspring. The life cycles are organized into three classes based on whether the partner species are unicellular or multicellular. The dotted lines distinguish groups from free-living species. Following group formation, there is some time needed for groups to grow via cell reproduction so that they can reproduce. Each multi-species group can reproduce through fragmentation and/or dissociation. In all cases of fragmentation, a multi-species group splits into smaller groups. We show an equal splitting but this is not required. In dissociation some set of the constituent species leave the multi-species group, with the potential to form new groups.

we choose to focus on whether each species is unicellular or multicellular because it is likely to be influential in group formation. If one species is already multicellular, then it has the capacity to form a group of cells that can give rise to new groups. Adding another species to this system might be easier due to the pre-existing multicellularity, i.e., multicellularity could provide a scaffold for other species to join and modify.

If we consider all possible combinations of two species that can be either unicellular or multicellular, we arrive at three possible routes to forming a multi-species multicellular group: 1) both species are unicellular, 2) one is unicellular and the other is multicellular, and 3) both species are multicellular (see Figure 17.1). Each of these initial configurations constitutes a class with representative features. We discuss these features below and provide examples of organismal multi-species entities that may have evolved from similar starting configurations.

17.3.1 CLASS 1, TWO UNICELLULAR SPECIES

In the first class, both species are unicellular and come together to form a multicellular group. There are many examples of organismal groups that likely started this way. For example, microbial syntrophies often feature interdependent, metabolic coupling between different unicellular microbial species (Morris et al., 2013; Schink, 2005;

Stams and Plugge, 2009). Unlike typical microbial communities, these syntrophies exhibit a high degree of interdependence, specificity, and coupling between species. Nonetheless, these syntrophies are quite far from paradigmatic organisms and have not evolved to be much more than mutualistic communities.

The extent to which syntrophic communities resemble populations as opposed to multicellular organisms depends in part on their physical structure. Some microbes establish physical connections between cells in order to share resources while others rely on diffusion in a shared environment (D'Souza et al., 2018). Since paradigmatic multicellular organisms, like plants and animals, use physically attached cells to construct tissues, it is tempting to consider multi-species communities that rely on physical connections as more organismal than those that rely on diffusion. Yet, if the species occupy an environment in which communication via diffusion is effective or even better than rigid physical connections, then the environment would limit the evolution of physical structures associated with more paradigmatic organisms.

This first class of multi-species entities is not limited to groups that resemble communities. The endosymbiosis that gave rise to eukaryotes would also fall in this category. Here, we are not equating the current, derived form of a eukaryotic cell with a multicellular organism. Instead, we are considering the earliest stage of the endosymbiosis when the two species were unicellular organisms and the endosymbiont could leave the host without threatening the survival of either species. At that initial stage, the two unicellular species would have formed a multicellular group. It may seem unusual to equate an endosymbiosis with a multicellular group, but the fact that one species is inside the other does not seem so different from a group of clonal cells in which there are interior cells completely surrounded by exterior cells, e.g., some of the volvocine algae (Kirk, 2005). Of course, in the case of the eukaryotic endosymbiosis the initial multicellular group did not stay multicellular but rather evolved into a more complex form of unicellularity that in some lineages later re-evolved multicellularity (often clonal).

On the one hand, it seems surprising that this class of multi-species multicellularity did not evolve into large, complex chimeric organisms. Since both species are unicellular and build the multicellular group from the bottom up, there is the potential to produce integrated structures that harness the functional capacities of the two species. If, by contrast, they had already evolved complex multicellular tissues prior to forming a group, then it might be difficult to combine them in an effective way. On the other hand, it could be that building a complex structure is difficult for two different unicellular species because of possible breakdowns in cooperation. The ecological and evolutionary time scales for unicellular organisms often overlap, which means cheating mutations can arise and disrupt cooperation (Ennis et al., 2000; Velicer et al., 2000). Cheating mutants also present a problem for clonal multicellularity, but kin selection can help to stabilize the multicellular group and maintain cooperation (Gilbert et al., 2007; Kuzdzal-Fick et al., 2011). Without kin selection, multi-species groups composed of unicellular species may be limited in the extent to which they can evolve large, complex structures—this constraint may be lessened if one of the species has already evolved multicellularity as there would be some stable structure on which to add. Ultimately, we lack observations of large complex, multicellular forms being built from two different unicellular species.

17.3.2 CLASS 2, A UNICELLULAR AND A MULTICELLULAR SPECIES

In the second class of multi-species group formation, a unicellular species is incorporated as part of an existing multicellular species. There are many possible scenarios that can lead to such a multi-species group. For example, a unicellular species may share the same environment as a multicellular species and begin to grow on its surface, or it may be internalized (either accidentally or as a parasite or food). Whatever the case, the multicellular species provides a structural—and possibly ecological—niche for the unicellular species. And since the multicellular species would have already evolved a way of reproducing its multicellular form, it would regularly occur and be available for interactions with unicellular species.

This class includes lichens, which are probably the best example of a multi-species chimeric organism. Lichens evolved through associations between multicellular fungal lineages and unicellular photobionts (algae and/or cyanobacteria). Even though the original nature of the relationship varied between parasitic to mutualistic (Gargas et al., 1995), the evolved relationship is fairly similar across lineages. The fungal partner provides an organized physical structure that surrounds and protects the photobionts from harsh environmental conditions, and the photobionts provide the fungi with energy (Honegger, 1998; Nash, 2008). When lichens reproduce sexually it is typically via the fungal partner, whose offspring must associate with a free-living photobiont to produce a new lichen (Bowler and Rundel, 1975). Thus, it is the multicellular fungal species that drives the repeated formation of groups.

Other organismal entities that evolved from this second class of group formation share a similar organization to each other. In general, the multicellular species constitutes most of the group's cells and maintains its body while the unicellular species is an endosymbiont or is at least kept localized to a specific section or organ. Examples include termites that maintain unicellular endosymbionts in their hindguts (Lombardo, 2008) and bobtail squid that grow populations of *Vibrio fischeri* in specialized light organs (McFall-Ngai, 2014; Ruby and Lee, 1998). The similar organization of this second class of multi-species entities could help maintain and regulate the inter-species relationship (Estrela et al., 2016). For instance, if the unicellular species lived on the outside surface of the multicellular species, then the partnership would be susceptible to disruption or invasion from other species in the environment. Alternatively, if the unicellular species were free to move around inside the multicellular species, this could interfere with other internal structures or functions of the multicellular species.

Unlike single-species multicellularity or the first class of multi-species multicellularity, the second class features different time scales between constituent parts. The time it takes a multicellular species to complete its life cycle is likely to be significantly longer than its partner unicellular species (Marbà et al., 2007). The difference in time scales may lead to the multicellular species evolving mechanisms to manage the evolution of its partner unicellular species. Without such mechanisms, the inter-species cooperation could be lost. For example, in the case of the bobtail squid, when a population of its *Vibrio fisheri* evolves to produce less light, the squid expels them and acquires a new population from the environment (McFall-Ngai, 2014; McFall-Ngai et al., 2012). Examples from this class that involve an endosymbiont often have intricate regulation that allows the multicellular species to control reproduction of its unicellular endosymbionts, preventing their loss or rampant growth (Lowe et al., 2016; Ratzka et al., 2012). Indeed, the need for multicellular species to manage their unicellular partners might point to why many examples from this class have specialized physical structures that contain the unicellular species.

17.3.3   Class 3, Two Multicellular Species

Finally, in the third class of multi-species groups, two multicellular species unite to form a single group. Since both species are multicellular, they have traits that can be co-opted to assist in evolving multi-species multicellularity. For instance, by virtue of being multicellular, both species would already have the ability to create groups, perhaps by secreting extracellular glues or maintaining cell-cell attachment following reproduction (Rainey and Kerr, 2010; Ratcliff et al., 2012). If the watermelon-pumpkin grafting had a mechanism for reproduction and gaining adaptations, it would form an organismal group belonging to this class, as would the mythological chimera (Knox and Fagles, 1990).

Both the grafting and the mythological chimera demonstrate a key challenge within this class. Since the two multicellular species have evolved to build complex structures, it can be difficult to combine them in an effective way. If we return to our grafting example, there are many ways to combine parts of watermelons and pumpkins that would not create a surviving, reproducing entity. Many representative examples of this class of group formation also do not feature intimate physical integration; instead, the multicellular species are mostly distinct. For example, an acacia and the protective ants it houses could be viewed as a member of this class (Janzen, 1966), as would a fig and the pollinating wasps it houses (Janzen, 1979; Weiblen, 2002). In both examples, cells of each multicellular species are kept segregated, and the amount of cell-cell contact between species is limited (i.e., most cells are only in contact with clonemates).

The acacia-ant and fig-wasp groups are examples of mutualisms that because of partner specificity and co-evolution may be considered more organismal than other multi-species communities. Yet, they may be limited in the extent of organismality they can achieve. As a point of comparison, we consider honeybee colonies, which can contain tens of thousands of multicellular organisms cooperating together to grow and sustain a hive (Seeley, 2009). Their scale, functional integration, and complexity have led them to be placed in a class of highly organismal groups called superorganisms (Seeley, 1989). Unlike the acacia-ant and fig-wasp mutualisms, superorganisms contain only a single multicellular species, and despite their comparative genetic homogeneity, they have evolved intricate mechanisms to control who reproduces so as to maintain cooperation (Ratnieks et al., 2006). Multicellular mutualisms may not be able to attain similar levels of reproductive control because it would need to emerge from the interaction between two genomes rather than being encoded in a single genome. Lack of reproductive control provides evolutionary opportunities for both genomes to exploit their partner species or abandon the partnership all together, which would constrain the amount of interdependence and functional integration that evolves between the species in this third class of multispecies groups.

17.4 MULTI-SPECIES GROUP REPRODUCTION

In the previous section, we discussed the different configurations of multi-species groups, but for these groups to be organismal they need some mode of reproduction. Across the different classes of multi-species multicellularity there are two basic modes of reproduction: 1) fragmentation and 2) cycles of dissociation and re-association (Libby and Ratcliff, 2021, see Figure 17.1). In the case of fragmentation, the multi-species group simply splits into smaller groups, each of which contains at least both species and possibly maintains some type of physical or community structure. In the second mode of reproduction, one or both species abandons the group and at some later point re-associates with the same partner species—though not necessarily the same individuals or lineage—to recapitulate the multi-species group. Importantly, both modes of group reproduction allow groups to produce group offspring and gain adaptations.

A key distinction between the two modes of reproduction lies in whether the multi-species partnership is maintained or temporarily severed (or, alternatively, whether the offspring group is a result of staying together or coming together (Tarnita et al., 2013)). If the relationship is severed, as is the case in dissociation, there is the risk that the two species will not be able to re-establish their partnership. If this risk is high enough then it will reduce the fitness of groups and interfere with the species evolving interdependence (Estrela et al., 2016). While there is no requirement that species are interdependent in organismal multi-species groups, it is a common trait of more organismal, clonal multicellular organisms, e.g., cells in different organs rely on one another for the organism as a whole to function. When faced with a recurring risk of not re-establishing a partnership, the different species are likely to maintain self-reliance or evolve more general reliance on species present in the environment (Estrela et al., 2016; Oliveira et al., 2014).

Although dissociation can sever the relationship between specific lineages, it may also facilitate some kinds of adaptation. Many clonal multicellular organisms reproduce through a single-cell bottleneck, which helps purge deleterious mutations, consolidate developmental changes, and maintain cooperation (Grosberg and Strathmann, 1998; Queller, 2000; Ratcliff et al., 2017). A cycle of dissociation and then re-association can function in a similar manner by reducing the size of group offspring and allowing species to swap members of the same species. Moreover, dissociation can be coupled with sexual reproduction to give partner species a chance to generate genetic variation. In lichens, when fungi reproduce sexually they dissociate from the lichens to disperse and form new lichen associations (Bowler and Rundel, 1975).

We note that the two modes of reproduction are not mutually exclusive, so the same multi-species group can use both to reproduce depending on the circumstances. For example, many lichens reproduce via fragmentation as part of an asexual reproductive life cycle and also dissociate their mycobionts (and possibly photobionts) as part of a sexual reproductive life cycle (Bowler and Rundel, 1975). Fragmentation allows for faster colonization of new niches, while dissociation increases genetic variation in lichens. Since the two modes of reproduction have different costs and benefits, lichens may incorporate both modes into their life cycles to better propagate the multi-species groups across generations.

At the onset of multi-species multicellularity, there is nothing that theoretically prevents multi-species groups from reproducing via either mode. But if we look at the different classes of multi-species groups, there appears to be a bias in favor of dissociation when at least one multicellular species is involved—especially if it has complex traits such as developmental programs or sexual reproduction. For example, in the acacia-ant system acacia trees reproduce through seeds, so in order to re-establish the relationship, ants must disperse and colonize new acacia trees (dissociation). Similarly, in the squid-*Vibrio* system, the squid reproduces via eggs, which means the symbiosis goes through cycles of dissociation and re-association (Nyholm and McFall-Ngai, 2004). The bias towards dissociation, however, is less apparent when both species are unicellular, e.g., microbial syntrophies can fragment or dissociate depending on environmental conditions and whether the species disperse (Cremer et al., 2012). Ultimately the prevalence of either mode of reproduction likely stems from the pre-existing life cycles of the partner species. Since complex multicellular species have a higher prevalence of dissociation-based life cycles, so too do their multi-species groups.

17.5 CONCLUSIONS

In the beginning of this chapter, we considered the peculiarity of plant grafting and the seeming rareness of multi-species chimeric organisms. To understand this rareness, we used a bottom-up approach informed by studies on the origins of multicellularity to explore how a group of multiple species might first form and then reproduce. A key idea in our approach is that these simple, reproducing multi-species groups are starting points from which something more organismal might evolve (or not). We organized multi-species groups into three classes depending on whether the partner species are unicellular or multicellular. For each class, we gave examples of existing multi-species entities that may have evolved from similar initial configurations. We then considered the two ways in which such groups could reproduce—either via fragmentation or dissociation—and weighed the benefits and prevalence of their use. In this last section, we draw a few conclusions on the different types of multi-species multicellularity and their potential to evolve a chimeric organism.

Firstly, there are a dearth of examples of multi-species entities high on the organismality spectrum that were formed by two unicellular species. Apart from the examples of the eukaryotic cell and perhaps a dozen plastid endosymbioses (Hackett et al. 2007), the vast majority of multi-species entities from this class are likely to be similar to microbial syntrophies, i.e., unicellular communities with some degree of cooperation. The turbulent dynamics of microbial communities in which environments change and new species frequently invade likely inhibit these communities from persisting and reliably reproducing with the same partners (Oliveira et al., 2014). The exceptions from this class, including the original eukaryotic cell as well as primary, secondary, and tertiary endosymbiotic origins of plastids, established persistent and reliable relationships through endosymbiosis. The extreme rarity of unicellular host-endosymbiont pairs (Lane, 2017) suggests that there are important physical constraints that make this an unlikely route to multi-species multicellularity. So, while this first class of multi-species groups seldom becomes anything more organismal than a community, it has the potential to produce highly integrated, organismal entities.

Compared to the first class of multi-species group formation, the second and third classes describe many more existing multi-species entities. Both of these classes involve a multi-species group formed by at least one multicellular species. Their evolutionary trajectories often result in one of two characteristic configurations: 1) a multicellular species with a unicellular endosymbiont and 2) two multicellular species with a mutualistic relationship. These examples represent fundamental and ubiquitous ecological relationships. Although the organismality of any of these partnerships depends on partner specificity as well as the cooperation/conflict elements of their relationship, some symbioses in which there is a strong interdependence between partner species have been considered to be very organismal (Estrela et al., 2016). Thus, multi-species organisms may be quite prevalent, and if so they were likely formed by a merger involving at least one multicellular species.

Though multi-species organisms may be widespread, they have different conformations than the chimeras presented in mythology. Mythological chimeras were often odd assortments of parts of creatures, e.g., an amalgamation of a lion's head, a goat's body, and a serpent's tail (Knox and Fagles, 1990). In some sense, this is similar to our grafting example of a watermelon fruit sewn on a pumpkin root. Yet, it is far removed from the best extant example of a multi-species chimera, lichens. As with other organismal representatives of the second and third classes, one species is contained within another. In lichens, the photobiont is surrounded by the multicellular fungi, in the squid-*Vibrio* system, the *Vibrio* are contained in specialized light organs in the squid, and in the fig-wasp system the wasp lives inside the fig fruit. The fact that one species is contained inside another establishes a more persistent relationship, which may facilitate the evolution of species-specific adaptations (Estrela et al., 2016).

Finally, we end by noting an interesting consequence to the prevalence of the dissociation mode of reproduction in multi-species multicellularity. By separating species in order to reproduce, there is the risk the species will not re-establish a group. This may disrupt the evolutionary trajectory of a multi-species group, but it could lead to the partner species forming new partnerships with other species. There are many biological examples of inter-species partnerships being lost and exchanged for others, e.g., the evolutionary history of lichens shows that the symbiosis has been lost and re-gained across diverse fungal lineages (Lutzoni et al., 2001; Nelsen et al., 2020). Such fluidity in forging new relationships may complement the evolution of new species to create new types of multi-species chimeric organisms, whose genomes contain an evolutionary legacy of previous inter-species cooperation as members of previous chimeras.

REFERENCES

Birchler, J. A., Auger, D. L., and Riddle, N. C. (2003). In search of the molecular basis of heterosis. The Plant Cell, 15(10):2236–2239.

Black, A. J., Bourrat, P., and Rainey, P. B. (2020). Ecological scaffolding and the evolution of individuality. Nature Ecology & Evolution, 4(3):426–436.

Bordenstein, S. R. and Theis, K. R. (2015). Host biology in light of the microbiome: ten principles of holobionts and hologenomes. PLoS Biol, 13(8):e1002226.

Bourrat, P. and Griffiths, P. E. (2018). Multispecies individuals. History and Philosophy of the Life Sciences, 40(2):1–23.

Bowler, P. and Rundel, P. (1975). Reproductive strategies in lichens. Botanical Journal of the Linnean Society, 70(4):325–340.

Bronstein, J. L., Alarcón, R., and Geber, M. (2006). The evolution of plant–insect mutualisms. New Phytologist, 172(3):412–428.

Callaway, R. M. (1995). Positive interactions among plants. The Botanical Review, 61(4):306–349.

Colegrave, N. (2002). Sex releases the speed limit on evolution. Nature, 420(6916):664–666.

Cremer, J., Melbinger, A., and Frey, E. (2012). Growth dynamics and the evolution of cooperation in microbial populations. Scientific Reports, 2:281.

Davis, A. R., Perkins-Veazie, P., Sakata, Y., Lopez-Galarza, S., Maroto, J. V., Lee, S. G., Huh, Y. C., Sun, Z., Miguel, A., King, S. R., et al. (2008). Cucurbit grafting. Critical Reviews in Plant Sciences, 27(1):50–74.

Douglas, A. E. and Werren, J. H. (2016). Holes in the hologenome: why host-microbe symbioses are not holobionts. mBio, 7(2):e02099–15.

Doulcier, G., Lambert, A., De Monte, S., and Rainey, P. B. (2020). Eco-evolutionary dynamics of nested Darwinian populations and the emergence of community-level heredity. eLife, 9:e53433.

D'Souza, G., Shitut, S., Preussger, D., Yousif, G., Waschina, S., and Kost, C. (2018). Ecology and evolution of metabolic cross-feeding interactions in bacteria. Natural Product Reports, 35(5):455–488.

Ennis, H. L., Dao, D. N., Pukatzki, S. U., and Kessin, R. H. (2000). *Dictyostelium* amoebae lacking an F-box protein form spores rather than stalk in chimeras with wild type. Proceedings of the National Academy of Sciences, 97(7):3292–3297.

Estrela, S., Kerr, B., and Morris, J. J. (2016). Transitions in individuality through symbiosis. Current Opinion in Microbiology, 31:191–198.

Gargas, A., DePriest, P. T., Grube, M., and Tehler, A. (1995). Multiple origins of lichen symbioses in fungi suggested by SSU rDNA phylogeny. Science, 268(5216):1492–1495.

Gestel, J. V., Vlamakis, H., and Kolter, R. (2015). Division of labor in biofilms: the ecology of cell differentiation. Microbial Biofilms, 67–97.

Gilbert, O. M., Foster, K. R., Mehdiabadi, N. J., Strassmann, J. E., and Queller, D. C. (2007). High relatedness maintains multicellular cooperation in a social amoeba by controlling cheater mutants. Proceedings of the National Academy of Sciences, 104(21):8913–8917.

Gilbert, S. F., Sapp, J., and Tauber, A. I. (2012). A symbiotic view of life: we have never been individuals. The Quarterly Review of Biology, 87(4):325–341.

Godfrey-Smith, P. (2009). Darwinian populations and natural selection. Oxford University Press.

Grosberg, R. K. and Strathmann, R. R. (1998). One cell, two cell, red cell, blue cell: the persistence of a unicellular stage in multicellular life histories. Trends in Ecology & Evolution, 13(3):112–116.

Hackett, J. D., Yoon, H. S., Butterfield, N. J., Sanderson, M. J., & Bhattacharya, D. (2007). Plastid endosymbiosis: sources and timing of the major events. Pp. 109–132 in Falkowski, P. G. and Knoll, A. H. Evolution of Primary Producers in the Sea.

Hay, M. E., Parker, J. D., Burkepile, D. E., Caudill, C. C., Wilson, A. E., Hallinan, Z. P., and Chequer, A. D. (2004). Mutualisms and aquatic community structure: the enemy of my enemy is my friend. Annual Review of Ecology, Evolution, and Systematics, 35.

Herron, M. D., Borin, J. M., Boswell, J. C., Walker, J., Chen, I. C. K., Knox, C. A., Boyd, M., Rosenzweig, F., and Ratcliff, W. C. (2019). De novo origins of multicellularity in response to predation. Scientific Reports, 9(1):2328.

Honegger, R. (1998). The lichen symbiosis—what is so spectacular about it? The Lichenologist, 30(03):193.

Janzen, D. H. (1966). Coevolution of mutualism between ants and acacias in central America. Evolution, 20(3):249–275.

Janzen, D. H. (1979). How to be a fig. Annual Review of Ecology and Systematics, 10(1):13–51.

Janzen, D. H. (1985). The natural history of mutualisms. The Biology of Mutualism, 40–99.

Kessin, R. H. (2001). *Dictyostelium*: Evolution, Cell Biology, and the Development of Multicellularity. Cambridge University Press.

Kirk, D. L. (2005). *Volvox*: A Search for the Molecular and Genetic Origins of Multicellularity and Cellular Differentiation. Cambridge University Press.

Knowlton, N. and Rohwer, F. (2003). Multispecies microbial mutualisms on coral reefs: the host as a habitat. The American Naturalist, 162(S4):S51–S62.

Knox, B. and Fagles, R. (1990). Homer, the Iliad. Trans. Robert Fagles. New York: Penguin.

Kuzdzal-Fick, J. J., Fox, S. A., Strassmann, J. E., and Queller, D. C. (2011). High relatedness is necessary and sufficient to maintain multicellularity in *Dictyostelium*. Science, 334(6062):1548–1551.

Lane, N. (2017). Serial endosymbiosis or singular event at the origin of eukaryotes? Journal of Theoretical Biology, 434:58–67.

Lee, J. M., Kubota, C., Tsao, S., Bie, Z., Echevarria, P. H., Morra, L., and Oda, M. (2010). Current status of vegetable grafting: diffusion, grafting techniques, automation. Scientia Horticulturae, 127(2):93–105.

Libby, E. and Rainey, P. B. (2013). A conceptual framework for the evolutionary origins of multicellularity. Physical Biology, 10(3):035001.

Libby, E. and Ratcliff, W. C. (2021). Lichens and microbial syntrophies offer models for an interdependent route to multicellularity. Lichenologist, 53:283–290. https://doi.org/10.1017/S0024282921000256

Lombardo, M. P. (2008). Access to mutualistic endosymbiotic microbes: an underappreciated benefit of group living. Behavioral Ecology and Sociobiology, 62(4):479–497.

Lowe, C. D., Minter, E. J., Cameron, D. D., and Brockhurst, M. A. (2016). Shining a light on exploitative host control in a photosynthetic endosymbiosis. Current Biology, 26(2):207–211.

Lutzoni, F., Pagel, M., and Reeb, V. (2001). Major fungal lineages are derived from lichen symbiotic ancestors. Nature, 411(6840):937–940.

Marbà, N., Duarte, C. M., and Agustí, S. (2007). Allometric scaling of plant life history. Proceedings of the National Academy of Sciences, 104(40):15777–15780.

McFall-Ngai, M. (2014). Divining the essence of symbiosis: insights from the squid-vibrio model. PLoS Biol, 12(2):e1001783.

McFall-Ngai, M., Heath-Heckman, E. A., Gillette, A. A., Peyer, S. M., and Harvie, E. A. (2012). The secret languages of coevolved symbioses: insights from the *Euprymna scolopes–Vibrio fischeri* symbiosis. In Seminars in Immunology, Volume 24, pages 3–8.

Michod, R. E., Bernstein, H., and Nedelcu, A. M. (2008). Adaptive value of sex in microbial pathogens. Infection, Genetics and Evolution, 8(3):267–285.

Moran, N. A. and Sloan, D. B. (2015). The hologenome concept: helpful or hollow? PLoS Biology, 13(12):e1002311.

Morris, B. E. L., Henneberger, R., Huber, H., and Moissl-Eichinger, C. (2013). Microbial syntrophy: interaction for the common good. FEMS Microbiology Reviews, 37(3):384–406.

Muñoz-Dorado, J., Marcos-Torres, F. J., García-Bravo, E., Moraleda-Muñoz, A., and Pérez, J. (2016). Myxobacteria: moving, killing, feeding, and surviving together. Frontiers in Microbiology, 7:781.

Nash, T. H. (2008). Lichen Biology. Cambridge University Press, 2 edition.

Nelsen, M. P., Lücking, R., Boyce, C. K., Lumbsch, H. T., and Ree, R. H. (2020). The macroevolutionary dynamics of symbiotic and phenotypic diversification in lichens. Proceedings of the National Academy of Sciences, 117(35):21495–21503.

Niklas, K. J. and Newman, S. A. (2013). The origins of multicellular organisms. Evolution & Development, 15(1):41–52.

Nyholm, S. V. and McFall-Ngai, M. (2004). The winnowing: establishing the squid–vibrio symbiosis. Nature Reviews Microbiology, 2(8):632–642.

Oliveira, N. M., Niehus, R., and Foster, K. R. (2014). Evolutionary limits to cooperation in microbial communities. PNAS, 111(50):17941–17946.

Queller, D. C. (2000). Relatedness and the fraternal major transitions. Philosophical Transactions of the Royal Society of London. Series B: Biological Sciences, 355(1403):1647–1655.

Queller, D. C. and Strassmann, J. E. (2009). Beyond society: the evolution of organismality. Philosophical Transactions of the Royal Society B: Biological Sciences, 364(1533):3143–3155.

Queller, D. C. and Strassmann, J. E. (2016). Problems of multi-species organisms: endosymbionts to holobionts. Biology & Philosophy, 31(6):855–873.

Rainey, P. B. and Kerr, B. (2010). Cheats as first propagules: a new hypothesis for the evolution of individuality during the transition from single cells to multicellularity. Bioessays, 32(10):872–880.

Ratcliff, W. C., Denison, R. F., Borrello, M., and Travisano, M. (2012). Experimental evolution of multicellularity. Proceedings of the National Academy of Sciences, 109(5):1595–1600.

Ratcliff, W. C., Herron, M., Conlin, P. L., and Libby, E. (2017). Nascent life cycles and the emergence of higher-level individuality. Philosophical Transactions of the Royal Society B: Biological Sciences, 372(1735):20160420.

Ratcliff, W. C., Herron, M. D., Howell, K., Pentz, J. T., Rosenzweig, F., and Travisano, M. (2013). Experimental evolution of an alternating uni-and multicellular life cycle in *Chlamydomonas reinhardtii*. Nature Communications, 4(1):1–7.

Ratnieks, F. L., Foster, K. R., and Wenseleers, T. (2006). Conflict resolution in insect societies. Annu. Rev. Entomol., 51:581–608.

Ratzka, C., Gross, R., and Feldhaar, H. (2012). Endosymbiont tolerance and control within insect hosts. Insects, 3(2):553–572.

Rokas, A. (2008). The molecular origins of multicellular transitions. Current Opinion in Genetics & Development, 18(6):472–478.

Rosenberg, E. and Zilber-Rosenberg, I. (2016). Microbes drive evolution of animals and plants: the hologenome concept. mBio, 7(2):e01395–15.

Ruby, E. G. and Lee, K.-H. (1998). The *Vibrio fischeri-Euprymna scolopes* light organ association: current ecological paradigms. Applied and Environmental Microbiology, 64(3):805–812.

Schink, B. (2005). Syntrophic associations in methanogenic degradation. In Molecular Basis of Symbiosis, pages 1–19. Springer.

Seeley, T. D. (1989). The honey bee colony as a superorganism. American Scientist, 77(6):546–553.

Seeley, T. D. (2009). The Wisdom of the Hive: The Social Physiology of Honey Bee Colonies. Harvard University Press.

Skillings, D. (2016). Holobionts and the ecology of organisms: multi-species communities or integrated individuals? Biology & Philosophy, 31(6):875–892.

Stams, A. J. and Plugge, C. M. (2009). Electron transfer in syntrophic communities of anaerobic bacteria and archaea. Nature Reviews Microbiology, 7(8):568.

Strassmann, J. E., Zhu, Y., and Queller, D. C. (2000). Altruism and social cheating in the social amoeba *Dictyostelium discoideum*. Nature, 408(6815):965–967.

Tarnita, C. E., Taubes, C. H., and Nowak, M. A. (2013). Evolutionary construction by staying together and coming together. Journal of Theoretical Biology, 320:10–22.

Van Gestel, J. and Tarnita, C. E. (2017). On the origin of biological construction, with a focus on multicellularity. Proceedings of the National Academy of Sciences, 114(42):11018–11026.

Velicer, G. J., Kroos, L., and Lenski, R. E. (2000). Developmental cheating in the social bacterium *Myxococcus xanthus*. Nature, 404(6778):598–601.

Weiblen, G. D. (2002). How to be a fig wasp. Annual Review of Entomology, 47(1):299–330.

Section 5

Synthesis and Conclusions

18 Conclusion
The Future of Multicellularity Research

William C. Ratcliff
School of Biological Sciences Georgia Institute
of Technology Atlanta, Georgia, USA

Peter L. Conlin
School of Biological Sciences Georgia Institute
of Technology Atlanta, Georgia, USA

Matthew D. Herron
School of Biological Sciences Georgia Institute
of Technology Atlanta, Georgia, USA

CONTENT

18.1   INTRODUCTION

In this final chapter, we examine some of the major outstanding questions and approaches for multicellularity research that we believe will define the intermediate-term future of the field. This is not meant to be an exhaustive list of potential future directions, but rather an overview of particularly compelling topics and questions that are well poised to make major contributions to our understanding of the evolution of multicellularity, based on the current conceptual and technical state of

DOI: 10.1201/9780429351907-23

the field. The list is inevitably affected by our backgrounds and biases, and we have no doubt that some important contributions will come from directions that we have not foreseen.

The chapter is organized around eight research areas that are not truly distinct—indeed, there are many interactions among these topics, and we expect insight will just as often come from the edges of these networks as the nodes.

18.2   PHILOSOPHICAL ISSUES IN MULTICELLULARITY

The evolution of multicellularity has historically been a rich topic for philosophical inquiry. The evolution of a new type of individual requires us to confront questions for which we in biology have long relied on heuristic solutions, but which break down when examining this major transition. Understanding the transition from unicellular organisms to multicellular organisms challenges us to define what organisms and biological individuals are (Clarke, 2010; Godfrey-Smith, 2013; Pradeu, 2016; Queller and Strassmann, 2009) and what it means to be multicellular (Rose and Hammerschmidt, 2021). Once we have these definitions in hand, how do we know when the transition to multicellularity has occurred? Is multicellularity even a binary state, such that an organism is either multicellular or not, or is it a continuum? These are challenging philosophical questions, and addressing them has broad value for biology as a whole, not just the evolution of multicellularity.

In Chapter 2, Maureen O'Malley challenges the notion that multicellularity is even a coherent category. Specifically, she argues that multicellular organisms are so diverse that they do not naturally fit under a single label—and indeed, there may be many more differences than similarities between, say, a cellular slime mold such as *Dictylostelium* that spends most of its life in a unicellular state, only ephemerally inhabiting a multicellular state, and an animal, which does the opposite. This has relevance beyond categorization: the way we frame scientific questions depends on how we see the differences among organisms. For example, the question "Why have aggregative multicellular organisms remained relatively simple?" (Márquez-Zacarías et al., 2021) assumes that this is a reasonable question to ask (*i.e.*, that developmental mode is an evolutionarily-salient factor among the many other differences between these organisms).

In addition to these questions about the evolution of multicellularity *per se*, the philosophy of biology is concerned with understanding the approaches that are used to study it. In Chapter 4, Merlijn Staps, Jordi van Gestel, and Corina Tarnita classify these approaches into two broad categories: bottom-up approaches that address particular evolutionary trajectories, considering the roles of the nature of the unicellular ancestor and the environmental factors imposing natural selection, and top-down approaches that aim to identify general principles by comparing the features of existing multicellular organisms. Recognizing that both approaches have strengths and limitations, they advocate for integrating, for example, the top-down, multilevel selection approach of Rick Michod (Chapter 3) with a bottom-up, mechanistic model of the emergence of multicellular life cycles. The multilevel selection framework, including its application to the evolution of multicellularity, has itself been a frequent point of intersection between biologists and philosophers of biology (Bourrat, 2015; Godfrey-Smith, 2013; Michod, 2005; Okasha, 2006).

Philosophical research has significant potential to contextualize and generalize macroevolutionary trends. We believe that philosophy will continue to play a key role in describing how multicellular organisms become units of selection, how multicellular traits arise, become heritable, and affect fitness, and how parts of organisms emerge and become entrenched within a multicellular context.

18.3   NATURAL HISTORY

One of the most important resources for the field of multicellularity is the only truly long-term natural experiments that we will ever have access to: lineages that have independently evolved multicellularity, in most cases hundreds of millions of years ago. Each such origin is a replicate natural experiment, and phylogenetic comparative methods make them useful for testing hypotheses. Most of what we know about multicellularity we've learned from these organisms, and yet we believe that we are only beginning to utilize this resource.

Future work will be critical for more accurately resolving the phylogenies of multicellular organisms. These will be essential for answering basic questions, for instance: How many times have both simple and complex multicellularity evolved? The most widely cited figure is 'at least 25 origins,' which comes from an analysis by King (2004), performed on a phylogeny developed by Baldauf (2003) and widely popularized in a landmark review by Grosberg and Strathmann (2007). Recent work, however, suggests that the true number is likely far higher. For example, multicellularity appears to have evolved in the green algae alone at least 25 times (Umen and Herron, 2021), and complex multicellularity appears to have convergently evolved from simple multicellular ancestors in the fungi 8–11 times using the same suites of genes (Nagy et al., 2018). How and when has cellular differentiation evolved in different lineages? Which phylum is sister to the remaining animals, sponges or ctenophores (Jékely and Budd, 2021)?

We still have much to learn about the natural diversity of less ecologically-dominant multicellular taxa. For example, recent work has shown that choanoflagellates, the closest known relative of animals, are capable of forming epithelia-like sheets of cells and can regulate phototactic swimming by manipulating the shape of their sheet-like bodies (Brunet et al., 2019). Another choanoflagellate species forms hollow spherical groups that contain a microbiome (Hake et al., 2021). Work with non-Metazoan Holozoans shows that they are capable of expressing a remarkable diversity of cellular phenotypes (e.g., flagellated cells, amoeboid cells, cysts and coenocytes) through time (Sebe-Pedros et al., 2017), lending support to the hypothesis that temporal cellular differentiation may have been co-opted for spatial cellular differentiation in animals (Brunet and King, 2017), as it likely was in the volvocine algae (Nedelcu and Michod, 2006). Within fungi, recent work has shown that large, complex multicellular structures do not necessarily require large genomes (Nguyen et al., 2017). It is clear that we are only beginning to appreciate the diversity of multicellular life on Earth, particularly among the "minor" multicellular taxa (Herron et al., 2013).

Improved sampling of extant multicellular organisms will provide deeper insight into the manifold routes, mechanisms, and constraints on this major transition. In particular, it will be helpful to know how much the cell biology of the unicellular ancestor dictates the way in which multicellularity subsequently evolves. For example, we still don't have a widely accepted explanation for why complex multicellularity has only evolved in eukaryotes (which is especially surprising given that bacterial multicellularity had a ~600 million to 1 billion year head start (Bengtson et al., 2017; Gibson et al., 2018; Schirrmeister et al., 2015)), though hypotheses linking eukaryotic transcriptional regulation to complex multicellularity have been proposed (de Mendoza et al., 2013; Petroll et al., 2021; Tarver et al., 2015). Groundbreaking work has shown that genes put to extensive use in multicellular processes (e.g., cellular adhesion, development, and cancer suppression) often predate this transition (King 2004; Nedelcu and Michod, 2006; Rokas, 2008; Ruiz-Trillo et al., 2008), suggesting that they were co-opted for novel multicellular use. Improved taxonomic sampling will refine our ability to infer how, when, and why genes have been co-opted for multicellular functionality, and how important this process is for the evolution of novel multicellular traits.

18.4   LIFE CYCLES: AN ORGANIZING PRINCIPLE FOR MULTICELLULAR ORIGINS

All multicellular organisms possess a life cycle, which characterizes their growth and reproduction. As Merljin Staps, Jordi van Gestel, and Corina Tarnita explain in Chapter 4, life cycles are a foundational concept for the origin of multicellularity, because they describe the manner in which groups of cells are generated and reproduce themselves (Bonner, 1965; Buss, 1987). Such group-level reproduction is an essential component in multicellular groups becoming Darwinian entities (Rainey and Kerr, 2010), and variation in the life cycle (*i.e.,* how they partition genetic variation among multicellular groups) has profound implications for the subsequent evolution of multicellularity (Hammerschmidt et al., 2014; Ratcliff et al., 2017; Staps et al., 2019).

Despite the centrality of life cycles for this major evolutionary transition, we know relatively little about how they arise. Experimental evolution with yeast (Koschwanez et al., 2013; Ratcliff et al., 2012) and green algae (Herron et al., 2019; Ratcliff et al., 2013) shows that life cycles can arise through growth followed by physical fracture. In both of these examples, the emergent life cycles can include unicellular genetic bottlenecks, efficiently partitioning genetic variation between groups. Ecology itself can act as a "scaffold" (Black et al., 2020)—creating an environment that favors a multicellular phase (e.g., a biofilm) followed by a unicellular phase (Hammerschmidt et al., 2014). Over time, the cycling between unicellular and multicellular states could be brought under developmental control (Black et al., 2020). Alternatively, life cycles may arise due to environment-specific cues (e.g., expression of aggregative proteins during starvation) (Dworkin, 1963; Mahadeo and Parent, 2006). Despite the plausibility of these different routes, we do not yet have a general theory for how life cycles arise, nor have we resolved the ancestral life cycle state of all extant clades of multicellular organisms. Further, we know little about how, when and why developmental processes modify multicellular life cycles, taking them from simple, stochastic beginnings to the robust and highly regulated processes that characterize most extant multicellular organisms.

The above framework raises an intriguing and largely unexplored hypothesis: the ecology of early multicellular life cycles may constrain the types of multicellularity that ultimately evolve (Pichugin and Traulsen, 2020). Some early multicellular life cycles provide a clear ecological advantage—for example, single cells aggregating to form groups when stressed. Such a behavior can provide group survival benefits (Smukalla et al., 2008) and may limit the costs associated with multicellularity (*i.e.*, resource diffusion slowing growth (Pentz et al., 2020) or limited dispersal (Queller and Strassmann, 2014)). Indeed, simple environmentally-dependent aggregation has evolved many times among otherwise unicellular organisms, leading to the ubiquity of microbial biofilms (Flemming and Wuertz, 2019), though in some cases, this simple life cycle has served as the basis for more significant multicellular innovation (see Chapters 5–8). Might we be missing modes of multicellularity simply because those early life cycles were not ecologically advantageous enough to persist? Future work will be necessary to disentangle this effect from the downstream evolutionary consequences of how these life cycles affect the evolution of multicellularity itself.

18.5   THE CRITICAL ROLE OF ORGANISMAL SIZE

As John Tyler Bonner made clear (Bonner, 2006), size is a universally important trait for multicellular organisms. The benefits of multicellularity stem in part from advantages of size derived from group formation, and many multicellular lineages have undergone selection to form larger, more mechanically-robust multicellular bodies at some point in their evolutionary history. A number of key questions remain unanswered regarding the evolution of larger size in early multicellular organisms, however.

Biophysically, how do larger organisms evolve? This is not a trivial question: multicellular organisms face biophysical stresses that act over evolutionarily novel length scales, and we have no reason to believe that early multicellular bodies are particularly robust. Cells growing within groups face a challenge—as cells divide, they exert strain on neighboring cells, which accumulates until it causes the group to fracture (Jacobeen et al., 2018). In order avoid fragmenting, organisms growing in groups must either stop dividing, reduce the accumulation of cell-cell strain, or evolve to tolerate this strain by becoming tougher. Extant organisms do all three. For example, aggregative organisms such as *Dictyostelium discoideum* typically do not divide in the multicellular phase, which is induced by starvation (Jang and Gomer, 2011). In some clonal multicellular organisms, such as plants, growth is often developmentally regulated to limit strain accumulation by producing new cells in parallel sheets (Jackson et al., 2017), or incorporating information about packing-induced strain into cellular division planes (Dupuy et al., 2010). Other mechanisms for increasing multicellular toughness include cell-cell adhesion (Abedin and King, 2010; El-Kirat-Chatel et al., 2015) and the entanglement of filamentous biological materials (Zou et al., 2009). We know relatively little about how biophysical toughness evolves prior to the evolution of developmental systems that coordinate strain reduction and increased toughness. Further work should examine the biophysical organizing principles for multicellularity and assess how they constrain the types of multicellular organisms that can evolve.

Is selection for larger size itself a driver of increased complexity? Size is costly, as it reduces access to extracellular resources (e.g., food and oxygen), providing a selective incentive to evolve traits that overcome these limitations (Bonner, 2006; Knoll, 2011). Examples of such traits include circulatory systems, low surface-area to volume morphologies, and even oxygen-binding proteins such as myoglobin. Many of these require the evolution of increased developmental regulation and multicellular integration. Despite the conceptual strength of this hypothesis, however, it has yet to be directly tested.

18.6   ORIGINS OF MULTICELLULAR DEVELOPMENT

Few topics are more important for the evolution of complex multicellular life than the origin of multicellular development. Developmental regulation of cells within a multicellular group (defined as the genetic, bioelectric, and biophysical mechanisms that allow cells to perform spatially and temporally explicit behaviors) is required for the expression of any reasonably sophisticated multicellular structure, yet we still know little about how development evolves *de novo*.

One school of thought argues that development is an outcome of multicellular adaptation. This may occur through the rewiring of phenotypic variation that was expressed through time in the unicellular ancestor (e.g., flagellar swimming, amoebal crawling, etc.) to be expressed in a spatially dependent manner (Brunet and King, 2017; Mikhailov et al., 2009; Sebe-Pedros et al., 2017), or the *de novo* evolution of cellular differentiation (Arendt, 2008). Alternatively, Stuart Newman and Ramray Bhat have proposed that the initial steps of development may have been non-adaptive. Specifically, they hypothesize that dynamical patterning modules may initially have arisen through the interactions of cellular behaviors and biophysical mechanisms that would have had novel developmental consequences once in a multicellular context. These could then have been refined by selection for improved multicellular functionality (Newman and Bhat, 2009).

Future work resolving how development has evolved in disparate lineages, and how it can evolve from scratch in initially undifferentiated groups of cells, will be critical for developing a comprehensive understanding of this process. Major questions remain: how critical is the evolution of the life cycle to the origin and ultimate evolution of development (Fortezza et al., 2021; see Chapter 4)? How and when will cell-cell communication be important during development? What is the relative importance of co-opting ancestral plasticity in cellular phenotype vs. evolving plasticity *de novo*? This work will require understanding not just the selective advantages of development, but also the mechanisms through which it arises (see Chapters 14–16). This research area will benefit greatly from interdisciplinary collaboration among theorists, natural historians, and groups leveraging emerging tools in synthetic biology/experimental evolution to directly test hypotheses.

18.7   ENVIRONMENTAL DRIVERS AND NICHE CONSTRUCTION

One of the most fundamental lines of inquiry in the field of multicellularity concerns when and why multicellularity evolved on Earth (Knoll, 2011; Chapters 6 and 10). This is critically important for developing a robust understanding of the environmental drivers of simple multicellularity and the subsequent evolution of more complex lineages. This is a challenging research topic because these transitions occurred in the deep past, and early multicellular lineages tend not to be well preserved in the fossil record. Even with exceptional fossils (such as those from the Doushantuo Formation, where even cellular organelles remain visible (Sun et al., 2020)), it can be difficult to place the fossils into phylogenetic context (Chen et al., 2014). Future work will be crucial for increasing not just the sampling resolution of ancient multicellular lineages, but resolving their broader Earth context, their local environment, and their biotic interactions. The latter may be especially important for resolving how organismal interactions (e.g., arms races, co-evolution) underlie the evolution of increased multicellular complexity (Sperling et al., 2013), and subsequently impact biogeochemical processes (Butterfield, 2018).

It is clear that multicellularity has fundamentally transformed Earth's surface and biogeochemical processes—for instance, plants alone account for approximately 80% of the biomass of all life on Earth by one estimate (Bar-On et al., 2018). Despite recent progress, we do not fully understand how multicellular organisms have affected the evolution of Earth's biogeochemical cycles and climate, or how multicellular niche construction has affected the subsequent evolution of multicellularity. For example, we often define our biomes by the dominant multicellular taxa in that environment (e.g., forests, grasslands, coral reefs, kelp forests), and nearly all of the organisms in these environments are there because of the niches created, directly or indirectly, by the dominant multicellular taxa. Plants give rise to herbivores and pollinators, which in turn give rise to predators, predators of the predators, parasites of all of these, hyperparasites, predators of the hyperparasites, and so on. Clearly, the evolution of multicellularity depends on the prior evolution of multicellularity—but how, why, and are these principles general or lineage-specific?

Finally, it's unclear to what extent Earth's environment has constrained the evolution of multicellularity. For example, how has environmental oxygen affected the evolution of multicellular size (Bozdag et al., 2021; Cole et al., 2020)? Prior to the rise of near modern oxygen levels in the Phanerozoic (Lyons et al., 2014), did low ocean/atmospheric oxygen constrain the evolution of tissues more than a few cells thick? How might environmental constraints interact with the origin of life cycles, the environment in which multicellular organisms could persist (Turner, 2021), the topology of multicellular groups (Yanni et al., 2020), and the subsequent evolution of multicellular development?

18.8 ASTROBIOLOGY: THE ULTIMATE GENERALIZATION

A comprehensive understanding of how, when, and why multicellularity has evolved on Earth should provide insight into a fundamental question in astrobiology: assuming cellular life has independently evolved elsewhere in the Universe, what is the probability that it would evolve to be multicellular? This is not quite the f_i of the Drake equation ("the fraction of planets with life that actually go on to develop intelligent life (civilizations)"), but it is probably a factor contributing to f_i, unless we are imagining intelligent unicells (as some science fiction authors have done).

What biotic and abiotic factors are critical for the initial transition to multicellularity and the subsequent evolution of complex multicellularity, and how might these apply to other planetary contexts? Is this process necessarily slow, or can it occur quickly given the right conditions? Understanding the evolution of multicellularity on Earth, and, in particular, developing robust theory from evolutionary first principles, should allow us to make strong astrobiological hypotheses. In addition to its importance for the field of astrobiology, this has deep significance for humanity, helping address a fundamental existential question: are we alone, or is sentient multicellular life cosmologically common?

18.9   KEY METHODS

Understanding the evolution of multicellularity will take a plurality of methods and disciplines. Here, we highlight some of the key approaches that will be crucial for research progress. These approaches can be roughly classified into three broad categories: comparative, experimental, and theoretical. Although there is some overlap, and although other classifications are certainly possible, we think this is a reasonable first pass at making sense of the diversity of methods.

Comparative methods often involve describing the differences between multicellular organisms and their closest unicellular relatives. Differences between the two must have evolved since they diverged from a (presumably unicellular) ancestor, and some of these differences will be causes and consequences of the transition to multicellularity. We are employing a broad definition of comparative methods, to include comparisons of morphology, physiology, biochemistry, genetics, and so on.

Of course, such comparisons are only meaningful when we understand the evolutionary relationships among the organisms we're comparing, so accurate phylogenies are crucial (see Chapters 5 and 9). There is simply no substitute for knowing how and when multicellularity evolved in extant lineages. As we describe in Section 18.3, improved phylogenetic inferences have often led to new insights into the evolution of multicellularity.

Paleontology and Earth science play critical roles in ground-truthing comparative methods. Fossils are indispensable in calibrating divergence time estimates in phylogenies, and they provide a unique snapshot into the phenotype of early multicellular life. Earth system science (e.g., isotope geochemistry) will allow for more precise inference of the environmental and planetary contexts of early multicellular evolution.

Although comparative methods are powerful tools for understanding the evolution of multicellularity, they do have some important limitations. In most cases, the closest unicellular relatives of extant multicellular organisms diverged from them hundreds of millions of years ago, and subsequent, uncharacterized changes in the unicellular lineages limit their utility as stand-ins for the unicellular ancestors. Experimental/directed evolution offers the unique opportunity to examine the origin of multicellularity directly and has been recently applied to fungi (Ratcliff et al., 2012), bacteria (Hammerschmidt et al., 2014), algae (Boraas et al., 1998; Herron et al., 2019; Ratcliff et al., 2013), and ichthyosporea (Dudin et al., 2021). In addition to examining how simple multicellularity can evolve in various lineages, this method can provide insight into how increasingly complex multicellularity evolves and allows one to directly test hypotheses that are otherwise experimentally intractable (e.g., how early life cycles or environmental conditions affect the evolution of multi-cellularity (Bozdag et al., 2021)). The trade-off for this approach is that it sacrifices relevance to particular historical origins of multicellularity for tractability and (possibly) generality. It can, in other words, tell us about how multicellularity <u>can</u> evolve, but not about how it <u>did</u> evolve.

Other experimental approaches lie at different points along this continuum. Some, for example detailed studies of genetic and developmental mechanisms, synergize well with comparative methods by fleshing out the differences between unicellular and multicellular organisms. Others, for example biophysics, may derive general principles that should be relevant to many or all origins of multicellularity.

Another promising experimental approach is synthetic biology. Despite its potential as a sandbox allowing virtually unlimited experimental possibilities, relatively little work has leveraged synthetic approaches to study multicellularity (but see Basu et al., 2005; Solé et al., 2018; Toda et al., 2018). We expect that this will change as methods for gene editing become increasingly routine in non-model organisms (Booth and King, 2020), and as techniques for engineering organoids allow for increasingly precise control on *in vitro* morphogenesis (Hofer and Lutolf, 2021).

Theory plays a key role in multicellularity research, complementing natural history (which tells us what has happened in this one run of Earth's timeline) and experiments with organisms (which let us test various hypotheses about when, why and how these things happened). Theory thus provides a formal method for summarizing our knowledge about evolutionary processes, allowing hypotheses to be clearly defined and subsequently tested. By the same token, theory and computation allow us to explore the evolution of multicellularity in a way that is unencumbered by the historical contingencies that are otherwise an unimpeachable constraint of real organisms living in a specific environment. Computational and theoretical approaches are thus an invaluable tool for exploration, hypothesis generation, and the identification and formulation of general principles.

One theoretical approach that has been crucial for understanding the evolution of multicellularity is that of multilevel selection. In Chapter 3, Rick Michod reviews the development of multilevel selection theory, especially as it applies to the evolution of multicellularity and of cellular differentiation, including his own considerable contributions to this application. Among the most influential of these is a collection of population genetic models based on life-history tradeoffs that address the evolution of a division of labor among cells, or cellular differentiation. In Chapter 12, Guilhem Doulcier, Katrin Hammerschmidt, and Pierrick Bourrat analyze the life-history model and its underlying assumptions, limitations, and interpretation as a critical step in the transformation of cell groups into multicellular organisms.

Just as the origins of multicellularity in plants, animals, and other taxa are particular cases of a broader category (but see Chapter 2), the evolution of multicellularity writ large is a particular case of the broader category of "major transitions" in evolution (Maynard Smith and Szathmáry, 1995) or evolutionary transitions in individuality (Michod and Roze, 1997). Much of the theory that has been developed for the evolution of multicellularity can be applied more broadly to the larger category, though just how much of the theory can be so applied, and to which transitions, are very much open questions (Calcott and Sterelny, 2011; Herron, 2021; McShea and Simpson, 2011; O'Malley and Powell, 2016).

Comparative, experimental, and theoretical approaches can all benefit from the development of new model systems. Model systems of extant multicellular organisms, such as *Dictyostelium discoideum*, *Myxococcus xanthus*, *Neurospora crassa*, *Volvox carteri*, and *Salpingoeca rosetta* have been crucial for generating and testing hypotheses about the evolution of multicellularity. Given how contingent each of these model systems is (each species is nested within a lineage that has independently evolved multicellularity in largely idiosyncratic ways), new model systems will significantly expand our suite of experimental possibilities.

18.10   CONCLUSION

There has never been a better time to work on the evolution of multicellularity. Rapid advances in life sciences technology have opened new avenues for research that would have been unimaginable a generation ago, and will no doubt continue to revolutionize progress. And, despite the deep history of the field, it has remained a small enough niche that major conceptual breakthroughs are not only possible but are a regular occurrence.

David Kirk (1998), quoting Jerome Gross, advised that the best path to a happy and productive scientific career is to find "a quiet backwater where there are lots of big fish to be caught, but not many people fishing." The evolution of multicellularity is much less of a backwater today than it was when Kirk wrote that, but it is still a relatively small pond, and we are convinced that there are indeed still big fish to be caught. We hope that this book has been useful to you, and we invite readers to reach out to any of the authors if you have questions about the evolution of multicellularity or are interested in wetting a line.

ACKNOWLEDGMENTS

W.C.R. was supported by NSF DEB-1845363 and a Packard Fellowship for Science and Engineering. P.L.C. was supported by a NASA Postdoctoral Program Fellowship and funding from the David and Lucile Packard Foundation. This material is based upon work while M.D.H. was serving at the National Science Foundation.

REFERENCES

Abedin, M., and King, N. (2010). Diverse evolutionary paths to cell adhesion. Trends in Cell Biology *20*, 734–742.

Arendt, D. (2008). The evolution of cell types in animals: emerging principles from molecular studies. Nature Reviews Genetics *9*, 868–882.

Baldauf, S.L. (2003). The deep roots of eukaryotes. Science *300*, 1703–1706.

Bar-On, Y.M., Phillips, R., and Milo, R. (2018). The biomass distribution on Earth. Proceedings of the National Academy of Sciences *115*, 6506–6511.

Basu, S., Gerchman, Y., Collins, C.H., Arnold, F.H., and Weiss, R. (2005). A synthetic multicellular system for programmed pattern formation. Nature *434*, 1130–1134.

Bengtson, S., Sallstedt, T., Belivanova, V., and Whitehouse, M. (2017). Three-dimensional preservation of cellular and subcellular structures suggests 1.6 billion-year-old crown-group red algae. PLoS Biology *15*, e2000735.

Black, A.J., Bourrat, P., and Rainey, P.B. (2020). Ecological scaffolding and the evolution of individuality. Nature Ecology & Evolution *4*, 426–436.

Bonner, J.T. (1965). Size and cycle- an essay on the structure of biology. American Scientist *53*, 488–494.

Bonner, J.T. (2006). Why size matters (Princeton, NJ: Princeton University Press).

Booth, D.S., and King, N. (2020). Genome editing enables reverse genetics of multicellular development in the choanoflagellate *Salpingoeca rosetta*. eLife *9*, e56193.

Boraas, M.E., Seale, D.B., and Boxhorn, J.E. (1998). Phagotrophy by a flagellate selects for colonial prey: a possible origin of multicellularity. Evolutionary Ecology *12*, 153–164.

Bourrat, P. (2015). Levels, time and fitness in evolutionary transitions in individuality. Philosophy and Theory in Biology *7*, 1–17.

Bozdag, G.O., Libby, E., Pineau, R., Reinhard, C.T., and Ratcliff, W.C. (2021). Oxygen suppression of macroscopic multicellularity. Nature Communications *12*, 2838.

Brunet, T., and King, N. (2017). The origin of animal multicellularity and cell differentiation. Dev Cell *43*, 124–140.

Brunet, T., Larson, B.T., Linden, T.A., Vermeij, M.J., McDonald, K., and King, N. (2019). Light-regulated collective contractility in a multicellular choanoflagellate. Science *366*, 326–334.

Buss, L.W. (1987). The evolution of individuality (Princeton, N.J.: Princeton, N.J.: Princeton University Press).

Butterfield, N. (2018). Oxygen, animals and aquatic bioturbation: an updated account. Geobiology *16*, 3–16.

Calcott, B., and Sterelny, K. (2011). The major transitions in evolution revisited (Cambridge, MA: MIT Press).

Chen, L., Xiao, S., Pang, K., Zhou, C., and Yuan, X. (2014). Cell differentiation and germ–soma separation in Ediacaran animal embryo-like fossils. Nature *516*, 238–241.

Clarke, E. (2010). The problem of biological individuality. Biological Theory *5*, 312–325.

Cole, D.B., Mills, D.B., Erwin, D.H., Sperling, E.A., Porter, S.M., Reinhard, C.T., and Planavsky, N.J. (2020). On the co-evolution of surface oxygen levels and animals. Geobiology *18*, 260–281.

de Mendoza, A., Sebé-Pedrós, A., Šestak, M.S., Matejčić, M., Torruella, G., Domazet-Lošo, T., and Ruiz-Trillo, I. (2013). Transcription factor evolution in eukaryotes and the assembly of the regulatory toolkit in multicellular lineages. Proceedings of the National Academy of Sciences *110*, E4858–E4866.

Dudin, O., Wielgoss, S., New, A.M., and Ruiz-Trillo, I. (2021). Regulation of sedimentation rate shapes the evolution of multicellularity in a unicellular relative of animals. bioRxiv, 2021.2007.2023.453070.

Dupuy, L., Mackenzie, J., and Haseloff, J. (2010). Coordination of plant cell division and expansion in a simple morphogenetic system. Proceedings of the National Academy of Sciences *107*, 2711–2716.

Dworkin, M. (1963). Nutritional regulation of morphogenesis in *Myxococcus xanthus*. Journal of Bacteriology *86*, 67–72.

El-Kirat-Chatel, S., Beaussart, A., Vincent, S.P., Flos, M.A., Hols, P., Lipke, P.N., and Dufrêne, Y.F. (2015). Forces in yeast flocculation. Nanoscale *7*, 1760–1767.

Flemming, H.C., and Wuertz, S. (2019). Bacteria and archaea on Earth and their abundance in biofilms. Nature Reviews Microbiology *17*, 247–260.

Fortezza, M.L., Rendueles, O., Keller, H., and Velicer, G.J. (2021). Hidden paths to endless forms most wonderful: ecology latently shapes evolution of multicellular development in predatory bacteria. bioRxiv, 2021.2006.2017.448787.

Gibson, T.M., Shih, P.M., Cumming, V.M., Fischer, W.W., Crockford, P.W., Hodgskiss, M.S., Wörndle, S., Creaser, R.A., Rainbird, R.H., and Skulski, T.M. (2018). Precise age of *Bangiomorpha pubescens* dates the origin of eukaryotic photosynthesis. Geology *46*, 135–138.

Godfrey-Smith, P. (2013). Darwinian individuals. In From Groups to Individuals: Evolution and Emerging Individuality, p. 17–36. Eds. Frederic Bouchard and Philippe Huneman. MIT Press.

Grosberg, R.K., and Strathmann, R.R. (2007). The evolution of multicellularity: a minor major transition? Annual Review of Ecology, Evolution, and Systematics *38*, 621–654.

Hake, K., West, P.T., McDonald, K., Laundon, D., De Las Bayonas, A.G., Feng, C., Burkhardt, P., Richter, D.J., Banfield, J., and King, N. (2021). Colonial choanoflagellate isolated from Mono Lake harbors a microbiome. bioRxiv, 2021.03.30.437421.

Hammerschmidt, K., Rose, C.J., Kerr, B., and Rainey, P.B. (2014). Life cycles, fitness decoupling and the evolution of multicellularity. Nature *515*, 75–79.

Herron, M.D. (2021). What are the major transitions? Biology & Philosophy *36*, 1–19.

Herron, M.D., Borin, J.M., Boswell, J.C., Walker, J., Chen, I.-C.K., Knox, C.A., Boyd, M., Rosenzweig, F., and Ratcliff, W.C. (2019). *De novo* origins of multicellularity in response to predation. Scientific Reports *9*, 2328.

Herron, M.D., Rashidi, A., Shelton, D.E., and Driscoll, W.W. (2013). Cellular differentiation and individuality in the 'minor' multicellular taxa. Biological Reviews *88*, 844–861.

Hofer, M., and Lutolf, M.P. (2021). Engineering organoids. Nature Reviews Materials *6*, 402–420.

Jackson, M.D., Xu, H., Duran-Nebreda, S., Stamm, P., and Bassel, G.W. (2017). Topological analysis of multicellular complexity in the plant hypocotyl. eLife *6*, e26023.

Jacobeen, S., Pentz, J.T., Graba, E.C., Brandys, C.G., Ratcliff, W.C., and Yunker, P.J. (2018). Cellular packing, mechanical stress and the evolution of multicellularity. Nature Physics *14*, 286.

Jang, W., and Gomer, R.H. (2011). Initial cell type choice in *Dictyostelium*. Eukaryotic Cell *10*, 150–155.

Jékely, G., and Budd, G.E. (2021). Animal phylogeny: resolving the slugfest of ctenophores, sponges and acoels? Current Biology *31*, R202–R204.

King, N. (2004). The unicellular ancestry of animal development. Developmental Cell *7*, 313–325.

Kirk, D.L. (1998). Volvox: A Search for the Molecular and Genetic Origins of Multicellularity and Cellular Differentiation, Vol 33 (Cambridge, UK: Cambridge University Press).

Knoll, A.H. (2011). The multiple origins of complex multicellularity. Annual Review of Earth and Planetary Sciences *39*, 217–239.

Koschwanez, J.H., Foster, K.R., and Murray, A.W. (2013). Improved use of a public good selects for the evolution of undifferentiated multicellularity. eLife *2*, e00367.

Lyons, T.W., Reinhard, C.T., and Planavsky, N.J. (2014). The rise of oxygen in Earth's early ocean and atmosphere. Nature *506*, 307–315.

Mahadeo, D.C., and Parent, C.A. (2006). Signal relay during the life cycle of *Dictyostelium*. Current topics in Developmental Biology *73*, 115–140.

Márquez-Zacarías, P., Conlin, P.L., Tong, K., Pentz, J.T., and Ratcliff, W.C. (2021). Why have aggregative multicellular organisms stayed simple? Current Genetics, 1–6.

Maynard Smith, J., and Szathmáry, E. (1995). The major transitions in evolution (NY: Oxford University Press).

McShea, D.W., and Simpson, C. (2011). The miscellaneous transitions in evolution. In: The Major Transitions in Evolution Revisited, p. 19–34. Eds. Brett Calcott and Kim Sterelny. MIT Press.

Michod, R. (2005). On the transfer of fitness from the cell to the multicellular organism. Biology and Philosophy *20*, 967–987.

Michod, R.E., and Roze, D. (1997). Transitions in individuality. Proceedings of the Royal Society of London Series B: Biological Sciences *264*, 853–857.

Mikhailov, K.V., Konstantinova, A.V., Nikitin, M.A., Troshin, P.V., Rusin, L.Y., Lyubetsky, V.A., Panchin, Y.V., Mylnikov, A.P., Moroz, L.L., Kumar, S., *et al.* (2009). The origin of Metazoa: a transition from temporal to spatial cell differentiation. Bioessays *31*, 758–768.

Nagy, L.G., Kovacs, G.M., and Krizsan, K. (2018). Complex multicellularity in fungi: evolutionary convergence, single origin, or both? Biological Reviews *93*, 1778–1794.

Nedelcu, A.M., and Michod, R.E. (2006). The evolutionary origin of an altruistic gene. Molecular Biology and Evolution *23*, 1460–1464.

Newman, S.A., and Bhat, R. (2009). Dynamical patterning modules: a "pattern language" for development and evolution of multicellular form. International Journal of Developmental Biology *53*, 693–705.

Nguyen, T.A., Cissé, O.H., Wong, J.Y., Zheng, P., Hewitt, D., Nowrousian, M., Stajich, J.E., and Jedd, G. (2017). Innovation and constraint leading to complex multicellularity in the Ascomycota. Nature Communications *8*, 1–13.

O'Malley, M.A., and Powell, R. (2016). Major problems in evolutionary transitions: how a metabolic perspective can enrich our understanding of macroevolution. Biology & Philosophy *31*, 159–189.

Okasha, S. (2006). Evolution and the Levels of Selection. Oxford University Press.

Pentz, J.T., Márquez-Zacarías, P., Bozdag, G.O., Burnetti, A., Yunker, P.J., Libby, E., and Ratcliff, W.C. (2020). Ecological advantages and evolutionary limitations of aggregative multicellular development. Current Biology *30*, 4155–4164. e4156.

Petroll, R., Schreiber, M., Finke, H., Cock, J.M., Gould, S.B., and Rensing, S.A. (2021). Signatures of transcription factor evolution and the secondary gain of red algae complexity. Genes *12*, 1055.

Pichugin, Y., and Traulsen, A. (2020). Evolution of multicellular life cycles under costly fragmentation. PLoS Computational Biology *16*, e1008406.

Pradeu, T. (2016). Organisms or biological individuals? Combining physiological and evolutionary individuality. Biology & Philosophy *31*, 797–817.

Queller, D.C., and Strassmann, J.E. (2009). Beyond society: the evolution of organismality. Philosophical Transactions of the Royal Society B: Biological Sciences *364*, 3143–3155.

Queller, D.C., and Strassmann, J.E. (2014). Fruiting bodies of the social amoeba *Dictyostelium discoideum* increase spore transport by *Drosophila*. BMC Evolutionary Biology *14*, 1–5.

Rainey, P.B., and Kerr, B. (2010). Cheats as first propagules: a new hypothesis for the evolution of individuality during the transition from single cells to multicellularity. Bioessays *32*, 872–880.

Ratcliff, W.C., Denison, R.F., Borrello, M., and Travisano, M. (2012). Experimental evolution of multicellularity. Proceedings of the National Academy of Sciences *109:5*, 1595–1600.

Ratcliff, W.C., Herron, M.D., Howell, K., Pentz, J.T., Rosenzweig, F., and Travisano, M. (2013). Experimental evolution of an alternating uni-and multicellular life cycle in *Chlamydomonas reinhardtii*. Nature Communications *4* (1), 1–7.

Ratcliff, W.C., Herron, M.D., Conlin, P.L., and Libby, E. (2017). Nascent life cycles and the emergence of higher-level individuality. Philosophical Transactions of the Royal Society B: Biological Sciences *372* (1735), 1–12.

Rokas, A. (2008). The origins of multicellularity and the early history of the genetic toolkit for animal development. Annual Review of Genetics *42*, 235–251.

Rose, C., and Hammerschmidt, K. (2021). What do we mean by multicellularity? The evolutionary transitions framework provides answers. Frontiers in Ecology and Evolution *9*, 1–6.

Ruiz-Trillo, I., Roger, A.J., Burger, G., Gray, M.W., and Lang, B.F. (2008). A phylogenomic investigation into the origin of Metazoa. Molecular Biology and Evolution *25*, 664–672.

Schirrmeister, B.E., Gugger, M., and Donoghue, P.C. (2015). Cyanobacteria and the Great Oxidation Event: evidence from genes and fossils. Palaeontology *58*, 769–785.

Sebe-Pedros, A., Degnan, B.M., and Ruiz-Trillo, I. (2017). The origin of Metazoa: a unicellular perspective. Nature Reviews Genetics *18*, 498–512.

Smukalla, S., Caldara, M., Pochet, N., Beauvais, A., Guadagnini, S., Yan, C., Vinces, M.D., Jansen, A., Prevost, M.C., Latgé, J.-P., Fink, G.R., Foster, K.R., Verstrepen, K.J. (2008). *FLO1* is a variable green beard gene that drives biofilm-like cooperation in budding yeast. Cell *135*, 726–737.

Solé, R., Ollé-Vila, A., Vidiella, B., Duran-Nebreda, S., and Conde-Pueyo, N. (2018). The road to synthetic multicellularity. Current Opinion in Systems Biology *7*, 60–67.

Sperling, E.A., Frieder, C.A., Raman, A.V., Girguis, P.R., Levin, L.A., and Knoll, A.H. (2013). Oxygen, ecology, and the Cambrian radiation of animals. Proceedings of the National Academy of Sciences *110*, 13446–13451.

Staps, M., van Gestel, J., and Tarnita, C.E. (2019). Emergence of diverse life cycles and life histories at the origin of multicellularity. Nature Ecology & Evolution *3*, 1197–1205.

Sun, W., Yin, Z., Cunningham, J.A., Liu, P., Zhu, M., and Donoghue, P.C. (2020). Nucleus preservation in early Ediacaran Weng'an embryo-like fossils, experimental taphonomy of nuclei and implications for reading the eukaryote fossil record. Interface Focus *10*, 20200015.

Tarver, J.E., Cormier, A., Pinzón, N., Taylor, R.S., Carré, W., Strittmatter, M., Seitz, H., Coelho, S.M., and Cock, J.M. (2015). MicroRNAs and the evolution of complex multicellularity: identification of a large, diverse complement of microRNAs in the brown alga *Ectocarpus*. Nucleic Acids Research *43*, 6384–6398.

Toda, S., Blauch, L.R., Tang, S.K., Morsut, L., and Lim, W.A. (2018). Programming self-organizing multicellular structures with synthetic cell-cell signaling. Science *361*, 156–162.

Turner, E.C. (2021). Possible poriferan body fossils in early Neoproterozoic microbial reefs. Nature *596*(7870), 87–91.

Umen, J., and Herron, M. (2021). The evolution of multicellularity in green algae. Annual Review of Genetics 55:603–632.

Yanni, D., Jacobeen, S., Márquez-Zacarías, P., Weitz, J.S., Ratcliff, W.C., and Yunker, P.J. (2020). Topological constraints in early multicellularity favor reproductive division of labor. eLife *9*, e54348.

Zou, L.N., Cheng, X., Rivers, M.L., Jaeger, H.M., and Nagel, S.R. (2009). The packing of granular polymer chains. Science *326*, 408–410.

Index

Note: Locators in *italics* represent figures and **bold** indicate tables in the text.